Behavioural Ecotoxicology

Ecological and Environmental Toxicology Series

Series Editors

Jason M. Weeks
National Centre for
Environmental Toxicology,
WRc-NSF, UK

Sheila O'Hare
Scientific Editor
Hertfordshire
UK

Barnett A. Rattner
Patuxent Wildlife
Research Center, USGS
Laurel, MD, USA

The fields of environmental toxicology, ecological toxicology and ecotoxicology are rapidly expanding areas of research. This explosion of interest within the international scientific community demands comprehensive and up-to-date information that is easily accessible to both professionals and an increasing number of students with an interest in these subject areas.

Books in the series will cover a diverse range of relevant topics ranging from taxonomically based handbooks of ecotoxicology to current aspects of international regulatory affairs. Publications will serve the needs of undergraduate and postgraduate students, academics and professionals with an interest in these developing subject areas.

The Series Editors will be pleased to consider suggestions and proposals from prospective authors or editors in respect of books for future inclusion in the series.

Published titles in the series

Environmental Risk Harmonization:
Federal and State Approaches to Environmental Hazards in the USA
Edited by Michael A. Kamrin (ISBN 0 471 972657)

Handbook of Soil Invertebrate Toxicity Tests
Edited by Hans Løkke and Cornelius A. M. van Gestel (ISBN 0 471 971030)

Pollution Risk Assessment and Management:
A Structured Approach
Edited by Peter E. T. Douben (ISBN 0 471 972975)

Statistics in Ecotoxicology
Edited by Tim Sparks (ISBN 0 471 96851 X CL, ISBN 0 471 972 991 PR)

Demography in Ecotoxicology
Edited by Jan E. Kammenga and Ryszard Laskowski (ISBN 0 471 490024)

Forecasting the Environmental Fate and Effects of Chemicals
Edited by Philip S. Rainbow, Steve P. Hopkin and Mark Crane (ISBN 0 471 491799)

Ecotoxicology of Wild Mammals
Edited by Richard F. Shore and Barnett A. Rattner (ISBN 0 471 974293)

Environmental Analysis of Contaminated Sites
Edited by Geoffrey I. Sunahara, Agnès Y. Renoux, Claude Thellen, Connie L. Gaudet and Adrien Pilon (ISBN 0 471 986690)

Forthcoming titles in the series

Predictive Ecotoxicology
Mark Crane (ISBN 0 471 984043)

PAHs: An Ecological Perspective
Edited by Peter Douben (ISBN 0 471 560243)

Behavioural Ecotoxicology

Edited by

Giacomo Dell'Omo
Institute of Anatomy,
University of Zürich, Switzerland

JOHN WILEY & SONS, LTD

Baffins Lane, Chichester,
West Sussex, PO19 1UD, UK

National 01243 779777
International (+44) 1243 779777
e-mail (for orders and customer service enquiries): cs-books@wiley.co.uk
Visit our Home Page on http://www.wileyeurope.com
or http://www.wiley.com

Other Wiley Editorial Offices

John Wiley & Sons, Inc., 605 Third Avenue,
New York, NY 10158-0012, USA

Wiley-VCH Verlag GmbH, Pappelallee 3,
D-69469 Weinheim, Germany

John Wiley & Sons Australia Ltd., 33 Park Road, Milton,
Queensland 4064, Australia

John Wiley & Sons (Asia) Pte Ltd., 2 Clementi Loop #02-01,
Jin Xing Distripark, Singapore 129809

John Wiley & Sons (Canada) Ltd., 22 Worcester Road,
Rexdale, Ontario M9W 1L1, Canada

Library of Congress Cataloguing-in-Publication Data

Behavioural ecotoxicology/edited by Giacomo Dell'Omo.
p.cm. — (Ecological and environmental toxicology series)
Includes bibliographical references.
ISBN 0-471-96852-8 (alk. paper)
1. Animal behavior. 2. Pollution — Environmental aspects. I. Dell'Omo, Giacomo. II. Ecological & environmental toxicology series.
QL751 .B3453 2002
591.5 — dc21 2001057372

British Library Cataloguing in Publication Data

A catalogue record for this book is available from the British Library

ISBN 0 471 96852 8

Typeset in 10/12pt Garamond from the author's disks by Laserwords Private Limited, Chennai, India
Printed and bound in Great Britain by Biddles Ltd, Guildford, Surrey
This book is printed on acid-free paper responsibly manufactured from sustainable forestry, in which at least two trees are planted for each one used for paper production.

Contents

List of Contributors

Prof. Elizabeth Adkins-Regan
Department of Psychology and Department of Neurobiology and Behavior, Cornell University, 218 Uris Hall, Ithaca, New York, NY 14853, USA. e-mail: er12@cornell.edu

Dr. Carmen Aguilar
Marine and Freshwater Biomedical Sciences Center, University of Wisconsin-Milwaukee, 600 East Greenfield Avenue, Milwaukee, WI 53204-2944, USA. e-mail: aguilar@uwm.edu

Dr. Erik Baatrup
Institute of Biological Sciences, University of Aarhus, Building 135, DK-8000, Aarhus C, Denmark. e-mail: erik.baatrup@biology.au.dk

Dr. Mace G. Barron
P.E.A.K. Research, 1134 Avon Lane, Longmont, Colorado 80302, USA. e-mail: mbarron@hotmail.com; macebarron@hotmail.com

Dr. Mark Bayley
Department of Zoology, Institute of Biological Sciences, University of Aarhus, DK-8000, Denmark. e-mail: mark.bayley@biology.au.dk

Dr. Windy A. Boyd
Department of Environmental Health Science, The University of Georgia, Athens, GA 30602, USA. e-mail: waboyd@usa.net

Dr. Sandra K. Brewer
U.S. Fish and Wildlife Service, New Jersey Field Office, Pleasantville, New Jersey, USA. e-mail: Sandra_Brewer@usgs.gov

Prof. Joanna Burger
Division of Life Sciences, Rutgers University, 604 Allison Road, Piscataway, New Jersey 08854-8082, USA. e-mail: burger@biology.rutgers.edu

Dr. Flavia Chiarotti
Section of Behavioural Pathophysiology, Laboratorio di Fisiopatologia di Organo E Sistema, Istituto Superiore di Sanità, Viale Regina Elena 299, I-00161 Rome, Italy. e-mail: chiarott@iss.it

Dr. Giacomo Dell'Omo
Division of Neuroanatomy and Behaviour, Institute of Anatomy, University of Zürich, Winterthurerstrasse 190, CH-8057 Zürich, Switzerland. e-mail: g.dellomo@anatom.unizh.ch

Susan C. Gardner
Centro de Investigaciones Biologicas del Noroeste, SC La Paz, BCS, CP23090, Mexico. e-mail: sgardner@cirbnor.mx

Peter L. Gibert
Graduate Group in Ecology, Department of Animal Sciences, One Shelds Avenue, University of California, Davis, CA 95616, USA. e-mail: gibertp@usa.net

Dr. John P. Giesy
Department of Zoology, Institute for Environmental Toxicology, Michigan State University, East Lansing, MI 48824, USA. e-mail: jgiesy@aol.com

Dr. Michael Gochfeld
Consortium for Risk Evaluation with Stakeholder Participation, and Graduate Program in Toxicology, Rutgers University, Piscataway, New Jersey 08854, USA. e-mail: gochfeld@eohsi.rutgers.edu

Dr. Christian E. Grue
US Geological Survey, Biological Resources Division, Washington Cooperative Fish and Wildlife Research Unit, School of Aquatic and Fisheries Sciences, Box 355020, University of Washington, Seattle, WA 98195, USA. e-mail: cgrue@u.washington.edu

Dr. Kurunthachalam Kannan
Department of Zoology, Institute for Environmental Toxicology, Michigan State University, East Lansing, MI 48824, USA. e-mail: kuruntha@msu.edu

Prof. Hans-Peter Lipp
Division of Neuroanatomy and Behaviour, Institute of Anatomy, University of Zürich, Winterthurerstrasse 190, CH-8057 Zürich, Switzerland. e-mail: hplipp@anatom.unizh.ch

Dr. Edward E. Little
US Geological Survey, Columbia Environmental Research Center, Columbia, Missouri 65201, USA. e-mail: edward_little@usgs.gov

Dr. Paola Palanza
Dipartimento di Biologia Evolutiva e Funzionale, Parma University, Parco area delle Scienze, 11/A, I-43100 Parma, Italy. e-mail: palanza@biol.unipr.it

David B. Peakall
Deceased

David H. Petering
Marine and Freshwater Biomedical Sciences Center, University of Wisconsin-Milwaukee, 600 East Greenfield Avenue, Milwaukee, WI 53204-2944, USA. e-mail: petering@uwm.edu

Randall Ryder
Department of Curriculum and Instruction School of Education, University of Wisconsin-Milwaukee, WI 53211, USA. e-mail: randall@uwm.edu

Dr. Tim Sparks
Centre for Ecology and Hydrology, Monks Wood, Abbots Ripton, Huntingdon, Cambridgeshire, PE28 2LS, UK. e-mail: t.sparks@nerc.ac.uk

Dr. Helen Thompson
Central Science Laboratory, Department for Environment, Food and Rural Affairs, Sand Hutton, York, Y04 1LZ, UK. e-mail: h.thompson@csl.gov.uk

Dr. Frederick vom Saal
Division of Biological Sciences, University of Missouri-Columbia, Columbia, MO 65211, USA. e-mail: vomSaalF@missouri.edu

Prof. Daniel N. Weber
Marine & Freshwater Biomedical Sciences Center, University of Wisconsin-Milwaukee, 600 East Greenfield Avenue, Milwaukee, WI 53204-2944, USA. e-mail: dweber@csd.uwm.edu

Prof. Phillip L. Williams
Department of Environmental Health Science, The University of Georgia, Athens, GA 30602, USA. e-mail: pwilliam@uga.cc.uga.edu

Series Foreword

BEHAVIOURAL ECOTOXICOLOGY

***Behaviour*— the way one conducts oneself; the treatment of others, the way in which a chemical substance works; the response of a person or animal to a stimulus; the pattern of activity when being observed.**

This is a book about behaviour and how behaviour is modified by environmental toxicants. It attempts to guide the reader though a hitherto uncharted terrain of measuring, compiling evidence and understanding the significance of, often subtle, chemical-induced changes in a behaviour pattern or trait. This book is divided into four main sections; first, the identification of pollutant-induced behavioural changes in different receptor organisms; second, an overview of the methods for measuring such changes in behaviour. Third, a series of case studies and examples of behavioural aberrations in the real environment and, finally, a look towards further understanding, teaching requirements and regulatory considerations.

This book has been written by a series of world experts and covers a diverse range of topics ranging from the fundamental mechanisms of behaviour to the statistical interpretation of observed changes. We as series editors are proud to have this book among the titles in the series. This book provides an overview of the complexity and interwoven nature of the subject area that is *Behavioural Ecotoxicology*. It will hopefully encourage greater understanding of the need to recognise what is normal behaviour to an organism and to understand the significance of pollutant-induced aberrations to that behaviour. Thus, we can construct better and more informed models to enable us to unravel this complex issue.

This latest contribution to the *Ecological and Environmental Toxicology Series* is welcomed and we wish to thank the editor for his considerable effort in delivering a text that brings to the fore an area of science often overlooked and misunderstood when interpreting ecotoxicology information.

This book crosses many different disciplines and will be of interest to a wide range of readers including ecologists, biologists, ethologists, toxicologists, managers, lawyers, risk assessors, behaviourists, and those merely curious.

Jason M. Weeks
Sheila O'Hare
Barnett A. Rattner

Editor's Preface

Spiders making abnormal orb webs, mosquitoes with altered oviposition behaviour, aquatic invertebrates with impaired escape reflexes or altered mating and feeding behaviour, fish with altered swimming behaviour, poor hunting ability and parental care in birds, rodents displaying abnormal territorial behaviour, male fish, alligators, and even panthers with feminised behaviours. These are just a few examples of a much longer list of behavioural alterations that have been observed in many animals inhabiting polluted environments. The number of species and the types of behavioural alterations observed are both increasing as is the evidence of a causal link to toxicants released by human activities.

The effects of pollutants on the behaviour of invertebrates, fish, and wildlife have been investigated in many studies over the last few decades. Reports of such studies can be found in journals that cover a variety of disciplines, ranging from behavioural neuroscience, toxicology, ecotoxicology, psychology, pharmacology, and conservation biology, to name just a few. However, to date, this considerable body of work has not been summarised and the behavioural approach to ecotoxicology issues has not been formalised. In fact, behavioural aspects have received little attention in many ecotoxicology books while an ecological approach that links effects on individuals to those on populations has also been disregarded by behavioural toxicologists and pharmacologists, even when they are dealing with toxicants of environmental concern.

However, in reality, the fields of behavioural studies and ecotoxicology have both already profited from interactions with each other. This book reviews when and how this fruitful relationship has occurred and will try to unify an approach across the two disciplines by highlighting the advantages, and subsequent consequences, of further development of this comprehensive and promising field.

Giacomo Dell'Omo

Introduction

WHY IS A BEHAVIOURAL APPROACH NEEDED TO STUDY THE EFFECTS OF ENVIRONMENTAL POLLUTANTS?

Behaviour is the cumulative interaction of a variety of biotic and abiotic factors that represents the animal's response to internal (physiological) and external (environmental, social) factors and relates one organism to another. Behaviour is adaptable, and can be modulated in type, intensity, and time of occurrence, and so is a major mechanism by which animals adapt to changes in their environment, including exposure to contaminants (Evans 1994). Behavioural effects can be the result of exposure either to neurotoxic chemicals that act on specific areas of the nervous system, or to contaminants that primarily affect other systems but, as a consequence, alter behaviour. Acute effects can be transient if caused by toxicants that degrade rapidly in the environment. However, many contaminants are persistent, resulting in long-term exposure for some species, and lipophilic compounds in particular can biomagnify through food-chains so that top predators are exposed to relatively high concentrations. Such chronic exposure can adversely affect behaviours involving sensation, perception, cognition, co-ordination, and motor function. This can lead to deficits in feeding, predator avoidance, social relationships and reproductive success that can affect individual survival and result in shift in population dynamics.

Toxicant-induced changes in behaviour may indicate toxicity (the failure of adaptive mechanisms) or conversely be the adaptive response of an animal to mitigate or obviate the potential effects of exposure. Behaviours reflecting toxicity are often overt but subtle adaptation can be mild and difficult to observe. Indeed, whether or not such effects are observable can depend on the sensitivity of the investigative approach: no-observed effect does not always mean no effect. In either cases, behavioural changes show that environmental factors, including chemical contamination, are impinging on the animal (Evans 1994).

In the perception of the public, a behavioural change that is suddenly observed in a wild species is considered *a priori* to be negative, even if it does not have direct, measurable deleterious effects. This is probably because the concept of adaptation of an organism is largely associated with time on a evolutionary scale. Thus, new rapid, adaptive responses, including changes in behaviour, are not regarded as driven by natural causes but are perceived as a consequence of an artificial alteration of the natural equilibrium. The interest of the public on these issues is raised in part by the recognition that animals and humans are equally vulnerable to exposure to chemicals and the pathways of exposure and mechanisms of effects are basically similar.

In the scientific community there is a growing awareness that understanding sub-lethal effects is critical for assessing environmental effects and for limiting environmental problems. Environmental insults can leave lasting effects, whether or not we have observed them (Matthews *et al.* 1996), and behavioural disruption in invertebrates, fish (Døving 1995), and wildlife could provide useful information for preventing more serious problems and maintaining healthy ecosystems. Behavioural alteration, although a relatively insensitive indicator of the degree of exposure, has the advantages of revealing both direct and indirect effects of pollutants and in some cases represents an early warning sign, since it is evidence that environmental contaminants cause functional deficits. There is also a growing interest, to abandon the classical laboratory toxicity tests that use lethal or sub-lethal endpoints and search for less invasive approaches that are more acceptable to the public and more informative of the ecological relevance of the pollutant-mediated effects (Fossi and Leonzio 1994). For example, the LD50 acute oral toxicity test (the so-called Test Guidelines 401, used until recently for calculating the dose of a chemical that would kill 50% of the animals after a single exposure), has been recently abolished from the Organisation for Economic Cooperation and Development (OECD) internationally accepted Test Guidelines (Environment Directorate 2001). It has been replaced with other tests that minimise the use of animals and use toxicity rather than lethal endpoints. The scientific community is also searching for less conventional (but standardized) animal models. In fact, although laboratory experiments are necessary for screening toxicants and for acquiring basic information, it can be dangerous to rely exclusively on data gathered from laboratory rodents to predict likely effects in wildlife and humans because the physiological responses to the same chemicals may differ markedly between species (Winneke and Lilienthal 1992, Watkinson and Gordon 1993, Dickerson *et al.* 1994).

The release of chemicals is only one of the many human activities that have an impact on the environment and wildlife. The effects are not as macroscopic and rapid as those produced by activities such as intensive agriculture, deforestation, water management, and recreation. However, on the long-term, the consequences of chemical contamination could be even more dramatic for the survival of species, including ourselves.

BACKGROUND

The use of behavioural parameters as an index of sub-lethal toxicity in wildlife was initially suggested by Warner and colleagues (1966). They listed three main advantages, namely that behaviourial change: i) is the final integrated result of biochemical and physiological processes and so a more comprehensive measure than a physiological or biochemical parameter; ii) is a sensitive indicator of sublethal toxicity; and iii) can be measured without harming the animal. As noted by Peakall (1996 and also Chapter 4 of this book), most of their arguments have resisted the challenges of other more analytical and technological approaches and have also adapted to new scenarios.

However, in those years, ecotoxicology was a young discipline and behaviour mainly supported more advanced, consolidated fields, such as pharmacology and toxicology. Behavioural toxicology, behavioural pharmacology, and psychopharmacology, as already-emerged disciplines, were instrumental to affirm the role of behaviour in neuroscience and for investigating the biochemical mechanisms of brain function and the effects of natural and artificial molecules. A book, consisting of selected papers presented at an international conference on environmental toxicity in 1972, dealt specifically with behavioural toxicology for the first time (Weiss and Laties, 1975). Later, the status of behavioural pharmacology was also upgraded (Seiden and Balster 1985) and reference to these two disciplines was frequent in the '80s (see e.g. Weiss and O'Donogue 1984, Annau, 1987, Cory-Sletcha 1989). However, throughout this period, behaviour had little to do with ecotoxicology and contributions apparently dealing with environmental issues were also much too dependent on conventional laboratory-based toxicology (Barthalmus and Rand 1980). As noted by Cairns (1992) "when the toxicologists added the prefix *eco* to the field of toxicology, so that it became *ecotoxicology*, they continued primarily to make the same measurements they had made before the name was changed... without any substantive change".

At the beginning of the '90s, concerns about environmental pollution, and the possible implications for ourselves, increased. This was partly because sufficient evidence had already accumulated on the deleterious effects of many pollutants, such as heavy metals, organochlorines, pesticides) and new proof had emerged that even very low doses of some substances (PCBs and other organic compounds) could interfere with endocrine function. This produced an awareness that the effects of environmental toxicants should be treated differently from those of pharmaceutical agents (Nelson 1990) and also encouraged a demand for more environmentally realistic test systems. From a behavioural perspective, this has meant the adoption of an "ethological approach" for studying behaviour (Silverman 1988) that is considered as more suitable for adding ecological relevance to laboratory-based studies (MacPhail *et al.* 1983, Parmigiani *et al.* 1988, Cohn and MacPhail 1990, Dixon *et al.* 1990, Dourish *et al.* 1995, Needelman 1995).

Little (1990) concluded that "observations of behaviour provide a unique toxicological perspective, one that links the biochemical and ecological consequences of environmental contamination. Chemically-induced changes of behaviour are linked to impaired biochemical, physiological or general metabolic processes within the animal. Since behaviour can be considered as the functional interface between the individual and the population, its ecological importance in population maintenance is intuitively obvious".

WHAT IS BEHAVIOURAL ECOTOXICOLOGY?

The study of behaviour in ecotoxicology, or behavioural ecotoxicology, is a comprehensive field which is the summation of many interacting disciplines. It can be considered as a combination of three main disciplines, ethology (the biological study of the natural behaviour of an animal; Tinbergen 1963), ecology (the study of the relationship between animals and their environment), and toxicology (the study of toxic agents) or pharmacology (which differs from toxicology in that it deals with agents that have known, reversible effects or a known mechanism of action).

Behavioural ecotoxicology could also be considered as the result of a number of other combinations between more specific fields, for example, between behavioural ecology and toxicology/pharmacology, between behavioural toxicology/pharmacology (also psychology and etho-pharmacology) and ecology, or ecotoxicology and ethology, among many others. The figure below simplifies these relationships and highlight its position with respect to these disciplines.

However, there could be differences in the functional meaning of these combinations. This is because each branch has its own specific features. For example, behavioural toxicology is limited to the study of effects of toxicants on brain functions and the approach is barely ethological. Similarly psychopharmacology, and behavioural pharmacology deal with the effects of drugs on single elements of the behavioural repertoire that can be controlled (in terms of occurrence and intensity), quantified, and replicated. On the other hand, etho-pharmacology is more oriented towards the study of the species' own "natural" behaviour (although the distinction between natural and non natural is not always useful; Dixon *et al.* 1990). Furthermore, toxicology is based on individuals whereas ecology investigates the effects at the population level. Ecotoxicologists are concerned with population but etho-ecotoxicologists should come back to the individual without losing the implications for populations.

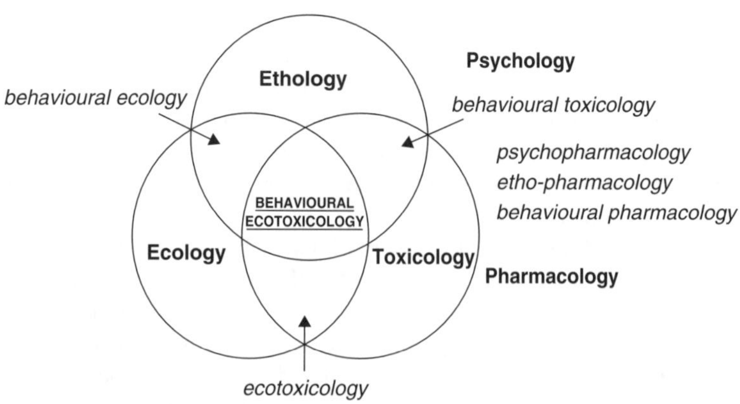

Figure 1 Relationships between behavioural ecotoxicology and other disciplines.

Perhaps it would be more appropriate not to force behavioural ecotoxicology into the rules of already-existing fields. It may be better instead to adopt a more flexible and wider interpretation of a field which requires a broad multidisciplinary comprehension of several processes.

Behavioural ecotoxicology does not deal with drugs but with substances that are present in natural compartments of the environment, such as soil, water and air. It requires ecotoxicologists to have a broad understanding of the biological basis of behaviour and apply an ethological approach to its study. This means that it is necessary to search for functional changes that have ecological relevance (that affect the capability to survive or adapt in the environment).

RATIONALE FOR THE BOOK

In his contribution for the Scientific Group on Methodologies for the Safety Evaluation of Chemicals (SGOMSEC) meeting held in 1994 in Rochester, New York, and concerning disrupted patterns of behaviour in wildlife, David Peakall concluded that "... a condensed review of behavioural (eco)toxicology requires a book..." (Peakall 1996). We hope to have moved with this book in the direction he suggested. In fact, in the plethora of ecological and behavioural books written to date, the synergy between these two disciplines has not yet been considered. While the use of behaviour as a measurable end-point has reached sophisticated levels in toxicology, neuroscience, and pharmacology and has been defined in many books, its use for assessing the effects of environmental toxicants on wild populations of animals has not been formalised and the literature is spread over various disciplines. This book was needed to resume and unify the various approaches, which are transversal across many disciplines, in a discipline on its own. However, this book is not the pretext for patenting a new discipline: behavioural ecotoxicology already exists.

AIM OF THE BOOK

The aspiration of this book is to synthesise the state of the knowledge on the impact of environmental pollutants on the behaviour of wild animals and to identify and discuss the implications of the behavioural approach for detecting exposure, investigating the effects of pollutants, and supporting the regulatory process in limiting the risk of exposure. The authors of the various chapters were invited to review each topic with examples from their own work and from the published literature in order to produce an overview of the state of our knowledge.

The book also includes chapters intended to introduce readers unfamiliar with behaviour or environmental toxicants, to these specific topics. The procedures employed to detect and to measure behavioural alterations are described for aquatic and terrestrial invertebrates and vertebrates. Other chapters discuss the relationships between behaviour and other bioindicators of exposure to pollutants, and show the importance of the behavioural approach in the regulatory process and education. The authors were identified on the basis of their contributions to specific fields of research. Such a selection was inevitably constrained by the number of chapters to be assigned and we are aware that other colleagues from the same countries or other regions of the world could have contributed to this book as well. Even if they do not figure among the authors, their contribution as a whole has furthered research in this field and has often been acknowledged by citations to their work throughout the book.

STRUCTURE OF THE BOOK

There were a number of ways for organising a book such as this, depending upon whether priority was given to ethological, toxicological, or ecological arguments. The chapters could have been grouped according to type/mechanisms of behaviour, category of pollutants, or habitat similarities. An organisation that took into account the increasing complexity of organisms (i.e. taxonomic) could also have been useful for zoologists. None of these criteria when considered alone was satisfactory, however. Given the multidisciplinarity of the arguments that encompass several fields, we ended up with a mixed criterion that has tried to take into account all the various possibilities simultaneously.

The book contains 12 chapters that can be treated as self-standing reviews for those interested in specific arguments. The chapters have different lengths that reflect the complexity and the relevance of the arguments that are considered. Behaviour is the main link across the chapters and it is considered from various perspectives. The first chapter by Christian Grue and colleagues is of key importance since it discusses the ecological relevance of behavioural alterations and the importance of an ecological context in determining the vulnerability of species to environmental pollutants. This chapter reviews the available data on fish and wildlife including biological (in terms of effects on reproduction and survival), ecological (as measured at population level), and regulatory aspects (current use or lack thereof in regulatory decision-making). It also discusses the main issues treated thereafter and could equally as well have been placed as a summation at the end of the book. However, we thought that it best served as an exhaustive introduction to most of the arguments treated later, including the relationships with other biomarkers, extrapolation from laboratory to field, and the implications for environmental monitoring and risk assessment.

The other chapters are organised in four topical sections. Namely, one introductory section, one concerning methods for measuring and analysing behaviour, a third describing case studies, and the last providing perspectives for the use of behaviour in the regulatory process and education. The first two chapters in the introductory section are propedeutic for those who are unfamiliar with behaviour (Chapter 2) or with the types and the properties of environmental contaminants (Chapter 3). In particular, Chapter 2 synthesises our current understanding on the mechanisms of behaviour, focussing on the role of neurotransmitters and hormones in determining the behaviour of an organism.. The list of behaviours and mechanisms is wide and is not limited just to those known to be susceptible to environmental pollutants. The behaviours considered are not only those strictly relevant for survival but also patterns that are less investigated or that are difficult to observe in the field. The intention was to provide the reader with a comprehensive list of behaviours that may be affected by contaminants and to stimulate interest in behaviours that are often neglected and are also dependent on the integrity of neuronal and hormonal pathways.

Chapter 3 categorises the environmental contaminants on the basis of the behaviours they affect. A detailed description of the chemical properties and the fate of chemicals in the environment is not given in this chapter but can be found in the literature. This chapter should be regarded as a tool for supporting the reader with basic information on contaminants. Chapter 4 discusses the relationships between behaviour and the available biomarkers of exposure. A common feature of behaviour and biomarkers is that both give information on the effects of pollutants and rarely if ever give information on the levels of contaminants present either in the environment or often in the animal itself. However, the aim of this chapter is not to force behaviour into the rules of more conventional biomarkers, nor to propose its use as a new biomarker but, instead, to review literature in which behaviour and other biomarkers were used at the same time and provide evidence of a link between these two sets of measures.

The following section of the book is dedicated to the measurements of behaviour and a general overview on data analysis (Chapter 7). Chapter 5 and 6 are complementary in that the first deals with basic behaviours and is mainly concerned with invertebrate species while the second concerns complex behaviours in vertebrates. The complexity of the behavioural responses increases with increasing complexity of the organisms, and measures of activity in aquatic and terrestrial invertebrates are among some of the simplest endpoints that can be quantified. Basic behaviours are mainly activity responses that reflect changes at any of the many integrated nervous pathways. Tests based on the analysis of basic behaviours are apical tests and are important at first-tier screening to collect evidence of changes. The identification of specific mechanisms or more subtle changes can be done later with selective tests suitable for single end-points. The observation of behaviour in the wild is not always feasible and thus animals should be brought to the laboratory for appropriate investigations. This can be done easily with aquatic organisms and invertebrate species that can be tested in laboratory analogues of real world microhabitats (artificial streams, meso and microcosms). However, this is not feasible with many terrestrial vertebrate species. Both these methodological chapters are necessarily based on a laboratory approach and, particularly in the case of Chapter 6, on methods imported from other disciplines. In theory, any element of the behavioural repertoire of a species has the potential of being affected by pollutants. In practice, only few can be observed, other are rare or elicited only under particular circumstances. Field observations of behavioural alterations can be important as early warning signals in suggesting an effect of pollutants but laboratory studies are a vital part of this approach because they are needed to ascertain the cause-effect relationship. In Chapter 6, methods and approaches such as those for the behavioural phenotyping of transgenic mice are discussed together with other approaches in semi-naturalistic settings and with non-conventional species.

The section on the case studies contains Chapters 8, 9, and 10. The first two chapters are comprehensive reviews on the behavioural effects of environmental contaminant in invertebrates and birds, respectively. These two groups, together with fish, are those more extensively investigated. Sedentary invertebrates are often advocated as biomonitors for local contamination whereas birds, particularly top predators where the contaminants of concern biomagnify through the foodchain, could reflect the health of entire ecosystems. Other classes of vertebrates have received less attention and have not required a full chapter. This reflects the status of the research in these groups. Reference to effects on amphibians can be found in Chapter 8, fish are considered in Chapters 1 and 11, and studies on reptiles and mammals feature in Chapter 10. This latter chapter is dedicated to endocrine disrupters their effects on reproduction and also on other minor, but no less ecologically-relevant, effects. Specific consideration is given to low doses, which are also likely to be relevant to humans. Exposure, can be critical during foetal development although effects may not be apparent until adulthood, and can have consequences for population structure and dynamic and implications for risk assessment.

The last section provides contribution in two main fields that could potentially benefit from behavioural ecotoxicology: the regulatory process (Chapter 11) and education (Chapter 12). The use of neurobehavioural tests on laboratory animals for the risk assessment process have been recommended for a long time (National Academy of Sciences 1975). However behavioural changes in wildlife exposed to contaminants are not currently used in risk assessment, apart from avoidance reactions in fish and lethargy and tremors in birds and mammals. The criteria for defining substances as harmful are important because these can drive regulatory programs and can have societal implication. An ethological approach can certainly add realism to the risk assessment. The last chapter provides examples of how, in practice, the interest for this discipline can be brought into schools and explores the possibility of using behavioural ecotoxicology demonstrations to educate students in the care of the environment.

DEDICATION

This book is dedicated to David Peakall, who contributed a chapter to this book on behaviour and biomarkers, and who sadly died in September 2001. He was one of the pioneers who, at the end of the 1960s, drew attention to the effects of DDT and its metabolites on the survival of many raptorial birds, and on the population consequences of exposure to such persistent contaminants. He provided the evidence of the causality linking eggshell thinning, one of the first biomarkers that arose from field observations, to DDE, a fact that contributed to the subsequent banning of these chemicals in North America and in many European countries. Later, his interest concentrated on biomarkers and their use as a tool in environmental assessment. He gave precious advice and support for the organisation of this work. Many of us are indebted to him.

ACKNOWLEDGEMENTS

I wish to thank Catalina Betancur for her precious help in finalising this project. Erik Baatrup, Jennifer Batty, and Mike Hooper supported the initial work. Charles Drews drew my attention to the potential of using behavioural ecotoxicology for educating students to the care of the environment. Giorgio Bignami and Richard Shore were always available with helpful advise.

This project was finalised during my work for the Swiss National Science Foundation and NCCR "Neural Plasticity and Repair".

REFERENCES

Annau Z (ed), (1986) *Neurobehavioral Toxicology*. John Hopkins University Press, Baltimore.

Barthalmus GT and Rand GM (1980) Behavioral toxicology. In *Introduction to Environmental Toxicology*, Guthrie FE and Perry JJ (eds), Blackwell Scientific Publications, Oxford, pp. 161-171.

Cairns JJr (1992) The threshold problem in ecotoxicology. *Ecotoxicology*, **1**, 3-16.

Cohn JC and MacPhail RC (1996) Ethological and experimental approaches to behaviour analysis: implications for ecotoxicology. *Environmental Health Perspectives*, **104** (suppl. 2), 299-305.

Cory-Slechta DA (1989) Behavioural measures of neurotoxicity. *NeuroToxicology*, **10**, 271-296.

Davis M (1982) Ethological approaches to behavioral toxicology. In *Nervous System Toxicology*, Mitchell CL (ed), Raven Press, New York, pp. 29-44.

Dickerson RL, Hooper MJ, Gard NW, Cobb GP and Kendall RJ (1994) Toxicological foundations of ecological risk assessment: biomarker development and interpretation based on laboratory and wildlife species. *Environmental Health Perspectives*, **102** (Suppl), 65-69.

Dixon AK, Fisch HU and McAllister KH (1990) Ethopharmacology: a biological approach to the study of drug-induced changes in behavior. *Advances in the Study of Behavior*, **19**, 170-204.

Dourish CT, Grewal SS, Shepherd JK, Stanhope KJ, Bill DJ and Fletcher A (1995) Benefit of ethological analysis of behaviour. *Trends in Pharmacological Science*, **16**, 260–261.

Døving KB (1991) Assessment of animal behaviour as a method to indicate environmental toxicity. *Comparative Biochemistry and Physiology C*, **100**, 247–252.

Environment Directorate, Joint Meeting of the Chemicals Committee and the Working Party of Chemicals, Pesticides and Biotechnology (2001) Guidance document on acute oral toxicity testing. *Organisation for Economic Co-operation and Development Series on Testing and Assessment*, 24.

Evans HL (1994) Neurotoxicity expressed in naturally occurring behavior. In *Neurobehavioral Toxicity. Analysis and Interpretation*, Weiss B and O'Donogue JL (eds), Raven Press, New York, pp. 111–136.

Fossi MC and Leonzio C (eds), (1994) *Nondestructive Biomarkers in Vertebrates*. Lewis, Boca Raton.

Little EE (1990) Behavioral toxicology: stimulating challenges for a growing discipline. *Environmental Toxicology and Chemistry*, **9**, 1–2.

MacPhail RC, Crofton KM, and Reiter LW (1983) Use of environmental challenges in behavioral toxicology. *Federal Proceedings of the American Society for Experimental Biology*, **42**, 3196–3200.

Matthews RA, Landis WG and Matthews GB (1995) The community conditioning hypothesis and its application to environmental toxicology. *Environmental Toxicology and Chemistry*, **15**, 597–603.

National Academy of Sciences (1975) *Principles for Evaluating Chemicals in the Environment*. National Academy of Sciences, Washington DC.

Needleman HL (1995) Behavioral toxicology. *Environmental Health Perspectives*, **103**, 77–79.

Nelson BK (1990) Origin of behavioral teratology and distinctions between research on pharmaceutical agents and environmental/industrial chemicals. *Neurotoxicology and Teratology*, **12**, 301–305.

Parmigiani S, Palanza P and vom Saal FS (1998) Ethotoxicology: an evolutionary approach to the study of environmental endocrine-disrupting chemicals. *Toxicology and Industrial Health*, **14**, 333–340.

Peakall D (1996) Disrupted pattern of behavior in natural populations as an index of ecotoxicity. *Environmental Health Perspectives*, **104** (suppl 2), 331–335.

Seiden LS and Balster RL (eds), (1985) *Behavioral Pharmacology: the Current Status*. Alan R. Liss, New York.

Silverman AP (1988) An ethologist's approach to behavioural toxicology. *Neurotoxicology and Teratology*, **10**, 85–92.

Tinbergen N (1963) On aims and methods of ethology. *Zhurnal für Tierpsychologie*, **20**, 410–433.

Warner RE, Peterson KK and Borgman L (1966) Behavioural pathology in fish: a quantitative study of sublethal pesticide toxication. *Journal of Applied Ecology*, **3** (Suppl), 223–247.

Watkinson WP and Gordon CJ (1993) Caveats regarding the use of the laboratory rat as a model for acute toxicological studies: modulation of the toxic response via physiological and behavioral mechanisms. *Toxicology*, **81**, 15–31.

Weiss B, Laties VG (eds), (1975) *Behavioural Toxicology*. Collection of papers presented at the Fifth Rochester International Conference on Environmental Toxicity, 1972. New York, Plenum Press.

Weiss B and O'Donogue JL (1994) *Neurobehavioral Toxicity. Analysis and Interpretation*. Raven Press, New York.

Winneke G and Lilienthal H (1992) Extrapolation from animals to humans: scientific and regulatory aspects. *Toxicology Letters*, **65**, 239–246.

Acronyms

Organisations and acts

AAAS	American Association for the Advancement of Science
AID	Agency for International Development
ASTM	American Society for Testing and Material
ATSDR	Agency for Toxic Substances and Disease Registry
CERCLA	Comprehensive Environmental Response, Compensation and Liability Act
CWA	Clean Water Act
EPA	Environmental Protection Agency
ESA	Endangered Species Act
FIFRA	Federal Insecticide, Fungicide and Rodenticide Act
FPA	Federal Power Act
IEPA	International Environmental Protection Act
IOBC	International Organisation for Biological and Integrated Control of Noxious Animals and Plants
NIEHS	National Institute for Environmental Health Sciences
NPDES	National Pollution Discharge and Elimination System
NRDA	Natural Resource Damage Assessment
NRC	National Research Council
NRDA	Natural Resource Damage Assessment
OECD	Organisation for Economic Cooperation and Development
OPA	Oil Pollution Act
TSCA	Toxic Substances Control Act
US	United States
WHO	World Health Organisation

Chemicals

DDD	dichlorodiphenyldichloroethane
DDE	2,2-bis (4-chlorophenyl)-1,1-dichloroethene, a major metabolite of DDT
DDT	dichlorodiphenyltrichloroethane
DDVP	dichlorvos
DES	diethylstilboestrol
DFP	diisopropyl-fluorophosphate
DMP	dimethyl phthalate

ED	endocrine disrupter
EPN	O-ethyl-O-(4nitrophenyl)phenylphosphotionate
HCB	hexachlorobenzene
MeHg	methylmercury
MT	metallothionein
NO	nitric oxide
OC	organochlorine
OP	organophosphate; organophosphorus
PAH	polycyclic aromatic hydrocarbon
PCB	polychlorinated byphenils
PCDD	polychlorinated dibenzo-*p*-dioxin; pentachlorodibenzo-*p*-dioxin
PCDF	polychlorinated dibenzofuran
PCP	pentachlorophenol
PE	polyethylene
PHAH	polyhalogenated aromatic hydrocarbon
PVC	polyvinyl chloride
SDS	sodium dodecyl sulphate
TCDD	tetrachlorodibenzo-*p*-dioxin
TCDF	tetrachlorodibenzofuran
TMP-4	dipiroxim, 1,1′-threemethylene-bis (4-formilpiridine bromide) dioxin, AChE reactivator
TPP	triphenyl phosphite

Hormones, enzymes, neurotransmitters

AAT	alanine aminotransferase
AChE	acetylcholinesterase
ACTH	adrenocortico-trophic hormone
Ah	aryl hydrocarbon
AHH	aryl hydrocarbon hydroxylase
ALAD	aminolevulinic acid dehydratase
AR	androgen receptor
ATP	adenosin tri-phosphate
AVP	vasopressin
AVT	vasotocin
AZT	azidotimidine
BChE	butyrylcholinesterase
BNST	bed nucleus of the *stria terminalis*
cAMP	cyclic adenaline monophosphate
CbES	carboxyl esterase
CCK	cholecystokinin
cGMP	cyclic guanidine monophosphate
ChE	cholinesterase
CNS	central nervous system

CREB	calcium response binding element protein
CRH	corticotropin releasing hormone
DHT	5α-dihydrotestosterone
DNA	deoxyribonucleic acid
E-G	oestradiol glucuronide
ER	oestrogen receptor
ERE	oestrogen response element
EROD	ethoxyresofurin-*O*-dealkylase
FSH	follicle-stimulating hormone
GABA	γ-aminobutyric acid (neurotransmitter)
GnRH	gonadotropin-releasing hormone
GSH	gonadal-stimulating hormone
HPA	hypothalamo-pituitary-adrenal
HPG	hypothalamo-pituitary-gonadal
LH	luteinising hormone
MFO	mixed function oxidase
mRNA	messenger ribonucleic acid
NAT	N-acetyltransferase
NMDA	N-methyl-D-aspartate
NPY	neuropeptide Y
PBAN	pheromone biosynthesis activating neuropeptide
PNS	peripheral nervous system
PR	progesterone receptor
RNA	ribonucleic acid
T-G	testosterone glucuronide
VIP	vasoactive intestinal polypeptide
5-HAT	serotonin

Others

AEC	adenylate energy charge
ALD	angular light distribution
ANOVA	analysis of variance
BAS	behavioural assessment scale
CR	clearance rate
CS	conditioned stimulus
CTA	conditioned taste aversion
CTMC	continuous time Markov chain
DRL	differential reinforcement of low-rate schedule
EC50	effective concentration of toxicant (in the diet or tissue) required to elicit a sublethal response in 50% of a test population
EODS	electric organ discharge
EPC	electronic particle counter
FI	fixed interval (schedule)

FR	fixed ratio (schedule)
GaPPS	*Gammarus pulex* precopula separation test
IIP-MF	intra/infrapyramidal mossy fiber
Ijs	infective juveniles
IQ	intelligence quotient
IST	induced separation test
ITI	intertrial interval
LC 50	median lethal concentration, the statistically derived concentration of toxicant in the diet that can be expected to kill 50% of the animal tested
LD 50	median lethal dose, the statistically derived single dose of a toxicant that can be expected to kill 50% of the animal tested
LOEC	lowest observed effect concentration
LTP	long-term potentiation
MANOVA	multivariate analysis of variance
MPOA	medial preoptic area
NOEC	no observed effects concentration
NOEL	no observed effect level
OPIDN	organophosphate-induced delayed neuropathy
PCA	principal component analysis
PEC	predicted environmental concentration
PICT	pollution-induced community tolerance model
QSAR	quantitative structure activity relationship
SAR	structure activity relationship
SCN	suprachiasmatic nucleus
SEM	mean standard error
SL	spontaneous locomotion
spp.	species
TAR	thoracic appendage rate (measured during the feeding activity of Dapnids)
TSD	temperature-dependent sex determination
UCS	unconditioned stimulus
wt	weight

1

On the Significance of Pollutant-induced Alterations in the Behaviour of Fish and Wildlife

CHRISTIAN E. GRUE[1], SUSAN C. GARDNER[2] AND PETER L. GIBERT[3]

[1]U.S. Geological Survey, Biological Resources Division, Washington Cooperative Fish and Wildlife, Research Unit, University of Washington, Seattle, WA, USA

[2]Centro de Investigaciones Biologicas del Noroeste, Mexico

[3]Department of Animal Sciences, University of California Davis, CA, USA

1.1 INTRODUCTION

Behaviour is the product of the interaction(s) of an organism with its external environment and represents the integration of underlying physiological processes and mechanisms, described in Chapter 2, with the environmental stimuli that trigger them and the evolutionary forces that have and continue to shape them. Real (1994) states that 'the only successful way to think about behaviour is to unite its internal mechanistic foundations with its external effects in specific ecological settings, and to recognise that the internal and external worlds of the organism are united through the action of natural selection and evolution', the union of which is the basis for the field of behavioural ecology (Krebs and Davies 1997).

The internal mechanisms underlying behaviour have been traditionally referred to as 'proximate' causes (Tinbergen 1963; Table 1) of an organism's actions. Included are factors that explain how particular actions or behaviours are effected. 'Ultimate' factors (Tinbergen 1963; Table 1), also referred to as 'ecological consequences' (Real 1994), explain why a particular behaviour exists: its evolutionary history, adaptive value or significance in terms of reproductive success (fitness).

Table 1.1 Proximate and ultimate causes of behaviour (after Alcock 1998: 5)

Proximate Causes
Genetic-developmental
Gene–environment interactions underlying the development of sensory-motor mechanisms
Sensory-motor mechanisms
Nervous system for detection of environmental stimuli
Hormone system for adjusting responsiveness to environmental stimuli
Skeletal-muscular systems for effecting responses
Ultimate Causes
Historical pathways leading to current behaviour
Evolutionary events from the origin of the trait to the present
Selective processes shaping history of a behavioural trait
Past and present value of the trait in terms of reproductive success

The internal processes that form the mechanistic basis of behaviour can be divided into five interacting components (Figure 1.1, after Real 1994): hormonal processes with a variety of cellular and molecular foundations, neurobiology and development, decision processes that promote the efficient use of resources (e.g., optimal foraging strategies [see Alcock 1998: 341–382) or the efficient processing of information (Giraldeau 1997); and psychophysical constraints that may limit memory, learning and perception. The ecological consequences of the resulting behaviour can be divided into four principal categories with a variety of interactions (Figure 1.1, after Real 1994): foraging, interspecific interactions, social organisation, and mating.

In this chapter, we explore the significance of pollutant-induced alterations in behaviour primarily in terms of their biological (e.g., manifestation into decreased reproduction or survival) and ecological (alterations in populations, community structure, or ecosystem function) effects as well as their role in the regulation of anthropogenic sources of pollution. Although we emphasise effects on the behaviour of fish and wildlife, we also examine our ability to extrapolate effects observed under laboratory conditions (often with surrogate species) to free-living individuals and their populations. In many cases, our understanding of the effects (or potential effects) of different pollutants on behaviour comes from pharmacological studies directed toward human health issues, and where appropriate we have included references to these findings. A thorough review of the literature is beyond the scope of this chapter and, where possible, we focus on examples within the literature of the last 10 years. Additional examples and/or more thorough discussion of specific effects of pollutants on the behaviour of exposed animals can be found in subsequent chapters. We conclude our discussion with recommendations and research needs necessary to more clearly define and integrate the roles of behavioural toxicology.

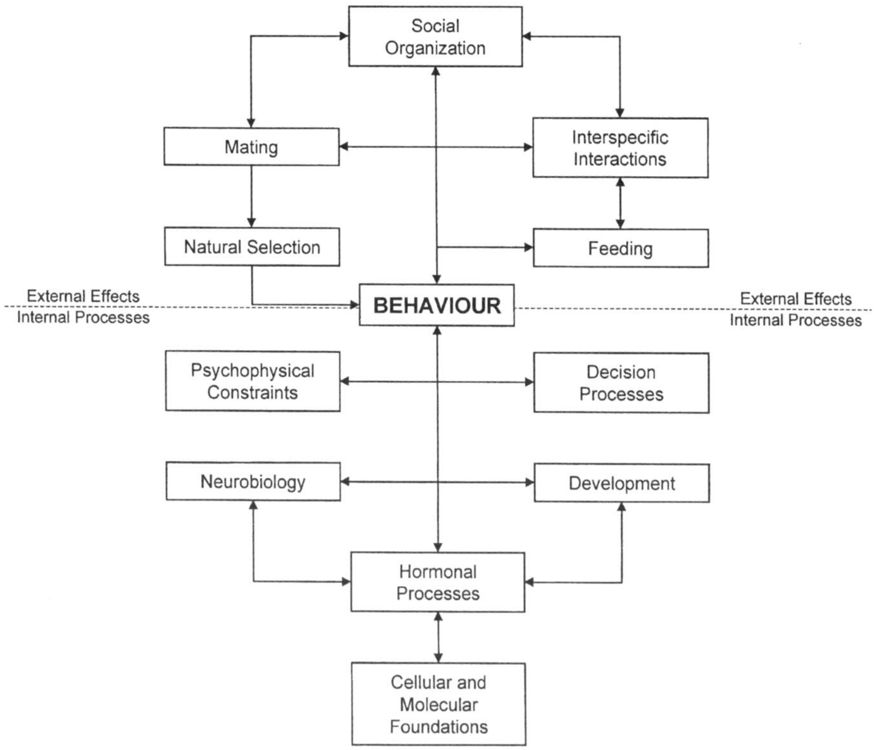

Figure 1.1 Interrelationships among the internal processes governing behaviour and the resulting external (ecological) effects on which natural selection acts to shape the organisms internal processes and behavioural responses (after Real 1994).

For the purposes of this chapter, we define behaviour in terms of both physical movements (flexion of limbs, running, flying) and the function or consequence (attacking, fleeing, nest building) (Immelmann and Beer 1989). Whereas one may be tempted to exclude the former from a discussion of the significance of pollutant-induced behavioural effects, many of the signs of intoxication associated with chemical exposure fall into this category and depending on frequency, severity and duration have the potential to reduce the survival or reproduction of affected individuals. We use the term pollutant (frequently used interchangeably with contaminant) to describe individual substances or mixtures (organic or inorganic) not normally associated with natural systems (the organism itself, its habitat, and the community and ecosystem of which it is a part) or in amounts that exceed normal/natural levels. Herein, we focus our discussion on the effects of chemicals introduced into the environment by man including natural substances whose concentrations in nature have been increased through anthropogenic means. Examples of the latter include sediment loading of streams associated with timber practices (Welch *et al.* 1998), heavy metal accumulation within terrestrial and aquatic habitats following mining operations (Blus *et al.* 1991, Henny *et al.* 1994a,b), the accumulation of selenium, boron and arsenic as a result of irrigation (Lemly *et al.* 1993), and the mobilisation of aluminium within lakes associated with acid rain (Sparling and Lowe 1996).

1.2 UNDERLYING ASSUMPTIONS AND 'PIOUS HOPES'

The primary assumptions underlying our chapter and the field of behavioural toxicology are: (1) most actions and behaviours exhibited by an animal have adaptive value (i.e., have been selected for), and (2) significant deviations from the normal range of responses to environmental stimuli will lower an organism's probability of survival or reproduction. Pollutant-induced alterations in behaviour may not only affect the individual, but may permeate ecological systems. In his recent synthesis, Real (1994) suggests that 'all ecological phenomena and patterns of community organisation can be viewed as the immediate consequences of individual actions and behaviors'. Although a tremendous amount of literature has dealt with documenting the effects of pollutants on behaviour of fish and wildlife, relatively few studies have linked these effects to reductions in survival or reproduction of free-living individuals. Not surprisingly, even less is known about how observed effects in individuals actually affect population viability and community structure.

In his 1985 review of the behavioural responses of birds to pollutants, Peakall coined the term 'pious hopes' to describe this expectation (implied biological significance) with respect to 'sublethal' pollutant-induced changes in physiology or behaviour within the concluding remarks of papers documenting these sublethal effects without demonstrating their effects on survival or reproduction. As noted later in our chapter, the lack of clear demonstrations of the biological or ecological significance of pollutant-induced behavioural effects has and continues to limit our ability to use this information within the regulatory framework (Grue 1994).

Not all behaviours need be explained by existing selection pressures as the latter are likely to vary with changes in ecological context through time (see discussion of the 'Panglosian paradigm' [Gould and Lewontin 1979] in Krebs and Davies 1997). Furthermore, Emlen *et al.* (1998) argue that natural selection cannot alone account for the remarkable ability of many organisms to adjust their physiology and behaviour 'appropriately' to day to day changes in their environment over a variety of scales (plasticity). These authors suggest that organisms, as complex structures, tend to self-organise in adaptive ways independent of the underlying genome, a process (the 'attractor hypothesis') that complements the neoDarwinian paradigm. Irrespective of the mechanisms underlying adaptation, the degree to which pollutant-induced alterations in behaviour are selected against will also depend on the environment (ecological context) in which they are elicited (see discussion in next section).

1.3 IMPORTANCE OF ECOLOGICAL CONTEXT

For the purpose of our chapter, we define ecological context as those factors, both intrinsic (within the organism) and extrinsic (within the organism's environment) that determine whether or not pollutant-induced alterations in behaviour will occur and if so, result in reduced survival or reproductive success.

1.3.1 INTRINSIC FACTORS

1.3.1.1 Inherent sensitivity

Inherent sensitivity varies among species and chemicals because of differences in the absorption (the amount of chemical entering the organism), translocation (movement of the chemical within the body), and biotransformation (metabolic activation, detoxication and excretion) of chemicals in relation to the presence of target and nontarget receptors and the chemical's affinity for the receptors (Stine and Brown 1996:11-15, Loomis and Hayes 1996:33-50). Differences among taxa in their inherent sensitivity to pollutants are important determinants of the magnitude of resultant effects, including alterations in behaviour. The most sensitive species are likely those in which effects will be manifested first and depending on the level of ambient contamination, may be the only species directly affected. Subsequent chemical-induced reductions in their reproduction or survival, or altered habitat use, may significantly alter community structure. For example, in their seminal paper, Hurlbert and his coworkers (Hurlbert *et al.* 1972) observed, among other effects, a rapid and persistent increase in phytoplankton populations in ponds treated with the organophosphorus insecticide, Dursban, in response to the chemical-induced mortality of herbivores ('cascading trophic interactions', Carpenter *et al.* 1985). Warner *et al.* (1993) note that 'when interacting species differ in their responses to the same gradient of an abiotic factor then the specific level of that factor ... can influence the outcome of species interactions, and patterns in communities will be determined by the synergistic effects of biotic and abiotic influences.' Furthermore, the authors note the ability of 'inferior competitors' to tolerate a more stressful environment which may, in some cases, give them a competitive advantage, or at least reduce the level of competition they experience. Examples of similar effects resulting from differential chemical-induced modifications of behaviour are few (see Section 1.8.2). The greater sensitivity in the avoidance response of rainbow trout (*Oncorhynchus mykiss*) to metal contamination compared with brown trout (*Salmo trutta*) may explain observed differences in the distributions of the two species in the Clark Fork River in Montana (Hansen *et al.* 1999).

Species can differ greatly in their sensitivity to pollutants and we refer the reader to reviews by Gaines (1969), Schafer (1972, 1994), Tucker and Leitzke (1979), Hudson *et al.* (1984), Mayer and Ellersiek (1986), Smith (1987), and most recently Eisler (2000). It is not unusual for the sensitivity of multiple species to a particular chemical to differ by 2–3 orders of magnitude, and differences may be as great as 4–5 orders of magnitude (Schafer 1994, Eisler 2000). Similar variability exists in the sensitivity of individual species among chemicals as well (e.g., Schafer 1994), supporting the conclusion that the probability of identifying the most sensitive species *a priori* is remote ('myth of the most sensitive species', Cairns 1986). Classic examples include the greater sensitivity of a variety of raptors and fish-eating birds to eggshell-thinning resulting from exposure to DDE compared to pheasants, quails and chickens (Peakall 1993) and, conversely, the high sensitivity of chickens, mustellids, and subhuman primates to specific PCB congeners (Eisler 2000).

Genetic polymorphism among individuals of the same species accounts for much of the intraspecific variation in sensitivity to xenobiotics (Eaton and Klaassen 1996). However, intraspecific sensitivity can also vary with the route of exposure; sex, age, and diet of the exposed animals; and ambient environmental conditions (Grue *et al.* 1983, Grue and Shipley 1984, Mineau 1991, Rattner and Heath 1995, Moser and Padilla 1998, Eisler 2000, and references therein). For example, mortality of adult mallards fed a diet of corn frequently died 10–14 days after ingesting a single lead shot, whereas similar birds on a nutritionally balanced commercial diet appeared normal after ingesting as many as 32 shot of the same size (Wobeser 1981). Additionally, sensitivity to chemicals can change within and between generations, particularly in response to repeated or chronic exposures (see discussion of tolerance and resistance in Section 1.3.1.3).

1.3.1.2 Behavioural modification of chemical exposure

The frequency, severity, and duration of chemical exposure in relation to inherent sensitivity of the organism are also important determinants of the magnitude of resultant alterations in behaviour. To a great extent, exposure is a function of habitat selection (contaminated vs. uncontaminated) and the ways an animal utilises the habitat(s) it occupies, its food habits and associated behaviours. An animal's normal behaviour can predispose or enhance its exposure to particular pollutants. The former is particularly true of foraging behaviours. Examples include the consumption of granular pesticides by seed-eating birds mistaking the granules for grit (Best and Gionfriddo 1994), mortality of magpies (*Pica pica*) associated with their consumption of hair from the backs of cattle treated with the insecticide famphur to control warbles (*Hypoderma* sp.) (Henny *et al.* 1985), ingestion of lead shot by waterfowl (Sanderson and Bellrose 1986), and the lethal exposure of Wilson's phalaropes (*Steganopus tricolor*) to fenthion, a mosquito larvicide, following application to wet-meadow habitat, most likely a result of dermal exposure to the chemical facilitated by the birds' circular swimming/foraging behaviour (DeWeese *et al.* 1983). Normal behaviours can also prevent or reduce an animal's exposure to pollutants.

Examples include the enhanced avoidance of cadmium by lake whitefish (*Coregonus clupeaformis*) as a result of their shoaling behaviour (McNicol *et al.* 1996), and the hulling of treated seeds by granivorous birds (Avery *et al.* 1997).

Timely regurgitation of foods containing noxious or harmful substances can significantly reduce their absorption. In birds, this has been documented in response to both naturally occurring toxins (Fink and Brower 1981, Kvitek 1991) and to anthropogenic contaminants (Hart and Thompson 1995, Yusufu 1996). Pascual *et al.* (1999) found that some rock doves (*Columba livia*) were able to regurgitate most (>60%) of the fonofos-treated seed consumed, reducing the dose of pesticide assimilated by 50%, from lethal to sublethal levels. Regurgitation reduced, but did not prevent, mortality among treated birds.

Conditioned food aversion has been found to be nearly universal among vertebrates (Riley and Tuck 1985). It is a learned response in which preference for a food may be reduced after only a single meal, even when the first symptoms of toxicity occur more than 2 h after ingestion. This aversion tends to be long lasting and may continue even when uncontaminated food is available if the animal is unable to discriminate between tainted and untainted prey. In birds, aversion can become associated with visual as well as gustatory cues so that referent prey are identified and avoided at a distance (Nicolaus *et al.* 1983). Although a majority of the work on conditioned food aversion in wildlife has involved intentional contamination of baits as a nonlethal method of predator control, such a response is likely to be common in agricultural areas where ingestion of pesticide tainted prey may cause illness, even if relatively minor or short-lived. Indeed, aversions to perfectly innocuous foods may develop if they are consumed prior to exposure to toxicants via some other route (Nicolaus and Lee 1999).

A recent study (Nicolaus and Lee 1999) of conditioned food aversion in free-living red-winged blackbirds (*Agelaius phoeniceus*) found birds consumed up to three meals of prey tainted with parathion, a common organophosphate (OP) pesticide, before completely refusing them for the remainder of the study (up to 28 days) even though parathion was no longer present. Levels of the pesticide were insufficient to produce any outward signs of illness or intoxication and caused significant inhibition of acetylcholinesterase (AChE) in only two-thirds of the high dose group (1.58–1.84 mg/kg, brain AChE activity 39–73% of controls). For most birds, inhibition did not meet a common criterion for biologically-significant exposure to cholinesterase (ChE) inhibitors (brain AChE inhibition >40%, Grue *et al.* 1991).

Other chemical cues may also significantly affect the extent and duration of exposure. An example comes from studies of presentation method and avian avoidance of treated food in which it was hypothesised that detection of higher vapor concentrations accounted for differences in food consumption from bowls and trays. Rock doves consumed significantly less fonofos-treated seed presented in 7-cm deep bowls than the same seed spread on 1-cm deep trays. The effect was lessened when the experiment was carried out in outdoor aviaries instead of indoor cages, presumably due to reduced accumulation of vapors allowed by increased air movement (Fryday *et al.* 1998).

1.3.1.3 Tolerance and resistance

Although incidents of acute poisoning and resultant wildlife mortality often attract the greatest attention, the widespread and frequent use of many agricultural and commercial chemicals creates the potential for much larger numbers of individuals and species to receive chronic low doses of contaminants. Responses to chronic exposure may remain constant over time, may increase (due to the cumulative effects of repeated injury, accumulation of the contaminant in tissues, or increased sensitivity), or may decrease. A decrease in response of individuals elicited by the same dose over time (but within their lifetimes) or, alternatively, the need for increasing doses to elicit the same response, is referred to as 'tolerance' in the pharmacological literature (Eaton and Klaassen 1996). For example, tolerance in man resulting from drug abuse is well known. After repeated exposure to morphine, a 100-fold increase in dose can be required to produce the original response, and a lethal dose of opium to a nonuser may produce minimal effects in an addict (McMillan 1999). Tolerance is also often used in the ecotoxicological literature to describe reduced sensitivity in individuals, populations and communities to long-term (multigenerational) exposure to environmental contamination through the removal (death) of individuals ('adaptation', Klerks and Weis 1987; 'inherited resistance', Spurgeon and Hopkin 2000) or species ('pollution-induced community tolerance model' ['PICT'], Blanck and Wangberg 1988) most vulnerable to direct toxic effects or chemical-induced changes in their environment. However, for our purposes, decreased sensitivity developed across multiple generations of species as a function of natural selection (genetic adaptation) will be considered 'resistance'. It is this latter process through which pesticide resistance has arisen in over 500 crop pests (Georghiou 1990). These processes as defined are not mutually exclusive (Clements 1999), and both have significant implications for the extrapolation of the results from laboratory studies to the field (see discussion by Klerks and Weis 1987).

A number of physiological mechanisms may be involved in the development of tolerance including increased metabolism and detoxification processes, sequestration of contaminants, and changes in levels of neurotransmitters, receptors and second messengers. However, in most cases, the underlying mechanisms are not well understood. One of the better studied examples is the development of tolerance to prolonged inhibition of ChE by many organophosphorus and carbamate pesticides in laboratory rodents (for reviews, see Costa *et al.* 1982, Bushnell *et al.* 1993). These nerve toxins act mainly by inhibiting AChE, the primary enzyme responsible for hydrolysis of the neurotransmitter acetylcholine at synaptic junctions. However, behavioural deficits and overt signs of toxicity frequently disappear, or partially so, well before AChE activity has returned to normal. This recovery is often explained by physiological changes such as the down-regulation of post-synaptic receptors; yet it also appears that animals can learn to compensate for the effects of hypercholinergia without demonstrable biochemical recovery, especially when there is strong motivation to do so (such as the avoidance of predators or pain or the procurement of food). For example, the differences in activity between bobwhite quail (*Colinus virginianus*) dosed with parathion and controls decreased substantially when a predator was present (Galindo *et al.* 1985). Findings such as these have led to speculation that as long as an animal retains some control over its behaviour, its response to intoxication will likely be adaptive and context dependent (Hart 1993). While adaptive responses may not fully compensate for contaminant effects, extrapolations from laboratory experiments in which behavioural responses are measured in the absence of natural environmental stressors may risk overestimating the impact of similar exposures in the wild.

Practice also plays a role in behavioural tolerance, defined here as the ability to regain essentially normal functional performance in spite of continued biochemical impairment. Thus, animals given the opportunity to repeat a task while intoxicated may learn to compensate for deleterious effects faster than those exposed for the same length of time, but without practice (Bignami *et al.* 1975, 1985, van Dongen and Wolthuis 1989). It should be noted, however, that functional recovery is rarely universal and may vary greatly with the demands of the specific task. For example, chronic administration of chlorpyrifos to rats initially reduced the accuracy of both learning and performance tasks in repeated trials using food reinforcement. After several weeks, accuracy of the performance task recovered completely while the learning task remained impaired (Cohn and McPhail 1997). The phenomena described above, variously termed the 'psychological', 'behaviorally-augmented', 'learned' (Bignami *et al.* 1975, 1985), or 'practice-related' (Wolthuis *et al.* 1990) components of tolerance, further complicate the measurement of behavioural effects and their extrapolation to the field.

Although we have focused on tolerance to anticholinesterases in mammals, similar interactions undoubtedly occur for a wide range of compounds and taxa. Development of tolerance has been shown in carp (*Catla catla*) exposed to the organophosphate dichlorvos (Venugopal *et al.* 1996), fathead minnows (*Pimephales promelas*) exposed to cadmium and copper, and rats exposed to the organochlorine (OC) insecticide lindane (Llorens *et al.* 1991).

While there has been some effort to understand the consequences of tolerance in aquatic systems, (e.g., Clements 1999), we are not aware of any studies that have examined the role of tolerance in mitigating contaminant-induced behavioural effects in terrestrial systems. Laboratory and controlled-field studies with wildlife have quantified behavioural effects following severe but 'sublethal' acute exposure to contaminants in relation to reproductive success or survival (e.g., Grue *et al.* 1982b, Galindo *et al.* 1985, Fryday *et al.* 1994, Dell'Omo and Shore 1996b). However, the effects of repeated or chronic doses, over a sufficient period to provide information on the possible contributions of tolerance effects have been over looked. Without such work, speculation that behavioural deficits seen in short-term studies are representative of effects from more chronic exposures in the wild is provocative, but premature (Galindo *et al.* 1985, Fryday *et al.* 1994, Dell'Omo and Shore 1996b).

1.3.1.4 Behavioural plasticity

Behavioural plasticity is the ability of a single genotype to produce more than one alternative, potentially adaptive, behaviour in response to environmental conditions (West-Eberhard 1989). An organism's ability (or inability) to alter its behaviour in response to contaminant-induced changes, whether internal or environmental, may significantly affect its chances for survival or reproduction. When considering how the plasticity of many behaviours may affect the magnitude and duration of contaminant-induced changes it may be useful to draw a distinction between direct and indirect effects. Direct effects are those in which the affected individual is actually (directly) exposed to a particular chemical or chemical mixture, whereas indirect effects result from chemical-induced changes in an animal's environment, but the animal itself may not have been directly exposed (discussed in greater detail in a subsequent section of this chapter).

Information on compensatory behavioural changes in animals directly exposed to contaminants is lacking, though there is some indication that such changes may occur. For example, when deprived of either visual (eyes) or infrared (facial pits) stimuli, rattlesnakes were able to maintain predatory performance by shifting to the intact sensory system (Kardong and Berkhoudt 1999). However, when forced to switch between chemosensory stimuli (olfactory and vomeronasal information) predatory performance declined. Likewise, rattlesnakes were not able completely to compensate for the absence of either modality, radiation reception (visual, infrared) or chemoreception (olfactory, vomeronasal), by reliance on the other. This complexity of response among sensory systems with similar nervous system organisation illustrates the difficulty in generalising behavioural responses within species, let alone to other taxa. Yet, it does establish a precedent for behavioural shifts in response to contaminant-induced sensory impairment.

Such impairment has been documented in a number of studies. In the case of ChE inhibitors, for example, vision in humans may be impaired within minutes following exposure to organophosphates in the form of vapor, dust or aerosol (Namba *et al.* 1971). Animals dosed with the carbamate insecticide, carbaryl, showed a decrease in response to electric shock (Sideroff and Santolucito 1972). Reischl and his coworkers (1975) reported a decrease in auditory detection in squirrel monkeys (*Saimiri sciureus*) dosed with parathion. Exposure to ChE inhibitors may also impair an animal's ability to differentiate between contaminated and uncontaminated diets (Bussiere *et al.* 1989).

Compensatory behavioural changes have been documented in response to changes in habitat conditions caused by environmental contaminants. For example, three species of tits (*Parus cristatus, Aegithalos caudatus* and *P. ater*) adopted less energy costly foraging methods within forests with thinned canopies and fewer arthropods caused by acidic deposition compared with unpolluted forests (Brotons *et al.* 1998).

Behaviours resulting from indirect effects, in which direct toxic effects on an individual affect the behaviour of an associated but unexposed individual, may either compensate for or extend the reach of contaminant effects. In species where both sexes share parental care, such as most birds, an unaffected pair member may compensate for a reduction in the attentiveness of its mate due to illness, contaminant effects, or predation. White *et al.* (1983), in a study in which one member of a laughing gull (*Larus atricilla*) pair received a single sublethal dose of parathion, found that the undosed member compensated for reduced care by its mate by incubating the clutch in its absence. In a similar study, female European starlings (*Sturnus vulgaris*) were given a single oral dose of dicrotophos on day 6 of incubation. Dosed birds spent less time incubating over the next 6 days while their mates compensated by spending more time on the nest relative to controls. Hatching success was not affected (Grue, unpublished data).

However, when the disruption occurs during the nestling phase the ability of the unaffected parent to compensate for its mate tends to vary greatly with the age of the nestlings. When one member of great and blue tit (*Parus major* and *P. caeruleus*) pairs was removed when their nestlings were at various ages, fledging success was unaffected if the nestlings were older than 6 days for blue tits and 8 days for great tits (the age at which the young reach approximately half their maximum weight). Nestlings younger than this critical age when one parent was removed had significantly higher mortality. All those younger than 3–4 days being fed by a single male died, most likely due to the additional stress of hypothermia since the male did not assume the brooding duties of the missing female. Both sexes increased their feeding frequency and the diversity of food delivered (Sasvari 1986). Grue *et al.* (1982b) found that widowed female European starlings given the same dose of dicrotophos as above made significantly fewer sorties to feed their young and remained away from their nests for longer periods of time. Nestlings of dosed birds lost an average of almost three times more weight than those of controls (9.3% vs. 3.2%) during the 24 hours following pesticide exposure.

In all of these examples only one member of a pair was dosed. In the perhaps more likely situation where both members of a pair are exposed, or in species where only one sex has primary responsibility for parental care, the potential effect of contaminant induced disruption of care might be catastrophic for the young, even in cases where the adults recover quickly and suffer no long term ill affects.

1.3.2 EXTRINSIC FACTORS

As previously defined, extrinsic factors are variables within an animal's environment that may mitigate or enhance the effects of pollutant-induced alterations on its behaviour on its survival or reproduction. Included are the level and duration of chemical exposure, interactions with other chemical stressors, the availability of resources, intra- and interspecific interactions (competition, predation and social aggregations), and environmental conditions.

1.3.2.1 Magnitude of chemical exposure

The level, frequency and duration of exposure (dose, Eaton and Klaassen 1996) relative to the inherent sensitivity of the organism and its metabolic (detoxification) capabilities largely determine the magnitude of the resultant changes in behaviour. The greater the dose, the more dramatic the changes in physiological function and behaviour(s) (dose response, Eaton and Klaassen 1996). Similarly, the greater the persistence of the chemical in the environment, the greater the probability of repeated doses in the absence of behavioural modifications of exposure (see previous section). However, the manifestation of behavioural effects may not always coincide with exposure. In some cases, the onset of behavioural effects may be delayed and/or persist long after exposure and signs of intoxication have ended. In other cases, behavioural effects may disappear, although chemical-induced physiological changes persist (see discussion of tolerance in previous section).

Chemicals that persist in the environment frequently also have properties that facilitate transfer within food chains and from parent to offspring, and therefore greater potential for significant delays between initial exposure and resultant behavioural effects. Some of these compounds (e.g., organochlorines) are also oestrogen mimics and their effects on endocrine systems may be additive (Colborn *et al.* 1993). Developmental feminisation of males and altered reproductive morphologies in females have been reported in alligators (Guillette *et al.* 1994) and birds (Fry and Toone 1981) exposed *in ovo* to dicofol and/or DDT and its metabolites. In some colonies of affected birds, it is not clear whether pairs consist of normal males with genotypic males feminised during development or whether they are all genotypic females showing abnormal behaviour (Fry *et al.* 1987, Fox 1992, Colburn *et al.* 1993). Similarly, more recent studies of second generation captive adult male and female American kestrels (*Falco sparverius*) from females exposed to dicofol indicated abnormal reproductive behaviour and reduced reproductive success (MacLellan *et al.* 1996, 1997). Fjeld *et al.* (1998) reported reduced foraging efficiency in grayling (*Thymallus thymallus*) 3 years after embryos were exposed to methylmercury during the first 10 days of development. Prenatal exposure to low levels of lead is associated with permanent impairment of cognitive function and abnormal behaviour in humans (Needleman *et al.* 1990). Similar effects have been reported following prenatal exposure to PCBs (Jacobson and Jacobson 1996).

Delayed and long-term physiological and behavioural effects are not restricted to chronic exposures or chemicals that persist in the environment. The 'intermediate syndrome' and 'organophosphate-induced delayed neuropathy' (OPIDN) are well-known examples of delayed effects following acute exposure to selected organophosphate (OP) pesticides, chemicals rapidly metabolised by warm-blooded animals and which are relatively short-lived in the environment. The former is a transient, but life-threatening myopathy in the neck and respiratory muscles of humans that develops several days after acute OP poisoning and appears to be associated with an OP-induced excess of acetylcholine (Eyer 1995). In contrast, OPIDN is a delayed (2–3 weeks) and prolonged (sometimes permanent) ataxia resulting from a single or repeated exposure to selected OPs that occurs in domestic chickens and man, but is not related to an accumulation of acetylcholine in nerve synapses (Sultatos 1994). The conditioned taste aversion (up to 28 days) in red-winged blackbirds following acute exposure to the OP parathion (Nicholas and Lee 1999), referred to earlier in this chapter, is another example, in this case a learned behaviour.

Ecological context is, therefore, more likely to change as the lag between exposure and the manifestation of behavioural effects (latency) increases. If effects are delayed, affected individuals may avoid having to cope concurrently with associated chemical-induced alterations in their environment. Or, delays may force exposed individuals to deal with alterations in behaviour at a time of greater challenges (e.g., mobilisation of organochlorines from fat to the brain of birds during migration, Stickel 1973).

1.3.2.2 Chemical interactions

It is likely that fish and wildlife are frequently exposed to multiple chemical stressors (for review, see Yang 1994). Unfortunately, little is known about the effects of chemical interactions on the severity of behavioural alterations let alone their physiological bases. [This is probably because of logistical and financial constraints and the current regulatory focus on individual chemicals (i.e., active ingredients, see Porter *et al.* 1999 and subsequent sections of this chapter).] Most of our knowledge of chemical interactions (additive effects, synergism, and potentiation; Eaton and Klaassen 1996) comes from human health related studies (e.g., drug interactions) and not behaviour, and even these studies are limited (Carpenter *et al.* 1998). However, there is reason to believe that chemical interactions play a significant role in altering the magnitude and/or persistence of behavioural effects in humans as well as fish and wildlife. For example, endocrine, immune and behavioural changes were observed in laboratory mice (*Mus musculus*) exposed to mixtures of agricultural chemicals (i.e., nitrates + individual pesticides) at concentrations detected in Wisconsin groundwater, but rarely when the chemicals were tested singly (Porter *et al.* 1999). The authors suggest that changes in thyroid hormone concentrations which were consistently associated with the mixtures and not individual chemicals may have been the mechanism underlying the observed effects. In addition, a number of recent studies have suggested a relationship between pesticide exposure and physiological and/or behavioural abnormalities in pesticide applicators, farmers and residents within areas of heavy pesticide use and their offspring (Garry *et al.* 1996, Parron *et al.* 1996, Cole *et al.* 1997). Whether or not exposure to multiple chemicals results in greater or lesser toxicity likely depends on the order in which the chemicals are encountered, their site(s) of action, and resultant effects on metabolic processes (chemical activation of toxic metabolites or detoxification processes, Hill 1995; Carpenter *et al.* 1998). For example, the toxicity to naive birds and laboratory rodents is additive when acutely challenged with combinations of ChE inhibitors (organophosphates and carbamates) and or organochlorines (Hill 1995). Synergistic effects when they occur in wildlife are usually less than a two-fold increase in toxicity and rarely more than three-fold, irrespective of whether exposure is via gavage (acute) or short-term dietary (Hill 1995).

Some investigators have avoided many of the logistical and financial constraints associated with studies of multiple chemical exposures by conducting studies of animals from polluted environments in the laboratory (e.g., Zhou and Weis 1998), or conducting studies in the polluted environments themselves (e.g., Rands 1985, Greig-Smith 1991), or both (Clements 1999). These approaches facilitate determinations of biological and ecological significance, but may or may not provide information of value within the current regulatory framework, in that chemical specific cause-effect relationships are frequently not examined.

1.3.2.3 Availability of resources

The availability (distribution and abundance) of critical resources may be important in determining whether or not chemical-induced behavioural effects ultimately reduce survival or reproduction. The greater the availability of a particular resource (food, cover, nest sites, territories, mates), the more likely that intra- and interspecific competition for these resources would be at a level at which behaviourally-impaired individuals could still meet their energetic and reproductive needs.

In most cases, affected individuals must not only cope with the physiological effects of chemical exposure, but also chemical-induced changes in their environment (indirect effects), particularly changes in the distribution and abundance of food resources. Frequently, these environmental changes are of a scale limiting the ability of affected individuals to move to undisturbed habitats, or site fidelity is maintained because alternative food sources remain, or because of expended energetic costs associated with advanced stages of reproduction. Environmental changes may be pronounced such as the alteration of the composition and structure of plant communities following herbicide applications resulting in loss of cover or nest sites (Best 1972), food resources

(Chiverton and Sotherton 1991), or both (Morrison and Meslow 1984, Anthony and Morrison 1985), or more subtle such as a reduction in only the abundance of the target species following application of an insecticide (Douthwaite and Fry 1982). Unfortunately, studies of the relationship between the degree behavioural alterations due to direct toxicity and varying availability of critical resources (indirect effects) are lacking.

We do know that reductions in target (and frequently nontarget) insects following insecticide applications can be severe; reductions of 50% to 100% have been reported (Barrett and Darnell 1967, Doane and Schaefer 1971, Hunter *et al.* 1984, Powell 1984, Pascual and Peris 1992, Pascual 1994, Fair *et al.* 1995, Martin *et al.* 1998). The magnitude of associated cumulative behavioural effects in wildlife (due to direct toxicity and chemical induced changes in the environment) varies among studies and appears to depend on the severity of the reduction of target insects and the availability of alternative prey. For example, Powell (1984) did not detect a change in prey selection or reproductive behaviour or success of red-winged blackbirds following a 50% reduction in eruciform larvae (primarily *Polia* spp.), the birds' primary food, within wet meadow habitat sprayed with fenthion to control mosquitoes. In comparison, the 100% mortality of the target lepidopteran larvae of the green moth (*Tortrix viridana*) within Spanish oak forests following application of malathion did not alter the reproductive success of blue tits (*Parus caeruleus*), but the birds did shift to the larvae of another lepidopteran (*Archips xylosteana*) whose abundance was unaffected by the insecticide (Pascual 1994). Similar responses have been observed in small mammals. White-footed mice (*Peromyscus leucopus*) compensated for the loss of lepidopteran prey following an application of diflubenzuron to control gypsy moths (*Lamantria dispar*) by using other food sources more (Seidel and Whitmore 1995).

These results are in contrast to the significant reductions in reproductive success of blue tits following an application of cypermethrin also to control green moth larvae in Spanish oak forests; however in this case, all potential arthropod food sources were eliminated (Pascual and Peris 1992). In this study, and the others described above, affected animals remained on site. We found only a few examples in the literature documenting off-site movements in response to contaminant-induced reductions in food resources, and in each a relatively small area was treated. For example, application of dimethoate to clover resulted in about a 90% reduction in insects compared to an unsprayed portion of the field. Within the 28 days postspray, a greater proportion of the composition of the small mammal community on the sprayed plots (ca. 1 ha) consisted of herbivores (prairie deer mice, *Peromyscus maniculatus*, and prairie voles, *Microtus ochrogaster*) than omnivores (house mice) compared to prespray conditions due to emigration of house mice and immigration of prairie voles (Barrett and Darnell 1967). Similarly, warblers appeared to leave portions of forested areas sprayed with carbaryl to utilise 'refugia' with a greater abundance of insect prey created by uneven application of the insecticide (only

30% of the area was 'effectively' sprayed, Hunter and Witham 1985). However, when the insecticide was applied to another area more uniformly the following year, no off-site movement of the birds was detected.

Shifts in foraging behaviour and/or habitat use may occur within contaminated areas. For example, in the study by Hunter and Witham (1985), warblers that utilised the sprayed forest patches shifted their foraging from the upper canopy of conifers to lower heights and to a greater variety of trees, presumably where arthropod abundance was greater. In another study, three species of tits (*Parus cristatus*, *Aegithalos caudatus* and *P. ater*) shifted to less energy costly foraging methods within forests with thinned canopies and fewer arthropods caused by acidic deposition from a coal fired power plant (Brotons *et al.* 1998). Vigilance was also greater within the polluted environment where the thin canopies provided less protection from predators. The degree to which alternative foods are contaminated may also be important. Alternative prey may be less contaminated, may serve as a continued source of chemical exposure, or possibly the primary source. Repellency, pesticide-induced anorexia, and conditioned taste aversions may also reduce the benefits of the availability of alternative prey (for discussion, see above).

1.3.2.4 Intra- and interspecific interactions

Contaminants can significantly alter competitive abilities of individuals directly through the direct toxic effects of sublethal exposure on their behaviour and/or indirectly through chemical-induced changes in their environment (for review, see subsequent sections of this chapter). Chemical-induced changes in competitive abilities may be enhanced or offset by concurrent chemical-induced changes in the availability of resources, reductions in the abundance or abilities of competitors, or both. Unfortunately, studies of the competitive abilities of individuals whose behaviour has been chemically altered in relation to chemical-induced changes in their environment are lacking. A few studies have documented changes in competitive interactions caused by the loss (poisoning) of a competitor. Hunter *et al.* (1986) documented a substantial overlap in the diets of juvenile black ducks (*Anas rubripes*) and fish within circumneutral ponds in Maine. On acidic ponds (pH = 4.5–4.8), ducklings spent less time searching and moving and more time feeding and resting and, as a result, grew faster than their counterparts on circumneutral ponds in response to the absence of fish competitors. Similarly, a low pH environment enabled larvae of the tree frog *Hyla femoralis* to compete more effectively with larvae of *Hyla gratiosa* because the latter suffered decreased survival at low pH (Warner *et al.* 1993). The intensity of intraspecific competition also varied with pH in *H. gratiosa*, but not *H. femoralis*. At low pH (the more stressful environment), survival of *H. gratiosa* was reduced as density increased, but there was no density dependent affect on survival at high pH (Warner *et al.* 1991).

Whether or not chemical-induced changes in behaviour ultimately result in decreased survival due to predation likely depends on the severity and duration of behavioural effects, the conspicuousness of affected individuals, and the abundance of predators. Although numerous studies have documented decreased abilities of prey given sublethal chemical exposures to avoid predators under laboratory conditions (see below and other chapters in this book), few have reported reduced survivorship in the field. (An exception may be avian responses to poisoned prey following application of insecticides, but here it is not clear whether the affected animals were dead, dying, or actually suffering from a sublethal dose.) In some cases, a significant numerical response by predators is unlikely because of their inherently low densities.

Social aggregations consisting of single species or mixed species may mitigate or enhance the impacts of chemical-induced alterations in behaviour. They may confer a number of benefits to their individual members in terms of survival and/or reproduction. Location and capture of prey may be enhanced by aggregations of foraging individuals (for review, see Alcock 1998:341 - 381). One can hypothesise that chemical-induced changes in foraging behaviour may be compensated for by the benefits afforded by the group as long as affected individuals can physically maintain their association with the group. However, benefits associated with foraging efficiency of social aggregations may come at an increased risk of severe chemical exposure. The largest die-offs of wildlife (i.e., birds) have been associated with flocking species because large numbers are simultaneously exposed with little warning of impending toxicity (Grue *et al.* 1983, Goldstein *et al.* 1999). In contrast, the shoaling behaviour of whitefish enhanced avoidance of cadmium exposure (McNicol *et al.* 1996).

Social aggregations have also been shown to convey greater protection from predation compared to single individuals. Whether or not aggregations mediate the potential increased risk of predation of individuals suffering chemical-induced changes in behaviour is not known. Bobwhite quail exposed to a sublethal dose of methyl parathion were unable to maintain their affiliation with their original covey, and in some cases attempted to affiliate with coveys with adjacent ranges (Buerger *et al.* 1991). In other cases, unexposed individuals within aggregations may suffer because of toxic effects on some members of the group, particularly behaviourally dominant individuals. For example, contaminants may have contributed to the death of a dominant male sperm whale (*Physeter macrocephalus*) and the subsequent stranding and death of three younger animals that maintained their social affiliation with the dominant male (Joiris *et al.* 1997). McEwen and Brown (1966) observed 'an unusual amount of intensive challenging and fighting' among territorial male sharp-tailed grouse (*Pedioecetes phasianellus*) following exposure to dieldrin and malathion, and in some cases, effective breeding behaviour was terminated.

1.3.2.5 Environmental conditions

Environmental conditions (defined here as abiotic factors excluding chemical contamination and interactions among pollutants) are important in shaping community structure and the interactions among species (Dunson and Travis 1991). Abiotic factors dictate how effectively an organism can function physiologically (and therefore, behaviourally) within a particular habitat, and whether or not its performance in that habitat is optimal (Walter 1973). If individuals or species differ in their responses to a particular gradient of an abiotic factor, changes in intra- or interspecific interactions can be expected. For example, Dunham *et al.* (1989) found that intraspecific competition intensified in iguanid lizards (*Scelopterus merriami*) when exposed to suboptimal temperatures. The classic studies by Park (1962) demonstrated differences in the competitive abilities between two species of flour beetle (*Tribolium* spp.) when exposed to flour under six different climatic conditions. The studies by Warner and her coworkers (1991, 1993) described above on the effects of pH on intra- and interspecific competition in larval amphibians are other examples. In fact, these studies suggest that the ability of normally inferior competitors to tolerate a more stressful environment frequently gives them a competitive advantage under these conditions. Changes in environmental conditions (e.g., weather) may also differentially alter the rates of feeding in predators and the abundance of prey (Kingsolver 1989).

Although it is well known that environmental conditions (e.g., temperature) can significantly alter exposure to and toxicity of contaminants (for recent review, see Rattner and Heath 1995), the role of environmental conditions (as defined here) in mitigating or enhancing the impacts of chemical-induced alterations in behaviour has received little attention. As noted above, environmental conditions may significantly alter intra- and interspecific interactions. The intensity of competition for food resources may be altered by the effects of abiotic factors on prey availability or the foraging intensity of competitors. Similarly, the risk of predation may be altered by the effects of abiotic factors on the activity and energetic demands of predators and the availability of alternative prey. In addition to these indirect effects, environmental conditions may ultimately dictate whether or not an organism's inability to correctly respond behaviourally to these conditions will reduce its chances of survival. For example, common tern (*Sterna hirundo*) chicks exposed to lead were less able to seek shelter (shade) from high temperatures and avoid heat stress than controls (Burger and Gochfeld 1985). Similarly, erigonid spiders (*Oedothorax apicatus*) exposed to the synthetic pyrethroid insecticide, deltamethrin, were less able to select between favourable (moist) and unfavourable (dry) habitat conditions (Everts *et al.* 1991). Furthermore, it has been shown that acute exposure to a variety of contaminants (e.g., metals and ChE inhibitors) causes a reduction in body temperature in birds and small mammals (Rattner and Franson 1983, Gordon *et al.* 1988, Gordon 1994). Animals suffering contaminant-induced hyperthermia select cool temperatures which in turn increase the hyperthermic effect of the contaminant(s) (Gordon *et al.* 1988, Gordon 1994). This response ('regulated hyperthermia', Gordon *et al.* 1988) may be protective in that, in many cases, the toxicity of contaminants to homeotherms increases with increasing body temperature (Gordon *et al.* 1988, Rattner and Heath 1995). However, high or low ambient temperatures may reduce survival if affected animals are behaviourally unable to select a more moderate (appropriate) environment (e.g., see Rattner and Franson 1983).

1.4 DIRECT BEHAVIOURAL EFFECTS

In view of the importance of the nervous system to most if not all of the physiological processes and mechanisms underlying behaviour and the fact that many pollutants, particularly pesticides, are neurotoxins, the potential for these chemicals directly to alter behaviour is high. For the purpose of our chapter, we define direct effects as behavioural changes observed in animals that have been exposed to a particular pollutant or chemical mixture. Behavioural responses resulting from the detection of pollutants in the environment are included even though the response(s) may prevent intoxication. An exhaustive review is beyond the scope of this chapter. We refer the reader to the most recent reviews available (Peakall 1985; Atchison *et al.* 1987; Grue *et al.* 1991, 1997; Zinkl *et al.* 1991; Burger 1995; Eisler 2000), as well as other chapters in this text. However, we have made an effort to include selected examples of the variety of ways pollutants have been shown directly to alter behaviour in fish and wildlife. The biological, ecological and regulatory significance of these behavioural effects is discussed in Section 1.8 of our chapter.

Knowledge of the direct effects of contaminants on animal behaviour varies among chemicals, chemical classes and among taxa (see reviews listed above). In comparison to other contaminants, the direct effects of lead and ChE inhibitors on behaviour have been well studied, undoubtedly because they are both nerve poisons and have been shown to pose significant threats to human health. In terms of taxa, the number of studies on the behavioural responses of fish, birds and laboratory rodents to contaminants exceed those for invertebrates, amphibians, reptiles and other species of mammals. For example, information on behavioural effects was generally lacking in recent reviews of the effects of contaminants on amphibians and reptiles (Sparling *et al.* 2000) and marine mammals (O'Shea 1999). Past emphases probably reflect economic considerations, historical differences in inherent sensitivity and exposure, human health concerns, ease of study and societal values (Hopkins 2000).

Behavioural responses to contaminant exposure may vary among species because of differences in ecological context (reviewed in Section 1.3) and/or behavioural repertoires. In addition, different chemicals or combinations of chemicals can elicit very different behavioural responses (e.g., Rice *et al.* 1997, Porter *et al.* 1999), and the direction of the response may vary with dose (e.g., hormesis, Janssen *et al.* 1993). As a result, generalisations are difficult in most cases (see Section 1.7). An exception are AChE inhibitors for which the 'general picture of acute ... intoxication is mainly one of depression, as shown by a reduction in a wide variety of behavioural outputs, both innate and learned

(Bignami *et al.* 1975:156; for recent examples, see Dell'Omo and Shore 1996a, Sparling *et al.* 1997). [We note that, in some cases, low doses of ChE inhibitors may stimulate activity (Brunet *et al.* 1997). In others, periods of hyperactivity may precede or follow periods of reduced activity (Hart 1993, Fryday *et al.* 1995).] Furthermore, exposure to ChE inhibitors has been shown to alter daily activity patterns (Brunet and Cyr 1992). Such chemical-induced changes in activity undoubtedly contribute to many of the more specific behavioural effects (e.g., reduced nest attentiveness and predator avoidance) described below, and are therefore, not reviewed separately here.

The vast majority of studies on the effects of contaminants on the behaviour of fish and wildlife have focused on acute exposures and short-term effects, and this is reflected in our review below. However, there is increasing concern about the effects of more chronic low level exposures, and that 'silent doses' (Needleman 1995), acute or chronic, may result in significant behavioural effects later in life or in subsequent generations (see Sections 1.3.2.1 and 1.6.2). Psychiatric disorders such as depression and its association with an increased risk of suicide in farm workers chronically exposed to OPs are an example of the former (Parron *et al.* 1996, Amr *et al.* 1997), although comparable data for fish and wildlife are lacking. A recent example of the latter is provided by Faulk *et al.* (1999). Parental exposure to DDT resulted in reductions in burst speed in response to a visual stimulus, response to a vibratory stimulus, and swimming activity in larval Atlantic croaker (*Micropogonias undulatus*), all of which are important in predator avoidance and/or foraging.

1.4.1 PERCEPTION OF ENVIRONMENTAL STIMULI

Within this category of chemical-induced behavioural effects, we focus on alterations in an animal's ability to detect natural chemical cues in its environment and subsequent behaviours, as well as behavioural responses to the detection of chemical contamination (e.g., avoidance or attraction). We note that exposure to contaminants can also affect an animal's ability to sense other stimuli. For example, mallard (*Anas platyrhynchos*) ducklings from hens fed a diet containing 0.5 ppm methyl mercury were less responsive to maternal calls than controls, but were hyper-responsive to a frightening stimulus (Heinz 1979). Larval bullfrogs (*Rana catesbeiana*) from sites heavily polluted by a coal-burning power plant were much less responsive to a physical stimulus than tadpoles from an unpolluted site (Hopkins *et al.* 2000). Similar results were reported by Raimondo *et al.* (1998).

1.4.1.1 Natural chemical cues

Several recent studies have examined the effects of aquatic toxicants on chemoreception. Moore and Waring (1996) demonstrated a significant effect on the olfactory mediated aspects of reproduction in mature male Atlantic salmon parr (*Salmo salar*) exposed to diazinon (an OP pesticide) at environmentally realistic concentrations. Exposed males were less responsive to prostaglandin, an odorant in mature males, and urine from ovulating females (pheromones), both important in synchronising reproductive physiology and behaviour. McCormick *et al.* (1998) suggested that pollutants may also interfere with the development of olfactory imprinting by juveniles and its expression (homing by adults) in Atlantic salmon. In a subsequent study, Sholtz *et al.* (2000) reported impaired olfaction in juvenile and adult Pacific salmon (chinook, *Oncorhynchus tshawytscha*) exposed to environmentally realistic concentrations of diazinon. Antipredatory responses in juveniles and homing in adults were disrupted. Similarly, the ability of snails (*Lymnaea palustris*) to detect and respond to chemical cues emanating from an extract of crushed conspecifics was impaired by exposure to heavy metals (Lefcort *et al.* 1999). Unexposed snails changed their behaviour in response to crushed snail extracts (indicative of threat of predation), whereas snails exposed to heavy metals did not. Saglio and Trijasse (1998) studied olfactory-mediated antipredatory responses in goldfish (*Carassius auratus*) exposed to the herbicides, atrazine or diuron. Significant decreases in grouping, sheltering, and/or changes in swimming orientation in the presence of a skin extract solution were observed in fish exposed to the herbicides. Instead of the normal avoidance of the skin extract, goldfish exposed to diuron were attracted to the alarm substance.

1.4.1.2 Contaminants

Chemical avoidance and attraction are well documented in fish exposed to a variety of contaminants and these responses, particularly avoidance, can occur at concentrations well below those at which other effects are observed (see Section 1.6.1). For example, Woodward *et al.* (1997) found that cutthroat trout (*Oncorhynchus clarki lewisi*) avoided copper and zinc even with a 90-day acclimation to the metals, but did not avoid water contaminated with cadmium or lead. Trout avoided zinc concentrations that were one sixth or less of those measured in the South Fork and lower Coeur d'Alene River basins in Idaho. In a similar study, rainbow trout (*O. mykiss*) avoided a mixture of metals (copper, cadmium, lead and zinc) at concentrations as low as 10% of that occurring in the Clark Fork River in Montana (Hansen *et al.* 1999). Steelhead (*Salmo gaidneri*) avoided domestic sewage effluent whereas chinook salmon (*O. tshawytscha*) and striped bass (*Morone saxatilis*) were attracted to the discharges from oil refineries (Smith and Bailey 1990).

Birds and mammals have been shown to avoid contaminated diets under laboratory (e.g., standardised toxicity test) conditions (e.g., Schafer *et al.* 1983, Linder and Richmond 1990, Luttik 1998; for a review of chemical repellents for birds, see Conover 1984 and references therein). Avoidance of contaminated diets may be induced through a variety of means that are not mutually exclusive. Chemicals may induce food avoidance during the act of feeding because the diet is distasteful (taste repellency, Conover 1984). Avoidance can also occur because of illness following ingestion of the chemical. It may be of short duration (i.e., a reduction in consumption of contaminated diets associated with a threshold dose which lasts only while the animal is exposed [chemical-induced anorexia, Grue 1982, Grue *et al.* 1997]), or be long lasting (conditioned aversion, e.g., conditioned taste aversion, CTA, Nicolaus and Lee 1999). The duration of the aversion appears to depend on the availability of alternative foods and the number of cues associated with the presence of the chemical (Conover 1984).

An animal's ability to avoid contaminated diet may be altered by the toxic actions of the chemical(s) in the food on sensory processes. For example, a variety of commonly used insecticides and herbicides were found to alter the electrophysiological response within the tongue of female Mongolian gerbils (*Meriones unguiculatus*) to 11 taste stimuli (Schiffman *et al.* 1995). In most cases, exposure to the pesticides resulted in a suppression of taste responses, but some exposures increased sensitivity to taste stimuli. However, there were no clear directional patterns within or among chemical classes or among taste stimuli. Some chemical concentrations inhibited responses to a given taste stimuli whereas others were stimulatory. Similarly, prior exposure to methyl parathion (6 mg/kg body weight) reduced the ability of 2-week-old northern bobwhite (*Colinus virginianus*) to discriminate between diets with the insecticide and untreated food (Bussiere *et al.* 1989). During the first day of the discrimination test, the OP-dosed quail actually preferred the treated diets, whereas controls avoided the treated food.

It is appropriate to reiterate the importance of ecological context in governing the magnitude of chemical exposure and resultant behavioural effects (see Section 1.3) by providing a few additional examples that relate to an animal's ability to perceive and/or avoid contaminants and illustrate the difficulties in making generalisations. Juvenile spot (*Leiostomus xanthurus*) which feed on benthic prey were tested in laboratory avoidance experiments using clean sediment and sediment with polycyclic aromatic hydrocarbons (Hinkle-Conn *et al.* 1998). Based on location and feeding strike patterns, spot did not avoid the contaminated sediment. Pairs of black oystercatchers (*Haematopus bachmani*) with territories that contained both oiled and unoiled foraging sites in Prince William Sound, Alaska, spent less time feeding on persistently oiled shorelines (Andres 1999). Mussel densities were similar among shorelines and therefore did not appear to influence differential use of the areas and adults were occasionally observed to reject prey items on the oiled sites. The author suggests that use of multiple foraging sites and the variety of available prey probably reduced the negative effects of occasionally ingesting oil-contaminated mussels (Andres 1999). Similarly, laboratory studies by Bennett (1989a,b) indicate that, in some cases, the ability of birds to discriminate among contaminated diets depends on the number and availability of treated and untreated alternatives, as well as the bird's ability to detect the chemical. For example, 14-day-old northern bobwhite were given one of four exposure scenarios (1 feeder with treated diet or negative control; 1:1, i.e., 1 feeder treated diet: 1 feeder negative control; 5:5 or 9:1) for several dietary concentrations of a variety of organophosphates and carbamates. Birds exposed to the 1:1 scenario were able to avoid ingesting lethal amounts of the chemicals, whereas birds given the 9:1 were not and suffered mortality similar to those in the single feeder scenario. And finally, recent studies McNicol and his coworkers (McNicol *et al.* 1999 and references therein) illustrate the role competing environmental gradients can play in the avoidance of chemical contamination. For example, lake whitefish strongly avoid cadmium in the absence of a light gradient (McNicol *et al.* 1999). However, when the fish encounter competing gradients of cadmium and light, their attraction to shaded conditions suppresses avoidance of toxic concentrations of the metal, even over an extended period (18.5 h).

1.4.2 MEMORY AND LEARNING

Memory of learned behaviours and cues is critical to the reproduction and survival of fish and wildlife. There are several lines of evidence linking contaminant exposure to impairments in memory and learning. However, in most cases, interpretation of this body of literature is difficult because of the number of different research paradigms used and the complex nature of the memory process. In addition, response deficits may be caused by other peripheral effects such as muscle weakness, an inability to perceive the conditioned stimulus, and/or a reduction in motivation (for review, see Cuomo *et al.* 1996, Kulig *et al.* 1996).

The effects of lead and ChE inhibitors on memory and learning in surrogates for man have received considerable attention. Needleman and his coworkers have demonstrated that extremely low levels of lead exposure ('silent' doses) can result in irreversible decrements in learning (Needleman *et al.* 1990, Needleman 1995). Acute and chronic exposure to ChE inhibitors can also disrupt cognitive processes (for review, see Grue *et al.* 1991), although, in most cases, effects are short-lived with recovery associated with the synthesis of new enzyme, spontaneous reactivation, or the development of tolerance (see Section 1.3.1.3). However, the extent to which such deficits occur in free-living fish and wildlife and their biological significance (i.e., effects on reproduction or survival) is largely unknown.

Burger and her coworkers have documented deficits in the memory of juvenile herring gulls (*Larus argentatus*) exposed to lead (as lead nitrate). Recognition of their caretaker was delayed in lead-exposed young compared with controls. Whereas untreated chicks demonstrated a preference for their caretaker by 5 days of age, lead-impaired chicks did so by day 10 or 14, depending on the dosage given (0.1 vs. 0.2 mg/g body weight, Burger 1990). Variations in their ability to recognise their caretaker were also related to differences in the timing of the dose. Birds exposed to lead on day 6 chose their caretaker correctly more often, but more slowly, than birds exposed to lead on day 2 (Burger and Gochfeld 1993). Common tern chicks given 0.2 mg of lead nitrate per kg body weight at 5 days of age were less able to respond quickly to a novel-feeding situation than controls (Burger and Gochfeld 1985). Following dosing chicks were conditioned to accept fish positioned with their head to the right of the birds for 25 days (four times per day) after dosing. The head position of the fish was then reversed. Lead-dosed chicks took four times longer to properly grab and swallow the first fish given to them in the new position compared with controls. Results of the two studies suggest lead-induced deficits in perception, memory, coordination and learning.

Cholinergically active compounds can affect memory in a number of ways (for review of existing models, see Deutsch 1983). ChE inhibitors may 'scramble' a strong memory by increasing synaptic conductance through an increase in acetylcholine at nerve synapses. Alternatively, they may not affect, or may even enhance, an old, faded memory by raising synaptic conductance within the retrievable range (Mineau *et al.* 1994a). For example, Vyas *et al.* (1995) exposed adult and juvenile migratory white-throated sparrows (*Zonotrichia albicollis*) to 256 ppm acephate in their diet for 14 days. After 7 days of exposure to the organophosphate, treated adults exhibited random activity and did not establish a preferred migratory orientation. However, treated juveniles, birds without previous migratory experience, orientated themselves correctly, suggesting that exposure to the chemical impaired memory of the migratory route in adults (Vyas *et al.* 1995). Mineau and his coworkers used food caching as a natural paradigm for studying the effects of sublethal exposure to anticholinesterases on memory in black-capped chickadees (*Parus atricapillus*). In contrast to the effects of acephate on the migratory orientation in white-throated sparrows noted above, chickadees orally dosed with the OP insecticide fenitrothion before or after caching were as, or more successful, at relocating stored food as controls (Mineau *et al.* 1994a). Results were similar to those of a companion study with the carbamate physostigmine (Mineau *et al.* 1994b). Reasons for the lack of memory impairment in the chickadees are not clear. The authors suggest that either the level of cholinergic activity required for cache site recall may be so low that the dose given was ineffective, or the signal — noise ratio in the memory trace is sufficiently high to offset any impairment (Mineau *et al.* 1994a).

1.4.3 THERMOREGULATION

In cold-blooded vertebrates (poikilotherms, ectotherms), temperature regulation is mainly behavioural. Examples include the 'shuttling' back and forth between shade and sun in reptiles and the selection of favourable water temperatures by amphibians and fish. Behavioural responses are also important in thermoregulation of young and adult warm-blooded vertebrates (homeotherms, endotherms). For example, neonatal birds and mammals (that lack the ability physiologically to maintain body temperature) will move to inanimate or animate sources of heat or shade when chilled or heat stressed, respectively (see review by Satinoff and Henderson 1977). Shading of young (altricial) brown pelicans (*Pelecanus occidentalis*) by their parents within exposed rookeries on desert islands and the nocturnal activity patterns and/or fossorial habits of desert rodents, are specific examples of behavioural responses to high temperatures (Bartholomew 1964). As noted by Satinoff and Henderson (1977), reflexive and operant responses to thermal stress are functionally and anatomically separate, and animals can compensate for deficits in one system through the mechanisms of the other. In fact, many studies have shown that when given a choice between behavioural or reflexive (autonomic) responses, the former is preferred probably because of the lower energy demands and greater efficacy (Gordon 1994).

Little is known about the effects of environmental contaminants on behaviours associated with regulating body temperature in animals. An exception is the effects of ChE inhibitors on thermoregulation in laboratory mammals and man, probably because of human health concerns related to use of these chemicals as pesticides and warfare agents. Studies indicate that in birds and mammals, inhibition of AChE activity of greater than 50% is associated with a significant but transient reduction in body temperature (hypothermia) (Rattner and Franson 1983, Gordon 1994, reviewed by Grue *et al.* 1997). In laboratory mammals, these reductions in core temperatures are associated with the selection of cooler ambient conditions (Gordon 1994), a behavioural response common to laboratory mice and rats following acute exposure to a variety of chemicals including heavy metals, solvents, pesticides and natural toxins (Gordon 1993:205–223). Therefore, underlying the marked hypothermia caused by ChE inhibitors is the natural response of mammals to lower their body temperature (metabolism) to minimise the adverse effects of toxins (Gordon 1993:205–223). This is in sharp contrast to physiological and behavioural responses to infection in fish, amphibians, birds and mammals in which warmer ambient temperatures are selected and core body temperatures are elevated (fever) (Satinoff 1979). Unfortunately, the interactions between the physiological and behavioural responses to chemical toxicity and those associated with infection are poorly understood.

In addition to inducing voluntary movements toward favourable ambient temperatures, contaminants may inhibit such movement. As noted above, movement to warmer or cooler temperatures is an important aspect of thermoregulation in neonatal birds and mammals. For example, common tern chicks given 0.2 mg of lead as lead nitrate per g body weight took longer to get to shade, spent less time there, and were out of the shade (once they reached it) more often than controls (Burger and Gochfeld 1985). Lead dosed chicks also often selected an object that did not provide shade and remained there instead of moving to the more appropriate alternative.

1.4.4 FEEDING

Contaminants can directly affect feeding in a number of ways and some have already been discussed such as sensory and motivational aspects (e.g., detection and avoidance of contaminated diets, Section 1.4.1.2). Here we focus on the effects of contaminants on the physical abilities and decision-making processes necessary for animals to locate, capture and consume their food. Effects on predator detection and avoidance are discussed in Section 1.4.6.1.

In some cases, impaired feeding abilities may be the result of chemical-induced morphological abnormalities. For example, swimming and feeding were impaired in gray treefrogs (*Hyla chrysoscelis*) that suffered malformations (e.g., lordosis, malformed legs, missing eyes) following exposure to methyl mercury or one of three agrichemicals (Britson and Threlkeld 1998). The authors concluded that tadpoles with these malformations are less likely to find the food resources necessary for growth and will remain in the larval stage for a longer period of time than normal. In other cases, impaired feeding may be the result of chemical-induced reductions in coordination and motor reflexes. Decreased begging by nestling songbirds following exposure to the OP pesticide dicrotophos, is one example, although the contribution of pesticide-induced anorexia is not clear (Grue and Shipley 1984). Similarly, herring gull chicks exposed to lead (100 mg/kg body weight as lead acetate) were less effective in stimulating their parents to regurgitate food by pecking their parents' bills than controls; treated chicks missed their parents' bill 16% of the time compared with 2% in controls. During feeding experiments, common tern chicks given a single intraperitoneal injection of lead nitrate were less able to manipulate fish presented to them than controls (Burger and Gochfeld 1985). Adult female house sparrows (*Passer domesticus*) given a single sublethal oral dose of the OP pesticide, chlorfenvinfos became clumsy and less able to husk awkward seeds without dropping them during the first 1.5 h after dosing and lost weight (Fryday *et al.* 1994). Similar changes have been seen in fish. Bluegill (*Lepomis macrochirus*) exposed to copper took significantly more time to handle their prey compared with controls with only a few exceptions, irrespective of prey density or exposure of prey to copper (Sandheinrich and Atchison 1989). Handling time did vary with prey size, with copper exposed fish having greater difficulty with small prey. Juvenile spot feed by obtaining sediment and manipulating it in their buccal cavity to remove meiobenthic prey. Cumulative processing time (a function of per-strike processing time and the number of strikes) decreased when the fish were exposed to sediment containing 122 ppm polycyclic aromatic hydrocarbons, largely because of a reduction in the number of strikes (Hinkle *et al.* 1998).

A number of studies have documented reductions in capture efficiency following exposure to contaminants. For example, foraging efficiency of grayling (*Thymallus thymallus*) exposed to methyl mercury during embryogenesis was

15% to 24% less than that of unexposed controls (Fjeld *et al.* 1998). Unexposed grayling caught two to six times more prey than exposed fish. Prey capture rates for adult mummichogs (*Fundulus heteroclitus*) collected from a site polluted with heavy metals and organics were significantly slower than those for fish from an unpolluted site (reviewed by Weis *et al.* 1999). Subsequent studies with mummichog larvae from the contaminated site indicated that young fish (3–4 months old) were more active and captured prey faster than larvae from the same clean site, primarily because of increased encounter rates between the larvae and their prey. However, by the time they were yearlings, prey capture rates declined. Differences between the polluted site and another control site, however, were not consistent.

Contaminants may not only affect the ability to handle and capture prey, but also decision-making processes underlying prey selection. For example, when simultaneously offered two sizes of daphnia (*Daphnia magna*), juvenile fathead minnows generally begin to switch from the older, larger and more difficult to catch daphnia to younger, smaller daphnia within 3 days. However, in feeding experiments, fish exposed to 0.5 ppm lead switched after 10 days and fish exposed to 1.0 ppm lead did not switch at all (Weber 1996). The increased energy required by lead exposed fish to capture the larger prey, and the fact that they satiated more quickly than controls resulted in smaller weight gains (Weber 1996).

1.4.5 INTRASPECIFIC INTERACTIONS

1.4.5.1 Communication

Communication is the process by which signals or cues from one animal influence the behaviour of another (for review, see Johnstone 1997). Therefore, communication is dependent on the ability of the 'signaller' to send the proper signal and that of the 'receiver' to 'accurately deduce the nature and intentions of the signaller' (Johnstone 1997). It represents the foundation for intraspecific interactions. Unfortunately, the effects of contaminants on the ability of fish and wildlife to communicate through vocalisation, movement, chemical cues, or other signals are poorly understood. Studies have largely focused on the signaller with little attention to the receiver (for exceptions, see Section 1.4.1.1). For example, sublethal exposure of birds to OP pesticides has been associated with marked declines in singing and displaying under both laboratory and field conditions (reviewed by Grue *et al.* 1997). In honey bees (*Apis mellifera*), 'dances' are an important means of communication. Bees exposed to insecticides perform unusual 'alarm' dances (reviewed by Cox and Wilson 1984). The 'wagtail' dance is used to communicate the location of nectar, water or a source of pollen. Upon returning from a feeding station, bees exposed to the organophosphate insecticide, parathion, performed a wagtail dance indicating the wrong location of the food source (Schricker and Stephen 1970). Movements (posturing, aggressive behaviours) are also important in predator avoidance and conveying behavioural dominance or submission (fitness) and there is evidence linking exposure to ChE inhibitors to enhancement of aggressive behaviour in both birds and small mammals (for review, see Grue *et al.* 1991). Urine marking by dominant males is a central feature in the reproductive strategy of many mammals, including mice. Vom Saal *et al.* (1995) observed that male mice exposed to *o,p*′-DDT and methoxychlor *in utero* marked their territories more frequently than controls and suggested that this change in behaviour might be associated with an increase in intermale aggression. The electric organ discharges (EODs) of tropical electric fish serve as a means of communication and perception of their surroundings. Thomas *et al.* (1996) found that the EOD of the tropical knifefish, *Apteronotus albifrons*, is extremely sensitive to cyanide and this was reflected in a decrease in discharge frequency and an increase in discharge waveform.

1.4.5.2 Social Organisation

Social organisation is a strategy to enhance communication, the location or capture of food, reproductive success and/or predator avoidance (see Section 1.3.2.4, Alcock 1998:555–601). Effects of contaminants on communication were reviewed in the previous section and, as in the case of communication, the effects of contaminants on social organisation represent a topic worthy of further study.

Fry and his coworkers have documented a variety of effects of OC pesticides, DDT and kelthane (Dicofol®) on the social interactions of birds. Abnormal female — female pairings in western gulls (*Larus occidentalis*) on the Channel Islands off the California Coast appear to have been due to DDT-induced feminisation of male embryos and sex ratios highly skewed toward females because of the failure of feminised males to reproduce (Fry *et al.* 1987). Feminisation of California (*Larus californicus*) and western gull embryos occurred following *in ovo* injections of DDT at concentrations comparable to those found in the eggs of western gulls on the islands (Fry and Toone 1981). Similar effects were observed in male American kestrels (*Falco sparverius*) from females exposed to *o,p*′-Dicofol (MacLellan *et al.* 1996). Second-generation males attempted fewer copulations when paired with comparable control females and were less likely to attain and maintain a high social rank in the presence of conspecifics when size and age of the birds were held constant due to reduced aggressiveness (MacLellan *et al.* 1997). McEwen and Brown (1966) observed the behaviour of two male sharp-tailed grouse on their breeding grounds following an acute oral dose of the OC dieldrin. One male given 10 mg per kg body weight was able to defend its territory for 3 days, but on the following day (probable time of maximum effect of the chemical) was repeatedly challenged and eventually defeated by an invading male. Similarly, another male (15.2 mg/kg) was also active for the first 3 days postdose, but on the fourth day remained on the edge of the central breeding area and occasionally ran through the central display area eliciting chases from other males. The interactions between contaminants and behavioural dominance in fish have been investigated in a series of studies by Henry and Atchison in which bluegill served as a model. In each of their studies with groups of fish with similar initial hierarchies, waterborne exposures to methyl parathion (Henry and Atchison 1984), copper (Henry and Atchison 1986), or cadmium and zinc (Henry and Atchison 1979), the most dominant and subordinate individuals were those most affected. Dominant individuals became more aggressive (threats and nips) and 'comfort movement behaviours' (S-jerks, fin flicks and partial jerks) increased in subordinates. Fish of intermediate rank were generally the least affected, the least active, and experienced the fewest social interactions. In none of these studies, did the authors report a change in social status among individual fish. The authors suggest that the observed responses reflect the fact that dominant and subordinate individuals are under the greatest social stress: dominants must expend energy to maintain their positions and subordinates must continuously avoid being attacked. An earlier study by Sparks *et al.* (1972) reported that subordinate bluegills died before dominant fish in acute toxicity tests with zinc. Placement of structures within the test tanks decreased social contact and resulted in more even effects across social ranks.

MacKenzie and Winston (1989) studied the effects of sublethal exposure to the OP pesticide diazinon on division of labour in honey bees. Surprisingly, few behaviours were affected (only 10% of the comparisons between treated and control bees were statistically significant). Significant effects were most often associated with alterations in the onset and duration of foraging and handling of nectar, but the direction of effects was not consistent. The effect of the OP parathion on the communication of locations of food resources among worker bees is discussed in Section 1.4.5.1.

Schooling in fish and flocking in birds are behavioural adaptations that facilitate communication, location of resources and/or predator avoidance. Fathead minnows exposed to cadmium (0.45 mg/l) for 48 h often changed direction within their school orientating themselves perpendicular to the axis of the school for several seconds before returning to their original position (Sullivan *et al.* 1978). In contrast, school cohesion ('uniformity and structure') in Atlantic silversides (*Menidia menidia*) exposed to low concentrations of copper was increased, despite a chemical-induced increase in swimming speed (Koltes 1985). Exposed fish decreased their rate of change of direction and nearest-neighbour distances, and swam more in parallel orientation than controls.

Evidence of direct effects of contaminants on flocking in birds is circumstantial and the little that is available is provided by a study of the effects of methyl parathion on the survival of free-living northern bobwhite quail (Buerger *et al.* 1991). Observations of three bobwhites given 4 or 6 mg of the organophosphate per kg body weight suggest an effect on covey integrity. One quail became separated from its covey for 3 days and was observed roosting on the periphery of another covey. Another bird was unable to maintain association with its covey for 5 days and attempted to join two other coveys with adjacent ranges. The third quail remained with its covey for 8 days after dosing, but then fluctuated between its covey and an adjacent one.

1.4.5.3 Reproduction

Under the previous subheadings of this section of our chapter, we have reviewed a variety of the known direct effects of contaminants on reproductive behaviour of fish and wildlife, and we will not repeat them here. The indirect effects of contaminants on reproductive behaviour are discussed in Section 1.5. Here, we focus on contaminant-induced changes in nest construction, nest defence, incubation and care of young by birds. We focus on these effects because they are the best documented of the effects in fish and wildlife not previously addressed (e.g., see Figure 1.2). Many of the studies described involve OP insecticides. This undoubtedly reflects the use of these chemicals during avian breeding seasons when insects, including pest species, are most abundant.

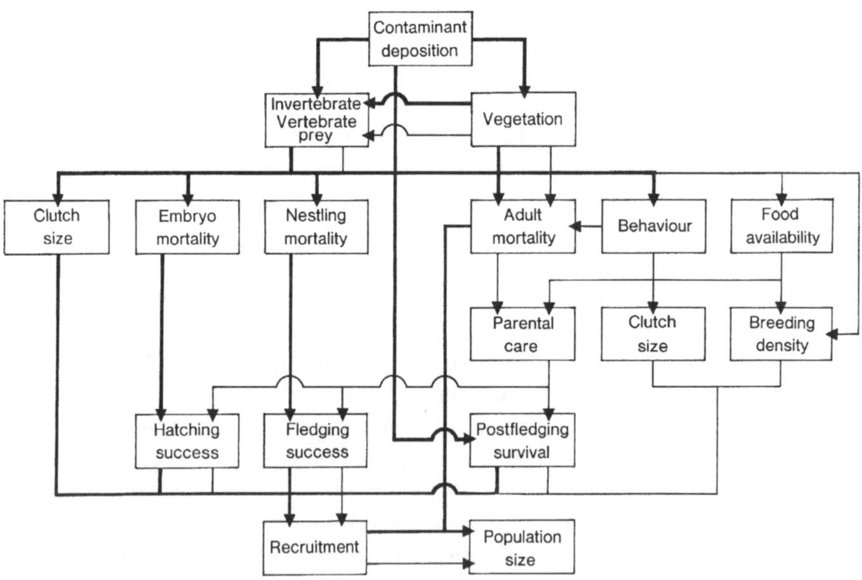

Figure 1.2 Potential direct (bold lines) and indirect (light lines) effects of contaminants on birds. The schematic may also be representative of potential effects on other taxa by changing to the appropriate habitat and reproductive terminology for fish, amphibians, reptiles or mammals (from Grue *et al.* 1986). Our review indicates that contaminant-induced changes in behaviour have been associated with most of these endpoints, although the amount of information varies among taxa.

Nest quality is important to reproductive success in birds. For example, the abundance of feathers in the lining of tree swallow (*Tachycineta bicolor*) nests is correlated with faster growth of young and a reduction in parasites. In addition, the size of the grass mat is correlated with increased clutch size (for review, see McCarty and Secord 1999a and references therein). Tree swallows exposed to PCBs along the Hudson River in New York State built nests that were smaller and of poorer quality (contained fewer feathers) than swallows from uncontaminated areas, effects that were independent of the age of the adults (McCarty and Secord 1999a).

Abandonment of territories, nests and young, and/or reductions in nest defence have been reported in adult free-living birds exposed to OPs, petroleum and organochlorines. Territory abandonment, in some cases due to poor defence, contributed to a significant reduction (ca. 30%) in the numbers of breeding white-throated sparrows following an application of the OP, fenitrothion, to control spruce budworm (*Choristoneura fumiferana*; Busby *et al.* 1990). Reduced nest defence in merlins (*Falco columbarius*) was correlated with exposure to organochlorines when exposure was expressed as the degree of eggshell thinning and residues in their eggs (Fyfe *et al.* 1976). A similar reduction in nest defence was reported in herring gulls exposed to organochlorines (Fox *et al.* 1978). McCarty and Secord (1999b) reported an unusually high occurrence of nest abandonment during incubation in tree swallows containing levels of PCBs that were among the highest reported for the species. Similarly, wedge-tailed shearwaters (*Puffinus pacificus*) orally dosed with weathered crude oil or to which oil was applied to their feathers abandoned their colony after dosing (i.e., during courtship) or their nests during incubation. Of the former which returned the following year, the 'predisposition to return and attempt to breed remained, but the cooperative effort required of both birds to hatch and raise a chick was evidently impaired' (Fry *et al.* 1986). More recently, Walton *et al.* (1997) reported an exceptionally low return rate of breeding kittiwakes (*Rissa tridactyla*) within a colony exposed to crude oil off the coast of Scotland the year before. Nest site and mate fidelity were also low. Observed effects were due to some adults not breeding for 1 or 2 years after the oil spill, and subsequently returning. More than 50% of hen mallards given 400 ppm of the OP methyl parathion, in their diet either abandoned their eggs or significantly reduced time incubating their eggs (i.e., left their nests for hours or days) while exposed to the chemical. Effects were greatest early in incubation (Bennett *et al.* 1991). The fact that chemical exposure frequently results in abandonment prior to egg laying or early in incubation may not only reflect toxicity, but also the minimisation of parental (energy) investment when intrinsic or extrinsic conditions are not favourable.

Other studies provide additional examples of contaminant-induced aberrant behaviour in birds caring for their eggs. Fox and his coworkers compared the incubation behaviour of herring gulls in colonies contaminated with organochlorines on Lake Ontario with that of birds in colonies less contaminated elsewhere.

Gulls in the Lake Ontario colonies spent less time on their eggs and, as a result, egg temperatures frequently dropped below the minimum necessary for survival and normal development. Within the Lake Ontario colonies, the total time birds were off nests that failed (no eggs hatched) was three times greater than that of successful nests (all eggs hatched). The time off nests was positively correlated with organochlorine residues in the eggs, an indicator of contaminant exposure in females. Exposure data for males were lacking (Fox *et al.* 1978). Captive female American kestrels whose mothers had been dosed with 5 mg of *o,p′* dicofol lost a greater number of eggs from their nests than controls. Eggs were considered lost if they were no longer in the nest box. The authors observed dicofol-dosed females eating their eggs or picking them up and dropping them on the cage floor, behaviours not observed in controls (MacLellan *et al.* 1996).

In species in which males and females share responsibilities for incubating the eggs, the ultimate effects of chemical-induced alterations in incubation behaviour on hatching success may depend on the level of intoxication of both caregivers. Less affected members of the pair may compensate for the aberrant behaviour of their mate. For example, one member of pairs of laughing gulls was given 6 mg of the OP parathion, per kg body weight at a time when they were incubating eggs. OP-dosed gulls spent significantly less time incubating their eggs during the second and third days after dosing. The member of the pair not exposed to the OP compensated for the inattentiveness of their mate, and as a result hatching success was not affected (White *et al.* 1983). In a similar study, female European starlings given 2.5 mg of dicrotophos per kg of body weight on day 6 of incubation, spent less time on their eggs than controls during the next 6 days (i.e., until hatching). However, their male partners (birds not exposed to the OP) spent more time incubating the eggs and hatching success was not affected (Grue, unpublished data). Dosages given to the gulls and starlings resulted in severe, but sublethal inhibition of brain AChE activity (ca. 50%).

In other species-contaminant scenarios, behavioural compensation for the inattentiveness of a mate may not occur or be insufficient. Butler *et al.* (1988) exposed one member of pairs of nesting Leach's storm-petrels (*Oceandodroma leucorhoa*) either internally or externally to crude oil or an oil-dispersant emulsion during the early or late stages of incubation or early chick rearing. Hatching success and chick survival were significantly reduced in the latter two exposure scenarios due to the temporary (several-day) abandonment of the nests by the treated member of the pair. In comparison, female red-winged blackbirds, a species in which the female incubates the eggs, took an average of about 2.5 h to return to their nests following a dose of 4.21 mg of methyl parathion which resulted in brain AChE inhibition of 35%. Once birds returned to their nests, behaviour was not significantly different from controls and hatching success was not affected (Meyers *et al.* 1990). Unfortunately, data for extended exposures are lacking.

A few studies have also reported altered parental care in birds exposed to contaminants. Free-living female European starlings given 2.5 mg of dicotophos per kg body weight made fewer sorties to feed their young than controls (Grue *et al.* 1982b). Nestlings of OP-dosed females lost weight compared with their control counterparts. Males were captured and moved to an aviary prior to dosing of the females and therefore could not compensate for the reduced attentiveness of their mates. Brewer *et al.* (1988) reported abandonment of broods by a female wood duck (*Aix sponsa*) and blue-winged teal (*Anas discors*) following aerial application of methyl parathion to a disked agricultural field and adjacent wetlands. Distraction display was reduced and distance flushed greater in willow ptarmigan (*Lagopus l. lagopus*) hens given 0.5 ul of cadmium per hour for 2 weeks compared with controls (Pedersen and Saether 1999). Neither difference was statistically significant, but survival of chicks associated with the cadmium-dosed hens was significantly lower than controls. Heinz *et al.* (1983) observed similar hyper-responsiveness (increased avoidance, i.e., distance moved when startled) in captive black duck ducklings after they and their mothers were fed 4 ppm cadmium.

Examples of aberrant behaviour in young birds suffering direct toxic effects of contaminants have already been described. Included are the reduction in begging of nestling songbirds exposed to OPs, the failure of common tern chicks dosed with lead to seek adequate shade, and the hyper-responsiveness of ducklings exposed to mercury or cadmium to a fright stimulus. Another example is provided by a recent study (Nocera and Taylor 1998) of the activity budgets of young common loons (*Gavia immer*) within a population with the highest mercury concentrations in North America. Elevated levels of mercury in the blood of the chicks was correlated with a decrease in the amount of time they spent brooding (riding on the back of an adult) and an increase in preening, and that the former was associated with a decrease in fledging rate. Energy expenditures by the chicks in the water were not compensated for by increases in feeding rates or begging.

In the review above, we focused on studies that have documented effects. However, we note that in some cases comparable studies have not detected similar behavioural responses for closely related species or chemicals. For example, observations by Fyfe *et al.* (1976) indicated that nest defence behaviour was reduced in merlins exposed to organochlorines, but not in prairie falcons (*Falco mexicanus*), although the latter's reproductive success was adversely affected at lower levels of contamination.

1.4.6 INTERSPECIFIC INTERACTIONS

1.4.6.1 Predator avoidance

Predator avoidance is a function of an animal's ability to minimise encounters with predators (avoid detection or detect predators before being attacked) and avoid capture once detected. We have already discussed many of the ways contaminants can affect behaviours important to predator avoidance (see Sections 1.4.1.1, 1.4.5.1 and 1.4.5.2). A review of the literature on fish (Mesa *et al.* 1994) indicates that prey impaired by any of a variety of stressors, including contaminants, are often more vulnerable to predation. In nine of the 10 studies reviewed involving contaminants (14 of the 15 chemicals), predators captured significantly more affected prey (referred to as 'substandard' by the authors). More recently, Weis and her coworkers have compared predator avoidance in adult and larval mummichogs from a creek in New Jersey polluted with a variety of contaminants to that of fish from less contaminated sites. Adult mummichogs from the contaminated creek were slow moving and, as a result, more vulnerable to predation by blue crabs (*Callinectes sapidus*) than conspecifics from less polluted sites (Smith and Weis 1997). However, the effects of contamination on the behaviour of larval mummichogs varied with age, as did predator avoidance. Young larvae (7 days posthatch) from the contaminated creek moved less, hid more and were less noticeable to predators (yearling mummichogs) than controls in which activity was greater (Zhou and Weis 1999). However, by the time they were 30 days old, larvae from the contaminated creek were more vulnerable to predation than control larvae of the same age because their ability to escape (swimming performance) was impaired. In a similar study with larvae from the same contaminated site, but with two control sites, Zhou and Weis (1998) found that, overall, exposure to methylmercury caused hyperactivity and an increase in vulnerability to predators. However, effects varied with age of the larvae, duration of exposure and site.

Increased vulnerability to predators has also been reported in birds exposed to organophosphate pesticides possibly in response to impaired sensory perception, coordination and motor skills, and reduced activity (for review, see Grue *et al.* 1991, Hunt *et al.* 1992). For example, northern bobwhite quail orally dosed with 8 mg of methyl parathion per kg body weight spent more time still than controls, and were more vulnerable to capture by a domestic cat (*Felis domesticus*; Galindo *et al.* 1985). Similarly, American kestrels captured house sparrows dermally exposed to fenthion instead of controls in 12 of 15 trials in which at least one bird was captured (Hunt *et al.* 1992). Five of the six sparrows exposed to the avicide and which exhibited aberrant behaviour were captured by kestrels. Evidence for preferential selection of affected free-living birds following operational applications of OPs is limited. Red-billed quelea (*Quelea quelea*) exhibiting aberrant behaviour (e.g., wing flapping) following application of the avicide, Queleatox® (60% fenthion), were 'conspicuous and quickly eaten by passing raptors' (Thomsett 1987). McEwen and Brown (1966) suggested that wild sharp-tailed grouse given a sublethal dose of malathion or dieldrin were 'more than normally vulnerable to predators'. The attractiveness of large numbers of intoxicated or moribund prey to avian predators following operational applications of OPs and other pesticides is well documented (White *et al.* 1979, Douthwaite 1982, Clements and Bale 1988, Mineau *et al.* 1999, Goldstein *et al.* 1999).

Aquatic and terrestrial arthropods provide additional examples of increased or decreased vulnerability of prey impaired by exposure to contaminants. The vulnerability of net-spinning caddisflies (*Chimarra* sp. and *Hydropsyche morosa*) to predation by the stonefly, *Paragnetina media*, was significantly greater in experimental streams contaminated with copper than controls (Clements *et al.* 1989). *H. morosa* was the preferred prey in the contaminated streams. The authors speculate that exposure to copper disrupted silk-spinning by this caddisfly resulting in abnormalities in their capture nets, and causing them to spend more time repairing their nets outside their retreats where they were more vulnerable to predators. More recently, Kiffney (1996) reported a similar outcome in artificial streams contaminated with a mixture of cadmium, copper and zinc. Similarly, exposure to the synthetic pyrethroid, deltamethrin, decreased walking speed in the erigonid spider, *Oedothorax apicatus*, which increased their likelihood of capture by carabid beetles (Everts *et al.* 1991). Exposure of carabid beetles (*Harpalus pennsylvanicus*) to imidacloprid, a nicotinic receptor agonist within the postsynaptic nerve membranes of insects, resulted in a variety of sublethal neurotoxic effects including paralysis, impaired walking and excessive grooming (Kunkel *et al.* 2001). Whereas affected beetles recovered within a few days in the laboratory, in the field they were highly vulnerable to predation by ants (*Formica schaufussi*). In contrast, predator-prey interactions in rotifers are largely mediated by contact through chemo- and mechanoreception with frequency of encounters determined by density, body size and swimming speed (reviewed by Preston *et al.* 1999). These authors found that the effects of sublethal exposure to pentachlorophenol significantly reduced swimming speed in some species, but not in others. In those in which swimming was impaired, encounters with predatory rotifers and numbers ingested declined relative to controls.

1.4.6.2 Competition

We noted earlier (see Section 1.3.3.4) that few studies have examined the effects of sublethal contaminant exposure on behaviour in relation to an animal's ability to compete with conspecifics or individuals of another species, i.e., head-to-head competition. An exception is a recent study by Fjeld *et al.* (1998) in which embryos of grayling were exposed to one of four concentrations of methylmercury (including control) during the first 10 days of development and then maintained in clean water for a period of 3 years. Fish within each treatment group were colour marked and eight fish from an exposed group and an equal number of controls were placed together in an aquarium, acclimated for 3 to 5 days and then allowed to compete for *Daphnia magna* (50/min) for 20 min. Fish were sacrificed and the number of *Daphnia* in their stomachs counted. In all of the comparisons, fish that were exposed to methylmercury as embryos caught significantly (16–50%) fewer prey than controls. Other exceptions are the study of the competitive ability of second-generation male kestrels from females orally dosed with o,p′-Dicofol® by MacLellan *et al.* (1997) and McEwen and Brown's (1966) controlled-field study of the survival and reproductive behaviour of sharp-tailed grouse given an oral dosage of malathion or dieldrin (see Section 1.4.5.2). Field studies, such as those by Powell (1984) and Busby *et al.* (1990) that document territorial defence or abandonment by exposed individuals also represent measures of competitive ability, although cause — effect relationships are more difficult to determine.

Documented changes in the competitive interactions among species exposed to contaminants have largely been the result of differential sensitivity, i.e., the loss (poisoning) of a competitor and not chemical-induced changes in behaviours. Examples include the effects of pH on competition between black duck ducklings and fish in Maine (Hunter *et al.* 1986) and larval tree frogs (*Hyla femoralis* and *H. gratiosa*) in the southeastern coastal plain of the United States (Warner *et al.* 1993) which were discussed in Section 1.3.3.4. We close our discussion of direct toxic effects with the results of a recent study by Lefcort and his coworkers which examined the interaction between heavy metals (cadmium, lead and zinc), predators (bluegill sunfish), and competition between tadpoles of the Columbia spotted frog (*Rana luteiventris*) and snails. This is appropriate because the study (Lefcort *et al.* 1999) is an excellent example of the complexities of nature, the relationship between behaviour and contaminant exposure, and how contaminants can interact with biotic factors to alter the outcome of competitive interactions.

Metal exposure, competition and chemoreception — In the absence of predators (fish odour) or snails, tadpoles fed more often within the detritus above the contaminated sediment and accumulated higher tissue concentrations of lead than snails. In the presence of fish odour, tadpoles hid and foraged more within plant refugia minimising their exposure to the metals. In contrast, snails accumulated higher tissue metal concentrations in the presence of fish odour and tadpoles, presumably because they shifted their feeding to the contaminated sediments in response to the greater competition with tadpoles within plant refugia. Both tadpoles and snails that accumulated high levels of metals lost their ability to detect chemical cues from predators (fish odour and crushed snails, respectively).

Growth and development — In the absence of fish odour, exposure to the metal-contaminated sediment was sufficient to adversely affect the growth and development of tadpoles, but not that of the snails. In contrast, snail biomass declined in the presence of fish odour and tadpoles, but not tadpoles alone, likely because of the effects of fish odour on habitat use by tadpoles and corresponding shift by snails to contaminated sediments.

1.5 INDIRECT BEHAVIOURAL EFFECTS

Indirect effects are defined here as behavioural responses to pollutant-induced changes in an animal's environment. They differ from direct toxic effects in that the behavioural response of interest cannot be attributed to direct toxicity in the affected animal. Included are responses to chemical-induced changes in the animal's habitat (e.g., food or cover) or intra- or interspecific interactions. Such effects can be difficult to document because of the large number of potential interactions among species and their environments, varying temporal and spatial scales for both chemical exposure and effects, and our incomplete understanding of natural systems. In addition, animals may be affected both directly and indirectly, and it is often difficult to separate these effects when assessing overall responses to a contaminant (e.g., see Figure 1.2). As a result, indirect effects of contaminants on fish and wildlife are poorly understood. A comprehensive review of the literature on the indirect effects of pollutants on the behaviour of fish and wildlife is beyond the scope of our chapter. Indirect effects are also briefly discussed in Sections 1.3.3.3 and 1.3.3.4.

1.5.1 HABITAT SELECTION

Animals select habitats based on the availability of requisites for survival and reproduction (e.g., cover, food and nest sites) as well as prior experience (i.e., successful reproduction, which we note may or may not be indicative of subsequent changes in habitat quality). The availability of critical resources is often influenced by intra- and interspecific interactions (competitive and predator — prey, for reviews see Bowers 1994, Alcock 1998:383 - 427, Fryxell and Lundberg 1998:82 - 115). Variation in the quality of the habitat(s) selected by individuals will ultimately govern their reproductive success and contribution to subsequent generations (i.e., their fitness). By altering the availability of requisites and or intra- or interspecific interactions, contaminants may alter the relative contributions of high fitness ('survival', 'source') and low fitness ('colonization', 'sink') habitats (Anderson 1970, Pulliam 1988; for discussion, see Bowers 1994, Lacher 1994). Fluctuations in species composition and density among habitats (chemically altered vs unaltered) through time may, therefore, be indicative of the quality of these habitats for specific species (for discussion of the 'ideal-free distribution', see Alcock 1998:384, Fryxell and Lundberg 1998:83). Densities and reproductive output within source habitats would be expected to remain relatively constant whereas that of sink habitats would be expected to experience wide fluctuations (Bowers 1994). The biological and ecological significance of chemical impacts on a species' habitat will likely depend on whether or not the habitat is a population source or sink (for discussion, see Lacher 1994). To date, the effects of contaminants relative to source vs. sink populations have received little attention.

Changes in species composition and density following the use of herbicides in terrestrial systems are well documented. For example, the use of herbicides to control weeds along the margins (conservation headlands) of cereal fields has been associated with significant reductions in the abundance of nonpest arthropods associated with the target plants (Chiverton and Sotherton 1991). [Herbicide-induced changes in these arthropod communities have been implicated in declines of grey partridge (*Perdix perdix*) populations; chick survival is reduced in treated headlands because of reductions in the nonpest arthropods (particularly Heteroptera) on which they depend for food (Potts 1977, Rands 1985).] Morrison and Meslow (1984) reported decreases in the numbers of avian species normally associated with shrubs and/or deciduous trees (e.g., MacGillivray's warbler, *Oporonis tolmiei*) one year following the application of glyphosate to remove plants competing with young conifers on clear cuts in western Oregon. Species (e.g., American goldfinch, *Carduelis tristis*) that prefer more open habitats increased in abundance. Similarly, numbers of Oregon voles (*Microtus oregoni*), a species that prefers grasses and forbs, increased one year after herbicide treatment (Anthony and Morrison 1985). Two years after treatment, numbers of both species declined coincident with the recovery of the shrubs.

Comparable studies with glyphosate in aquatic systems (freshwater wetlands) provide similar results. Application of the herbicide to create a mosaic of open-water and live and dead emergent vegetation (primarily cattails, *Typha* spp.) in wetlands in North and South Dakota resulted in an increase in use by diving and dabbling ducks (Solberg and Higgins 1993, Linz *et al.* 1996). However, the success of over-water nests was not as great as those surrounded by unfragmented emergent vegetation (Solberg and Higgins 1993). Surprisingly, other studies of the indirect effects of chemical-induced changes in aquatic vegetation are lacking, and this is reflected in a recent symposium on the biology, ecology and management of aquatic plants (Hydrobiologia 415:1–339, 1999) in which this issue was not addressed.

Insecticide-induced changes in invertebrate populations may also result in shifts in the composition of vertebrate communities. For example, Barrett and Darnell (1967) suggested that the shift in the composition of the small mammal community from one that was largely omnivorous to one that was primarily herbivorous resulted from an abrupt (90%) decline in insects within a clover field following the application of dimethoate. House mice (an omnivore) comprised 68% of the community prespray, but individuals subsequently emigrated from the study area and the species comprised only 37% of the community 4 weeks postspray. In contrast, prairie voles (a herbivore) immigrated into the study area after the insecticide was applied and comprised 44% of the community postspray compared to 13% at the beginning of the study. Numbers of, and per cent composition by, deer mice, largely a herbivore, remained unchanged.

1.5.2 HABITAT USE

As noted above, studies of the effects of chemical-induced changes in habitat on fish and wildlife have largely focused on documenting changes in species composition and density (i.e., habitat selection), and not modifications in behaviour (i.e., how an individual uses its habitat). Here, we focus on changes in the behaviour of animals related to chemical-induced changes in vegetative cover and food resources.

1.5.2.1 Vegetative cover

Alterations in vegetative composition or structure have also been shown to affect the behaviour of bird species. In their study of the effects of glyphosate on bird community structure in Western Oregon, Morrison and Meslow (1984) found that rufous-sided towhees (*Pipilo erythrophthalmus*) increased their use of deciduous trees (red alder, *Alnus rubra*) for foraging in an attempt to compensate for the loss of their preferred shrub habitat. The use of 2,4-D to control sagebrush (*Artemesia tridentata*) resulted in Brewer's sparrows (*Spizella breweri*) selecting large highly branched dead sagebrush plants for nest sites compared to much smaller live plants on untreated areas in an effort to compensate for the loss of foliage (Best 1972). Changes in behaviour varied among three species of tits (*Parus ater*, *P. cristatus* and *Aegithalos caudatus*) in response to reductions in needle densities on the pine trees in which the birds foraged as a result of pollution from a nearby coal-fired power plant. Although vigilance increased in all three species, most likely in response to a greater risk of predation in the thinned and more exposed outer canopy, only the coal tit (*P. ater*) altered its use of the trees. This species normally prefers the needled twigs at the outer edges of the canopy as its foraging substrate, but shifted to the more central and higher parts of trees in the polluted area (Brotons *et al.* 1998). Similarly, Hake (1991) reported increased scanning rates and flock sizes in willow tit (*Parus montanus*) populations in winter in response to needle loss within conifer forests in a polluted environment.

1.5.2.2 Food resources

Contaminant-induced changes in the abundance or distribution of food resources can alter the behaviour of fish and wildlife in a number of ways. As previously noted (Section 1.3.2.3), the severity of the resultant behavioural effects appears to depend on the magnitude of the chemical-induced change in food resources. Availability of alternative prey, the ability to utilise existing alternatives (prey specificity), as well as foraging efficiency can influence behavioural responses to chemical-induced reductions in food resources (see discussion below). In cases where chemical exposure results in an increase (usually temporary) in food availability, foraging efficiency may increase in the absence of a change in prey. For example, capture rates for grasshoppers and nonnative arthropods by nesting killdeer (*Charadrius vociferus*) increased within the 8 days following the application of carbaryl to control grasshoppers (Fair *et al.* 1995). Similarly, application of endosulfan to control tsetse flies (*Glossina morsitans*) resulted in easier capture of spray-debilitated fish by pied kingfishers (*Ceryle rudis*, Douthwaite 1982).

In situations where emigration may not be an appropriate trade-off (e.g., after considerable reproductive investment), animals may be forced to alter their diets because of chemical-induced reductions in preferred prey. In some cases, the quantity and/or quality of alternative foods may not alter survival or reproductive success. For example, grasshoppers accounted for the majority of the diet fed nestling chestnut-collared longspurs (*Calcarius ornatus*) prior to the application of Decis 5F, a broad-spectrum pyrethroid insecticide, to control grasshoppers. Following application of the insecticide, grasshopper density was reduced by 93% and adult longspurs switched to arthropod taxa (lepidopteran larvae) less affected by the insecticide application such that the overall biomass fed to their nestlings was not significantly reduced (Martin *et al.* 1998). Similarly, white-footed mice responded to the loss of lepidopteran larvae following an application of diflubenzuron to control gypsy moths (*Porthetria dispar*) by using other food sources without a change in stomach weights (Seidel and Whitmore 1995). However, in similar situations, but where alternatives are of reduced quantity or quality, the survival or reproduction of affected animals may be reduced. For example, growth rates of American black duck and mallard ducklings were significantly reduced on ponds aerially sprayed with carbaryl simulating overspray or drift of the insecticide during operational treatments to control spruce budworm (Hunter *et al.* 1984). Ducklings on the sprayed ponds spent more time searching for food and less time resting than ducklings on unsprayed ponds where invertebrate numbers and biomass were double that on the treated ponds. Similarly, studies by Munkittrick and his coworkers suggest metal contamination (primarily copper and zinc) of lakes from mining activity significantly reduced the diversity and abundance of aquatic invertebrates. Resultant reductions in the quantity and quality of prey were associated with increased foraging activity and decreased growth, fecundity and energy stores in white suckers (*Catostomus commersoni*, Munkittrick *et al.* 1991).

Animals may also switch prey in response to an increase in the availability of poisoned target and nonnative species. An example is shown in Figure 3. Adult tree sparrows (*Passer montanus*) fed their young primarily beetles and spiders prior to the application of an organophosphorus aphicide to adjacent cereal fields. Following the insecticide application, adults shifted to the moribund and readily available aphids without any adverse effects on the adults or their young (Greig-Smith 1991). Similarly, analyses of stomach contents of white-footed mice, short-tailed shrews (*Blarina brevicauda*) and red-backed voles (*Clethrionomys gapperi*) before and after an application of acephate to

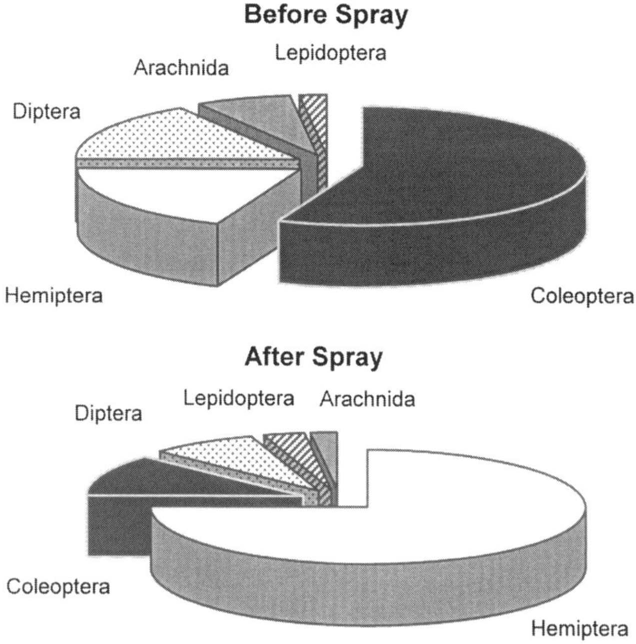

Figure 1.3 Diet of nestling tree sparrows (*Passer montanus*) before and after application of an aphicide on cereal fields (after Greig-Smith 1991).

control gypsy moths indicated a 70–400% increase in the number of arthropod fragments. The small mammals responded to the insect 'rain' from the forest canopy and/or affected ground-dwelling arthropods postspray (Stehn 1976).

It has been suggested that prey specialists may be particularly susceptible to contaminant-induced reductions in their prey (Bellocq *et al.* 1992). However, in some cases, foraging efficiency may compensate for severe reductions in preferred prey. For example Tennessee warblers (*Vermivora peregrina*), a spruce budworm specialist, were able to locate sufficient larvae to feed their young despite a 96% reduction in the numbers of caterpillars and pupae following application of an insecticide (tebufenozide) specific to juvenile lepidoptera (Holmes 1998). The authors suggested that the adults were able to locate refugia where the larvae were unaffected by the spray, such as the lower part of the canopy or small trees. As a result, adult females spent more time foraging and less time brooding their young than controls, but without adverse effects.

Changes in foraging niche have been reported in birds following the application of broad spectrum insecticides. For example, Hunter and Witham (1985) found that warblers (Parulinae) foraged lower in the canopy following an application of carbaryl to control spruce budworm when the canopy within the study areas was evenly sprayed (i.e., all areas treated). Warbler numbers within the sprayed areas did not decline postspray. However, in the previous year, when spray coverage was only about 30%, warbler numbers on the treated areas decreased, presumably as a result of the birds foraging in unsprayed refugia.

A number of authors have suggested out-of-area movement as a strategy to compensate for insecticide-induced reductions in prey (Moulding 1976, Hunter and Witham 1985, and references therein; Bellocq *et al.* 1992, Bishop *et al.* 2000), but it remains largely unconfirmed. An exception is the documented emigration of house mice from clover fields treated with the insecticide dimethoate in response to a 90% decrease in insect numbers postspray (Barrett and Darnell 1967).

Immigration in response to increased prey availability has also been reported. Bracher and Bider (1982) observed a significant increase in the presence of weasels (*Mustela erminea*) in fir forests treated with the insecticide aminocarb to control spruce budworm. The weasels were attracted to the increase in avian activity on/near the forest floor because of insecticide-induced reductions in prey in the canopy above.

1.5.3 INTRASPECIFIC INTERACTIONS

Controlled-field studies, those in which free-living animals are given a specific dosage of a chemical, provide some examples of how chemical exposure can affect intraspecific interactions, i.e., behaviour of unexposed individuals. Male sharp-tailed grouse given a sublethal oral dosage of either dieldrin or malathion exhibited abnormal behaviour on their breeding grounds. For example, 4 days postdose, a dieldrin-exposed male that had previously maintained a central territory on the breeding grounds, remained on the edge of the breeding grounds, did not display, but occasionally ran through the central display area eliciting challenges by the males defending their territories (McEwen and Brown 1966). Grue *et al.* (1982b) studied the care of nestlings by free-living female European starlings given a severe, but sublethal oral dose of the OP insecticide, dicrotophos, one that resulted in about 50% inhibition of brain ChE activity. Males were captured and moved to an aviary prior to the capture and dosing of their mates. OP-dosed females made fewer sorties to feed their young and remained away from their nests for longer periods of time than comparable controls and their nestlings suffered greater weight losses. In a comparable study with laughing gulls, in which one member of a pair was given a single sublethal oral dose of the OP parathion (ca. 50% brain ChE inhibition), the unexposed gull compensated for the chemical-induced changes in the behaviour of its partner (White *et al.* 1983). During the second and third days after treatment, OP-dosed gulls spent less time on their nests than controls, but the overall incubation of eggs by affected pairs did not differ from that of controls. Similarly, Grue and his co-workers found that free-living European starlings given the same sublethal oral dose of dicrotophos as that in their earlier study, made fewer visits to their nests to incubate their eggs and remained on their nests longer than controls. As in the case of the gulls, male starlings compensated for the lack of attentiveness of their OP-dosed females, and no differences in hatching success were detected between the two treatment groups (Grue, unpublished data; Grue *et al.* 1991).

Examples of comparable effects resulting from ambient environmental contamination are few, although they undoubtedly occur. For example, exposure of western gulls to DDT is believed to be responsible for the poor reproductive success of birds breeding off the southern California coast in the 1970s (Fry and Toone 1981). Feminisation of adult males is believed to have resulted in the observed sex ratio highly skewed to females and female-female pairings among the excess females due to a lack of true males. Another, but less conclusive example, is the stranding and subsequent death of three young sperm whales that maintained their affiliation with a dominant male which may have become stranded and died because of contaminant exposure (Joiris *et al.* 1997).

Studies by Warner and her coworkers on the effects of low (acidic) pH on hylid tadpoles provide examples of changes in competitive abilities of conspecifics within contaminated environments, although not specifically behavioural responses. Intraspecific competition among tadpoles of the barking tree frog (*Hyla gratiosa*) was accentuated at low pH and resulted in decreased survival or size at metamorphosis (Warner *et al.* 1991).

1.5.4 INTERSPECIFIC INTERACTIONS

1.5.4.1 Predator — prey relationships

Previously (see Section 1.5.2.2), we reviewed how chemical-induced changes in vegetative cover or food resources can alter interactions among predators and their prey. Predators may alter their foraging strategies in response to chemical-induced increases or decreases in prey availability. In the case of the former, foraging efficiency may increase (if only temporarily) and may or may not be associated with a shift in foraging substrate, prey, or out-of-area movement. In the case of the later, foraging efficiency may be sufficient to obtain adequate food, or if not, predators may change foraging substrate and/or switch to more abundant prey (which may or may not be of similar quality), or leave the immediate area. In some cases, changes in foraging strategy or emigration may not be possible (e.g., because of prey specificity or significant reproductive investment), and reproductive success may suffer. To minimise redundancy, we highlight a few additional examples of some of these effects.

Chemical-induced changes in vegetative cover have been associated with shifts in the abundance of predators and prey. For example, the presence of weeds in cereal crop headlands helps to maintain the diversity of arthropod species within the fields. The use of herbicides to control weeds within headlands has been associated with decreases in the numbers of predatory arthropods and an increase in the frequency of cereal aphid outbreaks. In untreated areas, Carabid beetles (which prey on aphids) consumed a greater number and diversity of prey items than beetles in herbicide-treated areas, and they contained more eggs (Chiverton and Sotherton 1991). Similarly, Greig-Smith (1991) reported that the overall potential for natural control of cereal pests, particularly summer aphids, was reduced with repeated annual (prophylactic, 'full insurance') pesticide applications. Many predators were reduced in number by a combination of toxic effects and changes in the availability of their prey, and as a result, pests became more liable to sudden, unpredictable outbreaks.

Studies of the effects of insecticides used to control the green moth in Spanish oak forests by Pascual and his coworkers further illustrate the importance of alternative prey in mediating the effects of chemical-induced reductions in prey availability. The use of the synthetic pyrethroid insecticide, cypermethrin, resulted in 100% mortality of lepidopteran larvae, the primary food of blue tits, as well as other arthropods, at a time when the birds' eggs were hatching or young nestlings were present. As a result of food shortages, reproductive success of the tits on the treated plots was poor (35% of nests successful), whereas that on the unsprayed controls was good (100% success, Pascual and Peris 1992). In contrast to these results, reproductive success of the tits was not affected by the application of the organophosphate insecticide, malathion, to control the moth. In this case, mortality of the target pest was nearly 100%, but other arthropods were unaffected and the food supply for the tits was similar to or higher than that on unsprayed plots (Pascual 1994).

> '... in these kinds of studies it is more important to assess the abundance of arthropods remaining alive after treatment than the degree of arthropod mortality caused by the insecticide' (Pascual 1994).

1.5.4.2 Competition

We could only find a few examples in which chemical-induced changes in an animal's environment altered the level of competition an animal experiences or its competitive abilities. Barrett and Darnell (1967) suggested that the immigration of prairie voles following an application of dimethoate to clover fields was facilitated by a reduction in interspecific competition with house mice. Numbers of the mice decreased as they left the treated plots in response to insecticide-induced reductions in insect prey. Bracher and Bider (1982) observed a decrease in the activity of toads (*Bufo americanus*) following the application of aminocarb to a fir forest to control spruce budworm. The authors postulated that the combined effect of the insecticide-induced mortality of invertebrates in the sprayed area, plus increased competition by birds and other insectivores for the remaining live arthropods on or near the forest floor made foraging difficult for the toads.

Studies by Hunter and Warner and their co-workers provide more definitive examples and both involve changes in competitive abilities associated with decreased pH. Hunter *et al.* (1986) studied the growth and behaviour of young black ducks on ponds in Maine which were physically similar, but had markedly different acidity. Surprisingly, ducklings on the more acid ponds (pH 4.5-4.8) spent less time searching and moving and more time feeding and resting and, as a result, grew faster than ducklings on circumneutral ponds. The authors hypothesised that acidification may have actually increased food availability for ducklings by extirpating fish competitors. Fish exclosure experiments indicated that the presence of fish decreased invertebrate abundance and diet composition data demonstrated that there was substantial overlap in the diets of brook trout (*Salvelinus fontinalis*) and ducklings. Similarly, in a low pH environment (4.5), larvae of the pinewoods tree frog (*Hyla femoralis*) were better able to compete with larvae of the barking tree frog (*H. gratiosa*) because the latter was more susceptible to the toxic effects of increased acidity (Warner *et al.* 1993). Under less acid conditions, tadpoles of *H. gratiosa* exerted a marked competitive effect on the *H. femoralis*; larval period was increased and size at metamorphosis was reduced.

1.6 IS BEHAVIOUR A SENSITIVE INDICATOR OF EXPOSURE AND EFFECT?

One of the reasons behavioural endpoints are attractive as biomarkers[1] of direct chemical exposure and/or effect is that, in some cases, they are more sensitive than other endpoints, either in terms of chemical concentration, response time, or both. Here, we compare the magnitude of chemical exposure associated with behavioural effects with that associated with lethality and detectable changes in biochemistry and physiology. We conclude this section with a comparison of the chemical specificity of biomarkers and the relationship between chemical specificity and ecological relevance. We have excluded indirect effects from our discussion because little is known about the relative sensitivities of biochemical, physiological, and behavioural markers in response to chemical-induced changes in an animal's environment compared to direct exposure, and because of the inherent difficulties in separating the contributions of direct and indirect effects to observed responses (e.g., see Busby *et al.* 1990). Changes in behaviour associated with indirect effects of pollutants are reviewed in Section 1.5.

1.6.1 BEHAVIOUR VS. LETHALITY

It is not surprising that a wide variety of behavioural alterations occur in animals exposed to 'sublethal' concentrations of pollutants. However, it is surprising that relatively little effort has been devoted to comparing the actual levels of exposure associated with behavioural abnormalities and death. In most cases, comparisons are only possible with data from studies conducted with different species and/or test conditions (Atchison *et al.* 1987).

Ventilatory behaviour in fish has been shown to be a sensitive indicator of exposure to a variety of contaminants with changes observed at chemical exposures (incorporating concentration and response time) estimated to be 10–30% of 96-h median lethal concentrations (LC_{50}s, Diamond *et al.* 1990). Swimming capacity (often defined as the maximum velocity which can be maintained for a given time) and activity (or pattern) are frequently even more sensitive markers with chemical induced alterations detected at 0.5–5% of corresponding LC_{50}s (Little and Finger 1990).

Fish are frequently able to detect and avoid chemicals in water at concentrations equal to or lower than the lowest concentration at which effects are observed during standard chronic toxicity tests (LOECs, Atchison *et al.* 1987). The latter values themselves are often 1–2 orders of magnitude lower than 96-h LC_{50}s (Atchison *et al.* 1987), and therefore, avoidance responses may be the most sensitive of the behavioural endpoints in fish associated with contaminant exposure or effect studied to date.

Few data are available to adequately compare chemical concentrations that elicit other behavioural endpoints in fish with those that result in mortality. Included here are foraging behaviour and predator — prey interactions (for review, see Sandheinrich and Atchison 1990), which, like avoidance responses, are attractive 'sublethal' endpoints because of their apparent greater ecological relevance than other responses (e.g., ventilatory behaviour). [It is important to note that the value of a particular behavioural endpoint may not need to be the ease by which one can relate the response to ecological effects, but its ability to detect unacceptable levels of environmental contamination in real time (see Section 1.6.3).] Foraging behaviour and predator — prey interactions appear to be more sensitive than mortality, but less sensitive than ventilatory behaviour with chemical-induced changes in many cases occurring at LOECs and concentrations of about 25% of LC_{50}s of various durations (Sandheinrich and Atchison 1990). In a more recent study, lake charr (*Salvelinus namaycush*) exposed to 0.5 and 5.0 ug/l of cadmium captured fewer prey or made fewer attacks, respectively, compared to controls (Kislalioglu *et al.* 1996). Median-lethal concentrations (96-h) for cadmium and adult fish in the same genus (brook trout) are 2.4 and 4.1 ug/l (reviewed by Eisler 2000).

The development and use of behavioural endpoints as indices of contamination in terrestrial wildlife lags behind that of aquatic organisms. In many cases, investigators have selected one or only a few sublethal dosages at levels expected to result in significant alterations in behaviour (Hart 1993, Burger 1995). As a result, behavioural effects associated with low levels of contaminant exposure have been poorly studied. For example, Grue and his co-workers conducted a series of studies on the sublethal effects of exposure to OP pesticides on songbirds using the European starling and dicrotophos, as models, and a dosage of 2.5 mg/kg body weight (for review, see Grue *et al.* 1997). This dosage was selected because it was sublethal under laboratory conditions and resulted in brain ChE inhibition of about 50%, a level at which significant behavioural effects might be expected (Hart 1993, Grue *et al.* 1997). Observed effects included reductions in activity and reproductive behaviours (e.g., singing, displaying, and nest attentiveness; reviewed in Grue *et al.* 1997). This dosage represented about 30% of the 24-h median lethal dose (LD_{50}) for dicrophos in adult starlings (Grue and Shipley 1984). Similarly, Kendall and his coworkers have conducted a number of studies with bobwhite quail given dosages of the OP methyl parathion that resulted in sublethal brain ChE inhibition of $\geq$40%. For example, Galindo *et al.* (1985) found that the activity of quail given a dosage of 8 mg/kg was significantly reduced and subsequent predation was greater than controls, but lower dosages had little effect on predator avoidance. The authors reported the LD_{50} for methyl parathion in bobwhite to be 10 mg/kg. Hart (1993) reported subtle behavioural effects at much lower levels of brain ChE inhibition in starlings exposed to the OP chlorfenvinphos. Starlings suffering brain ChE inhibition of about 12% spent less time standing on one leg than controls; the associated dosage (3 mg/kg) was $\leq$30% of the 24-h LD_{50} (>10 mg/kg, Hart and Westlake 1986).

The effects of lead on a variety of endpoints in common tern and herring gull chicks have been studied by Burger and Gochfeld (reviewed in Burger 1995 and Chapter 10 of this book). Chicks were given dosages of 100 and 200 mg lead (as lead acetate)/kg which elicited a variety of behavioural effects including reduced coordination, begging for food, and thermoregulatory behaviour. The dosages selected for study represent approximately 10 and 20 percent of the 17-day LD_{50} for lead acetate in common tern chicks, respectively (Burger and Gochfeld 1988).

More recent studies provide additional comparisons with other terrestrial species and behavioural endpoints. For example, Walton *et al.* (1997) reported abnormal breeding behaviour in kittiwakes 1 or 2 years after sublethal exposure to crude oil following the wreck of the Braer oil tanker off the coast of Scotland in 1993. Although data on the actual exposure of the birds to the oil is not known, 133 kittiwakes were among the seabirds found dead after the spill (8.6% of the birds collected) and several haematological parameters differed between birds in the exposed colony vs. those from control sites. Moser and her coworkers (Moser *et al.* 1995) examined the neurobehavioural effects of 10 known toxicants in female laboratory rats. In their acute studies with pesticides, behavioural effects were observed at dosages usually <10% of the corresponding LD_{50} value. Similarly, Dhawan *et al.* (2000) compared the sediment concentrations of five metals (24-h EC_{50}s for interstitial pore water) that altered the movement of the soil nematode, *Caenorhabditis elegans*, with the respective LC_{50}s. In each case, significant changes in locomotion occurred at metal concentrations less than 10% of the median lethal concentration.

1.6.2 BEHAVIOUR VS. BIOCHEMICAL AND PHYSIOLOGICAL ENDPOINTS

As noted by Peakall (1996), chemical-induced behavioural effects have generally not been shown to be more sensitive than biochemical or physiological endpoints. The severity of behavioural effects frequently varies with the magnitude of the underlying biochemical and physiological changes. However, as in the case of comparisons between levels of chemical exposure that elicit behavioural effects and mortality, one must be cautious in making generalisations. Comparisons between the sensitivities of biochemical, physiological and behavioural endpoints are difficult because few studies have made such comparisons and as a result the number of chemicals, exposure scenarios and endpoints for which information is limited. Additionally, there is the potential for test conditions to differentially influence responses (Triebskorn *et al.* 1997).

Measures of behavioural endpoints are routinely used in neurotoxicological risk assessment and screening where it is recognised that subtle impairments such as memory dysfunction are often among the earliest and most sensitive indicators of neurotoxicant exposure and frequently precede signs of neuropathology (Evangelista de Duffard and Duffard 1996). Stephens *et al.* (1995) reported reduced neuropsychological performance in sheep farmers who suffered long-term exposure to sheep dips containing OPs and concluded that the subtle effects observed were unlikely to manifest themselves as clinical symptoms. Similarly, lead at 'silent doses' produces deficits in psychological functions in children including intelligence, perception, attention, language function, and possibly social adjustment (Needleman 1995). Other behavioural effects that manifest themselves after clinical symptoms have disappeared include OPIDN and the 'intermediate syndrome', both associated with acute exposure to OPs (see Section 1.3.2.1).

Outside the field of medicine, several studies have also shown that behavioural endpoints can, in some cases, be as sensitive as biochemical and physiological markers. In a study of the chronic (60-day) toxicities of four organic compounds to rainbow trout fry, behaviour (activity, body orientation, startle response) was comparable in sensitivity to indices of growth in determining no observed effect levels (NOELs). However, the behavioural endpoints did not provide any significant advantages over simple measures of length and weight (Passino-Reader *et al.* 1995). Triebskorn and her coworkers (1997) compared the sensitivities of biochemical (heat shock proteins), physiological (liver ultrastructure), and behavioural (swimming behaviour) markers in brown trout exposed to water from two streams with different levels of contamination and for various periods of time (13–21 weeks). The markers were equally sensitive in separating the two streams, with the exception of the heat shock proteins in which the response was water temperature dependent. Bryan *et al.* (1995) found growth in bluegill exposed to cadmium over a period of 21–22 days to be a more sensitive endpoint than foraging behaviours. For example, attack rates of exposed bluegill varied with size of the prey (*Daphnia*) and duration of exposure, whereas, growth was significantly reduced at all cadmium concentrations tested (small and medium *Daphnia* only). Porter *et al.* (1999) compared endocrine, immune and behavioural (aggression and exploratory behaviour) responses of male laboratory mice or deer mice exposed chronically (22 - 103 days) to a variety of groundwater pollutants, both singly and as mixtures. Although the individual chemicals rarely affected any of the endpoints, mixtures did, and changes in aggressive behaviour were frequently consistent with changes in the most sensitive endpoint — immune response, whereas exploratory behaviour ('holepoke assay') was not affected by any chemical or chemical combination. The reported abnormal breeding behaviour in kittiwakes 1 or 2 years after sublethal exposure to crude oil (Walton *et al.* 1997) described above is another example in which responses in both physiological and behavioural endpoints were equally sensitive indicators of exposure and effect.

Inhibition of brain ChE to $<40-60\%$ of normal has frequently been associated with behavioural changes and overt signs of toxicity in birds and mammals exposed to OPs (for reviews, see Grue *et al.* 1991, 1997). At lower levels of exposure, when animals do not appear to be distressed, more careful observations have revealed subtle changes in behaviour. As noted above, Hart (1993) found alterations in posture of European starlings when brain ChE activity was 88% of normal; $\leq 80\%$ of normal brain ChE is the criteria for diagnosing exposure to anticholinesterases in birds (Grue *et al.* 1991). In a test of the hypothesis that such changes were due to impaired balance or coordination, Fryday *et al.* (1994) detected reductions in the seed-handling efficiency of house sparrows when brain ChE was less than about 70% of normal. Marmosets (*Callithrix jacchus*) given doses of OPs or carbamates (ChE inhibitors) that caused no measurable physiological changes beyond enzyme inhibition (as low as 23% in marmosets exposed to physostigmine) displayed decreased hand-eye and visual discrimination performance and appeared to have difficulty making choices (Wolthuis *et al.* 1995). In comparison, a recent review of the relationship between AChE inhibition and the occurrence of sublethal effects (e.g., swimming capacity [stamina]) in estuarine fish suggests that these effects are often only observed when enzyme inhibition approaches lethal levels ($>70\%$, Fulton *et al.* 2001). This appears to be frequently true for behavioural effects associated with exposure to ChE inhibitors in other species of fish as well (Zinkl *et al.* 1991). These results are in sharp contrast to avoidance responses in fish that are often elicited at chemical concentrations 1 - 2 orders of magnitude lower than 96-h LC_{50}s (Atchison *et al.* 1987).

As noted in Section 1.3.1.3, the development of tolerance associated with subacute or chronic chemical exposures can result in a disassociation between biochemical and physiological markers of exposure and behavioural effects. For example, Blick *et al.* (1994) used the performance of a compensatory tracking task by rhesus monkeys (*Macaca mulatta*) to estimate the behavioural effects of repeated low doses of soman, an OP, such as might be experienced by military personnel under combat conditions. Performance deficits were detected following acute exposure resulting in 65–70% inhibition of plasma ChE, but not clinical signs of intoxication. However, measures of plasma ChE activity were of limited value in predicting deficits after chronic exposure because performance deficits comparable to those following acute exposure did not manifest themselves until inhibition of plasma ChE exceeded 85%.

1.6.3 TIME TO RESPONSE

The greater sensitivity of behavioural endpoints (in terms of dosage and response time) to contaminants compared to lethality (e.g., see Rice *et al.* 1997) has resulted in a number of investigators promoting the use of behavioural responses in aquatic organisms as candidates for 'real-time' water quality monitoring systems. Such systems would provide early warnings of, and screening assessments for, chemical hazards. At the present time, the most likely candidate appears to be ventilatory behaviour in fish in which significant changes can be detected within 60 min (usually within 15 min) of exposure to various contaminants at 10% to 30% of the associated 96-h LC_{50}s (Diamond *et al.* 1990). Other taxa and endpoints appear to be more suited to decreasing the time necessary to conduct laboratory or *in situ* bioassays ('rapid screening bioassays', Charoy and Janssen 1999). For example, Janssen and his co-workers have been investigating swimming behaviour in the rotifer, *Brachionus calyciflorus*, as a candidate for such bioassays. Comparisons with results from standard acute toxicity tests and other rapid screening assays for copper suggest concentrations necessary to reduce swimming speed and/or swimming periods in the rotifer to 50% of controls within 6 hours (6-h EC_{50}) were the lowest reported. However, 6-h EC_{50}s for pentachlorophenol and lindane for the rotifer were three to >100 times higher than results for tests with other species, but the latter took longer to conduct. The 6-h EC_{50}s for the rotifer were at least as sensitive as the corresponding 24-h LC_{50}s (Charoy and Janssen 1999). A more promising behavioural bioassay, because of its greater simplicity, may be one of the two described by Blockwell *et al.* (1998) which uses the precopulatory guarding behaviour of the amphipod *Hyalella azteca*. The 'indirect precopulatory pair separation' test can be conducted in 4 hours and tests with lindane indicate the bioassay is more sensitive than the 24-h LC_{50} for neonates of the amphipod, but not as sensitive as the corresponding 48-h LC_{50}.

1.6.4 CHEMICAL SPECIFICITY AND ECOLOGICAL RELEVANCE

The ecological relevance of biomarkers is generally considered to be a function of their degree of biological complexity (Adams *et al.* 1989, Adams 1990).

Unfortunately, as they become more complex, biomarkers tend to become less contaminant-specific and the response measured more difficult to attribute to a specific cause (Peakall 1994). In this respect, behaviour is no different from other measures taken at the individual level or higher in that responses are rarely diagnostic of the effects of a specific chemical or class of chemicals. Attempts to improve the specificity of behavioural measures have had some success (see Section 1.8.1.1). For example, Drummond *et al.* (1986) compared the behavioural and morphological responses of fathead minnows to 139 chemicals. By applying discriminant function analysis to the behavioural data, they were able to separate a subset of the chemicals according to probable modes of action. Others have had similar success using distinct behaviours or sets of behavioural and morphological responses to distinguish among various contaminants (Drummond and Russom 1990, Rice *et al.* 1997). These studies suggest that behaviour may be useful in monitoring effluents for toxicity and in comparing and predicting the mode of action of new or unknown contaminants (see Sections 1.8.1.1 and 1.8.1.2.2).

The noninvasive nature of behavioural observations makes them potentially invaluable in situations where more traditional biomarkers are not possible, either because they require sacrificing the animal or capture and handling would cause undue stress to sensitive species. This is particularly true for endangered or difficult to capture species for which there are not good surrogates. In addition, 'nondestructive biomarker approaches' enable the same individual or population to be monitored over time (Fossi 1998).

Additionally, an animal's behaviour represents the cumulative product of interactions occurring across biochemical, cellular and individual levels within the context of a variety of biotic and abiotic factors. This integration of multiple variables through time may allow behavioural endpoints to provide meaningful information when other methods would fail. For example, under conditions of fluctuating chemical inputs and water flows, as is frequently encountered in rivers and streams, assessment of contaminant concentrations through chemical analyses is likely to provide highly variable results of questionable ecological relevance (Wendt-Rasch *et al.* 1998). As an alternative, several studies have utilised the net-spinning behaviour of hydropsychid caddisfly larvae to monitor water quality (e.g., Wendt-Rasch 1998; Tessier *et al.* 2000a, b). A variety of pollutants, including metals, organochlorines, and anti-cholinesterase pesticides have been shown to increase the frequency of structural deformities in the capture nets of *Hydropsyche* larvae, both in the laboratory and the field. While responses vary with species and contaminant, at least one combination (*H. slossonae* exposed to 2,4-dichlorophenol) has yielded chronic EC_{50}s at least as sensitive as conventional toxicity tests (Tessier *et al.* 2000b). In a related study, Tessier *et al.* (2000c) found that not only were net anomalies sensitive indicators of malathion exposure, but also their occurrence was strongly correlated with AChE activity (Spearman coefficient, $r = 0.98$).

Several authors (e.g., Triebskorn *et al.* 1997, Fossi 1998) suggest that the effectiveness of biomarkers in environmental assessments (i.e., probability of detecting effects and establishing their ecological relevance) would be enhanced through the use of multiple endpoints representing different levels of biological organisation. This approach, referred to as the 'Multiple Response Concept' (Depledge 1994), attempts to address concerns inherent to biomarkers (i.e., differences in sensitivity, chemical specificity, transiency of response and ecological relevance), particularly under field conditions in which organisms are likely exposed to a variety of stressors, both abiotic and biotic. An example is the study by Triebskorn *et al.* (1997) in which heat shock protein expression, changes in liver ultrastructure and alterations in swimming behaviour were simultaneously examined in brown trout exposed to water from two polluted streams for various lengths of time and at two water temperatures. They found that the three markers, representing responses at the biochemical, cellular and individual levels of biological organisation, respectively, were more coincident with one another in the highly polluted stream compared to the moderately polluted stream in which the markers showed little relation to one another. The authors conclude that this was primarily due to the differential capacities of the biomarkers to integrate stress responses over time. More specifically, the induction of stress proteins proved to be a very sensitive, rapid responding, general marker that integrated stress over a short period of time, but one with low ecological relevance and high temperature dependency, and therefore best applied in conjunction with structural (cellular) endpoints. The latter were also sensitive general markers, but slow in responding with moderate ecological relevance. The behavioural endpoint used (swimming performance) was a sensitive, rapidly responding, general marker of both short- and long-term pollution, with the greatest ecological relevance. As a result, Thriebskorn and her co-workers recommend that environmental assessments incorporate a suite of biomarkers, ones that are specific and general, slow and rapid-responding, and short- and long-term integrators of stress, and some with high ecological relevance. Furthermore, Fossi (1998) notes that the goal in using biomarkers in environmental assessments is not to provide a quantitative evaluation of levels of a toxic substance to which an organism is exposed, but an indication of the health of the organism or its population within its environment. Conceptually, such an evaluation can be extended to the health of the community and ecosystem to which the species belongs (e.g., see Fossi 1998).

1.7 'LEAPS OF FAITH' – EXTRAPOLATING FROM THE LABORATORY TO THE FIELD

A basic premise of the regulatory applications of environmental toxicology is that observations from controlled laboratory studies can be used to predict the responses of the same or other species under field conditions. With the implementation of the Environmental Protection Agency's (EPA) 'New Paradigm' for assessing the risks pesticides pose to wildlife (i.e., reliance on standardised laboratory tests for predicting environmental effects without the benefit of field studies), our ability to extrapolate from the laboratory to the field assumes even greater significance (for discussion, see Section 1.8.3.1). However, it is unlikely that controlled laboratory studies can predict all of the potential exposure-effect scenarios in the field (for recent review, see Blus and Henny 1997). Whether or not contaminant effects manifest themselves in the field is a function of a variety of intrinsic and extrinsic factors (see Section 1.3 for discussion of the importance of ecological context). As a result, field studies have served to document problems that have 'slipped through the system' (Blus and Henny 1997). Heinz *et al.* (1994) suggest that laboratory tests need not try to mimic exactly all the conditions existing in nature, but more importantly, we understand how the conditions of the laboratory bias extrapolations to the field. For example, one could argue that laboratory toxicity studies that fail to incorporate natural stressors would underestimate hazards to free-living animals. Or, the hazard to free-living animals may be less because they are more strongly motivated to compensate for the effects of intoxication than their laboratory counterparts (Bignami *et al.* 1975, Galindo *et al.* 1985, Hart 1993). Test subjects in the laboratory lacking, for example, the immediate threat of predation or food shortages, may simply have little to gain from the extra effort needed to maintain functional normalcy. In addition, because behaviour is context dependent it is likely that natural variation in behaviour in the field will mask subtle chemical-induced changes. We note here that these contexts can vary significantly geographically and with them the behavioural repertoires of species (behavioural ecotypes, Foster and Endler 1999) adding to the difficulties of extrapolating from the context of the laboratory to that in the field.

Do behavioural effects observed under laboratory conditions manifest themselves in the field? In this section of our chapter, we review the extent to which chemical-induced changes in behaviour observed under laboratory conditions have also been documented in free-living conspecifics as well as the difficulties inherent in extrapolations among species. The translation of chemical-induced behavioural effects in individuals to higher levels of biological organisation is discussed in Section 1.8. Because experimental systems represent a continuum from the single species toxicity tests to multispecies mesocosms, we treat all such constructs under the rubric of laboratory studies rather than draw a distinction at some arbitrary level of realism.

1.7.1 EXTRAPOLATION WITHIN SPECIES

Few studies have simultaneously compared the inherent sensitivity of captive and free-living conspecifics to chemical exposure (e.g., as measured by a LD_{50}) or biochemical, physiological and behavioural endpoints. Results of these limited studies suggest that responses at molecular, cellular and organismal levels may be more similar among captive and free-living individuals than anticipated when the animals are given equivalent chemical exposures. However, effects that are 'sublethal' in the laboratory may ultimately reduce survival or reproduction in the field. For example, Grue *et al.* (1982a) compared body weights and brain ChE activity in captive and free-living adult female European starlings following a severe but sublethal exposure to the OP dicrotophos. Responses were similar except that captive controls lost less weight than their wild counterparts, probably because the latter were caring for young. Similarly, Buerger *et al.* (1994) found that LD_{50} estimates for methyl parathion were similar for wild and captive populations of northern bobwhite when the birds were penned after dosing. However, wild caught birds released after dosing suffered statistically greater predation than controls (survival probabilities of 0.84 vs. 0.96 over the 3-year study; Buerger *et al.* 1991). Evidence of altered covey integrity suggested that behavioural effects contributed to the increase in mortality. Increased susceptibility to predation had been reported earlier in laboratory studies with bobwhite given a similar 'sublethal' dose of methyl parathion (Galindo *et al.* 1985). Using similar dose-release field trials, Brewer *et al.* (1996) reported a 7% greater mortality in wild bobwhite given 21 mg/kg of the OP terbufos compared to results from laboratory toxicity tests with laboratory-reared bobwhite.

A recent study by Matz and her coworkers (Matz *et al.* 1998) further illustrates the importance of ecological context in extrapolating of laboratory results to the field. They compared results from a controlled field study of the effects of the OP azinphos-methyl on northern bobwhite chicks to those from a risk assessment based on laboratory dietary toxicity data for young bobwhite and estimated residues on foliage and in insects. Survival, brain AChE activity and growth rates in the field were significantly reduced at application rates predicted to be safe, based on the laboratory toxicity data and nomogram estimate of residues of the insecticide. Chemical exposure in the field was believed to be greater than that in the laboratory because of dermal contact and inhalation. Although not a study of behavioural effects, behavioural responses such as food repellency and conditioned aversion, altered time-activity budgets, and impaired predator avoidance were also suggested as factors contributing to the discrepancy between predicted and observed effects.

Studies that have combined field and laboratory observations to investigate the effects of pollutants on natural systems (cause-effect) offer additional comparisons between responses in the laboratory and those in the field. As part of study of the effects of a spill of fuel oil on the salt-marsh crab, *Uca pugnax*, Krebs and Burns (1977) found that the escape response of crabs exposed in the field was uncharacteristically slow and the crabs were lethargic and showed loss of equilibrium. Laboratory experiments supported the field observations: the escape response times of crabs ingesting oiled sediments more than doubled. Similarly, Dicks (1973) used a combination of field and laboratory observations to study the effects of crude oil on the limpet (*Patella vulgata*). He found crude oil caused detachment of the limpets from rocks through the toxic effects of the oil on foot muscles, and as a result, the limpets were more susceptible to predation before they could successfully reattach. More recently, Weis and her co-workers have applied this approach to studies of a polluted estuary in New Jersey, with an emphasis on predator-prey relationships (for review, see Weis *et al.* 1999). Laboratory studies indicated that exposure to contaminated sediments, water and grass shrimp (*Palaemonetes pugio*) from the estuary impaired the predatory and predator-avoidance abilities of mummichogs. These effects corresponded to an observed decrease in the growth and longevity of the fish, and an increase in number and size of grass shrimp in the estuary; the latter were shown not to be as affected by the pollution.

Additional comparisons of chemical-induced behavioural responses in captive and free-living conspecifics come from studies of the attraction and avoidance responses of aquatic invertebrates and fish to contaminants. For example, Burris *et al.* (1990) demonstrated that the stream-dwelling operculate snail, *Elimia clavaeformis*, can detect relatively low levels of contaminants, both in the laboratory and *in situ*, and will disperse downstream in response to contamination, whereas its natural tendency is to disperse upstream in uncontaminated streams. Similarly, juvenile steelhead (*Salmo gairdneri*) avoided sewage effluent in laboratory studies, whereas juvenile chinook salmon were attracted to the effluent from an oil refinery; responses also noted in limited field observations of migrating adults (Smith and Bailey 1990). And finally, the greater avoidance of mixtures of metals, simulating water pollution in the upper Clark Fork River in Montana, by rainbow trout compared to brown trout in the laboratory was in agreement with population responses in the river (Hansen *et al.* 1999). We refer the reader to the review by Atchison *et al.* (1987) for additional examples.

To some extent, these comparisons suggest that laboratory studies do not necessarily have to be complex to predict responses of specific taxa in the field — just well designed and executed. Additionally, as suggested by others (Rand 1985, Cohn and MacPhail 1996, Dell'Omo and Shore 1996b, Weis *et al.* 1999), the combination of laboratory and field studies provides a powerful approach to identify cause-effect relationships and facilitate the extrapolation of responses observed in the laboratory to natural systems.

1.7.2 EXTRAPOLATION AMONG SPECIES

Historically, toxicologists have focused on identification of the pharmacological dose of a compound associated with specific effects in a small number of test species in order to determine the potential hazards to human health. These test animals often represent highly inbred strains whose responses relative to humans have been well characterised through many years of study. Yet even with these few species of intentionally reduced genetic diversity, it is standard practice to incorporate uncertainty factors. These usually represent an order of magnitude reduction from no observed adverse effect levels (NOELs) for each level of extrapolation (e.g., from human to sensitive human, animal to human, subchronic to chronic exposure; Hattis *et al.* 1996).

Similarly, fish and wildlife species of concern to regulatory agencies are often limited in their availability or their ability to adapt to captivity, and regulators must frequently rely on studies with alternative species (surrogate, indicator, or sentinel; see Schafer 1994 for definitions). Investigations have generally been confined to relatively few species that make up standard test batteries. As a result, available comparative data on the effects of pesticides and other environmental contaminants on fish and wildlife are quite limited. For example, a 1994 review concluded that only about 1% (less than 10) of North American wild bird species have been the subject of detailed comparative toxicological studies (Schafer 1994).

Therefore, one is frequently presented with the problem of extrapolating effects observed in the laboratory (generally in the form of a NOEL, LD_{50} or LC_{50}) to the field and/or from one species to another in the absence of direct empirical data. Several methods have been employed to account for variation in sensitivity among species in estimating acceptable levels of environmental contamination. These include simple weight-based extrapolation (Urban and Cooke 1986), the incorporation of scaling factors (Davidson *et al.* 1986, Mineau *et al.* 1996), and more complex statistical methods. The latter seek to provide estimates, confidence limits and safety factors based on species sensitivity (LD_{50}) distributions – especially when the extrapolation must be made from small data sets (Aldenberg and Slob 1993, Luttik and Aldenberg 1997, Roman *et al.* 1999, Aldenberg and Jaworska 2000, Mineau *et al.* 2001).

Reviewing the available information for birds, Schafer (1994) found that the species and family sensitivity distributions for six chemicals (for which there were sufficient data) varied from just over one to nearly three orders of magnitude. The assumption that certain avian species are consistently among the most sensitive to chemical intoxication (and hence likely to be protective of other species) was shown to be unfounded. Schafer concluded that there is little justification for the use of any one species as a surrogate for another or as an indicator of a larger taxonomic group. Furthermore, his comparisons suggest that extrapolation of responses for a particularly species among chemicals or for a particular chemical among species can at best be expected to be within a factor of two to three orders of magnitude in either direction along an exposure gradient. We note that many of the problems that have 'slipped through the system' (Blus and Henny 1997) have been associated with failures adequately to account for either differential sensitivity among species (e.g., DDE-induced eggshell thinning) or behaviours that result in greater than predicted chemical exposure (e.g., mortality of magpies, following ingestion of hair from cattle treated with famphur).

In view of the potential variability in inherent sensitivities, chemical exposures, behavioural repertoires and ecological contexts among species, successful extrapolations of chemical-induced behavourial effects appears unlikely except for closely-related species within a specific ecological context for which a biomarker is available to assess equivalency of chemical exposure. Although numerous studies have been conducted on the effects of pollutants on the behaviour of a variety of taxa (see Sections 1.4 and 1.5 for review), variation in toxicant, methods and behaviours quantified makes meaningful comparisons difficult, if not impossible in most cases. An exception may be the ChE inhibitors, chemicals for which a significant amount of data have been collected on their effects on the behaviour of birds, laboratory rodents and nonhuman primates. The biomarker, AChE, is a marker of exposure as well as effect; and the enzyme alters what has been referred to as 'the only substance that can influence every physiological or behavioural response thus far examined' (Meyers in Russell 1982:439). Additionally, general patterns of recovery in brain AChE activity following acute exposure to OPs appear to be similar among species of birds and mammals (Grue *et al.* 1997), irrespective of whether they are captive or free-living (Roberts *et al.* 1988). In the case of the carbamates, enzyme inhibition is reversible, and recovery can occur very rapidly (spontaneous reactivation; O'Brien 1976).

A large number of the signs of intoxication associated with exposure to ChE inhibitors are shared by birds, laboratory mammals and man (Grue *et al.* 1991). Acute exposure to these chemicals is frequently associated with a reduction in activity in the laboratory and in the field. For example, Grue and Shipley (1981) reported reduced activity (flying, foraging, and singing and displaying) in male European starlings following a single oral dose of the OP, dicrotophos. These findings agree with reports of 'behavioural slumps' in a variety of species of free-living birds following operational applications of these chemicals to control insects in agricultural fields (Edwards and Graber 1968) and forests (Busby *et al.* 1990). The similarity of laboratory and field observations has led to concerns of the effects of chemical-induced reductions in activity on survival (e.g., predator avoidance) and reproductive success (e.g., parental care) of free-living individuals (for review, see Grue *et al.* 1991, 1997), as well as potential biases in field monitoring techniques dependent on activity to enhance detection of affected individuals (Grue and Shipley 1981, Mineau and Peakall 1987, Fryday *et al.* 1996). Reductions in the consumption of clean food has also been observed in captive birds and mammals following acute exposure to these chemicals, although it is not clear to what extent the responses are the result of pesticide-induced anorexia or conditioned aversion or to what extent comparable effects occur in the field (for review, see Grue *et al.* 1991, 1997).

As in the case of extrapolations from the laboratory to the field for particular species, our ability to extrapolate observations from the laboratory to multiple species in the field depends on our understanding of the factors that govern inherent sensitivity, chemical exposure, and behaviour. Such an understanding can only be achieved through a combination of laboratory and field studies (Grue 1994, Blus and Henny 1997). Furthermore, the role of chemical-induced alterations in behaviour (and 'sublethal effects' in general) in regulatory decision-making will depend on our ability to demonstrate links between behavioural abnormalities and alterations in the reproduction or survival of free-living fish and wildlife — links that to date have largely been missing.

1.8 ANSWERING THE 'SO WHAT' QUESTION

> ... 'all ecological phenomena and patterns of community organization can be viewed as the immediate consequences of individual actions and behaviours ...'
> Real (1994)

With the publication of *Silent Spring* (Carson 1962) and subsequent increase in the public's awareness of the connectivity between man's actions and human and environmental health, environmental legislation was enacted in the 1970s that continues to serve as the primary stimulus and framework for studies of the effects of environmental contaminants on fish and wildlife (see Section 1.8.3.1). Since that time, contaminants have been shown to affect 'virtually every aspect of behaviour of terrestrial and aquatic organisms' (Little 1990, briefly reviewed in previous sections of this chapter and in more detail in the chapters that follow). In spite of this, and the theoretical basis for the biological and ecological significance of behaviour (Real 1994), debate has ensued over the value of many of these studies (and 'sublethal' effects in general), primarily because few have linked behavioural effects to reductions in the reproduction or survival of free-living individuals, let alone effects at higher levels of biological organisation: populations, communities and ecosystems (Cairns 1983, Peakall 1985, 1996; Heinz 1989, Calow 1994, Clements and Kiffney 1994, Grue 1994).

In this section of our chapter, we examine contaminant-induced changes in behaviour in terms of their biological, ecological and regulatory significance. We define biological significance as the use of behavioural responses to contaminants in elucidating underlying biochemical, physiological mechanisms and processes, their use in monitoring environmental health, and their manifestation into decreased reproduction or survival in exposed individuals. Ecological significance is defined as the role of contaminant-induced changes in behaviour in altering populations, community structure, and ecosystem function. Regulatory significance is defined as the extent to which behaviour is or could be incorporated into the regulatory framework protecting human and environmental health.

1.8.1 BIOLOGICAL SIGNIFICANCE

1.8.1.1 Mechanisms and processes

Behavioural responses to contaminants can be used to help identify underlying biochemical, physiological mechanisms and processes. For decades, investigators have used the behavioural effects of anticholinesterases as tools to elucidate the role of peripheral and central cholinergic systems (e.g., Bignami *et al.* 1975, Moser and Padilla 1998, and references therein). More recent applications include the use of differences in behavioural effects to identify the mechanisms of action of ortho- and coplanar-substituted PCBs and dioxin (TCDD) on the central nervous system (Schantz *et al.* 1996, 1997, and references therein). In tricoptera larvae, a variety of chemicals have been shown to alter their net-spinning behaviour as reflected in specific anomalies in the capture-net they build (Tessier *et al.* 2000a,b,c, and references therein). In the case of 2,4-dichlorophenol, midline abnormality was the first to be observed and tended to reflect the physiological stress associated with polar narcosis effects. A more significant anomaly, termed chaotic net, appeared to be related to the uncoupling effects of oxidative phosphorylation, as confirmed by a significant decrease in larval ATP levels (Tessier *et al.* 2000b).

Conversely, knowledge that particular behavioural abnormalities are indicative of exposure to specific chemicals or classes of chemicals would greatly enhance the predictive capabilities of behavioural endpoints in environmental monitoring. For example, with respect to human health, behavioural endpoints are used, or are being recommended, as a screening tool to detect and characterise potential neurotoxic effects of chemicals in humans (for recent comprehensive review, see Environmental Health Perspectives [1999] Vol. 104, Supplement 2) and the US EPA has established testing guidelines for a variety of neurobehavioural endpoints (US EPA 1991). Rice *et al.* (1997) examined the behavioural effects of a number of chemicals with different modes of action on medaka (*Oryzias latipes*). In most cases, chemicals with different modes of action elicited a distinct behaviour or set of behavioural responses: abnormal swimming (permethrin), forward posturing (chlorpyrifos), loss of equilibrium and convulsions (strychnine), anesthesia (phenol), and all were observed at concentrations below the 48-h LC_{50}. Studies by Drummond and his coworkers are exploring the use of 'behavioral toxicity syndromes' (i.e., hypoactivity, hyperactivity and physical deformity) in fathead minnows to classify chemicals according to their mode of action. To date, their work (Drummond and Russom 1990) suggests that it may be possible to identify chemicals at least to a few classes based on behavioural responses which would be useful in focusing subsequent chemical analyses to identify the specific chemical(s) involved.

1.8.1.2 Monitoring environmental health

1.8.1.2.1 Incidental observations

Incidental observations of aberrant behaviour in humans and fish and wildlife can lead to the discovery of important environmental health problems associated with chemical contamination. For example, after the first childhood lead poisoning incident was identified, it was believed that if lead did not kill the child, then he or she suffered no long-term effects. Only through observations of behavioural alterations of the children who had recovered from the disease (reflected, e.g., in high rates of failure in school), was it determined that neurological effects had occurred (Needleman 1995). Similarly, reproductive abnormalities and skewed sex ratios of gulls on Santa Barbara Island might have remained unrecognised, had it not been for observations of unusual female-female pairings in the field (Fry and Toone 1981). In a tragic example, there were various ecological and behavioural changes that had occurred prior to the outbreak of Minamata disease as a result of sublethal methyl mercury exposure in Kyushu, Japan. Domestic cats in the village had showed behavioural symptoms similar to those subsequently manifested in human victims, such as ataxic gait, slowness, unsteady movement and hysterical running patterns, the latter resulting in many of the cats jumping into the sea and drowning. However, there is no evidence that these behavioural warning signs had been taken seriously by the local authorities (Watanabe and Satoh 1996).

1.8.1.2.2 Environmental surveillance and targeted monitoring

We define environmental surveillance as routine monitoring to identify unanticipated contamination and effects (e.g., the National Water Quality Assessment Program, Gilliom *et al.* 1995), and targeted monitoring as that focused on specific, known contaminant situations (e.g., effluents, Natural Resource Damage Assessments [NRDAs], mitigation or restoration). Behavioural endpoints have often been demonstrated to be more sensitive (and therefore may be more protective) indicators of contaminant exposure than the traditional toxicity test endpoint, mortality, but often less sensitive than biochemical measures (e.g., the enzymes ALAD for lead and ChEs for OPs and carbamates) (see Section 1.6). Furthermore, behaviour represents 'the final integrated result of a diversity of biochemical and physiological processes', conveys greater ecological significance than these processes, and can usually be quantified without direct physical harm to the organism (Peakall 1996). As a result, the recent literature is replete with studies in which the authors suggest the biological responses measured (in many cases behavioural) may serve as biomarkers[1] of environmental contamination.

Fossi (1998) provides a conceptual framework for the application of biomarkers in assessing environmental health. She identifies the goals of environmental monitoring using biomarkers as the identification and prediction of early signs of exposure to contaminants or their effects to avoid unacceptable ecological consequences. Biomarkers may be classified as markers of *exposure* and/or *effects*, and as *specific* or *general* in terms of their chemical specificity (Peakall and Shugart 1993). However, with the exception of a few biochemical and physiological markers, the effectiveness of most potential biomarkers in the field, including behavioural endpoints, has yet to be demonstrated (Peakall 1996). [We refer readers to the recent series of articles on biomarkers in Ecotoxicology (1999, Vol. 8:327-416) and references therein.] Issues limiting the use of behaviour as a marker (and biomarkers in general) are: logistical constraints (e.g., monitoring behaviour the field [Peakall 1996]), standardisation of methods, and their validation (e.g., determination of cause-effect relationships in both laboratory and the field) (Little 1990, Munkittrick and McCarty 1995). Munkittrick and McCarty (1995) suggest that 'the development of the biomarker approach for investigating organism responses to environmental stressors has decreased the emphasis on validating extrapolations' both in terms of exposure and resultant effects at higher levels of biological organisation (see Section 1.8.3 for regulatory implications).

Although we do not envision use of behavioural endpoints in environmental surveillance within the near future, studies suggest that they may be a valuable addition to targeted monitoring efforts and standardised toxicity tests (for review, see Environmental Toxicology and Chemistry [1990], Vol. 9[1]). For example, Smith and Bailey (1990) examined the responses of species of anadromous fish to effluent from two oil refineries and domestic sewage water under laboratory conditions. Chinook salmon the only species tested at the first refinery, showed a strong attraction to the effluent at all dilutions tested. Similarly, steelhead and striped bass (*Morone saxatilis*) were attracted to the effluent of the second refinery, whereas steelhead showed a strong avoidance of the sewage water. Limited field observations appeared to corroborate the laboratory results. Adult salmon were observed in the discharge channel of the second refinery even though spawning habitat was absent, and adult steelhead were only observed in tributaries above the sewage outflow during the no-discharge period. The authors also observed fin loss or damage in the striped bass exposed to the effluent of the second refinery in the laboratory and in the field, indicating that the fish attracted to the effluent remained for a sufficient time to suffer toxic effects. Additional examples of avoidance responses of fish to contaminants associated with changes in community structure are given in Section 1.8.3.

1.8.1.2.3 Standardised toxicity tests

As also noted by Atchison *et al.* (1987), we do not see behavioural tests replacing existing standardised toxicity tests, but instead using behavioural endpoints to improve the interpretation and ecological relevance of test results. Of the behavioural endpoints evaluated to date, avoidance/preference responses, such as those noted in the previous section, appear to have the greatest potential to complement the existing suite of standardised aquatic and terrestrial toxicity tests (i.e., how behavioural responses to a particular chemical in water or food may alter the chemical exposure an organism receives in the laboratory and the field [Linder and Richmond 1990]). For example, monitoring the feeding behaviour during median lethal dietary toxicity (LC_{50}) tests has been recently suggested as a means of assessing repellency of test chemicals to birds (Luttik 1998). Similar tests have been suggested for small mammals (Linder and Richmond 1990). Mineau *et al.* (1994c) suggest that behavioural endpoints (feeding, reproductive behaviours) may also improve the standardised avian reproduction test by better assessing sublethal toxicity in parents, particularly from compounds known to be neurotoxic. Both the US EPA and the Organization for Economic Cooperation and Development (OECD) are currently considering these recommendations.

1.8.1.3 Reproduction and survival

Our review identified a relatively small number of studies that have demonstrated a relationship between contaminant-induced behavioural alterations and a change in an organism's fitness (reproductive success or survival). Unfortunately, the majority of the evidence for the significance of behavioural changes is incomplete (for review, see Section 1.4 and other chapters in this text) and further research is needed to document the importance of these changes to the fitness of affected organisms. Studies which not only demonstrate a relationship between contaminant-induced behavioural alterations and a change in an organism's fitness, but also population, community or ecosystem effects are reviewed in Section 1.8.2.

Several of the studies that have documented reductions in reproductive success associated with aberrant behaviour involve seabirds exposed to petroleum. Fry *et al.* (1986) reported increased nest abandonment and reduced incubation of eggs by wedge-tailed shearwaters (*Puffinus pacificus*) exposed to crude oil either on their plumage or internally. Return rates and reproductive success of birds exposed to external applications of 2.0 ml of oil were reduced 1 year after exposure. Similarly, decreased reproductive success in Leach's storm petrels (*Oceanodroma leucorhoa*) exposed externally and internally to crude oils and dispersants was probably associated with temporary abandonment of nesting burrows by treated adults, but in this case, reproductive performance the following year appeared normal (Butler *et al.* 1988). More recently, Walton *et al.* (1997) examined the effects of an oil spill on the breeding behaviour of kittiwakes and observed that many birds did not breed in the first and second years immediately following the spill. This decrease in return rate was not a result of low survival, but rather adults missing one or more breeding attempts.

Reductions in nest attentiveness and subsequent reductions in reproductive success of have also been observed in a other bird species following sublethal exposures to chemicals other than petroleum. For example, behavioural responses of breeding white-throated sparrows following an aerial application of the OP fenitrothion included territory abandonment, disruption of normal incubation and clutch desertion which were thought to have contributed to the observed reduction in reproductive success (Busby *et al.* 1990), although the authors could not rule out the potential influence of chemical-induced reductions in food abundance and prey item contamination as factors influencing nestling mortality. Similarly, reductions in nest defense and incubation associated with sublethal organochlorine exposures were believed to be sufficient to partially account for the lower reproductive success of Lake Ontario herring gulls compared to less contaminated colonies elsewhere (Fox *et al.* 1978; Mineau *et al.* 1984).

There are also few examples of reductions in survival as a result contaminant-induced behavioural alterations. In a frequently cited study, sharp-tailed grouse exposed to sublethal dosages of dieldrin or malathion under field conditions were more vulnerable to predation than controls as a result of decreased alertness and flight capacity, although the authors acknowledge that the data were too limited to be conclusive (McEwen and Brown 1966). Additional support for their findings can be found in more recent studies. Buerger *et al.* (1991) reported greater predation of wild bobwhite quail receiving a severe dose of methyl parathion that was sublethal under laboratory conditions (Galindo *et al.* 1985) compared to wild controls. Similar results were reported in a comparison of the toxicity of terbufos to laboratory-reared and wild bobwhite (Brewer *et al.* 1996) in which the mortality of the wild birds was 7% greater than that predicted for their laboratory counterparts. This difference was believed to be largely due to predation by avian and mammalian predators. The findings of these studies are further supported by laboratory studies indicating greater predation of birds given sublethal exposures of OP pesticides compared to controls (Gallindo *et al.* 1985, Hunt *et al.* 1992).

1.8.2 ECOLOGICAL SIGNIFICANCE

There are few demonstrations of the translation of chemical-induced changes in behaviour seen in the laboratory or the field to changes in populations, community structure or ecosystem function. The study by Krebs and Burns (1977) is frequently cited as an example (Rand and Petrocelli 1985, Smith and Logan 1997). The authors studied the effects of water and sediment contaminated with number 2 fuel oil on the behaviour and population dynamics of the fiddler crab (*Uca pugnax*) in a salt marsh over a period of 7 years. They concluded that impairment of locomotor ability and other behaviours (observed in the field and confirmed in the laboratory) accounted for the persistent reduction in populations of the crab through increased predation and, as a result, energy flow within the contaminated marsh, other salt marshes and coastal waters was altered.

More recently, Weis *et al.* (1999) studied predatory behaviour of mummichogs and an important prey species, the grass shrimp, in an estuary contaminated with organic chemicals, mercury and lead. Impairments in foraging behaviour detected in the fish from contaminated sites appear to have accounted for their reduced growth and longevity compared with those from uncontaminated sites. In contrast, the prey species had a greater population density and a larger size-frequency distribution at the polluted site, apparently because of reduced predation pressure.

If the individuals within a population are able to detect contaminants and actively avoid contaminated areas, then this will be reflected in the relative abundance of this species within the community. This may be especially meaningful if interspecies differences in the sensitivity to the pollutant exist. Hansen *et al.* (1999) suggest that the greater sensitivity in the avoidance response of rainbow trout as compared to brown trout to metal contamination under simulated field conditions in the laboratory may be responsible for the observed differences in the observed distributions and relative abundance of these two species within the Clark Fork River in Montana. Similarly, reductions in populations of native cutthroat trout (*O. clarki lewisi*) in the South Fork of the Coeur d'Alene River in Idaho appear to be due in part to their avoidance of heavy metal (cadmium, copper, lead and zinc) contamination (Woodward *et al.* 1997).

Difficulties in performing studies at levels of biological organisation above the individual and demonstrating effects, particularly behavioural effects, against the backdrop of high and unpredictable variability of natural systems undoubtedly contributes to the small number of studies at the population, community and ecosystem levels (Kulig *et al.* 1996). Historical differences in approaches of toxicologists and ecologists (Figure 1.4) may contribute as well. Clements and Kiffney (1994) attribute the small number of studies to a 'lack of interest among environmental toxicologists for assessing effects at the community and ecosystem levels' (and we would argue 'population level' also). This assessment may be harsh in view of the precedence of making regulatory decisions based on standardised procedures based largely on the individual and the relative lack of such procedures for higher levels of biological organisation (e.g., see Urban and Cook 1986, Nabholz *et al.* 1997, Touart and Maciorowski 1997) and it also ignores the increased logistics, cost and comparative lack of knowledge of ecosystem processes (Calow 1994), societal considerations (Beyer and Heinz 2000), and the relative youth of the field of ecotoxicology. Additionally, it ignores the fact that theoretical ecologists have been 'notoriously reluctant to provide substantial guidance . . . on just what parameters to measure in natural systems to validate predictions based on laboratory exposures' (Cairns 1989).

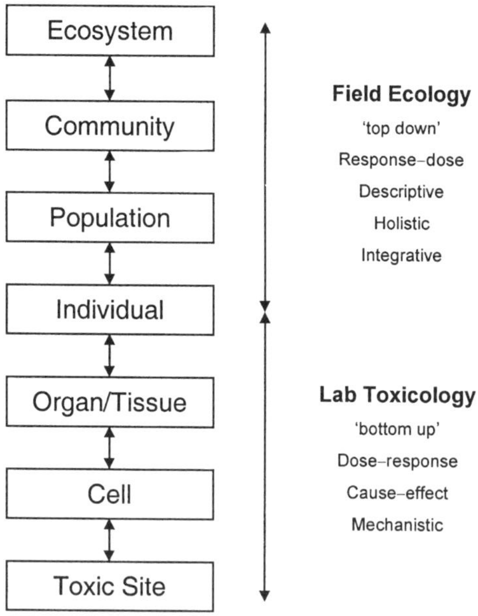

Figure 1.4 Differences between the toxicological and ecological approaches to studying effects of pollutants within the hierarchy of biological organisation (after Munkittrick and McCarty 1995).

Toxicology has its foundation in the individual-based human health focus of early pharmacological studies ('bottom-up', 'dose-response', 'reductionist') in which the primary objective was identifying the dose associated with a specific effect level (Clements and Kifffney 1994, Munkittrick and McCarty 1995). In comparison, ecology has as its origins descriptive natural history and biological sciences through which ecologists have attempted first to identify an effect or response and then search for the cause ('top down', 'response-dose'; Munkittrick and McCarty 1995). As noted by Munkittrick and McCarty (1995), both approaches currently rely heavily on information collected on the individual and may be better referred to as 'middle-down' (toxicology), 'middle-up' (ecotoxicology) 'middle-out' (both). Although the latter has formed the basis for regulatory decisions in the past, the authors suggest dependence on this approach is limiting the evolution of ecotoxicology because difficulties in linking cause-effect, extrapolating from the laboratory to the field, dealing with chemical mixtures, and extrapolating individual responses to the population are not being adequately addressed. The same argument could be made for application of behavioural and other 'sublethal' effects to regulatory decision making, except to add that in most cases the necessary links to the individual (reproduction or survival) have not been adequately demonstrated either. Reasons for this are discussed in greater detail in Sections 1.8.3.1 and 1.8.3.2.

1.8.3 REGULATORY SIGNIFICANCE

1.8.3.1 Past to present

Data requirements and criteria for pesticide registration in the United States under the Federal Insecticide, Fungicide and Rodenticide Act (FIFRA) have been major factors in both stimulating and limiting the advancement of aquatic-, wildlife-, and ecotoxicology.[2] Environmental concerns primarily associated with organochlorine pesticides (e.g., biomagnification and eggshell thinning) raised questions about our ability to predict effects in nature based on the results of standardised laboratory testing. The late 1970s to early 1990s was a time period in which greater focus was placed on the study of chemical exposure and resulting direct effects (including sublethal effects) in the natural world (mesocosms, controlled-field and field) (Kendall and Akerman 1992). The need to develop and validate appropriate field methodologies and endpoints translated into unprecedented funding from private (registrants) and public (federal government) sources (a time period we refer to as the 'Golden Years'). Unfortunately, potential methodologies were developed faster than regulators could validate them, and mesocosm, controlled-field and field studies served to heighten awareness of the difficulties in determining dose - response (cause - effect) outside the laboratory and interpreting the significance of observed effects in the field because of the inherent complexity of natural systems (e.g., see Section 1.3 of this chapter). Frustrations on the part of the regulators and registrants as a result of resultant delays in regulatory decision making led to a significant paradigm shift in aquatic and terrestrial hazard assessment in the United States (Fisher 1992).

Prior to 1992, most ecological risk assessments (in reality hazard assessments as no true probability determinations were made) employed primarily standardised laboratory toxicity tests within a tiered framework: terrestrial — median lethal dosages or dietary concentrations ⇒ reproductive toxicity tests ⇒ simulated or actual field testing; aquatic — median lethal concentrations ⇒ aquatic invertebrate and fish life-cycle tests ⇒ simulated (mesocosm) or actual field (pond) studies (Touart and Maciorowski 1997). In effect, assessments at the lower tiers in which established risk criteria were exceeded (Urban and Cook 1986) were viewed as hypotheses to be critically tested in simulated or actual field conditions (Fite *et al.* 1988, Touart 1988, Pratt *et al.* 1997). In 1992, the US EPA made the decision not to require the higher tiers (simulated or actual field testing) to support 'risk' assessments except in unusual circumstances (Fisher 1992, Touart and Maciorowski 1997). Although behavioural effects (and most sublethal effects) have not been included in the 'risk' assessment process before or after 1992, the decision to remove simulated and actual field studies from the regulatory decision-making process makes the incorporation of such effects even more unlikely.

There is precedence for including behavioural effects within the existing regulatory framework(s). According to Little (1990), behavioural testing received legal status in the USA in 1986 with the adoption of certain behavioural tests for National Resource Damage Assessments (NRDAs) required by the Comprehensive Environmental Response, Compensation, and Liability Act (CERCLA) of 1980. Here, biological responses including behaviour are admissible as irrefutable evidence of injury and carry the same legal weight as water quality criteria of the US EPA and action levels of the US Food and Drug Administration (US FDA) (NRDA 1986). NRDA acceptance criteria used in selecting biological responses as evidence of injury are demonstrated cause-effect in the laboratory and the field, the response is practical to measure, and yields scientifically valid results (NRDA 1986). Behavioural responses meeting these criteria include clinical behavioural signs of toxicity (e.g., lethargy or tremors) and avoidance reactions of fish (USDI 1987). Although the criteria call for field verification, unlike FIFRA and Toxic Substances Control Act (TSCA), a link to impaired reproduction or survival is not required (Little 1990).

In a number of countries, functional measures, especially behavioural endpoints, are used routinely, or are being recommended, to detect and characterise potential neurotoxic effects of chemicals in humans (Weiss 1988, Cory-Slechta 1989, Evangelista de Duffard 1996, Hattis *et al.* 1996, Dick and Ahlers 1998). A comprehensive recent review of this topic was published in 1996 in a series of papers in Environmental Health Perspectives (Vol. 104, Supplement 2). The USA EPA has established testing guidelines for a variety of neurobehavioural endpoints (US EPA 1991) and these data are used by other regulatory agencies in the USA including the FDA, Occupational Safety and Health Administration, and Consumer Product Safety Commission. Many of the issues facing the use of behavioural effects in protecting human health are similar to those in using behavioural effects to protect fish and wildlife, except that in the former field validation is not required and, as in NRDA, neither is a link to impaired reproduction or survival. Common issues include the standardisation and validation of test methodologies (Little 1990, Cohn and MacPhail 1996, Kulig *et al.* 1996).

1.8.3.2 Future — The search for the missing links

> '. . . to date there is no clear evidence in wildlife that behavioral changes caused by pollutants are a serious threat to populations' (Peakall 1996).

Ecotoxicologists and regulators must decide what is realistically needed to sufficiently 'complete the circle of evidence' (demonstration that sublethal effects seen in the laboratory occur in the field, and that these effects actually cause or contribute to reduced reproduction or survival in the field; Heinz 1989), such that behavioural endpoints can be included in the regulatory process. Heinz (1989) suggests that 'just a few such studies [demonstrating these links] would greatly increase the weight of all of the rest [referring to studies solely documenting sublethal effects].' Sandheinrich and Atchsion (1990) note that the predictive potential of behavioural data will not be fully realised until we can relate behavioural changes to ecological effects. Little (1990) suggests that 'a few monumental field studies will be needed to confirm what we as behavioural toxicologists continually assume to be intuitively obvious: that changes in behaviour are reflected at the population and community levels.'

We believe it is impractical to determine *a priori* the effort required to adequately test our 'leaps of faith' and 'pious hopes'. As noted by Peakall (1996), there are a myriad of difficulties in conducting the necessary studies. However, instead of dismissing the vast majority of existing studies as inadequate, we need to identify those chemicals, taxa and environmental situations for which a strong case can be made currently or with the least amount of additional effort, keeping in mind that it may be unrealistic to fill in all of the missing links (i.e., 'a weight of evidence approach'). For example, studies have shown that migratory orientation is altered in adult white-throated sparrows exposed to sublethal levels of the OP acephate (Vyas *et al.* 1995). However, it would be logistically difficult to assess the effect of altered orientation on the birds' survival or subsequent reproductive success.

If one adopts a 'weight of evidence' approach to completing the 'circle of evidence' and concurs with Beyer and Heinz (2000) that demonstrating harmful effects at levels of biological organisation above the individual (or possibly local populations) should not be a prerequisite for regulatory action, then the existing data on the severity of 'sublethal' (including behavioural) effects associated with brain AChE inhibition of $\geq$40% in birds exposed to OP pesticides in the field should be considered 'life threatening' (Busby *et al.* 1991), and for regulatory purposes, be considered an example of completing the 'circle of evidence' (see Section 1.7; Grue *et al.* 1991, 1997; Mineau 1998). We believe a regulatory threshold of 40% may be appropriate when applied to levels of enzyme inhibition in those species believed to be at greatest risk based on sensitivity, potential exposure, and population status. Here endpoints deal with actual exposure in sampled individuals vs. documentation of effects in the field (Busby *et al.* 1991). Criteria for the proportion of exposed populations that would have to be exceeded would need to be developed taking into account potential sampling biases (Mineau and Peakall 1987, Grue *et al.* 1997).

Behaviour and behavioural endpoints can and will play an increasingly important role in ecotoxicology and ultimately the regulatory process irrespective of our ability to 'satisfactorily' complete the 'circle of evidence'. Knowledge of the normal behaviours of nonnative species is critical to adequately predict contaminant exposure as part of the risk assessment framework (US EPA 1992; see discussion in Section 1.3.1.2) and in the selection of appropriate surrogates for effects data from the laboratory (standardized toxicity tests) or the field (Tiebout and Brugger 1995, Kendall *et al.* 1996). Behavioural responses (e.g., avoidance/preference) can facilitate the interpretation of the results of standardised toxicity tests (see Section 1.8.1.2.3). Behavioural endpoints will continue to help elucidate underlying biochemical mechanisms and physiological processes in humans, and fish and wildlife (see Section 1.8.1.1). Although we do not believe that environmental surveillance focused on behavioural endpoints is likely to be a realistic approach to assessing environmental quality, we do envision greater improvement in, and application of, standardised behavioural tests in monitoring environmental quality through their use as sensitive, biological (early warning) indicators of human health and water quality, and as accepted endpoints for NRDAs (i.e., targeted monitoring efforts). At the same time, behavioural anomalies reported in the field through other than targeted monitoring efforts must also be taken seriously because of the low probability of detecting them in the first place (see Section 1.3), and consequently their potential as indicators of environmental stress.

1.9 FINAL THOUGHTS

Over the last 30 years, hundreds of studies have documented changes in the behaviour of fish and wildlife following sublethal exposure to environmental contaminants, many of them reviewed in this chapter (Section 1.4) and elsewhere in this text. Although the theoretical basis for their biological and ecological significance is well-rooted in neo-Darwinian theory ('pious hopes', Peakall 1985), very few (see Section 1.8.1.3) have demonstrated a direct link between chemical-induced changes in behaviour and impaired reproduction or survival. Nor are we aware of many instances where behavioural abnormalities in and of themselves in the field have served as signals of contaminant problems or cases where behavioural alterations have translated into changes in community structure or ecosystem structure or function (see Section 1.8.2). Although disappointing, we do not find this surprising given, for example, the complexity of natural systems (see Section 1.3) and the relative youth of the field of ecotoxicology. Critics of this body of literature, and sublethal effects in general, suggest that unless the links between these effects and reductions in reproduction or survival can be more clearly and frequently demonstrated, these effects may be 'considered little more than interesting academic observations, or at best, early warning signals that heavier exposure might cause lethal effects' (Heinz 1989). Furthermore, recent changes in the paradigm used by the US EPA to evaluate the risks pesticides pose to the environment under FIFRA (discussed in the previous two sections) make it even less likely that the 'circle of evidence' will be completed (demonstrated to the satisfaction of critics) in the near future or that 'sublethal' effects (and we would argue community and ecosystem effects) will be incorporated into the regulatory process. Thus, we find ourselves in a regulatory dilemma in that the studies necessary adequately to evaluate the extent to which sublethal effects are sublethal, our ability to extrapolate from the laboratory to the field, and the extent to which 'ecological' risk assessments will reflect natural systems lack the regulatory stimulus responsible, in large part, for advancing the science and raising these issues.

Although one can appreciate the logistical concerns (expediency and cost) and scientific issues (e.g., lack of validation [regulatory history] of new field monitoring techniques and resultant difficulties in data interpretation), we hope that the US EPA will (as suggested by Touart and Maciorowski 1997) re-evaluate its 1992 paradigm and policy shift. Given the present regulatory dilemma (discussed above), the infancy of ecotoxicology, ecological modeling and ecological risk assessment, and many studies/commentaries that question our current ability adequately to assess/regulate under the existing paradigm, 'traditional risk assessments based on toxicity testing must be viewed as hypotheses to be critically tested ...' (Pratt *et al.* 1997). A balance must be obtained between expediency in regulatory decision-making and validation of the assessment methodologies employed. The use of conditional registrations (used in Western Europe, but not elsewhere) and implementation of targeted field monitoring to assess efficacy of exposure risk mitigation (already part of the existing paradigm, but to our knowledge not utilised) would help to address these concerns, and related to chemical-induced behavioural effects (and sublethal effects in general) provide the necessary stimulus and framework through which their significance and role in the regulatory process can also be adequately assessed.

Ultimately, the significance of contaminant-induced behavioural abnormalities (and sublethal effects in general) may not be decided by members of the scientific and regulatory communities, but by society in general (Heinz 1989, Beyer and Heinz 2000), and in particular, human health concerns.[3] For example, the scientific debate over the significance of endocrine disrupting chemicals in the environment continues despite evidence of feminisation (physiological, morphological and behavioural) in a variety of fish and wildlife (Sumpter 1998, Tyler *et al.* 1998). However, recent suggestions (Lemonick 2000) that these chemicals may be responsible for, or contributing to, early maturation (physiological, morphological and behavioural) in human females (Herman-Giddens *et al.* 1997) will undoubtedly heighten society's concerns over these and other sublethal effects of contaminants in both humans and fish and wildlife.

ACKNOWLEDGEMENTS

Financial support for the preparation of this chapter was provided by the U.S. Geological Survey, Washington Cooperative Fish and Wildlife Research Unit and the Centro de Investigaciones Biologicas del Noroeste, SC, La Paz, BCS, Mexico. The Unit is supported by the USGS, University of Washington, and the Washington Departments of Ecology, Fish and Wildlife, and Natural Resources. We thank Andy Hart, Gary Heinz and Pierre Mineau for their input and reviews of earlier drafts, David Tipton for assistance in obtaining and organising the literature, and Martin Grassley for finalising the figures. The senior author also thanks his family, his graduate students, the editor and the publisher for their support and patience during the preparation of the manuscript.

NOTES

[1]Depledge (1994) defined an ecotoxicological biomarker as a 'biochemical, cellular, physiological, or behavioral change that can be measured in a tissue, biological fluid, whole organism, or population, providing evidence of exposure to, and/or toxic effects of, one or more contaminants ...' Biomarkers may be classified as markers of *exposure* and/or *effects*, and as *specific* or *general* in terms of their chemical specificity (Peakall and Shugart 1993).

[2]The Toxic Substances Control Act (TSCA), also administered by the US EPA, regulates chemicals other than pesticides 'which may present an unreasonable risk of injury to health or the environment' or 'will be produced in substantial quantities, and ... enters or may reasonably be anticipated to enter the environment in substantial quantities ...' (Nabholz *et al.* 1997). Under TSCA, chemicals (30,000 since 1979; 80,000 total) are 'innocent until proven guilty' by the regulatory agency, in contrast to FIFRA, where the registrant is responsible for 'refuting the presumption of risk' (Nabholz *et al.* 1997). Because requests for toxicological data must be justified under TSCA *a posteriori*, its toxicological assessments are largely based on chemical structure activity relationships (SARs) and rarely (<5%) environmental toxicity testing (i.e., in most cases, single species acute toxicity tests) (Nabholz 1997).

[3]We note that human health issues, specifically the occurrence of DDT and its metabolites in human milk and concerns over carcinogenicity were major factors in the ban of DDT in the United States in 1972 (Dunlap 1981:202, Peterle 1991:203, Sullivan 1999:559).

REFERENCES

Adams SM (1990) Status and use of bioindicators for evaluating effects of pollutant stress in fish. In *Biological Indicators of Stress in Fish*, Adams SM (ed.), American Fisheries Society, Bethesda, MA, pp. 1-8.

Adams SM, Shepard KL, Greeley MS, Jimenez BD, Ryon MG, Shugart LR, McCarthy JF and Hinton DE (1989) The use of bioindicators for assessing the effects of pollutant stress on fish. *Marine Environmental Research*, **28**, 459-464.

Alcock J (1998) *Animal behavior-An Evolutionary Approach*. 6th edn, Sinauer Associates, Inc. Sunderland, MA.

Aldenberg T and Jaworska JS (2000) Uncertainty of the hazardous concentration and fraction affected for normal species sensitivity distributions. *Ecotoxicology and Environmental Safety*, **46**, 1-18.

Aldenberg T and Slob W (1993) Confidence limits for hazardous concentrations based on logistically distributed NOEC toxicity data. *Ecotoxicology and Environmental Safety*, **25**, 48-63.

Amr MM, Halim ZS and Moussa SS (1997) Psychiatric disorders among Egyptian pesticide applicators and formulators. *Environmental Research*, **73**, 193-199.

Anderson PK (1970) Ecological structure and gene flow in small mammals. *Symposia of the Zoological Society-London*, **26**, 299-325.

Andres BA (1999) Effects of persistent shoreline oil on breeding success and chick growth in black oystercatchers. *Auk*, **116**, 640-650.

Anthony RG and Morrison ML (1985) Influence of glyphosate herbicide on small-mammal populations in western Oregon. *Northwest Science*, **59**, 159–168.

Atchison GJ, Henry MG and Sandheinrich MB (1987) Effects of metals on fish behavior: A review. *Environmental Biology of Fishes*, **18**, 11–25.

Avery ML, Fischer DL, and Primus TM (1997) Assessing the hazard to granivorous birds feeding on chemically treated seeds. *Pesticide Science*, **49**, 362–366.

Barrett GW and Darnell RM (1967) Effects of dimethoate on small mammal populations. *American Midland Naturalist*, **77**, 164–175.

Bartholomew GA (1964) The roles of physiology and behavior in the maintenance of homeostasis in the desert environment. In *Thermoregulation*, Satinoff E (ed), Dowden, Hutchinson & Ross, Inc., Stroudsburg, PA, pp. 50–73.

Bellocq MI, Bendell JF and Cadogan BL (1992) Effects of the insecticide *Bacillus thuringiensis* on *Sorex cinereus* (masked shrew) populations, diet, and prey selection in a jack pine plantation in northern Ontario. *Canadian Journal of Zoology*, **70**, 505–510.

Bennett RS (1989a) Role of dietary choices in the ability of bobwhite to discriminate between insecticide-treated and untreated food. *Environmental Toxicology and Chemistry*, **8**, 731–738.

Bennett RS (1989b) Factors influencing discrimination between insecticide-treated and untreated foods by northern bobwhite. *Archives of Environmental Contamination and Toxicology*, **18**, 697–705.

Bennett RS, Williams BA, Schmedding DW and Bennett JK (1991) Effects of dietary exposure to methyl parathion on egg laying and incubation in mallards. *Environmental Toxicology and Chemistry*, **10**, 501–507.

Best LB (1972) First-year effects of sagebrush control on two sparrows. *Journal of Wildlife Management*, **36**, 534–544.

Best LB and Gionfriddo JP (1994) House sparrow preferential consumption of carriers used for pesticide granules. *Environmental Toxicology and Chemistry*, **13**, 919–925.

Beyer WN and Heinz G (2000) Implications of regulating environmental contaminants on the basis of wildlife populations and communities. *Environmental Toxicology and Chemistry*, **19**, 1703–1704.

Bignami G, Giardini V and Scorrano M (1985) Behaviorally augmented versus other components in organophosphate tolerance — The role of reinforcement and response factors. *Fundamental and Applied Toxicology*, **5**, S213–S224.

Bignami G, Rosic N, Michalek H, Milosevic M and Gatti GL (1975) Behavioral toxicity of anticholinesterase agents: methodological, neurochemical, and neuropsychological aspects. In *Behavioral Toxicology*, Weiss B and Laties VG (eds), Plenum Press, New York, pp. 155–215.

Bishop CA, Ng P, Mineau P, Quinn JS and Struger J (2000) Effects of pesticide spraying on chick growth, behavior, and parental care in tree swallows (*Tachycineta bicolor*) nesting in an apple orchard in Ontario, Canada. *Environmental Toxicology and Chemistry*, **19**, 2286–2297.

Blanck H and Wangberg S (1988) Induced community tolerance in marine periphyton established under arsenate stress. *Canadian Journal of Fisheries and Aquatic Sciences*, **45**, 1816–1819.

Blick DW, Weathersby FR Jr, Brown GC and Murphy MR (1994) Behavioral toxicity of anticholinesterases in primates: effects of daily repeated soman exposure. *Pharmacology Biochemistry and Behavior*, **48**, 643–649.

Blockwell SJ, Maund SJ and Pascoe D (1998) The acute toxicity of lindane to *Hyalella azteca* and the development of a sublethal bioassay based on precopulatory guarding behavior. *Archives of Environmental Contamination and Toxicology*, **35**, 432–440.

Blus LJ and Henny CJ (1997) Field studies on pesticides and birds: unexpected and unique relations. *Ecological Applications*, **7**, 1125–1132.

Blus LJ, Henny CJ, Hoffman DJ, and Grove RA (1991) Lead toxicosis in tundra swans near a mining and smelting complex in northern Idaho. *Archives of Environmental Contamination and Toxicology*, **21**, 549-555.

Bowers MA (1994) Use of space and habitats by individuals and populations: Dynamics and risk assessment. In *Wildlife Toxicology and Population Modeling — Integrated Studies of Agroecosystems*, Kendall RJ and Lacher TE Jr (eds), Lewis Publishers, Boca Raton, FL, pp. 109-122.

Bracher GA and Bider JR (1982) Changes in terrestrial animal activity of a forest community after an application of aminocarb (Matacil®). *Canadian Journal of Zoology*, **60**, 1981-1997.

Brewer LW, Driver CJ, Kendall RJ, Zenier C and Lacher TE Jr (1988) Effects of methyl parathion in ducks and duck broods. *Environmental Toxicology and Chemistry*, **7**, 375-379.

Brewer RA, Calock LL, Hooper MJ, Brewer LW, Cobb GP and Kendall RJ (1996) Toxicity, survivability, and activity patterns of northern bobwhite quail dosed with the insecticide terbufos. *Environmental Toxicology and Chemistry*, **15**, 750-753.

Britson CA and Threlkeld ST (1998) Abundance, metamorphosis, developmental, and behavioral abnormalities in *Hyla chrysoscelis* tadpoles following exposure to three agrochemicals and methyl mercury in outdoor mesocosms. *Bulletin of Environmental Contamination and Toxicology*, **61**, 154-161.

Brotons L, Magrans M, Ferrus L and Nadal J (1998) Direct and indirect effects of pollution on the foraging behavior of forest passerines during the breeding season. *Canadian Journal of Zoology*, **76**, 556-565.

Brunet R and Cyr A (1992) The impact of dimethoate on rhythms of three granivorous bird species. *Agriculture, Ecosystems and the Environment*, **41**, 327-336.

Brunet R, Girard C and Cyr A (1997) Comparative study of the signs of intoxication and changes in activity level of red-winged blackbirds (*Agelaius phoeniceus*) exposed to dimethoate. *Agriculture, Ecosystems and Environment*, **64**, 201-209.

Bryan MD, Atchsion GJ and Sandheinrich MB (1995) Effects of cadmium on the foraging behavior and growth of juvenile bluegill, *Lepomis macrochirus*. *Canadian Journal of Fisheries and Aquatic Sciences*, **52**, 1630-1638.

Buerger TT, Kendall RJ, Mueller BS, DeVos T and Williams BA (1991) Effects of methyl parathion on northern bobwhite survivability. *Environmental Toxicology and Chemistry*, **10**, 527-532.

Buerger TT, Mortenson SR, Kendall RJ and Hooper MJ (1994) Metabolism and acute toxicity of methyl parathion in pen-reared and wild northern bobwhites. *Environmental Toxicology and Chemistry*, **13**, 1139-1143.

Burger J (1990) Behavioral effects of early postnatal lead exposure in herring gull (*Larus argentatus*) chicks. *Pharmacology, Biochemistry & Behavior*, **35**, 7-13.

Burger J (1995) A risk assessment for lead in birds. *Journal of Toxicology and Environmental Health*, **45**, 369-396.

Burger J and Gochfeld M (1985) Early postnatal lead exposure: behavioral effects in common tern chicks (*Sterna hirundo*). *Journal of Toxicology and Environmental Health*, **16**, 869-886.

Burger J and Gochfeld M (1988) Lead and behavioral development: effects of varying dosage and schedule on survival and performance of young common terns (*Sterna hirundo*). *Journal of Toxicology and Environmental Health*, **24**, 173-182.

Burger J and Gochfeld M (1993) Lead and behavioral development in young herring gulls: effects of timing of exposure on individual recognition. *Fundamental and Applied Toxicology*, **21**, 187-195.

Burris JA, Bamford MS and Stewart AJ (1990) Behavioral responses of marked snails as indicators of water quality. *Environmental Toxicology and Chemistry*, **9**, 69-76.

Busby DG, White LM and Pearce PA (1990) Effects of aerial spraying of fenitrothion on breeding white-throated sparrows. *Journal of Applied Ecology*, **27**, 743–755.

Busby DG, White LM and Pearce PA (1991) Brain acetylcholinesterase activity in forest songbirds exposed to a new method of UULV fenitrothion spraying. *Archives of Environmental Contamination and Toxicology*, **20**, 25–31.

Bushnell PJ, Pope CN and Padilla S (1993) Behavioral and neurochemical effects of acute chlorpyrifos in rats: tolerance to prolonged inhibition of cholinesterase. *Journal of Pharmacology and Experimental Therapeutics*, **266**, 1007–1017.

Bussiere JL, Kendall RJ, Lacher TE Jr and Bennett RS (1989) Effect of methyl parathion on food discrimination in northern bobwhite (*Colinus virginianus*). *Environmental Toxicology and Chemistry*, **8**, 1125–1131.

Butler RG, Harfenist A, Leighton FA and Peakall DB (1988) Impact of sublethal oil and emulsion exposure on the reproductive success of Leach's storm-petrels: short and long-term effects. *Journal of Applied Ecology*, **25**, 125–143.

Cairns J (1983) Are single species toxicity tests alone adequate for estimating environmental hazard? *Hydrobiologia*, **100**, 47–57.

Cairns J (1986) The myth of the most sensitive species. *BioScience*, **36**, 670–672.

Cairns J (1989) Applied ecotoxicology and methodology. In *Aquatic Ecotoxicology: Fundamentals, Concepts and Methodology*, Volume II, Boudou A and Ribeyre F (eds), CRC Press, Boca Raton, FL.

Calow P (1994) Ecotoxicology: What are we trying to protect? *Environmental Toxicology and Chemistry*, **13**, 1549.

Carpenter DO, Arcaro KF, Bush B, Niemi WD, Pang S and Vakhria DD (1998) Human health and chemical mixtures: an overview. *Environmental Health Perspectives*, **106**, 1263–1270.

Carpenter SR, Kitchell JF and Hogsdon JR (1995) Cascading trophic interactions and lake productivity. *BioScience*, **35**, 634–639.

Carson R (1962) *Silent Spring*, Houghton-Mifflin, Boston, MA.

Charoy C and Janssen CR (1999) The swimming behaviour of *Brachionus calyciflorus* (rotifer) under toxic stress. II. Comparative sensitivity of various behavioural criteria. *Chemosphere*, **38**, 3247–3260.

Chiverton PA and Sotherton NW (1991) The effects on beneficial arthropods of the exclusion of herbicides from cereal crop edges. *Journal of Applied Ecology*, **28**, 1027–1039.

Clements RO and Bale JS (1988) The short-term effects on birds and mammals of the use of chlorpyrifos to control leatherjackets in grassland. *Annals of Applied Biology*, **112**, 41–47.

Clements WH (1999) Metal tolerance and predator–prey interactions in benthic macroinvertebrate stream communities. *Ecological Applications*, **9**, 1073–1084.

Clements WH and Kiffney PM (1994) Assessing contaminant effects at higher levels of biological organization. *Environmental Toxicology and Chemistry*, **13**, 357–359.

Clements WH, Cherry DS and Cairns J Jr (1989) The influence of copper exposure on predator-prey interactions in aquatic insect communities. *Freshwater Biology*, **21**, 483–488.

Cohn J and MacPhail RC (1996) Ethological and experimental approaches to behavior analysis: implications for ecotoxicology. *Environmental Health Perspectives*, **104** (Suppl. 2), 299–305.

Cohn J and MacPhail RC (1997) Chlorpyrifos produces selective learning deficits in rats working under a schedule of repeated acquisition and performance. *Journal of Pharmacology and Experimental Therapeutics*, **283**, 312–320.

Colborn T, vom Saal FS and Soto AM (1993) Developmental effects of endocrine-disruptive chemicals in wildlife and humans. *Environmental Health Perspectives*, **101**, 378–384.

Cole DC, Carpio F, Julian J, Leon N, Carbotte R and de Almeda H (1997) Neurobehavioral outcomes among farm and nonfarm rural Ecuadorians. *Neurotoxicology and Teratology*, **19**, 277-286.

Conover WR (1984) Response of birds to different types of food repellents. *Journal of Applied Ecology*, **21**, 437-443.

Cory-Slechta DA (1989) Behavioral measures of neurotoxicity. *Neurotoxicology*, **10**, 271-296.

Costa LG, Schwab BW and Murphy SD (1982) Tolerance to anticholinesterase compounds in mammals. *Toxicology*, **25**, 79-97.

Cox RL and Wilson WT (1984) Effects of permethrin on the behavior of individually tagged honey bees, *Apis mellifera* L. (Hymenopetera: Apidae). *Environmental Entomology*, **13**, 375-378.

Cuomo V, De Salvia MA, Petruzzi S and Alleva E (1996) Appropriate endpoints for the characterization of behavioral changes in developmental toxicology. *Environmental Health Perspectives*, **104** (Suppl. 2), 307-315.

Davidson IWF, Parker JC and Beliles RP (1986) Biological basis for extrapolation across mammalian species. *Regulatory Toxicology and Pharmacology*, **6**, 211-237.

Dell'Omo G and Shore RF (1996a) Behavioral and physiological effects of acute sublethal exposure to dimethoate on wood mice, *Apodemus sylvaticus* [I — Laboratory studies]. *Archives of Environmental Contamination and Toxicology*, **31**, 91-97.

Dell'Omo G and Shore RF (1996b) Behavioral effects of acute sublethal exposure to dimethoate on wood mice, *Apodemus sylvaticus* [II — Field studies on radio-tagged mice in a cereal ecosystem]. *Archives of Environmental Contamination and Toxicology*, **31**, 538-542.

Depledge M (1994) The rational basis for the use of biomarkers as ecotoxicological tools. In *Nondestructive Biomarkers in Vertebrates*, Fossi MC and Leonzio C (eds), Lewis Publishers, Baton Raton, FL, pp. 272-295.

Deutsch JA (1983) The cholinergic synapse and the site of memory. In *The Physiological Basis of Memory*, Deutsch JA (ed), Second edition, Academic Press, New York, pp. 367-386.

DeWeese LR, McEwen LC, Settimi LA, and Deblinger RD (1983) Effects on birds of fenthion aerial application for mosquito control. *Journal of Economic Entomology* **76**, 906-911.

Dhawan R, Dusenbery DB and Williams PL (2000) A comparison of metal-induced lethality and behavioral responses in the nematode *Caenorhabditis elegans*. *Environmental Toxicology and Chemistry*, **19**, 3061-3067.

Diamond JM, Parson MJ and Gruber D (1990) Rapid detection of sublethal toxicity using fish ventilatory behavior. *Environmental Toxicology and Chemistry*, **9**, 3-11.

Dick RB and Ahlers H (1998) Chemicals in the workplace: incorporating human neurobehavioral testing into the regulatory process. *American Journal of Industrial Medicine*, **33**, 439-453.

Dicks B (1973) Some effects of Kuwait crude oil on the limpet, *Patella vulgata*. *Environmental Pollution*, **5**, 219-229.

Doane CC and Schaefer PW (1971) Aerial application of insecticides for control of the gypsy moth: With studies effects on non-target insects and birds. *Connecticut Agricultural Experiment Station Bulletin*, **724**, VV.

Douthwaite RJ (1982) Changes in pied kingfisher (*Ceryle rudis*) feeding related to endosulfan pollution from tsetse fly control operations in the Okavango Delta, Botswana. *Journal of Applied Ecology*, **19**, 133-141.

Douthwaite RJ and Fry CH (1982) Food and feeding behavior of the little bee-eater *Merops pusillus* in relation to tsetse fly control by insecticides. *Biological Conservation*, **23**, 71-78.

Drummond RA and Russom CL (1990) Behavioral toxicity syndromes — a promising tool for assessing toxicity mechanisms in juvenile fathead minnows. *Environmental Toxicology and Chemistry*, **9**, 37–46.

Drummond RA, Russom CL, Geiger DL and DeFoe DL (1986) Behavioral and morphological changes in fathead minnow (*Pimephales promelas*) as diagnostic endpoints for screening chemicals according to modes of action. In *Aquatic Toxicology and Environmental Fate*, Poston TM and Purdy R (eds), *ASTM STP 921*, American Society for Testing and Materials, Philadelphia, **9**, 415–435.

Dunham AE, Grant BW and Overall KL (1989) Interfaces between biophysical and physiological ecology and the population ecology of terrestrial vertebrate ectotherms. *Physiological Zoology*, **62**, 335–355.

Dunlap TR (1981) DDT — *Scientists, Citizens, and Public Policy*, Princeton University Press, Princeton, NJ.

Dunson WA and Travis J (1991) The role of abiotic factors in community organization. *American Naturalist*, **138**, 1067–1091.

Eaton DL and Klaassen CD (1996) Principles of Toxicology, In *Casarett and Doull's Toxicology — The Basic Science of Poisons* (5th edn), Klaassen CD (ed), McGraw-Hill, New York, pp. 13–33.

Edwards WR and Graber RR (1968) Responses of avians to methyl parathion in a hayfield. In *20th Illinois Custom Spray Operators Training School — Summary of Presentations*, University of Illinois, College of Agriculture, Illinois Natural History Survey, Urbana, IL, pp. 53–59.

Eisler R (2000) *Handbook of Chemical Risk Assessment — Health Hazards to Humans, Plants and Animals*. Volumes 1–3, Lewis Publishers, Boca Raton, FL.

Emlen JM, Freeman DC and Graham JH (1998) How organisms do the right thing: The attractor hypothesis. *Chaos*, **8**, 717–725.

Evangelista De Duffard AM and Duffard R (1996) Behavioral toxicology, risk assessment, and chlorinated hydrocarbons. *Environmental Health Perspectives*, **104** (Suppl. 2), 353–360.

Everts JW, Willemsen I, Stulp M, Simons L, Aukema B and Kammenga J (1991) The toxic effect of deltamethrin on linyhiid and erigonid spiders in connection with ambient temperature, humidity, and predation. *Archives of Environmental Contamination and Toxicology*, **20**, 20–24.

Eyer P (1995) Neuropsychopathological changes by organophosphorus compounds — A review. *Human and Experimental Toxicology*, **14**, 857–864.

Fair JM, Kennedy PL and McEwen LC (1995) Effects of carbaryl grasshopper control on nesting killdeer in North Dakota. *Environmental Toxicology and Chemistry*, **14**, 881–890.

Faulk CK, Fuiman LA and Thomas P (1999) Parental exposure to *ortho, para*-dichlorodipenyltrichloroethane impairs survival skills of Atlantic croaker (*Micropogonias undulatus*) larvae. *Environmental Toxicology and Chemistry*, **18**, 254–262.

Fink LS and Brower LP (1981) Birds can overcome the cardenolide defense of monarch butterflies in Mexico. *Nature*, **291**, 67–70.

Fisher LJ (1992) Decisions on the ecological, fate, and effects task force [and program guidance on ecological risk management]. Memorandum [29 October] to D Camp, Director, Office of Pesticide Programs, Assistant Administrator, Office of Prevention, Pesticides, and Toxic Substances, US Environmental Protection Agency, Washington, DC.

Fite EC, Turner LW, Cook NJ and Stunkard C (1988) Guidance document for conducting terrestrial field studies. EPA 540/09-88-109, National Technical Information Service, Springfield, VA.

Fjeld E, Haugen TO and Vollestad LA (1998) Permanent impairment in the feeding behavior of grayling (*Thymallus thymallus*) exposed to methylmercury during embryogenesis. *The Science of the Total Environment*, **213**, 247-254.

Fossi MC (1998) Biomarkers as diagnostic and prognostic tools for wildlife risk assessment: Integrating endocrine-disrupting chemicals. *Toxicology and Industrial Health*, **14**, 291-309.

Foster SA and Endler JA (1999) Thoughts on geographic variation in behavior. In *Geographic Variation in Behavior — Perspectives on Evolutionary Mechanisms*, Foster SA and Endler JA (eds), Oxford University Press, New York, pp. 287-307.

Fox GA (1992) Epidemiological and pathological evidence of contaminant-induced alterations in sexual development in free-living wildlife. In *Chemically induced alterations in sexual and functional development: The wildlife/human connection*, Colborn T and Clemens C (eds), Princeton Scientific Publishing, Princeton, NJ, pp. 147-148.

Fox GA, Gilman AP, Peakall DB and Anderka FW (1978) Behavioral abnormalities of nesting Lake Ontario herring gulls. *Journal of Wildlife Management*, **42**, 477-483.

Fry DM and Toone KC (1981) DDT-induced feminization of gull embryos. *Science*, **213**, 922-924.

Fry DM, Toone CK, Speich SM and Peard RJ (1987) Sex ratio skew and breeding pattern of gulls: demographic and toxicological considerations. *Studies in Avian Biology*, **10**, 26-43.

Fry DM, Swenson J, Addiego LA, Grau CR and Kang A (1986) Reduced reproductive of wedge-tailed shearwaters exposed to weathered Santa Barbara crude oil. *Archives of Environmental Contamination and Toxicology*, **15**, 453-463.

Fryday SL, Chandler-Morris SA and Hart ADM (1998) Presentation method and the avoidance of fonofos-treated seed by captive birds. *Bulletin of Environmental Contamination and Toxicology*, **61**, 448-454.

Fryday SL, Hart ADM and Dennis NJ (1994) Effects of exposure to an organophosphate on the seed-handling efficiency of the house sparrow. *Bulletin of Environmental Contamination and Toxicology*, **53**, 869-876.

Fryday SL, Hart ADM and Langton SD (1996) Effects of exposure to an organophosphorus pesticide on the behavior and use of cover by captive starlings. *Environmental Toxicology and Chemistry*, **15**, 1590-1596.

Fryday SL, Hart ADM and Marczylo TH (1995) Effects of sublethal exposure to an organophosphorus pesticide on the flying performance of captive starlings. *Bulletin of Environmental Contamination and Toxicology*, **55**, 366-373.

Fryxell JM and Lundberg P (1998) *Individual Behavior and Community Dynamics*, Chapmam & Hall, London.

Fulton MH and Key PB (2001) Acetylcholinesterase inhibition in estuarine fish and invertebrates as an indicator of organophosphorus insecticide exposure and effects. *Environmental Toxicology and Chemistry*, **20**, 37-45.

Fyfe RW, Risebrough RW and Walker W II (1976) Pollutant effects on the reproduction of prairie falcons and merlins of the Canadian prairies. *Canadian Field Naturalist*, **90**, 346-355.

Gaines TB (1969) Acute toxicity of pesticides. *Toxicology and Applied Pharmacology*, **14**, 515-534.

Galindo JC, Kendall RJ, Driver CJ and Lacher TE Jr (1985) The effect of methyl parathion on susceptibility of bobwhite quail (*Colinus virginianus*) to domestic cat predation. *Behavioral and Neural Biology*, **43**, 21-36.

Garry VF, Schreinemachers D, Harkins ME and Griffith J (1996) Pesticide appliers, biocides, and birth defects in rural Minnesota. *Environmental Health Perspectives*, **104**, 349-399.

Georghiou GP (1990) Overview of insecticide resistance. In *Managing resistance to agrochemicals*, Green MB, Le Baron HM and Moberg WK (eds), American Chemical Society, Washington, DC, pp. 18-41.

Gilliom RJ, Alley WM, and Gurtz ME (1995) Design of the National Water-Quality Assessment Program: Occurrence and distribution of water-quality conditions, *US Geological Survey Circular* 1112.

Giraldeau LA (1997) The ecology of information use. In *Behavioral Ecology - An Evolutionary Approach*, Krebs JR and Davies NB (eds.), 4th edn, Blackwell Science Ltd, London, pp. 42-68.

Goldstein MI, Lacher TE Jr, Woodbridge B, Bechard MJ, Canavelli SB, Zaccagnini ME, Cobb GP, Scollon EJ, Tribolet R and Hooper MJ (1999) Monocrotophos-induced mass mortality of Swainson's hawks in Argentina, 1995-96. *Ecotoxicology*, **8**, 201-214.

Gordon CJ (1993) *Temperature Regulation in Laboratory Rodents*, Cambridge University Press, Cambridge.

Gordon CJ (1994) Thermoregulation in laboratory mammals and humans exposed to anticholinesterase agents. *Neurotoxicology and Teratology*, **16**, 427-453.

Gordon CJ, Mohler FS, Watkinson WP and Rezvani AH (1988) Temperature regulation in laboratory mammals following acute toxic insult. *Toxicology*, **53**, 161-178.

Gould SJ and Lewontin RC (1979) The spandrels of San Marco and the Panglossian paradigm: a critique of the adaptionist programme. *Proceedings Royal Society London, Series B*, **205**, 581-598.

Greig-Smith PW (1991) The Boxworth experience: Effects of pesticides on the fauna and flora of cereal fields. In *The Ecology of Temperate Cereal Fields*, Firbank LG, Carter N, Darbyshire JF and Potts GR (eds), 32nd Symposium of the British Ecological Society with the Association of Applied Biologists, University of Cambridge, Blackwell Scientific Publications, Oxford, pp. 333-371.

Grue CE (1982) Response of common grackles to dietary concentrations of four organophosphate pesticides. *Archives of Environmental Contamination and Toxicology*, **11**, 617-626.

Grue CE (1994) Modeling the effects of pesticides on wildlife: The need to evaluate our leaps of faith and pious hopes. In *Wildlife Toxicology and Population Modeling: Integrated Studies of Agroecosystems*, Kendall RJ and Lacher TE Jr (eds), Lewis Publishers, Boca Raton, FL, pp. 263-267.

Grue CE, DeWeese LR, Mineau P, Swanson GA, Foster JR, Arnold PM, Huckins JN, Sheehan PJ, Marshall WK and Ludden AP (1986) Potential impacts of agricultural chemicals on waterfowl and other wildlife inhabiting prairie wetlands: An evaluation of research needs and approaches. *Transactions North American Wildlife and Natural Resources Conference*, **51**, 357-383.

Grue CE, Fleming WJ, Busby DG and Hill EF (1983) Assessing hazard of organophosphate pesticides to wildlife. *Transaction of the North American Wildlife and Natural Resources Conference*, **48**, 200-220.

Grue, CE, Gibert PL and Seeley ME (1997) Neurophysiological and behavioral changes in non-target wildlife exposed to organophosphate and carbamate pesticides: Thermoregulation, food consumption, and reproduction. *American Zoologist*, **37**, 369-388.

Grue CE, Hart ADM and Mineau P (1991) Biological consequences of depressed brain cholinesterase activity in wildlife. In *Cholinesterase-inhibiting Insecticides — Their Impact on Wildlife and the Environment*, Mineau P (ed), Elsevier Science Publishers BV, Amsterdam, pp. 152-209.

Grue CE, Powell GVN and Gorsuch CH (1982a) Assessing effects of organophosphates on songbirds: Comparison of a captive and free-living population. *Journal of Wildlife Management*, **46**, 766-768.

Grue CE, Powell GVN and McChesney MJ (1982b) Care of nestlings by wild female starlings exposed to an organophosphate pesticide. *Journal of Applied Ecology*, **19**, 327-335.

Grue CE and Shipley BK (1981) Interpreting population estimates of birds following pesticide applications: Behaviour of male starlings exposed to an organophosphate pesticide. *Studies in Avian Biology*, **6**, 292-296.

Grue CE and Shipley BK (1984) Sensitivity of nestling and adult starlings to dicrotophos, and organophosphate pesticide. *Environmental Research*, **35**, 454-465.

Guillette LJ Jr, Gross TS, Masson GR, Matter JM, Percival FH and Woodward AR (1994) Developmental abnormalities of the gonad and abnormal sex hormone concentration in juveniles alligators from contaminated and control lakes in Florida. *Environmental Health Perspectives*, **102**, 680-688.

Hake M (1991) The effects of needle loss in coniferous forests in south-west Sweden on the winter foraging behaviour of willow tits *Parus montanus*. *Biological Conservation*, **58**, 357-366.

Hansen JA, Woodward F, Little EE, DeLonay AJ and Bergman HL (1999) Behavorial avoidance: Possible mechanism for explaining abundance and distribution of trout species in a metal impacted river. *Environmental Toxicology and Chemistry*, **18**, 313-317.

Hart ADM (1993) Relationships between behavior and the inhibition of acetylcholinesterase in birds exposed to organophosphorus pesticides. *Environmental Toxicology and Chemistry*, **12**, 321-326.

Hart ADM and Thompson HM (1995) Significance of regurgitation in avian toxicity tests: Bulletin of *Environmental Contamination and Toxicology*, **54**, 789-796.

Hart ADM and Westlake GE (1986) Acetylcholinesterase activity and the effects of chlorfenvinphos in regions of the starling brain. *Bulletin of Environmental Contamination and Toxicology*, **37**, 774-782.

Hattis D, Glowa J, Tilson H and Ulbrich B (1996) Risk assessment for neurobehavioral toxicity: SGOMSEC Joint Report. *Environmental Health Perspectives*, **104** (Suppl. 2), 217-226.

Heinz GH (1979) Methylmercury: Reproductive and behavioral effects on three generations of mallard ducks. *Journal of Wildlife Management*, **43**, 394-401.

Heinz GH (1989) How lethal are sublethal effects? *Environmental Toxicology and Chemistry*, **8**, 463-464.

Heinz GH, Brody M and Blus LJ (1994) How valuable are the results of models and laboratory studies when extended to field situations? In *Wildlife Toxicology and Population Modeling: Integrated Studies of Agroecosystems*, Kendall RJ and Lacher TE Jr (eds), Lewis Publishers, Boca Raton, FL, pp. 551-555.

Heinz GH, Haseltine SD and Sileo L (1983) Altered avoidance behavior of young black ducks fed cadmium. *Environmental Toxicology and Chemistry*, **2**, 419-421.

Henny CJ, Blus LJ, Kolbe EJ, and Fitzner RE (1985) Organophosphate insecticide (Famphur) topically applied to cattle kills magpies and hawks. *Journal of Wildlife Management*, **49**, 648-658.

Henny CJ, Blus LJ, Hoffman DJ, and Grove RA (1994a) Lead in hawks, falcons and owls downstream from a mining site on the Coer d'alene Rover, Idaho. *Environmental Monitoring and Assessment*, **29**, 267-288.

Henny CJ, Hallock RJ, and Hill EF (1994b) Cyanide and migratory birds at gold mines in Nevada, USA. *Ecotoxicology*, **3**, 45-58.

Henry MG and Atchison GJ (1979) Influence of social rank on the behaviour of bluegill, *Lepomis macrochirus* Rafinsque exposed to sublethal concentrations of cadmium and zinc. *Journal of Fish Biology*, **15**, 309-315.

Henry MG and Atchison GJ (1984) Behavioral effects of methyl parathion on social groups of bluegill (*Lepomis macrochirus*). *Environmental Toxicology and Chemistry*, **3**, 399-408.

Henry MG and Atchison GJ (1986) Behavioral changes in social groups of bluegills exposed to copper. *Transactions of the American Fisheries Society*, **115**, 590-595.

Herman-Giddens ME, Slora EJ, Wasserman RC, Bourdony CJ, Bhapkar MV, Koch GG and Hasemeier CM (1997) Secondary sexual characteristics and menses in young girls seen in office practice: A study from the Pediatric Research in Office Settings Network. *Pediatrics*, **88**, 505-512.

Hill EF (1995) Organophosphorus and carbamate pesticides. In *Handbook of ecotoxicology*, Hoffman DJ, Rattner BA, Burton GA Jr and Cairns J Jr (eds), Lewis Publishers, Boca Raton, FL, pp. 243-274.

Hinkle-Conn C, Fleeger JW, Gregg JC and Carman KR (1998) Effects of sediment-bound polycyclic hydrocarbons on feeding behavior in juvenile spot (*Leiostomus xanthurus* Lacepede: Pisces). *Journal of Experimental Marine Biology and Ecology*, **227**, 113-132.

Holmes SB (1998) Reproduction and nest behavior of Tennessee warblers *Vermivora peregrina* in forests treated with Lepidoptera-specific insecticides. *Journal of Applied Ecology*, **35**, 185-194.

Hopkins WA (2000) Reptile toxicology: Challenges and opportunities on the last frontier in vertebrate ecotoxicology. *Environmental Toxicology and Chemistry*, **19**, 2391-2393.

Hopkins WA, Congdon J and Ray JK (2000) Incidence and impact of axial malformations in larval bullfrogs (*Rana catesbeiana*) developing in sites polluted by a coal-burning power plant. *Environmental Toxicology and Chemistry*, **19**, 862-868.

Hudson RH, Tucker RK and Haegele MA (1984) *Handbook of toxicity of pesticides to wildlife* (2nd edn) US Department of Interior, Fish and Wildlife Service Resource Publication **153**, Washington, DC.

Hunt KA, Bird DM, Mineau P and Shutt L (1992) Selective predation of organophosphate-exposed prey by American kestrels. *Animal Behavior*, **43**, 971-976.

Hunter ML Jr and Witham JW (1985) Effects of a carbaryl-induced depression arthropod abundance on the behavior of Parulinae warblers. *Canadian Journal of Zoology*, **63**, 2612-2616.

Hunter ML Jr, Jones JJ, Gibbs KE and Moring JR (1986) Duckling responses to lake acidification: Do black ducks and fish compete? *Oikos*, **47**, 26-32.

Hunter ML Jr, Witham JW and Dow H (1984) Effects of a carbaryl-induced depression in invertebrate abundance on the growth and behavior of American black duck and mallard ducklings. *Canadian Journal of Zoology*, **62**, 452-456.

Hurlbert SH, Mulla MS and Willson HR (1972) Effects of an organophosphorus insecticide on the phytoplankton, zooplankton, and insect population of freshwater ponds. *Ecological Monographs*, **42**, 269-299.

Immelmann K and Beer C (1989) *A Dictionary of Ethology*, Harvard University Press, Cambridge, MA, p. 27.

Jacobson JL and Jacobson SW (1996) Intellectual impairment in children exposed to polychlorinated biphenyls *in utero*. *New England Journal of Medicine*, **355**, 783-789.

Janssen CR, Ferrando Rodrigo MD and Persoone G (1993) Ecotoxicological studies with the freshwater rotifer *Brachionus calyciflorus* - I. Conceptual framework and applications. *Hydrobiologia*, **255/256**, 21-32.

Johnstone RA (1997) The evolution of animal signals. In *Behavioural Ecology - An Evolutionary Approach*, Krebs JR and Davies NB (eds), (4th edn), Blackwell Science Ltd, Oxford, pp. 155-178.

Joiris CR, Holsbeek L, Bossicart M and Tapia G (1997) Mercury and organochlorines in four sperm whales stranded on the Belgian coast, November 1994. *Bulletin de l'Institut Royal des Sciences Naturelles de Belgique Biologie*, **67** (Suppl.), 69-73.

Kardong KV and Berkhoudt H (1999) Rattlesnake hunting behavior: Correlations between plasticity of predatory performance and neuroanatomy. *Brain, Behavior and Evolution*, **53**, 20-28.

Kendall RJ and Akerman J (1992) Terrestrial wildlife exposed to agrochemicals: An ecological risk assessment perspective. *Environmental Toxicology and Chemistry*, **11**, 1727-1749.

Kendall RJ, Lacher TE Jr, Bunck C, Daniel B, Driver C, Grue CE, Leighton F, Stansley W, Watanabe PG and Whitworth M (1996) An ecological risk assessment of lead shot exposure in non-waterfowl avian species: Upland game birds and raptors. *Environmental Toxicology and Chemistry*, **15**, 4-20.

Kiffney PM (1996) Main and interactive effects of invertebrate density, predation, and metals on a Rocky Mountain stream macroinvertebrate community. *Canadian Journal of Fisheries and Aquatic Sciences*, **53**, 1595-1601.

Kingsolver JG (1989) Weather and the population dynamics of insects: Integrating physiological and population ecology. *Physiological Zoology*, **62**, 314-334.

Kislalioglu M, Scherer E and McNicol RE (1996) Effects of cadmium on foraging behavior of lake char, *Salvelinus namaycush*. *Environmental Biology of Fishes*, **46**, 75-82.

Klerks PL and Weis JS (1987) Genetic adaptation to heavy metals in aquatic organisms: A review. *Environmental Pollution*, **45**, 173-205.

Koltes KH (1985) Effects of sublethal copper concentrations on the structure and activity of Atlantic silverside schools. *Transactions of the American Fisheries Society*, **114**, 413-422.

Krebs CT and Burns KA (1977) Long-term effects of an oil spill on populations of the salt-marsh crab *Uca pugnax*. *Science*, **197**, 484-487.

Krebs JR and Davies NB (eds.) 1997 *Behavioral Ecology — An Evolutionary Approach*, (4th edn), Blackwell Science Ltd, London, 456 pp.

Kulig B, Alleva E, Bignami G, Cohn J, Cory-Slechta D, Landa V, O'Donoghue J and Peakall D (1996) Animal behavioral methods in neurotoxicity assessment: SGOMSEC Joint Report. *Environmental Health Perspectives*, **104** (Suppl. 2), 193-204.

Kunkel BA, Held DW and Potter DA (2001) Lethal and sublethal effects of bendiocarb, halofenozide, and imidacloprid on *Harpalus pennsylvanicus* (Coleoptera: Carabidae) following different modes of exposure in turfgrass. *Journal of Economic Entomology*, **94**, 60-67.

Kvitek RG (1991) Sequestered paralytic shellfish poisoning toxins mediate glaucous-winged gull predation on bivalve prey. *Auk*, **108**, 381-392.

Lacher TE Jr (1994) Direct and indirect pathways in spatially structured habitats. In *Wildlife Toxicology and Population Modeling — Integrated Studies of Agroecosystems*, Kendall RJ and Lacher TE Jr (eds), Lewis Publishers, Boca Raton, FL, pp. 139-143.

Lefcort H, Thomson SM, Cowles EE, Harowicz HL, Livaudais BM, Roberts WE and Ettinger WF (1999) Ramifications of predator avoidance: predator and heavy-metal-mediated competition between tadpoles and snails. *Ecological Applications*, **9**, 1477-1489.

Lemly AD, Finger SE, and Nelson MK (1993) Sources and impacts of irrigation drainwater contaminants in arid wetlands. *Environmental Toxicology and Chemistry*, **12**, 2265-2279.

Lemonick MD (2000) Teens before their time. *Time*, October 30, 2000, 67-74.

Linder G and Richmond ME (1990) Feed aversion in small mammals as a potential source of hazard reduction for environmental chemicals: Agrochemical case studies. *Environmental Toxicology and Chemistry*, **9**, 95-105.

Linz GM, Blixt DC, Bergman DL and Bleier WJ (1996) Response of ducks to glyphosate-induced habitat alterations in wetlands. *Wetlands*, **16**, 38-44.

Little EE (1990) Behavioral toxicology: Stimulating challenges for a growing discipline. *Environmental Toxicology and Chemistry*, **9**, 1-2.

Little EE and Finger SE (1990) Swimming behavior as an indicator of sublethal toxicity in fish. *Environmental Toxicology and Chemistry*, **9**, 13-19.

Llorens J, Sunol C, Tusell JM and Rodriguezfarre E (1991) Evidence for acute tolerance to the behavioral effects of lindane — Concomitant changes in regional monoamine status. *Neurotoxicology*, **12**, 697-706.

Loomis TA and Hayes AW (1996) *Loomis's Essentials of Toxicology*, (4th edn), Academic Press, New York.

Luttik R (1998) Assessing repellency in a modified avian LC_{50} procedure removes the need for additional tests. *Ecotoxicology and Environmental Safety*, **40**, 201-205

Luttik R and Aldenberg T (1997) Extrapolation factors for small samples of pesticide toxicity data: Special focus on LD50 values for birds and mammals. *Environmental Toxicology and Chemistry*, **16**, 1785-1788.

MacKenzie KE and Winston ML (1989) Effects of sublethal exposure to diazinon on longevity and temporal division of labor in the honey bee (Hymenoptera: Apidae). *Journal of Economic Entomology*, **82**, 75-82.

MacLellan KNM, Bird DM, Fry DM and Cowles JL (1996) Reproductive and morphological effects of *o, p'*-Dicofol on two generations of captive American kestrels. *Archives of Environmental Contamination and Toxicology*, **30**, 364-372.

MacLellan KNM, Bird DM, Shutt LJ and Fry DM (1997) Behavior of captive American kestrels hatched from *o, p'*-Dicofol exposed females. *Archives of Environmental Contamination and Toxicology*, **32**, 411-415.

Martin PA, Johnson DL, Forsyth DJ and Hill BD (1998) Indirect effects of the pyrethroid insecticide deltamethrin on reproductive success of chestnut-collared longspurs. *Ecotoxicology*, **7**, 89-97.

Matz AC, Bennett RS and Landis WG (1998) Effects of azinphos-methyl on northern bobwhite: a comparison of laboratory and field results. *Environmental Toxicology and Chemistry*, **16**, 1364-1370.

Mayer FL Jr and Ellersiek MR (1986) *Manual of acute toxicity: Interpretation and database for 410 chemicals and 66 species of freshwater animals*. US Department of the Interior, Fish and Wildlife Service Resource Publication, **160**, Washington, DC.

McCarty JP and Secord AL (1999a) Nest-building behavior in PCB-contaminated tree swallows. *Auk*, **116**, 55-63.

McCarty JP and Secord AL (1999b) Reproductive ecology of tree swallows (*Tachycineta bicolor*) with high levels of polychlorinated biphenyl contamination. *Environmental Toxicology and Chemistry*, **18**, 1433-1439.

McCormick SD, Hansen LP, Quinn TP and Saunders RL (1998) Movement, migration, and smolting of Atlantic salmon (*Salmo salar*). *Canadian Journal of Fisheries and Aquatic Sciences*, **55**, 77-92.

McEwen LC and Brown RL (1966) Acute toxicity of dieldrin and malathion to wild sharp-tailed grouse. *Journal of Wildlife Management*, **30**, 604-611.

McMillan DE (1999) Neurobehavioral toxicology and addiction. In *Introduction to Neurobehavioral Toxicology: Food and Environment*, Niesink RJM, Jaspers RMA, Kornet LMW, van Ree JM and Tilson HA (eds), CRC Press, Boca Raton, FL, pp. 2-26.

McNicol RE, Begout-Anras ML and Scherer E (1999) Influence of light preferences on the avoidance responses of lake whitefish, *Coregonus clupeaformis*, to cadmium. *Environmental Biology of Fishes*, **55**, 295-306.

McNicol RE, Scherer E, and Gee JH (1996) Shoaling enhances cadmium avoidance by lake whitefish, *Coregonus clupeaformis*. *Environmental Biology of Fishes*, **47**, 311-319.

Mesa MG, Poe TP, Gadomski DM and Petersen JH (1994) Are all prey created equal? A review and synthesis of differential predation on prey in substandard condition. *Journal of Fish Biology*, **45** (Suppl. A), 81-96.

Meyers SM, Cummings JL and Bennett RS (1990) Effects of methyl parathion on red-winged blackbird (*Agelaius phoeniceus*) incubation behavior and nesting success. *Environmental Toxicology and Chemistry*, **9**, 807-813.

Meyers SM, Marden BT, Bennett RS and Bentley R (1992) Comparative response of nestling European starlings and red-winged blackbirds to an oral administration of either dimethoate or chlorpryrifos. *Journal of Wildlife Diseases*, **28**, 400-406.

Mineau P (1991) Difficulties in the regulatory assessment of cholinesterase-inhibiting insecticides. In *Cholinesterase-inhibiting Insecticides - Their Impact on Wildlife and the Environment*, Mineau P (ed.), Elsevier Science Publishers BV, Amsterdam, The Netherlands, pp. 278-300.

Mineau P (1998) Biomarkers: Are there linkages to ecological effects? In *Multiple Stresses in Ecosystems*, Cech JJ Jr, Wilson BW and Crosby DG (eds), Lewis Publishers, Boca Raton, FL, pp. 91-100.

Mineau P, Barill A, Collins BT, Duffe J, Joerman G and Luttik R (2001) Pesticide acute toxicity reference values for birds. *Reviews in Environmental Contamination and Toxicology*, **170**, 13-74.

Mineau P, Boag PT and Beninger RJ (1994a) Effects of fenitrothion on memory for cache-site locations in black-capped chickadees. *Environmental Toxicology and Chemistry*, **13**, 281-290.

Mineau P, Boag PT and Beninger RJ (1994b) The effects of physostigmine and scopolamine on memory for caches in the black-capped chickadee. *Pharmacology, Biochemistry and Behavior*, **49**, 363-370.

Mineau P, Boersma DC and Collins B (1994c) An analysis of avian reproduction studies submitted for pesticide registration. *Ecotoxicology and Environmental Safety*, **29**,1-25.

Mineau P, Collins BT and Baril A (1996) On the use of scaling factors to improve interspecies extrapolation of acute toxicity in birds. *Regulatory Toxicology and Pharmacology*, **24**, 24-29.

Mineau P, Fletcher MR, Glaser LC, Thomas NJ, Brassard C, Wilson LK, Elliott JE, Lyon LA, Henny CJ, Bollinger T and Porter SL (1999) Poisoning of raptors with organophosphorus and carbamate pesticides with emphasis on Canada, U.S. and U.K. *Journal of Raptor Research*, **33**, 1-37.

Mineau P, Fox GA, Nordstrom RJ, Weseloh DV, Hallet DJ and Ellenton JA (1984) In *Toxic Contaminants in the Great Lakes*, Nriagu JO and Simmons MS (eds), John Wiley & Sons, Ltd, Chichester, pp. 425-452.

Mineau P and Peakall DB (1987) An evaluation of avian impact assessment techniques following broad-scale forest insecticide sprays. *Environmental Toxicology and Chemistry*, **6**, 781-791.

Moore A and Waring CP (1996) Sublethal effects of the pesticide diazinon on olfactory function in mature male Atlantic salmon parr. *Journal of Fish Biology*, **48**, 758-775.

Morrison ML and Meslow CE (1984) Effects of the herbicide glyphosate on bird community structure, western Oregon. *Forest Science*, **30**, 95-106.

Moser V and Padilla S (1998) Age-and gender-related differences in the time course of behavioral and biochemical effects produced by oral chlorpyrifos in rats. *Toxicology and Applied Pharmacology*, **149**, 107-119.

Moser VC, Cheek BM and McPhail RC (1995) A multidisciplinary approach to toxicological screening: III. Neurobehavioral toxicity. *Journal of Toxicology and Environmental Health*, **45**, 173-210.

Moulding JD (1976) Effects of a low-persistence insecticide on forest bird populations. *Auk*, **93**, 692-708.

Munkittrick KR and McCarty LS (1995) An integrated approach to aquatic ecosystem health: Top-down, bottom-up or middle-out? *Journal of Aquatic Ecosystem Health*, **4**, 77-90.

Munkittrick KR, Miller PA, Barton DR and Dixon DG (1991) Altered performance of white sucker populations in the Manitouwadge chain of lakes is associated with changes in benthic macroinvertebrate communities as a result of copper and zinc contamination. *Ecotoxicology and Environmental Safety*, **21**, 318-326.

Nabholz JV, Clements RG and Zeeman MG (1997) Information needs for risk assessment in EPA's Office of Pollution Prevention and Toxics. *Ecological Applications*, **7**, 1094-1098.

Namba T, Nolte CT, Jackrel J and Grab D (1971) Poisoning due to organophosphate insecticides - Acute and chronic manifestations. *American Journal of Medicine*, **50**, 475-492.

Natural Resource Damage Assessments: Final Rule (1986) *Federal Register*, **51**, 27674-27753.

Needleman HL (1995) Behavioral toxicology. *Environmental Health Perspectives*, **103**, 77-79.

Needleman HL, Schell A, Bellinger D, Leviton A and Alfred EN (1990) The long-term effects of exposure to low doses of lead in childhood: An 11-year follow-up report. *New England Journal of Medicine*, **322**, 83-88.

Nicolaus LK and Lee H (1999) Low acute exposure to organophosphate produces long-term changes in bird feeding behavior. *Ecological Applications* **9**, 1039-1049.

Nicolaus LK, Cassel JF, Carlson RB and Gustavson CR (1983) Taste-aversion conditioning of crows to control predation on eggs. *Science*, **220**, 212-214.

Nocera JJ and Taylor PD (1998) *In situ* behavioral response of common loons associated with elevated mercury (Hg) exposure. *Conservation Ecology*, **2**, http://www.consecol.org/Journal/vol2/iss2/art10

O'Brien RD (1976) Acetylcholinesterase and its inhibition. In *Insecticide Biochemistry and Physiology*, Wilkinson CF (ed), Academic Press, New York, pp. 271-296.

O'Shea TJ (1999) Environmental contaminants and marine mammals. In *Biology of Marine Mammals*, Reynolds JE III and Rommel SA (eds), Smithsonian Institution Press, Washington, DC, pp. 485-564.

Park T (1962) Beetles, competition, and populations. *Science*, **138**, 1369-1375.

Parron T, Hernandez AF and Villaneuva E (1996) Increased risk of suicide with exposure to pesticides in an intensive agricultural area. A 12-year retrospective study. *Forensic Science International*, **79**, 53-63.

Pascual JA (1994) No effects of forest spraying of malathion on breeding blue tits (*Parus caeruleus*). *Environmental Toxicology and Chemistry*, **13**, 1127-1131.

Pascual JA, Hart ADM, and Fryday SL (1999) Incidence of lethal bird poisoning reduced by regurgitation of pesticide-treated food. *Environmental Toxicology and Chemistry*, **18**, 247-253.

Pascual JA and Peris SJ (1992) Effects of forest spraying with two application rates of cypermethrin on food supply and on breeding success of the blue tit (*Parus caeruleus*). *Environmental Toxicology and Chemistry*, **11**, 1271-1280.

Passino-Reader DR, Berlin WH and Hickey JP (1995) Chronic bioassays of rainbow trout fry with compounds representative of contaminants in Great Lakes fish. *Journal of Great Lakes Research*, **21**, 373-383.

Peakall DB (1985) Behavioral responses of birds to pesticides and other contaminants. *Residue Reviews*, **96**, 45-77.

Peakall DB (1993) DDE-induced eggshell thinning: an environmental detective story. *Environmental Review*, **1**, 13-20.

Peakall DB (1994) The role of biomarkers in environmental assessment. 1. Introduction. *Ecotoxicology*, **3**, 157-160.

Peakall DB (1996) Disrupted patterns of behavior in natural populations as an index of ecotoxicity. *Environmental Health Perspectives*, **104** (Suppl. 2), 331-335.

Peakall DB and Shugart L (1993) *Biomarkers: Research and Application in the Assessment of Environmental Health*. NATO ASI Series, Springer-Verlag, New York.

Pedersen HC and Saether M (1999) Effects of cadmium on parental behaviour in free-living willow ptarmigan hens. *Ecotoxicology*, **8**, 1-7.

Peterle TJ (1991) *Wildlife Toxicology*, Van Nostrand Reinhold, New York.

Porter WP, Jaeger JW and Carlson IH (1999) Endocrine, immune, and behavioral effects of aldicarb (carbamate), atrazine (triazine) and nitrate (fertilizer) mixtures at groundwater concentrations. *Toxicology and Industrial Health*, **15**, 133-150.

Potts GR (1977) Population dynamics of the grey partridge: overall effects of herbicides and insecticides on chick survival rates. *Proceedings of the Congress of Game Biologists*, **13**, 203-211.

Powell GVN (1984) Reproduction by an altricial songbird, the red-winged blackbird, in fields treated with the organophosphate insecticide fenthion. *Journal of Applied Ecology*, **21**, 83-95.

Pratt JR, Melendez AE, Barreiro R and Bowers NJ (1997) Predicting the ecological effects of herbicides. *Ecological Applications*, **7**, 1117-1124.

Preston BL, Snell TW and Dusenbery DB (1999) The effects of sublethal pentachlorophenol exposure on predation risk in freshwater rotifer species. *Aquatic Toxicology*, **47**, 93-105.

Pulliam HR (1988) Sources, sinks and population regulation. *American Naturalist*, **132**, 652-661.

Raimondo SM, Rowe CL and Congdon JD (1998) Exposure to coal ash impacts swimming performance and predator avoidance in larval bullfrogs (*Rana catesbeiana*). *Journal of Herpetology*, **32**, 289-292.

Rand GM (1985) Behavior. In *Fundamentals of Aquatic Toxicology*, Rand GM and Petrocelli SR (eds), Hemisphere Publishing Corporation, Washington, DC, pp. 221-263.

Rands MRW (1985) Pesticide use on cereals and the survival of grey partridge chicks: a field experiment. *Journal of Applied Ecology*, **22**, 49-54.

Rattner BA and Heath AG (1995) Environmental factors affecting contaminant toxicity in aquatic and terrestrial vertebrates. In *Handbook of Ecotoxicology*, Hoffman DJ, Rattner BA, Burton GA Jr and Cairns J Jr (eds), Lewis Publishers, CRC Press Inc., Boca Raton, FL, pp. 519535.

Rattner BA and Franson JC (1983) Methylparathion and fenvalerate toxicity in American kestrels: acute physiological responses and effects of cold. *Canadian Journal of Physiology and Pharmacology*, **62**, 787-792.

Real LA (1994) How to think about behavior: An introduction. In *Behavioral Mechanisms in Evolutionary Ecology*, Real LA (ed), University of Chicago Press, Chicago, IL, pp. 1-8.

Reischl P, Van Gelder GA and Karas GG (1975) Auditory detection behavior in parathion-treated squirrel monkeys (*Saimiri sciureus*). *Toxicology and Applied Pharmacology*, **34**, 88-101.

Rice PJ, Drewes CD, Klubertanz TM, Bradbury SP and Coats JR (1997) Acute toxicity and behavioral effects of chlorpyrifos, permethrin, phenol, strychnine, and 2,4-dinitrophenol to 30-day-old Japanese medaka (*Oryzias latipes*). *Environmental Toxicology and Chemistry*, **16**, 696-704.

Riley AL and Tuck DL (1985) Conditioned food aversions: a bibliography. In *Experimental Assessments and Clinical Applications of Conditioned Food Aversions*, Braveman NS and Bronstein P (eds.), New York Academy of Sciences, New York, pp. 381-437.

Roberts DK, Silvy NJ and Bailey EM Jr (1988) Brain acetylcholinesterase activity recovery following acute methyl parathion intoxication in two feral rodent species: comparison to laboratory rodents. *Bulletin of Environmental Contamination and Toxicology*, **41**, 26-35.

Roman G, Isnard P and Jouany J-M (1999) Critical analysis of methods for assessment of predicted no-effect concentration. *Ecotoxicology and Environmental Safety*, **43**, 117-125.

Russell RW (1982) Cholinergic system in behavior: the search for mechanisms of action. *Annual Reviews in Pharmacology and Toxicology*, **22**, 435-463.

Saglio P and Trijasse S (1998) Behavioral responses to atrazine and diuron in goldfish. *Archives of Environmental Contamination and Toxicology*, **35**, 484-491.

Sanderson GC and Bellrose FC (1986) A review of the problem of lead poisoning in waterfowl. Illinois Natural History Survey Special Publication 4, Urbana, IL.

Sandheinrich MB and Atchison GJ (1989) Sublethal copper effects on bluegill, *Lepomis macrochirus*, foraging behavior. *Canadian Journal of Fish and Aquatic Sciences*, **46**, 1977-1985.

Sandheinrich MB and Atchison GJ (1990) Sublethal toxicant effects on fish foraging behavior: empirical vs. mechanistic approaches. *Environmental Toxicology and Chemistry*, **9**, 107-119.

Sasvari L (1986) Reproductive effort of widowed birds. *Journal of Animal Ecology*, **55**, 553-564.

Satinoff E (1979) Drugs and thermoregulatory behavior. In *Body Temperature — Regulation, Drug Effects, and Therapeutic Implications*, Lomax P and Schonbaum E (eds), Marcell Dekker, Inc., New York, pp. 151-182.

Satinoff E and Hendersen R (1977) Thermoregulatory behavior. In *Handbook of Operant Behavior*, Honig WK and Staddon JER (eds), Prentice-Hall, Inc., Englewood Cliffs, NJ, pp. 153-173.

Schafer EW Jr (1972) The acute oral toxicity of 369 pesticidal, pharmaceutical and other chemicals to wild birds. *Toxicology and Applied Pharmacology*, **21**, 315-330.

Schafer EW Jr (1994) Comparative avian toxicology: What is its role in predicting and monitoring the effects of agricultural pesticides? In *Wildlife Toxicology and Population Modeling: Integrated Studies of Agroecosystems*, Kendall RJ and Lacher TE Jr (eds), Lewis Publishers, Boca Raton, FL, pp. 321-337.

Schafer EW Jr, Bowles WA and Hurlbut J (1983) The acute oral toxicity, repellency, and hazard potential of 998 chemicals to one or more species of wild and domestic birds. *Archives of Environmental Contamination and Toxicology*, **12**, 355-382.

Schantz SL, Seo B-W, Moshtaghian J and Amin S (1997) Developmental exposure to polychlorinated biphenyls or dioxin: do changes in thyroid function mediate effects on spatial learning? *American Zoologist*, **37**, 399-408.

Schantz SL, Seo B-W, Moshtaghian J, Peterson RE and Moore RW (1996) Effects of gestational and lactational exposure to TCDD or coplanar PCBs on spatial learning. *Neurotoxicology and Teratology*, **18**, 305-313.

Schiffman SS, Suggs MS, Abou Donia MB, Erickson RP and Nagle HT (1995) Environmental pollutants alter taste responses in the gerbil. *Pharmacology Biochemistry and Behavior*, **52**, 189-194.

Sholtz NL, Truelove NK, French BL, Berejikian BA, Quinn TP, Casillas E and Collier TK (2000) Diazinon disrupts antipredator and homing behaviors in chinook salmon (*Oncorhynchus tshawytscha*). *Canadian Journal of Fisheries and Aquatic Sciences*, **57**, 1911-1918.

Schricker B and Stephen WP (1970) The effect of sublethal doses of parathion on honeybee behaviour. I. Oral administration and the communication dance. *Journal of Apicultural Research*, **9**, 141-153.

Seidel GE and Whitmore RC (1995) Effects of Dimilin® application on white-footed mouse populations in a central Appalachian forest. *Environmental Toxicology and Chemistry*, **14**, 793-799.

Sideroff SI and Santolucito JA (1972) Behavioral and physiological effects of the cholinesterase inhibitor carbaryl (1-naphthyl methylcarbamate). *Physiology and Behavior*, **9**, 459–462.

Smith EH and Bailey HC (1990) Preference/avoidance testing of waste discharges on anadromous fish. *Environmental Toxicology and Chemistry*, **9**, 77–86.

Smith EH and Logan DT (1997) Linking environmental toxicology, ethology, and conservation. In *Behavioral Approaches to Conservation in the Wild*, Clemmons JR and Buchholz R (eds), Cambridge University Press, Cambridge, pp. 277–302.

Smith GJ (1987) *Pesticide use and toxicology in relation to wildlife: Organophosphorus and carbamate compounds*. US Department on the Interior, Fish and Wildlife Service Resource Publication **170**, Washington, DC.

Smith GM and Weis JS (1997) Predator-prey relationships in mummichogs (*Fundulus heteroclitus*): The effects of living in a polluted environment. *Journal of Experimental Marine Biology and Ecology*, **209**, 75–87.

Solberg KL and Higgins KF (1993) Effects of glyphosate herbicide on cattails, invertebrates, and waterfowl in South Dakota wetlands. *Wildlife Society Bulletin*, **21**, 299–307.

Sparks RE, Waller WT and Cairns J Jr (1972) Effect of shelters on the resistance of dominant and submissive bluegills (*Lepomis macrochirus*) to a lethal concentration of zinc. *Journal of the Fisheries Research Board-Canada*, **29**, 1356–1358.

Sparling DW and Lowe PT (1996) Environmental hazards of aluminum to plants, invertebrates, fish, and wildlife. *Reviews of Environmental Contamination and Toxicology*, **145**, 1–127.

Sparling DW, Linder G and Bishop CA (eds) (2000) *Ecotoxicology of Amphibians and Reptiles*. SETAC Press, Pensacola, FL.

Sparling DW, Lowe TP and Pinkney AE (1997) Toxicity of Abate® to green frog tadpoles. *Bulletin of Environmental Contamination and Toxicology*, **58**, 475–481.

Spurgeon DJ and Hopkin SP (2000) The development of genetically inherited resistance to zinc in laboratory-selected generations of earthworms *Eisenia fetida*. *Environmental Pollution*, **109**, 193–201.

Stehn RA (1976) Feeding response of small mammal scavengers to pesticide-killed arthropod prey. *American Midland Naturalist*, **95**, 253–256.

Stephens R, Spurgeon A, Calvert IA, Beach J, Levy LS, Berry H and Harrington JM (1995) Neuropsychological effects of long-term exposure to organophosphates in sheep dip. *The Lancet*, **345**, 1135–1139.

Stickel LF (1973) Pesticide residues in birds and mammals. In *Environmental pollution by pesticides*, Edwards CA (ed), Plenum Press, London, pp. 254–312.

Stine KE and Brown TM (1996) *Principles of Toxicology*, Lewis Publishers, Boca Raton, FL.

Sullivan JF, Atchison GJ, Kolar DJ and McIntosh AW (1978) Changes in the predator-prey behavior of fathead minnows (*Pimephales promelas*) and largemouth bass (*Micropterus salmoides*) caused by cadmium. *Journal of the Fisheries Research Board-Canada*, **35**, 446–451.

Sullivan TFP (ed) (1999) *Environmental Law Handbook*, Government Institutes, Inc., Rockville, MD.

Sultatos LG (1994) Mammalian toxicology of organophosphorus pesticides. *Journal of Toxicology and Environmental Health*, **43**, 271–289.

Sumpter JP (1998) Xenoendocrine disrupters Environmental impacts. *Toxicology Letters*, **102–103**, 337–342.

Tessier L, Boisvert JL, Vought LBM, and Lacoursière JO (2000a) Anomalies on capture nets of *Hydropsyche slossonae* larvae (Trichoptera; Hydropsychidae) following a sublethal chronic exposure to cadmium. *Environmental Pollution*, **108**, 425–438.

Tessier L, Boisvert JL, Vought LBM, and Lacoursière JO (2000b) Effects of 2,4-dichlorophenol on the net-spinning behavior of *Hydropsyche slossonae* larvae (Trichoptera; Hydropsychidae), an early warning signal of chronic toxicity. *Ecotoxicology and Environmental Safety*, **46**, 207-217.

Tessier L, Boisvert JL, Vought LBM, and Lacoursière JO (2000c) Anomalies on capture nets of Hydropsyche slossonae larvae (Trichoptera; Hydropsychidae), a potential indicator of chronic toxicity of malathion (organophosphate insecticide). *Aquatic Toxicology*, **50**, 125-139.

Thomas M, Florion A, Chretien D and Terver D (1996) Real-time biomonitoring of water contamination by cyanide based on analysis of the continuous electric signal emitted by a tropical fish: *Apteronotus albifrons*. *Water Research*, **30**, 3083-3091.

Thomsett (1987) Raptor deaths as a result of poisoning Quelea in Kenya. *GABAR*, **2**, 38.

Tiebout HM and Brugger KE (1995) Ecological risk assessment of pesticides for terrestrial vertebrates: Evaluation and application of the U.S Environmental Protection Agency's quotient model. *Conservation Biology*, **9**, 1605-1618.

Tinbergen N (1963) On aims and methods of ethology. *Zeitschrift fur Tierpsychologie*, **20**, 410-433.

Touart LW (1988) *Aquatic mesocosm tests to support pesticide registrations*. Hazard Evaluation Division Technical Guidance Document EPA 540/09-88-035. National Technical Information Service, Springfield, VA.

Touart LW and Maciorowski AF (1997) Information needs for pesticide registration in the United States. *Ecological Applications*, **7**, 1086-1093.

Triebskorn R, Köhler HR, Honnen W, Schramm M, Adams SM and Müller EF (1997) Induction of heat shock proteins, changes in liver ultrastructure, and alterations of fish behavior: are these biomarkers related and are they useful to reflect the state of pollution in the field? *Journal of Aquatic Ecosystem Stress and Recovery*, **6**, 57-73.

Tucker RK and Leitzke JS (1979) Comparative toxicology of insecticides for vertebrate wildlife and fish. *Pharmacology and Therapeutics*, **6**, 167-220.

Tyler CR, Jobling S and Stumpter JP (1998) Endocrine disruption in wildlife: a critical review of the evidence. *Critical Reviews in Toxicology*, **28**, 319-361.

Urban DJ and Cook NJ (1986) *Hazard Evaluation Division standard evaluation procedure: ecological risk assessment*. EPA 540/09-86-167, US Environmental Protection Agency, Office of Pesticide Programs, Washington, DC.

US Department of the Interior (1987) *Injury to Fish and Wildlife Species*. Type B Technical Information Document, National Technical Information Service, PB88-100169, Springfield, VA.

US Environmental Protection Agency (1991) *OPPTS Test Guidelines: Neurotoxicity*. National Technical Information Service, PB91-154617, Springfield, VA.

van Dongen CJ and Wolthuis OL (1989) On the development of behavioral tolerance to organophosphates I: behavioral and biochemical aspects. *Pharmacology Biochemistry and Behavior*, **34**, 473-481.

Venugopal G, Srinivas A, Irwin C and Waghray S (1996) Influence of pesticide (DDVP) on chemoresponses of freshwater carp, *Catla catla* - A behavioural study. *Journal of Aquaculture in the Tropics*, **11**, 83-89.

vom Saal FS, Nagel SC, Palanza P, Boechler M, Parmigiana S and Welshons WV (1995) Estrogenic pesticides: binding relative to estradiol in MCF-7 cells and effects of exposure during fetal life on subsequent territorial behavior in male mice. *Toxicology Letters*, **77**, 343-350.

Vyas NB, Kuenzel WJ, Hill EF and Sauer JR (1995) Acephate affects migratory orientation of the white-throated sparrow (*Zonotrichia albicollis*). *Environmental Toxicology and Chemistry*, **14**, 1961-1965.

Walter H (1973) *Vegetation of the Earth in Relation to Climate and Ecophysiological Conditions*, Springer-Verlag, New York.

Walton P, Turner CMR, Austin G, Burns MD and Monagahan P (1997) Sub-lethal effects of an oil pollution incident on breeding kittiwakes *Rissa tridactyla*. *Marine Ecology Progress Series*, **155**, 261-268.

Warner SC, Dunson WA and Travis J (1991) Interaction of pH, density, and priority effects on the survivorship and growth of two species of hylid tadpoles. *Oecologia*, **88**, 331-339.

Warner SC, Travis J and Dunson WA (1993) Effect of pH variation on interspecific competition between two species of hylid tadpoles. *Ecology*, **74**, 183-194.

Watanabe C and Satoh H (1996) Evolution of our understanding of methyl mercury as a health threat. *Environmental Health Perspectives*, **104** (Suppl. 2), 367-379.

Weber DN (1996) Lead-induced metabolic imbalances and feeding alterations in juvenile fathead minnows (*Pimephales promelas*). *Environmental Toxicology and Water Quality*, **11**, 45-51.

Weis JS, Smith GM and Zhou T (1999) Altered predator/prey behavior in polluted environments: Implications for fish conservation. *Environmental Biology of Fishes*, **55**, 43-51.

Weiss B (1988) Behavior as an early indicator of pesticide toxicity. *Toxicology and Industrial Health*, **4**, 351-360.

Welch EB, Jacoby JM and May CW (1998) Stream quality. In Naiman RJ and Bilby RE (eds) *River Ecology and Management — Lessons from the Pacific Coastal Ecoregion*, Springer, New York, pp. 69-94.

Wendt-Rasch L, Vought LB-M and Woin P (1998) Effects of fenvalerate on the net-spinning behaviour of *Hydropsyche siltalai* (Döhler) (Trichoptera; Hydropsychidae). *Hydrobiologia*, **382**, 53-61.

West-Eberhard MJ (1989) Phenotypic plasticity and the origins of diversity. *Annual Review of Ecology and Systematics*, **20**, 249-278.

White DH, King KA, Mitchell CA, Hill EF and Lamont TG (1979) Parathion causes secondary poisoning in a laughing gull breeding colony. *Bulletin of Environmental Contamination and Toxicology*, **23**, 281-284.

White DH, Mitchell CA and Hill EF (1983) Parathion alters incubation behavior of laughing gulls. *Bulletin of Environmental Contamination and Toxicology*, **31**, 93-97.

Wobeser GA (1981) *Diseases of Wild Waterfowl*, Plenum Press, New York.

Wolthuis OL, Groen B, Busker RW and Vanhelden HPM (1995) Effects of low-doses of cholinesterase-inhibitors on behavioral performance of robot-tested marmosets. *Pharmacology Biochemistry and Behavior*, **51**, 443-456.

Wolthuis OL, Philippens IHCHM and Vanwersch R (1990) On the development of behavioral tolerance to organophosphates III: behavioral aspects. *Pharmacology Biochemistry and Behavior*, **35**, 561-565.

Woodward DF, Goldstein JN and Farag AM (1997) Cutthroat trout avoidance of metals and conditions characteristic of a mining waste site: Coeur d' Alene River, Idaho. *Transactions of the American Fisheries Society*, **126**, 699-706.

Yang RSH (1994) Introduction to the toxicology of chemical mixtures. In *Toxicology of Chemical Mixtures*, Yang RSH, Academic Press, San Diego, CA, p. 1-10.

Yusufu SD (1996) Behaviour of starlings (*Sturnus vulgaris*) eating food treated with the insecticide methiocarb. *Discovery and Innovation*, **8**, 69-72.

Zhou T and Weis JS (1998) Swimming behavior and predator avoidance in three populations of *Fundulus heteroclitus* larvae after embryonic and/or larval exposure to methylmercury. *Aquatic Toxicology*, **43**, 131-148.

Zhou T and Weis JS (1999) Predator avoidance in mummichog larvae from a polluted habitat. *Journal of Fish Biology*, **54**, 44–57.

Zinkl JG, Lockhart WL, Kenny SA and Ward FJ (1991) The effects of cholinesterase-inhibiting insecticides on fish. In *Cholinesterase-inhibiting Insecticides- Their Impact on Wildlife and the Environment*, Mineau P (ed), Elsevier Science Publishers BV, Amsterdam, pp. 233–254.

2

Mechanisms of Behaviour

ELIZABETH ADKINS-REGAN[1] AND DANIEL N. WEBER[2]
[1]*Department of Psychology and Department of Neurobiology and Behaviour, Cornell University, NY, USA*
[2]*Marine & Freshwater Biomedical Sciences Center, University of Wisconsin–Milwaukee, WI, USA*

2.1 CONCEPTS AND PRINCIPLES OF NEURAL AND HORMONAL MECHANISMS OF BEHAVIOUR

A full understanding of behaviour is required to answer four basic questions about the ultimate and proximate causes of behaviour (Tinbergen 1963): (1) what is its function (adaptive value)? (2) what is its phylogenetic history? (3) what is its ontogeny (development in the individual)? and (4) what are its underlying physiological mechanisms? Progress on all four fronts has been significant, but nowhere more so than in addressing this last question. Discoveries of chemical mechanisms of behaviour occur with increasing frequency, spurred on in part by the rapid growth of modern neuroscience. Summarising what is known about mechanisms of behaviour is therefore not an easy task. This chapter will first briefly present some general principles and concepts about physiological mechanisms in most animals. It will then provide more detailed information about mechanisms associated with particular behaviours, giving disproportionate (but not exclusive) attention to work with nondomesticated animals. The emphasis will be on neurochemical and hormonal mechanisms of ecologically-relevant behaviour, because these are most directly related to understanding how many anthropogenic chemicals in the environment might directly affect the life of the animals in the wild.

Behavioural Ecotoxicology. Edited by Giacomo Dell'Omo.

2.1.1 MECHANISMS OF BEHAVIOUR ARE A MIXTURE OF RELATIVELY CONSERVED (GENERAL) AND RELATIVELY TAXON SPECIFIC (DIVERSE) CHARACTERS

Some mechanisms are very similar across a wide array of taxa and are remarkably invariant in the face of great ecological and morphological diversity. Such mechanisms may have evolved relatively early in the history of multicellular animal life and may also reflect a limited number of workable solutions to basic problems of living, given the physicochemical realities of the earth. Knowledge derived from a few species may be more widely applicable.

2.1.1.1 Key aspects of the structure and function of nervous systems

Some features of the overall organisational plan of the vertebrate nervous system are remarkably similar from lampreys to primates (Figure 2.1) (Butler and Hodos 1996, Striedter 1998). All vertebrates have a dorsally positioned and segmentally structured spinal cord and brainstem, the same major brain divisions (hindbrain or rhombencephalon, midbrain or mesencephalon, and forebrain or prosencephalon, with the latter including diencephalon and telencephalon), many of the same major dorsal ascending (sensory) and ventral descending (motor) pathways, and predominantly contralateral (opposite) control of each side of the body by the brain. Many brain structures, even telencephalic structures, in one class such as mammals have likely homologues in other vertebrate brains.

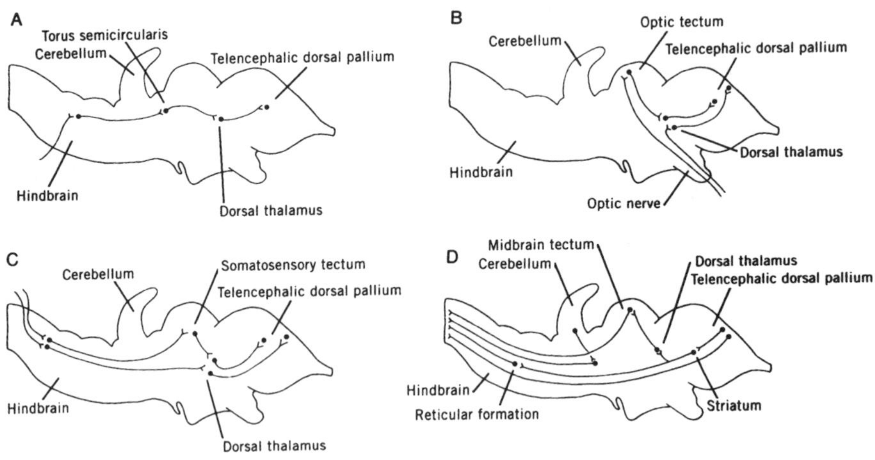

Figure 2.1 Conserved sensory and motor pathways in vertebrate brains. (A) Ascending auditory pathways; (B) the two major ascending visual pathways; (C) the two major ascending somatosensory pathways; (D) some of the descending motor pathways. From Butler and Hodos (1996) *Comparative Vertebrate Neuroanatomy*, © 1996 Wiley-Liss. Reprinted by permission of Wiley-Liss, Inc., a subsidiary of John Wiley & Sons, Inc.

Many but not all invertebrate nervous systems are also segmented (Figure 2.2). However, the single or double chain of ganglia running from anterior to posterior is ventrally rather than dorsally positioned. As in vertebrates, enlargement of the anterior end of the nervous system to receive sensory information from the head region and send descending signals to more posterior motor circuits is characteristic of a number of invertebrate groups.

Many features of the physiology of individual neurons are highly conserved across both invertebrates and vertebrates. This reflects the considerable similarity between neurons in the basic chemical machinery responsible for the

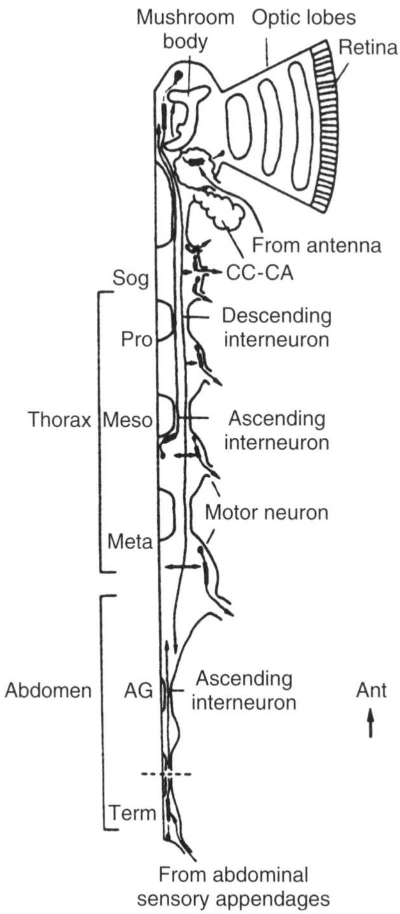

Figure 2.2 One half of an insect nervous system, showing the anterior cerebrum (brain) and the ventral chain of segmented ganglia (abbreviated Sog, Pro, Meso, Meta, AG, and Term) joined by paired connectives. CC-CA = corpora cardiaca-corpora allata. Original source: Figure 1a in Bicker and Menzel (1989) *Nature* (vol. **337**, pp. 33–39), reprinted with permission, © 1989 Macmillan Magazines Ltd.

transmission of information from one end of a neuron to the other. Regardless of the animal and regardless of whether the neuron is located in an insect head ganglion, a bird brain, or a fish spinal cord, sodium, potassium, chloride and calcium ion channels direct the movement of ions into and out of the neuron such that electrochemical signals (graded or action potentials) are generated and propagated along the neuron (Purves *et al.* 1997). These ion channels are protein molecules. As their amino acid sequences are determined, it is increasingly clear that the sequence similarity across animal phyla is substantial.

Synaptic transmission at chemical synapses also occurs in similar ways across taxa, but in combination with intraspecific differences depending on the type of neuron, its location in the nervous system, and the location of the synapse on the sending and receiving neurons. When graded or action potentials reach the presynaptic portion of one neuron, a chemical signal crosses the synaptic cleft to the postsynaptic (receiving) portion of the next neuron or the muscle fibre. Figure 2.3 shows the processes involved in synaptic transmission at a classical synapse (for example, an axo-dendritic synapse) using a classical neurotransmitter such as acetylcholine. All of these processes are potential points of pharmacological/toxicological disruption. Disruption will have a different effect depending on which process is altered. For example, blocking of synthesis will decrease activity at the synapse, while blocking of enzymatic destruction will increase activity. In addition, the processes have different temporal dynamics. As a result, it is not uncommon for the short-term effect of a toxicant affecting synaptic function to be rather different or even opposite from its long-term effect. Glial cell physiology and biochemistry may provide additional points of disruption (Jordan 1999, Laming *et al.* 2000).

2.1.1.2 Chemical regulators of neuronal activity

Internal chemical regulators, including neuromodulators as well as classical neurotransmitters and hormones, that are known or thought at this time to be involved in ecologically-relevant behaviours through their actions on the nervous system are listed in Table 2.1. The boundaries between these categories are not clearly defined in either vertebrates and invertebrates. Some vertebrate neural tissue can circumvent the blood–brain barrier and release products (neurohormones) into the circulation; most insect hormones are neurohormones (Truman 1992).

A glance at Table 2.1 quickly reveals the large (and still increasing) number of neuropeptides that have been discovered. Many of these fall into families based on amino acid sequence resemblance, such as the vasopressin–oxytocin family of nonapeptides (named for the predominant mammalian members of the family) or the gonadotropin releasing hormone (GnRH) family of decapeptides. Some of these families are very widely distributed, occurring in both invertebrates and vertebrates (e.g., the vasopressin–oxytocin family the GnRH family and the tachykinin family that includes substance P) (Lunt and Olsen

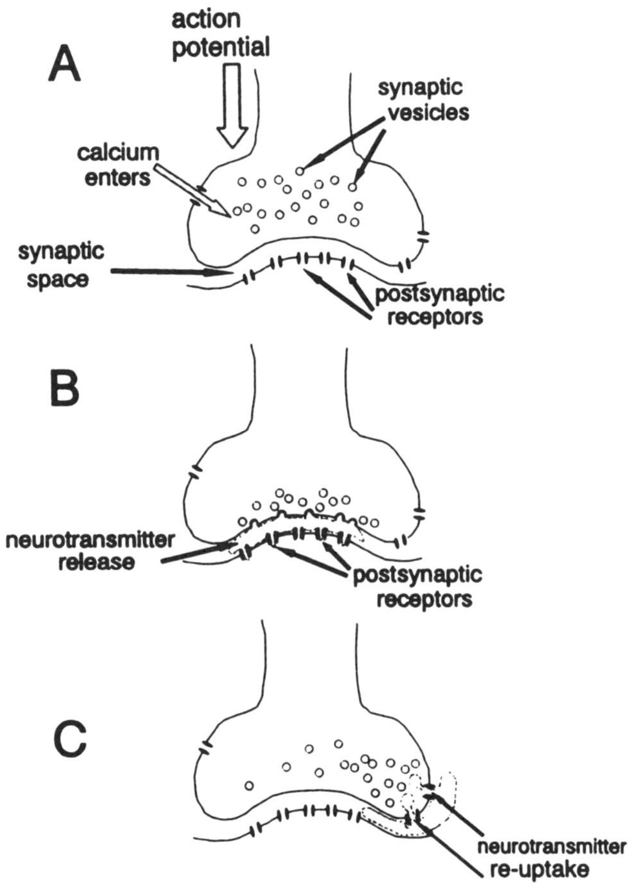

Figure 2.3 Transmission at a chemical synapse. The neurotransmitter is synthesised from precursors (which themselves may either be synthesised by the animal or derived from its food) and transported to the axon terminals, where it is stored in synaptic vesicles (A). When potentials reach these terminals (A) an influx of calcium ions occurs that then through a series of molecular events causes the vesicles to exocytose, releasing the neurotransmitter into the synaptic cleft (B). The neurotransmitter diffuses to the postsynaptic neuronal membrane where it binds to receptors (B). Excess neurotransmitter diffuses away, is enzymatically destroyed, or is taken back up (recycled) by transporters in the presynaptic membrane (C) for future use, or to generate a feedback signal about whether to increase or decrease the amount of neurotransmitter synthesised. From Butler and Hodos (1996) *Comparative Vertebrate Neuroanatomy*, © 1996 Wiley-Liss. Reprinted by permission of Wiley-Liss, Inc., a subsidiary of John Wiley & Sons, Inc.

1988, Gäde *et al.* 1997, Fernald and White 1999). Where sequence similarity is high, administration of one form of the peptide or of an antagonist is likely to have some effect across a wide range of species. The nature of the effect (the function thus altered) will not necessarily be similar, however, because similar molecules can be employed for different purposes in different taxa.

Table 2.1 Internal chemical regulators of behaviour. The list includes those neurotransmitters, neuromodulators, neurohormones, and hormones involved in ecologically-relevant behaviour discussed in this chapter. Sources: Butler and Hodos (1996); Coast and Webster (1998); Ewer *et al*. (1997); Lunt and Olsen (1988); Nijhout (1994); Norris (1996); Strand (1999)

Chemical	Acronym	Type of Chemical	Comments
Acetylcholine	ACh		
Norepinephrine (noradrenalin)	NE (NA)	catecholamine	
Dopamine	DA	catecholamine	
Histamine		biogenic amine	
Serotonin (5-hydroxytryptamine)	5-HT	indoleamine	
Octopamine	OA	indoleamine	invertebrates only
Melatonin	MT (MEL)	indoleamine	
γ-Aminobutyric acid	GABA	amino acid	
Glutamate	GLU	amino acid	
Prolactin	PRL	protein	
Growth hormone	GH	protein	
Vasopressin	AVP (VP)	nonapeptide	
Vasotocin	AVT (VT)	nonapeptide	
Oxytocin	OT (OXY)	nonapeptide	
Gonadotropin releasing hormone	GnRH	decapeptide	
Corticotropin releasing hormone	CRH	peptide	
Adrenocorticotrophic hormone	ACTH	peptide	
Cholecystokinin	CCK	peptide	
Vasoactive intestinal polypeptide	VIP	peptide	
Neuropeptide Y	NPY	peptide	
Substance P	SP	tachykinin peptide	
β-Endorphin	B-E	opioid peptide	
Bombesin		peptide	
Insulin		peptide	
Leptin		peptide	
Eclosion hormone	EH	peptide	arthropods
Ecdysis triggering hormone	ETH	peptide	arthropods
Crustacean cardioactive peptide	CCAP	peptide	arthropods
Prothoracicotrophic hormone	PTTH	peptide	arthropods
Proctolin	PROC	peptide	arthropods
Egg laying (bag cell) hormone		peptide	molluscs
Oestradiol-17β	E_2	steroid	
Progesterone	P	steroid	
Testosterone	T	steroid	
5α-Dihydrotestosterone	DHT	steroid	
11-Ketotestosterone	11-ketoT	steroid	teleost fishes
Corticosterone	CORT	steroid	
Cortisol		steroid	
20-Hydroxyecdysone		ecdysteroid	arthropods
Juvenile hormone	JH	terpenoid	arthropods
Thyroxine	T_4		
Triiodothyronine	T_3		
Prostaglandins	PG	eicosanoid	
Nitric oxide	NO		

Steroids, like peptides, are numerous and widely distributed. They are small lipophilic molecules that easily cross blood–brain barriers. The exact set of steroid molecules produced varies with taxon and, within taxon, with source. For example, vertebrate gonadal and adrenocortical (or interrenal) tissues do not produce the exact same set of steroids; androgens, estrogens, and progestins predominate in the steroid products of the former, whereas glucocorticoids and mineralocorticoids predominate in steroid products of the latter. Each particular steroid is structurally identical regardless of the species producing it, unlike amino acid based molecules. Testosterone refers to a particular androgen that is the same molecule in codfish, crocodiles, cuckoos and cats.

Some chemical regulators of neural activity are produced by neuronal cell groups that have characteristic and somewhat conserved neuroanatomical distributions. Several neurotransmitter systems such as serotonin (5-HT) and dopamine have similar distributions across vertebrates (see Brown 1994 for figures illustrating these distributions). Neurons producing the peptides GnRH and AVP (or AVT) tend to be located in similar areas in different vertebrate classes (Figure 2.4; Moore and Lowry 1998).

It is unlikely that any of the chemicals in Table 2.1 will have a specific (one-to-one) relation to behaviour; i.e., a neurotransmitter, neuromodulator, or hormone does not function in only one type of behaviour, and a single behaviour is not based on one chemical regulator (Orchard *et al.* 1993, Weiger 1997). Rather, multiple chemical systems regulate behaviours, each system is

PRIMARY SENSORY	SECONDARY SENSORY SENSORIMOTOR	MOTIVATIONAL-AUTONOMIC	MOTOR
● OLFACTORY NERVE	✱ OLFACTORY TUBERCLE	✱ CINGULATE CORTEX	▲ VOCAL CONTROL AREAS
● OLFACTORY BULB	▲ STRIATUM	✱ HIPPOCAMPUS	▲ MOTOR CRANIAL NUCLEI
● ACC. OLFACTORY BULB	✱ THALAMUS	✱ AMYGDALA	▲ SPINAL NUCLEI:
● RETINA	✱ TECTUM	✱ SEPTUM	(BULBOCAVERNOSUS)
	✱ TORUS SEMICIRCULARIS	✱ HABENULA	(FLEXOR CARPI RADIALIS)
	✱ CENTRAL GREY	✱ PREOPTIC AREA	(STERNORADIALIS)
	✱ CEREBELLUM	✱ HYPOTHALAMUS	
	▲ LAT, VESTIBULAR NUCLEUS		
	▲ SPINAL N. OF V		
	▲ N. SOLITARY TR.		
	▲ SPINAL CORD-DORSAL HORN		

● LHRH PATHWAYS
▲ STEROID-CONCENTRATING CELLS
✱ BOTH ● + ▲

Figure 2.4 The distribution of GnRH (LHRH) containing neural pathways and steroid concentrating cells in vertebrates. See also Figure 2.7. Original source: Figure 10 in Demski (1984) *American Zoologist* (vol. **24**, pp. 809–830). Reproduced by permission.

involved in several behaviours, and each chemical regulator is functionally versatile. Each act, however simple, results from a mix of excitatory and inhibitory inputs, each of which is modulated by multiple chemicals; the balance between them determines the output (Katz and Harris-Warrick 1999). These modulatory chemicals provide the flexibility and temporal and spatial coordination necessary for adaptive behaviour sequences (Bicker and Menzel 1989). The different chemical regulatory systems interact with each other at several levels. Neurotransmitters may be found in close anatomical association with another neurotransmitter or with one or more neuromodulators (Salter and De Koninck 1999). In some brain regions, neurotransmitters and neuromodulators are regulated by steroids (Micevych and Hammer 1995). Peptide and protein hormones involved in reproduction often require prior steroid priming in order to act (Norris 1996).

The chemicals that affect neuronal activity in behaviourally relevant ways include not only the internal chemical regulators in Table 2.1, but also chemicals from other animals such as pheromones. Some of these signalling molecules belong to the same chemical categories that appear in Table 2.1.

2.1.1.3 Mechanisms by which chemicals regulate neuronal activity

Chemical regulators of behaviour are normally present in minuscule amounts. Their mechanisms of action, however, involve cascades of events that amplify the signal. This is a general characteristic of biological signalling systems. This property helps explain how a physically tiny stimulus can have such a big behavioural effect — how a few molecules of an odour can stimulate directed migration in salmon.

Receptors for chemical regulators of neuronal activity are proteins. They are not identical across taxa, but nonetheless they often have marked amino acid sequence similarity indicating an ancient origin (Darlison and Richter 1999). It is increasingly common to find that there are multiple receptor types for a single chemical regulator. This could provide a basis for functional diversity. Angers *et al.* (1998) attribute the multiplicity of behavioural functions of 5-HT to different receptor subtypes that are coupled to different second-messenger systems. It is unlikely, however, that even if multiple receptors serve different functions, we will find a one-to-one relation between a receptor and a behaviour. At the receptor level, as at the hormonal and neuromodulator level, these systems interact. For example, the proteins produced when intracellular steroid receptors stimulate gene transcription include neurotransmitter and peptide receptors (Micevych and Hammer 1995).

Some neurotransmitter receptors (for example, the nicotinic acetylcholine (ACh) receptor and the $GABA_A$ receptor) are ionotropic (ligand-gated) receptors. Such receptors are closely linked to ion channels, enabling the neurotransmitter to have a very rapid effect on postsynaptic neural activity. Other receptors (for example, the muscarinic ACh receptor and receptors for the catecholamines and indoleamines) are metabotropic receptors which affect neurons through sequences of biochemical events involving ubiquitous highly conserved second messenger systems such as cAMP or cGMP (Purves *et al.* 1997). These receptors produce effects that are slower but longer lasting and that do not necessarily consist of alterations in neuronal firing.

Most neuropeptide receptors are members of the G-protein-coupled receptor superfamily, as are the receptors for some protein based hormones such as pituitary luteinising hormone (Figure 2.5) (Darlison and Richter 1999). Second

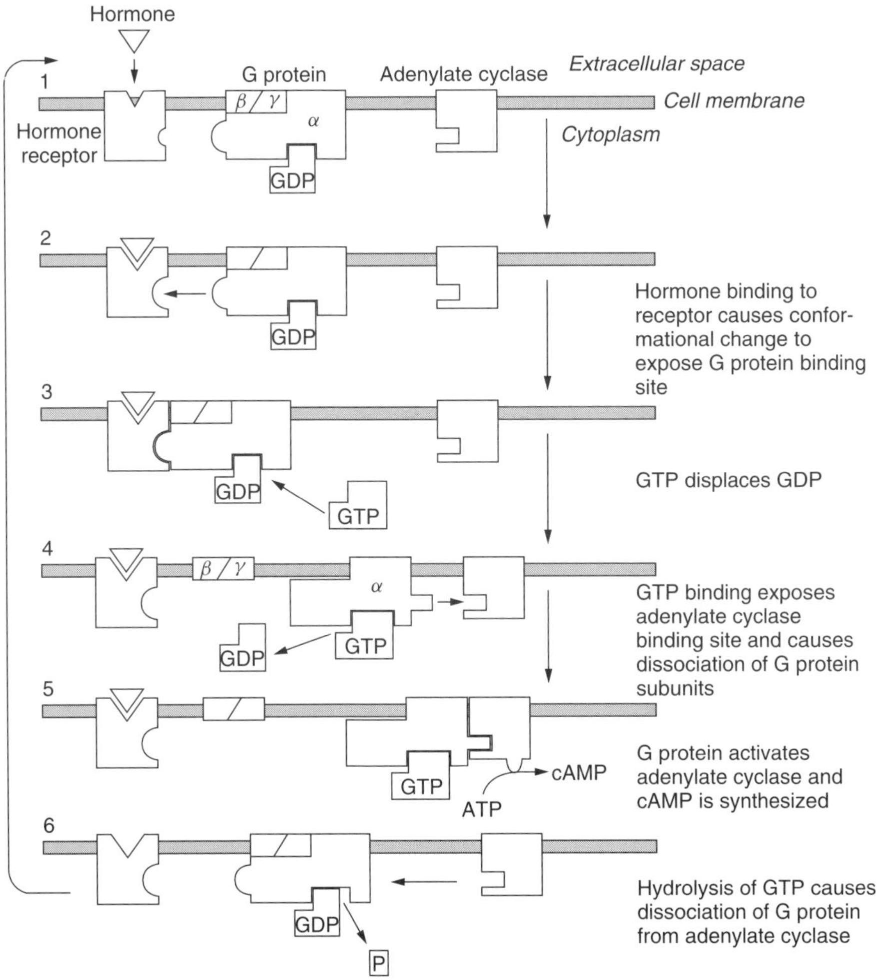

Figure 2.5 A typical second-messenger-mediated hormonal response pathway using cAMP. Original source: Figure 2.3 in Nijhout (1994) *Insect Hormones*. © 1994 by Princeton University Press, reprinted by permission.

messenger systems mediate the effects that occur when these ligands bind to their receptors (Norris 1996).

Steroids have diverse actions on neural tissue, ranging from rapid changes in firing rates to long-term structural changes. The same general classes of actions are taxonomically very widespread, occurring in both vertebrates and invertebrates (Nijhout 1994, Micevych and Hammer 1995, Weeks and Levine 1995, Gäde *et al.* 1997, Zakon 1998, Woolley 1999). For some actions, steroids (and also thyroid hormones) bind to members of the intracellular (nuclear) receptor family of ligand-activated gene transcription factors (Figure 2.6). Through this mechanism they stimulate expression of mRNA and production of peptides or proteins, including ion channels and receptors for neurotransmitters, neuromodulators and steroid receptors (the latter are themselves steroid regulated) (Harding 1992, Micevych and Hammer 1995, Young and Crews 1995). The anatomical distribution of these receptors is best known for the sex steroids,

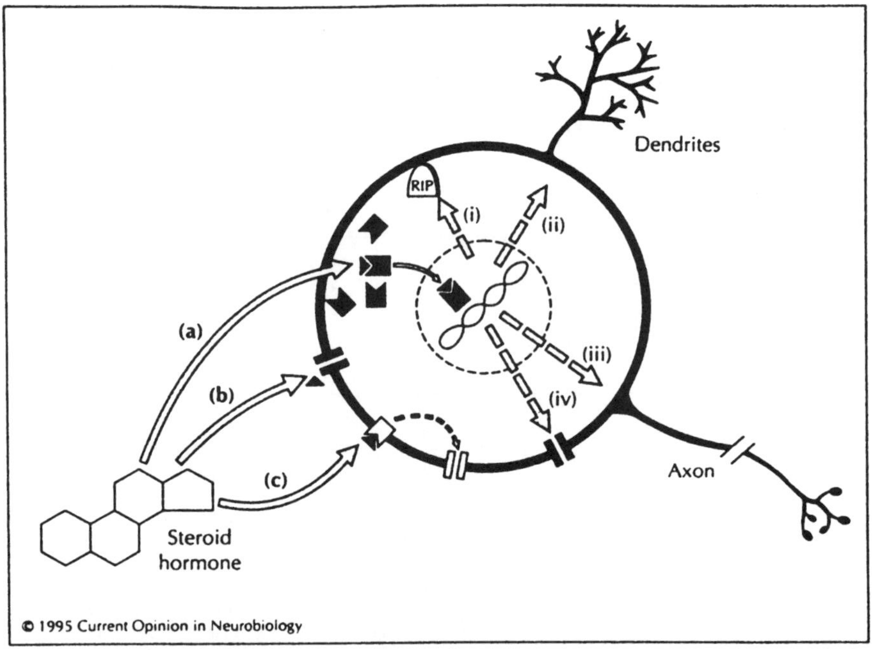

Figure 2.6 Mechanisms of steroid action on neurons. (a) Steroid hormones bind to intracellular receptors that alter gene expression, resulting in (i) programmed cell death, (ii) growth or regression of dendrites, (iii) changes in synaptic function, and (iv) synthesis or modulation of ion channels. Two classes of nongenomic effects are also shown: (b) direct binding of steroids to ion channels, and (c) binding of steroids to steroid receptors on the cell surface and modulation of ion channels via second messengers. The cell surface and intracellular receptors are structurally dissimilar, and intracellular steroid receptor antagonists do not block nongenomic steroid actions (Schmidt *et al*. 2000). Reprinted from Weeks and Levine (1995) *Current Opinion in Neurobiology* (vol. **5**, pp. 809–815), © 2000, with permission from Elsevier Science.

is remarkably similar in all vertebrate classes, and is similar for androgen, oestrogen and progesterone receptors (Figure 2.7). This process of gene transcription and production of new proteins takes time, and thus many effects of steroids on behaviour take several days or weeks to occur. More recently, evidence has accumulated in both vertebrates and invertebrates that in addition to these intracellular steroid receptors, there are membrane-level mechanisms, including receptors for steroid action (Figure 2.6). Some of these could be rapid, and could account for behavioural responses to a change in steroid level that are reported to occur in seconds or minutes (Moore and Orchinik 1991).

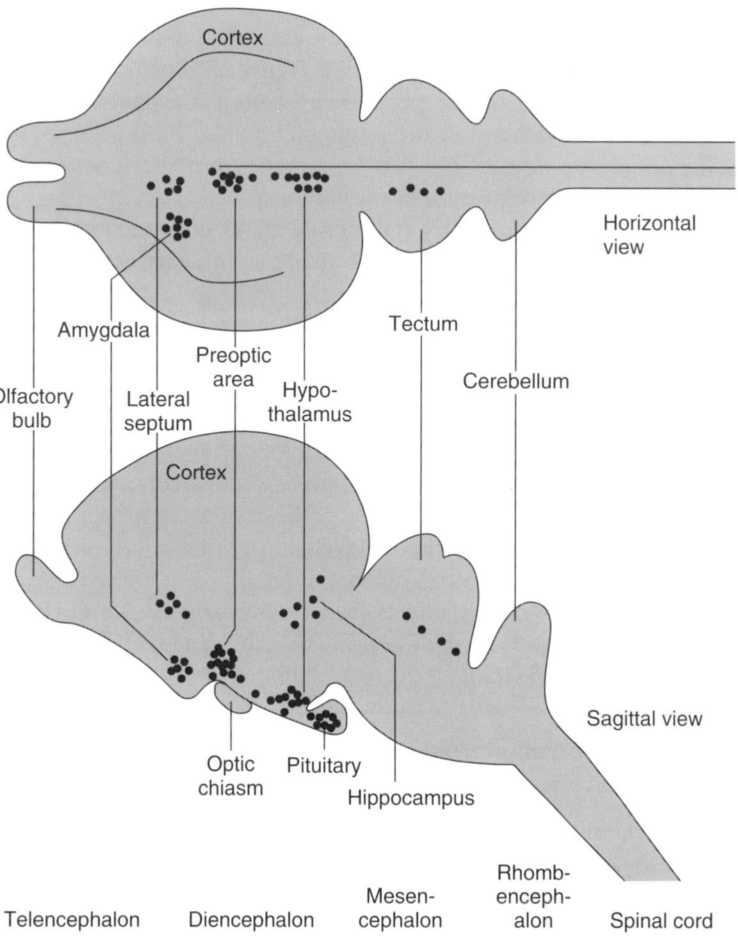

Figure 2.7 Distribution of cells concentrating sex steroids in vertebrates. The ability to concentrate sex steroids is an indication of the presence of intracellular steroid receptors. See also Figure 2.4. Original source: Figure 5.17 in Nelson (2000), *An Introduction to Behavioral Endocrinology*, 2nd edn, © Sinauer Associates, reproduced by permission.

2.1.1.4 Control of endocrine secretions relevant to behaviour by peptides from the brain (anterior nervous system)

In order for behaviour to occur at the right time, place, and social context, hormonal mechanisms regulating behaviour must themselves be regulated by the external physical and social environment. Information about the environment accesses brains via the sensory systems, but how does it access hormones produced elsewhere in the body? This important problem was solved for mammals by two discoveries: (1) the brain (more specifically a portion of the hypothalamus in the diencephalon) regulates the anterior pituitary trophic hormones that in turn regulate the secretions of the peripheral endocrine glands; (2) peptide neurohormones produced by the hypothalamus are the chemical mechanism for regulation of the anterior pituitary (Raisman 1997, Scharrer and Scharrer 1963). Peripheral hormones then feed back onto the hypothalamus and pituitary to adjust hormone production. Nonmammalian vertebrates also have hypothalamic regulation of the pituitary, and also have an array of similar hypothalamic neuropeptides, but less is known about the functional roles of these neuropeptides in regulation of the pituitary.

Of these several brain - pituitary - endocrine gland axes, the two most closely linked to behaviour are the HPG (hypothalamo-pituitary-gonadal) and HPA (hypothalamo-pituitary-adrenal or interrenal) axes (Figure 2.8a). There is much interest in the detailed anatomical localisation and chemical regulation of the hypothalamic cells producing GnRH and corticotropin releasing hormone (CRH), because these are the peptides serving as the final common paths from the brain for these two axes that are so important for reproduction and responses to environmental challenges (Aguilera 1998, Ball 1993, Wendelaar-Bonga 1997). A great variety of neurochemicals are involved in regulation of the pituitary (Figure 2.8b). Any changes in these or their receptors will affect behaviour indirectly, via alterations in levels of peripheral circulating steroids, in addition to any direct effects on behaviour.

There are some important parallels to the vertebrate HPG and HPA axes in insects. The insect brain is a major neuroendocrine organ, and steroidogenesis is regulated by the brain (Gäde *et al.* 1997, Nijhout 1994).

2.1.1.5 Prohormones: steroid metabolism and synthesis at neural targets

Hormones travel in the circulation to their targets, including targets in the brain and other parts of the nervous system. There they may undergo conversion (metabolism) through the action of localised enzymes from the prohormone form to a related but more potent molecule before binding to receptors. The expression and activity of these enzymes are a reminder that hormone targets are dynamic, not passive (Nijhout 1994). As important mechanisms for regulation of how much active hormone a neuronal target actually sees, these enzyme activities are potentially important points for behavioural disruption.

These prohormone conversion processes have been extensively investigated for steroid hormones. In vertebrates testosterone may be converted in the nervous system to 5α-dihydrotestosterone by the enzyme 5α-reductase, a process called 5α-reduction, or to 17β-oestradiol (hereafter referred to simply as oestradiol) by aromatase, an enzyme in the cytochrome P450 family, a process called aromatisation. These vertebrate testosterone metabolising enzymes are somewhat

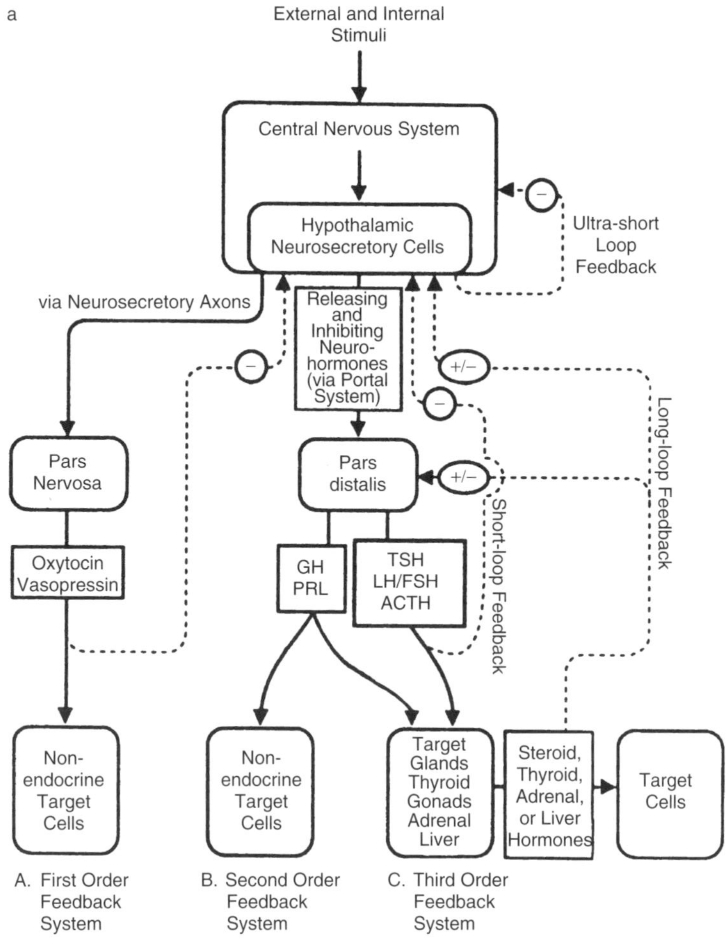

Figure 2.8 (a) The hypothalamo-pituitary axes of mammals. Many features of these axes are seen in other vertebrates. (b) (next page) Neurochemistry of hypothalamic control of pituitary gonadotropins in mammals. It is likely that some of these same regulators function similarly in other vertebrates. TSH, thyroid stimulating hormone; LH/FSH, luteinising hormone/follicle stimulating hormone; END, endogenous opiates; ARC, arcuate nucleus; POA, preoptic area; OC, optic chiasm. See Table 2.1 for other abbreviations. Original source: Figures 4-2 and 4-20B in Norris (1996), *Vertebrate Endocrinology*, © 1996 by Academic Press, reproduced by permission of the publisher. All rights of reproduction in any form reserved.

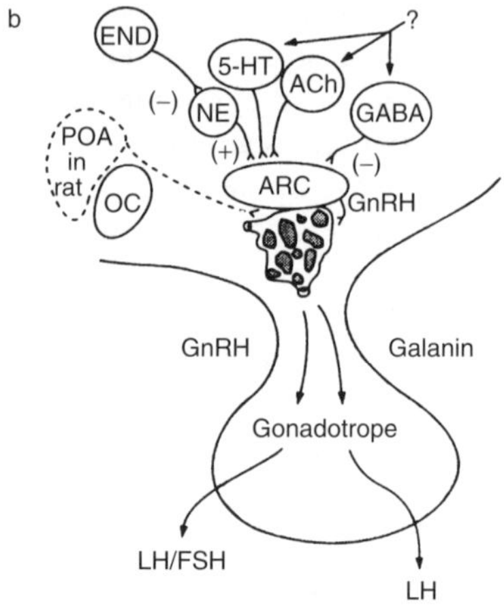

Figure 2.8 (*continued*)

region specific and are an essential part of the biochemical pathways whereby sex steroids stimulate behaviour. Highest levels of enzyme activity tend to be found in brain areas known to be involved in reproductive behaviour, often (but not always) in the same regions where steroid receptors are found (Figure 2.7) (Celotti *et al.* 1997, Schlinger and Arnold 1995, Silverin *et al.* 2000). In some regions these enzymes are themselves steroid regulated (Balthazart and Ball 1998, Roselli *et al.* 1997, Takase *et al.* 1999).

One important consequence of brain cell metabolism and synthesis of steroids is that brain steroid levels and blood steroid levels are somewhat independent. Brain produced oestrogens contribute to male behaviours in many species that have very low circulating estradiol levels (see Section 2.7.1). Some evidence even suggests that neurons or glia can synthesise steroids from cholesterol (Robel and Baulieu 1995). This makes it even riskier than usual to use correlations (or their absence) between circulating hormones and behaviour to reach conclusions about causal hormonal mechanisms of behaviour. The occurrence of a specific behaviour (e.g., territorial defence) at a time of year when gonads are regressed and circulating sex steroid levels are low allows one to conclude that gonadal hormones are probably not stimulating the behaviour, but does not allow one to conclude that sex steroids are not stimulating the behaviour (Schlinger *et al.* 1992). Nor can it be assumed that circulating steroids have come from the gonads or adrenals, because brain produced steroids can cross the blood–brain barrier and show up in the circulation (Schlinger and Arnold 1995).

2.1.2 DIVERSE, TAXON-SPECIFIC MECHANISMS

Pharmacological disruption of neural and hormonal mechanisms is likely to affect several behaviours in a broad array of species, and results obtained with one species, even a domestic animal, may generalise to others. Mixed in with these conserved features underlying behaviour are a diverse array of others that are more taxonomically limited. Their disruption is likely to have more circumscribed consequences, and results from one taxon cannot be safely generalised to distantly related others.

Some diversities in mechanisms are likely to be adaptations related to a particular way of life (morphology or ecological setting). For example, many insects grow within a rigid exoskeleton of hardened cuticle and thus need mechanisms to regulate the behaviours required to get out of the cuticle during moulting (ecdysis) and out of the pupa at the final metamorphosis (eclosion). This may explain why the ecdysteroids are not found in vertebrates, why insect juvenile hormones do not resemble chemically the hormones of vertebrates, and why eclosion hormone (a polypeptide) does not resemble other peptide families (Truman 1992).

Taxonomic differences in exact amino acid sequences are typical of protein and peptide neurochemicals and hormones (such as prolactin and CRH). This is why versions derived from one taxon may not be effective when administered to another. These variations in sequence structure reflect the slow accumulation of genetic changes since the time of initial phylogenetic divergence. Whether these changes are selectively neutral or reflect biologically significant changes in function due to selective pressures is not easy to determine.

While any particular steroid is the same in all species, there is nonetheless some diversity in exactly which steroidogenic pathways occur (which synthesising enzyme systems are present) and where. In goldfish (*Carassius auratus*), Gelinas and Callard (1993) found retinal, optic tectal and retinotectal neurons that contained aromatase, suggesting a role in regulation of reproduction-related visual functions. Many fishes, but not other vertebrates, produce the androgen 11-ketotestosterone, which is not aromatizable (Brantley *et al.* 1993). At young ages and during the nonbreeding season, birds have a very active 5β-reduction pathway whereby testosterone is converted to 5β-dihydrotestosterone, an androgen that is apparently physiologically and behaviourally inactive (Balthazart 1989). Levels of the same circulating sex steroid may differ markedly between species, with behavioural consequences that will depend on whether there are also associated differences in steroid receptor levels.

Sexual selection has considerable potential to produce and maintain diverse taxon-specific mechanisms. Thus, many mechanisms involved in social behaviour and social signalling show interesting diversity (Jacobs 1996). For example, the telencephalon of oscine passerine birds (songbirds) contains a set of interconnected nuclei involved in the learning and production of song (see Section 2.6.3.2). These nuclei are impressively large, sexually dimorphic in size and dedicated to a surprising degree to a single behaviour whose main function seems to be to attract mates and repel rivals. Unlike corresponding regions of the telencephalon of nonpasserines, most of them also contain sex steroid receptors (Metzdorf *et al.* 1999). This song system seems to be a specialty of the passerine lineage (Brenowitz and Kroodsma 1996).

2.1.3 NEURAL DEVELOPMENT AND PLASTICITY; LEARNING

The developing nervous system of vertebrates and complex invertebrates has considerable structural and functional plasticity (Purves *et al.* 1997). This plasticity is associated with certain kinds of behavioural flexibility and also with the capacity to repair certain types of damage (Brauth *et al.* 1991). In adulthood, the processes underlying this plasticity (such as neurogenesis and axon outgrowth) tend to occur to a lesser extent or in more circumscribed parts of the nervous system. The adult nervous system is surprisingly plastic nonetheless. For example, neurons in the cortex are capable of functionally relevant change at the level of biochemistry (alterations in excitability) and structure (alterations in cell size, number of synapses, dendritic arborisation, number of dendritic spines). Such changes can occur in response to ecologically-relevant conditions, as when avian forebrain song system nuclei change size with the seasons (Brenowitz and Kroodsma 1996). Some neurogenesis also occurs in adult nervous systems. In mammals and birds it is limited to a few telencephalic areas such as the hippocampus (Gould *et al.* 1999), but in fish there may be continual and more widespread adult neurogenesis (Zupanc 1999).

Neural plasticity (biochemical, structural, or both) is thought to be involved in behavioural change through experience, including change due to learning. Even simple animals and nervous systems show modification through experience in the form of habituation, sensitisation and classical (Pavlovian) conditioning, and are capable of modifying their behaviour in response to the rewards (positive stimuli) and punishments (negative stimuli) that result from their actions. Nervous systems and animals have memories for what has gone on before, as reflected in long-lasting behavioural change. These highly conserved processes are almost certainly adaptive (fitness-enhancing), enabling animals to track aspects of the environment that change too rapidly and are too local to be predictable on the time scale of genetic evolution.

Other forms of change through experience are more taxon and behaviour specific, such as the exceptional memory for stored food locations shown by some corvid birds (Balda and Kamil 1998) or the honeybee's ability to remember which flowers to visit at what times of day (Gould 1996). Some of the best studied cases are highly constrained forms of learning that occur during critical (sensitive) periods in development (learning by instinct — Marler 1991) such as song learning in birds. There is much interest in the chemical mechanisms that open and close these critical periods, whether steroids are among these mechanisms, and whether steroid effects on developmental plasticity are mediated by neurotrophins (e.g., McAllister *et al.* 1999).

The neurochemical and hormonal mechanisms involved in both general forms of learning such as Pavlovian conditioning and more specialised (and often developmentally restricted) forms such as song learning are currently the target of a very active research effort (e.g., Nottebohm 1991). One major focus is N-methyl-D-aspartate (NMDA) receptors, a subtype of glutamate receptor. Glutamate is an abundant and widespread neurotransmitter. NMDA receptors have the interesting property that they are activated only when both the ligand (glutamate) and prior postsynaptic membrane depolarisation are present together (i.e., when there is a co-occurrence of activity pre- and postsynaptically). NMDA receptors are critically involved in long-term potentiation (LTP), which in turn is hypothesised to be the neurophysiological basis of some kinds of associative learning (Ito 1989). There is increasing interest in determining how neurotrophins are involved in adult neural plasticity (McAllister *et al.* 1999). Parts of the brain that seem especially plastic or are thought for other reasons to be important for learning are receiving much attention, such as the visual cortex, hippocampus and amygdala of mammals, the hippocampus and neostriatum of birds, and the mushroom bodies of insects (Capaldi *et al.* 1999; see also Section 2.4.2). Disruption of mechanisms responsible for critical periods has great potential for permanent effects on behaviour.

2.1.4 HORMONAL BASES OF BEHAVIOUR: CONCEPTS AND PRINCIPLES

Many ecologically-relevant behaviours are dependent to some degree on hormones from endocrine glands or other tissues. Research focusing on mating, aggression and parental behaviour has yielded important concepts and principles that may generalise to other behaviours (especially other social behaviours) and other hormones. Three in particular deserve discussion here.

First, sex steroids can act both centrally and peripherally to alter behaviour (for example, to increase mating in a sexually inactive animal) (Nelson 2000). They can act: (a) on peripheral sense organs or sensory processes, such as the olfactory system, so that a steroid exposed animal is more sensitive (has lower thresholds) to stimuli (Cardwell *et al.* 1995, Halem *et al.* 1999); (b) on motor organs, such as the avian syrinx or the muscles controlling penile movements (Luine *et al.* 1980, Rand and Breedlove 1992); (c) on the central nervous system (brain and spinal cord), to facilitate not only central sensory and motor mechanisms but also motivation, arousal, discrimination of appropriate from inappropriate stimuli, etc.; (d) to change the stimuli given off by the animal itself, such as odours, which then determine whether other individuals, such as females, approach and provide a stimulus for behaviour.

Second, there is a relation between the amount of behaviour seen and the amount of circulating hormone only in certain circumstances. Mating behaviour, for example, is clearly sex steroid dependent in many species, disappearing when the gonads are removed and reappearing when gonadal steroids are replaced (see Section 2.7.1). If different dosages of hormone are given, the behaviour that results will often follow a threshold (step) function, not a graded relationship such as an S-curve (Albert *et al.* 1993). On the other hand (and contrary to what is often assumed), if behaviour level (frequency or quality) and circulating steroid level are measured in a population, more often than not there will be no significant positive correlation. There are many possible reasons for this, including: (a) the step function relationship (all the animals' steroid levels are above threshold); (b) brain produced steroids that stimulate mating (see Section 2.2.1.5); (c) circulating steroid levels that vary quite a bit from moment to moment, so that a single measure is not a sufficiently well-integrated signal to be meaningful at the target; and (d) the dependence of the behaviour on the social context, for example, the dominance status of the animal (Zumpe and Michael 1996).

Third, hormones can have both organisational and activational effects on behaviour. Organisational effects occur during development, are long-lasting or permanent, and establish some aspect of the adult behavioural phenotype. Activational effects occur at puberty and adulthood, are temporary (disappearing if the hormone goes away), and do not change the behavioural phenotype. The boundaries between these two categories are fuzzy (Arnold and Breedlove 1985), and the concepts may not apply to all vertebrates (Crews 1993). Nonetheless they remain an important foundation of much research in hormones and behaviour, with the potential to shed light on the mechanisms of alternative male types in reptiles and fish (Moore *et al.* 1998; see Section 2.6.1.1) and insect morphs and castes (Nijhout 1994, Truman 1992).

2.1.5 SOCIAL INFLUENCES ON HORMONES

Behaviour can alter hormone levels, and thus the hormone-behaviour relationship is two-way (Wallen and Schneider 2000). It is well known that social stimuli such as defeat, separation from the group, or the sudden appearance of strangers are powerful activators of the glucocorticoid stress response (HPA axis) in vertebrates (Schreck 1981, Sapolsky 1992). Reproductive hormones are also responsive to social stimuli via the HPG axis, e.g., exposure of males to female cues and successful interactions with other males can cause dramatic and surprisingly rapid elevations in circulating testosterone (Wingfield and Marler 1994). Physical contact with older honey bee workers delays endocrine and behavioural development in younger workers (Huang *et al.* 1998). Stimuli such as odours from other individuals can alter hormone levels even in the absence of the individuals themselves, e.g., substrate material containing male musk shrew (*Suncus murinus*) odours rapidly causes an increase in GnRH in the brains of females exposed to it (Rissman *et al.* 1997). Courtship activities stimulate ovarian development and ovulation in females of several species (Cheng 1993, Wingfield and Farner 1993). Because behavioural interactions can alter hormone levels, they can also alter hormone sensitive features of the nervous system and of individual neurons (Francis *et al.* 1993). The ability of steroids to act as gene transcription regulators and to mediate neural plasticity means that the consequences of these acute hormonal surges could be profound.

2.1.6 POLYMORPHISM AND SEXUAL DIMORPHISM

Most multicellular animals reproduce sexually, and many of these are sexually dimorphic. In some species one sex comes in more than one form (e.g., alternative male types in fish — Henson and Warner 1997). Thus, the mechanisms that determine the forms or morphs are also mechanisms determining behavioural type, and their disruption will impact not only body type but also many ecologically-relevant behaviours.

The neurochemical and hormonal mechanisms producing adult behavioural types are best understood for vertebrate sexual dimorphism, especially in mammals and birds (see also Section 2.8). The process of sexual differentiation begins with gonadal differentiation but thereafter depends importantly on organisational and activational effects of hormones produced by the differentiated gonads on the body (Adler *et al.* 1985, Gerall *et al.* 1992, Short and Balaban 1994). These effects include organisational actions of steroids (androgens, oestrogens and their locally produced metabolites) on neural development and adult activational effects on neural responsiveness. The result is adult sex differences in certain parts of the nervous system (e.g., the diencephalic preoptic area) and in certain behaviours (e.g., mating behaviour), with the former thought to be the cause of the latter. Neural outcomes include sex differences in neurochemistry (e.g., steroid enzyme or receptor levels), neuron numbers, fine structure of neurons (dendritic arborisation), and overall sizes of brain regions (Panzica *et al.* 1996, Cooke *et al.* 1998).

The magnitude of these neural sex differences varies considerably depending on the species and the brain region. Some are moderate and values for males and females overlap, that is, the phenotype is not categorically dimorphic. Such quantitative sex differences do, however, help us understand why males and females often differ in sensitivity to hormones and other chemicals, including drugs. Other neural sex differences are large, with little or no overlap in male and female values, and thus are truly dimorphic, e.g., the male-female differences in the sizes of the some of the telencephalic song system nuclei in the zebra finch (*Taeniopygia guttata*) (Nottebohm and Arnold 1976).

Although many characters are sexually differentiated to some degree, it is uncommon, at least in vertebrates, to find mechanisms that are themselves sex-specific. Hormone levels often differ between the sexes, but it is rare to find a hormone that is present in one sex and not the other (Gäde *et al.* 1997, Norris 1996). Nor do quantitative sex differences in mechanisms necessarily generalise from one species to another or correspond to our intuitions. Females can have as much or more circulating testosterone as males, and males can have as much or more circulating oestradiol as females (Schlinger and Arnold 1995, Staub and DeBeer 1997).

2.2 BEHAVIOURAL RHYTHMS

The ability accurately and consistently to predict a regularly-occurring environmental change and anticipate the necessary behavioural adjustments is critical to the survival of a population. The sensitivity of the physiological mechanisms of these behavioural rhythms varies temporally with age, season or time of day. Depending on the exact temporal relationship, similar environmental cues, or *zeitgebers*, may have radically different effects (Figure 2.9). A given cue, for example photoperiod, may be critical to induce reproductive behaviours for an adult yet have no effect on a juvenile.

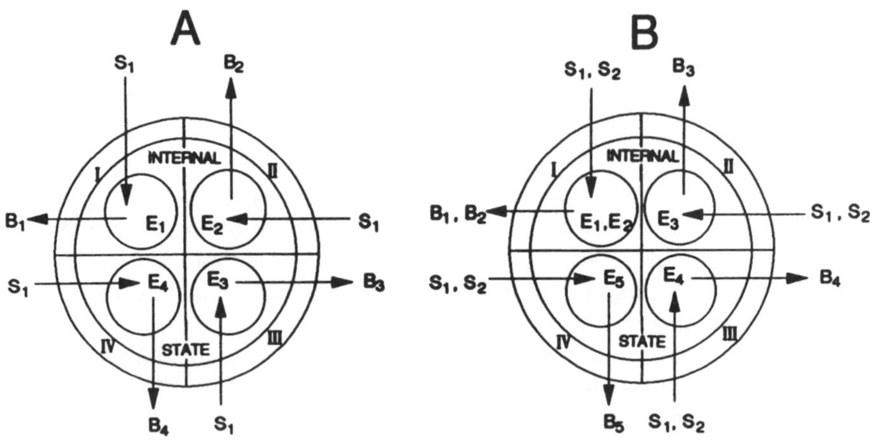

Figure 2.9 Temporal interactions between the external environment and internal state of the individual fish. I to IV = different seasons; S_1 and S_2 = different environmental stimuli; B_1 to B_5 = different behavioural responses to specific environmental stimuli; E_1 to E_5 = different internal states upon which environmental stimuli act and which provide the fundamental mechanisms that direct seasonal-specific behaviours. (A) A single environmental stimulus inducing a single behavioural response. (B) Multiple environmental stimuli inducing multiple behavioural responses. Original source: Weber and Spieler (1994) 'Behavioural mechanisms of metal toxicity in fishes,' in *Aquatic Toxicology: Molecular, Biochemical, and Cellular Perspectives*. Reprinted with permission, © CRC Press, Boca Raton, FL.

2.2.1 CIRCADIAN RHYTHMS

Many behavioural rhythms persist in the absence of changing external cues to time of day (i.e., they free-run), one of many lines of evidence that they are generated by one or more endogenous circadian oscillators (pacemakers or clocks) (Aschoff 1981) based on cellular and molecular level processes. Circadian oscillators can be entrained by any cue that varies on a daily basis, e.g., the daily light-dark cycle.

In addition to locomotor activity and foraging rhythms, many fish species display diurnal variation in breeding and parental care behaviours, e.g., fathead minnow (*Pimephales promelas*), medaka (*Oryzias latipes*), female convict cichlids (*Cichlasoma nigrofasciatum*), and rainbow cichlids (*Herotilapia multispinosa*) (Denny 1987, Weber and Spieler 1987, Reebs and Colgan 1991, Weber 1993).

Some behavioural rhythms, e.g., entrainment to feeding at specific times of day in crayfish (*Procambarus clarkii*) (Fernandez de Miguel and Arechiga 1994), are quite malleable in terms of ease of phase shifting. Maintaining the phase relationship of this associative behaviour involves specific neuronal pathways for long-term potentiation (LTP). Aquatic microcrustacea, e.g., *Daphnia* spp., undergo daily vertical migrations to feed upon algae that correlate with light levels, a rhythm entrainable by the nonspecific response to chemicals, or kairomones, released by many planktivorous fish species, and assist the planktons' ability to avoid visual predators (von Elert and Loose 1996). Work by Spieler and coworkers (Spieler and Noeske 1984, Spieler *et al.* 1987, Spieler and Clougherty 1989) has shown that a single daily meal can synchronise a wide variety of rhythms, e.g., locomotor, agonistic and circulating cortisol rhythms, and may be a stronger stimulus than the light-dark cycle, even in free-running goldfish (Spieler and Clougherty 1989). While mammalian circadian locomotor activity rhythms have a dual control, i.e., an anticipatory feeding phase that is entrained by meal feeding with the remaining portion tightly linked to the photoperiod, goldfish apparently have no such splitting. In Atlantic salmon (*Salmo salar*), temperature alone, independent of photoperiod, is sufficient to phase shift feeding patterns. Apparently, the salmonid retina is more sensitive at the colder temperatures thus facilitating night-time feeding (Fraser *et al.* 1993) and the accompanying changes in social interactions.

2.2.1.1 Oscillator and entrainment mechanisms

In rodents, the suprachiasmatic nuclei (SCN) in the diencephalon form a primary circadian oscillator; there may be a homologue with a similar role in other vertebrate groups (Norgren and Silver 1989, Underwood 1992, Wilczynski *et al.* 1993). Although some authors have suggested that fishes, like mammals, may have an SCN (Prosada *et al.* 1982), the lack of clear evidence for its existence in fishes leads others to place more emphasis on the role of the pineal gland (Underwood 1990). Some but not all vertebrate pineals are themselves circadian oscillators, continuing to produce melatonin in a rhythmic manner even when isolated (Underwood 1992, Falcon 1999, Okimoto and Stetson 1999). The eyes of some animals are oscillators even when disconnected from the rest of the nervous system (Underwood *et al.* 1990).

The circadian oscillator in the eye of *Aplysia* is entrainable by changing light cycles (Jacklet 1974, Eskin 1979) and appears to be controlled by extraocular neural pathways (Block *et al.* 1974). For some vertebrates whose behavioural rhythms are entrained by the daily light-dark cycle, one entrainment mechanism is a subdivision of the visual system that begins in the retina, projects directly to the hypothalamus (to the SCN in those species that have one) and continues from there indirectly to the pineal (Norgren and Silver 1989, Morin 1994). There are interesting differences between vertebrate taxa (even between species in the same class or order) in the importance of the eyes and pineal for entrainment by light; in birds and reptiles, extraretinal diencephalic photoreceptors are involved (Underwood 1992, Saldanha *et al.* 1994, Falcon 1999). In hagfish (*Eptatretus burgeri*), which have no pineal, only ablations of the ventromedial part of the telencephalon-diencephalon with sections of the hypothalamus resulted in arrhythmic activity (Underwood 1990). The hypothalamus may play a role in organising rhythmic behaviours in eel (Popek 1983).

2.2.1.2 Neurochemical and hormonal mechanisms

In several species serotonin participates in circadian rhythmicity. In *Aplysia* the eye possesses membrane receptors specific for serotonin and also synthesises serotonin (Corrent *et al.* 1982); serotonin is capable of entraining ocular circadian rhythms, but does so more slowly than light (Corrent and Eskin 1982). Injections of serotonin in the snail *Helix aspersa* intensify the associative daily rhythms of activity, feeding and sexual behaviour (Adamo and Chase 1991). In mammals the SCN is innervated by serotonergic and NPY-ergic fibres as well as by glutamatergic fibres from the retina; pharmacological manipulations of serotonin levels alter circadian rhythms in hamsters (*Mesocricetus auratus*) (Morin 1994).

The hormone most often associated with circadian rhythms is the indoleamine melatonin (Cassone 1990), although opioids may also be involved (Kavaliers 1989). Melatonin is a pineal hormone in vertebrates, but its synthesis is not restricted either to the pineal (Underwood *et al.* 1990) or to vertebrates (Morita 1990). Levels of melatonin are highest during the night in both nocturnal and diurnal vertebrates; light inhibits N-acetyltransferase (NAT), an essential enzyme for melatonin synthesis. In species with pineals that are circadian oscillators, pineals not only show this passive response to the light-dark cycle, but can continue showing rhythmic changes in NAT without rhythmic changes in light.

While some fishes may have an endogenous oscillator that controls pineal NAT activity, others may not (Underwood 1990, Okimoto and Stetson 1999).

Pinealectomy and melatonin administration affect circadian activity rhythms in a number of vertebrates (Cassone 1990, Underwood 1990), although they generally seem to have a stronger influence in lizards and birds (Underwood 1992, Gwinner *et al.* 1997). Pinealectomy in fishes has a variable effect on rhythmicity depending on the species, including significant alterations in free-running periods, bifurcation of the circadian rhythm, or arrhythmicity, which may be due to the wide distribution of melatonin receptor sites (Martinoli *et al.* 1991). Several other hormones, including thyroid hormones and gonadal sex steroids, are also known to influence the period of circadian locomotor rhythms (Nelson 2000).

2.2.2 ANNUAL RHYTHMS

Seasonal patterns in behaviour, e.g., reproductive behaviour, social organization or hibernation (diapause in insects), occur which are under various external and internal controls (Aschoff 1981, Lam 1983, Underwood 1992, Wingfield and Farner 1993, Nijhout 1994, Houck and Woodley 1995). Environmental cues, e.g., changes in daylength (photoperiod), temperature, rainfall/humidity, or the behaviour of conspecifics, that predict the availability of critical resources for reproduction, induce many seasonal changes in reproductive hormones via the HPG axis or its invertebrate analogues (see Sections 2.2.4 and 2.2.5) (Wallen and Schneider 2000). Where peaks in behaviour coincide with elevated hormones, some of these relationships are probably causal (see Sections 2.6 and 2.7), as has been shown for seasonal changes in the electric organ discharges (EODs) of some weakly electric fish (Zakon 1996).

For many terrestrial species, especially those living in the temperate zone, daylength is the most important predictive cue (the coarse tuner), with other cues such as temperature playing a supplemental role (the fine tuners) by, for example, determining what the critical daylength will be (Wallen and Schneider 2000). The circadian clock is often used to measure daylength, and so the mechanisms of the photoperiodic control of the HPG axis overlap to some extent with those of circadian clocks and their entrainment by light, with similar taxonomic diversity in the role of the eyes, extraretinal photoreceptors, the pineal and melatonin (Underwood 1992, Goldman and Nelson 1993, Saldanha *et al.* 1994). The brain itself varies seasonally in response to changing photoperiod; in several species of birds, seasonal changes in the size of brain regions (such as the song system nuclei or the hippocampus) can be mimicked by changing the photoperiod experimentally (Tramontin and Brenowitz 2000).

Little is known about the pathways whereby other environmental cues such as rainfall, humidity, or social cues regulate behaviour on a seasonal basis. The information about these cues likely alters the GnRH system of neurons, possibly via brain peptides such as AVT and VIP (Maney *et al.* 1999). In frogs (*Hyla* spp.) and ring doves (*Streptopelia* sp.) some progress has been made establishing the neural pathways involved in the stimulating effect of courtship and mate attraction vocalisation input signals on gonadal status of listeners (Wilczynski *et al.* 1993, Cheng and Zuo 1994) (see also Section 2.2.5).

While many seasonal changes in behaviour occur because of changes in gonadal hormone production, not all do. Animals with dissociated reproduction (spermatogenesis and high testosterone occurring at a different time of year than mating behaviour) seem to have annual rhythms in reproductive behaviour that are not produced by activation of the HPG axis (Crews 1984, Mendonça 1987). Even in species where reproductive behaviour and high circulating gonadal hormone levels do coincide, other seasonally varying mechanisms may also be involved. These could include nongonadal hormones, brain steroid receptors, activity of brain steroid synthesising and metabolising enzymes, peptide neuromodulators, and sensitivity to neuromodulators (Moore and Marler 1988, Pasmanik and Callard 1988, Saldanha *et al.* 1994, Wood *et al.* 1995, Takase *et al.* 1999). Some of this seasonal variation could occur independently of changes in circulating gonadal hormones (i.e., as a more direct response to changing photoperiod).

2.3 SPATIAL BEHAVIOUR

2.3.1 DISTRIBUTION WITHIN THE HABITAT, SHORT DISTANCE MIGRATIONS AND RANGING

Attainment of the adult state is often accompanied by movement away from the preadult habitat and into the rather different adult habitat. The adults may then return to the juvenile habitat to mate and lay or fertilise eggs. These habitat shifts that accompany life stage changes are regulated by hormones such as thyroxine, corticosteroids, CRH and prolactin (Rankin 1991, Norris 1996).

Ecdysis (moulting), pupation and eclosion (final moult and emergence from the pupa) have important consequences for the use of space by insects (Weeks *et al.* 1989). Considerable progress has been made in determining the mechanisms of these processes and their accompanying behaviours in the tobacco hornworm (Truman 1992, Ewer *et al.* 1997). In aquatic invertebrates, temporal and spatial distribution is often regulated by patterns of oxygen tension (Syed and Winlow 1991, Hermann and Bulloch 1998). Parasitic invertebrates provide a unique model for understanding mechanisms of short-distance migration (Sukhedo 1992). Migration patterns of marine invertebrates can involve a variety of sensory systems including the visual system (Ringelberg 1969, Nevitt *et al.* 1995, Svane and Dolmer 1995, Schultz and Clark 1997), rheotactic responses, sensitivity to superficial microbial film layers (Orlov 1997) and pheromones (Clare *et al.* 1994).

Locomotor behaviour has major consequences for spatial and temporal distributions. Evidence in several invertebrates from different phyla suggests an involvement of octopaminergic and serotonergic neurons in the regulation of stereotyped locomotor sequences (Willard 1981, Bicker and Menzel 1989, Orchard *et al.* 1993, Goldberg *et al.* 1994). In vertebrates steroid hormones influence locomotor activity levels (Wade 1976, Wada 1984, Cash and Holberton 1999).

Male birds of several species increase their breeding season territory and home range sizes when treated with testosterone (see Section 2.6.1.1). Also, intrauterine position predicts the home range size of free-ranging female house mice (*Mus musculus*), an effect which is likely to be due to organisational effects of sex steroids (Zielinski *et al.* 1992).

2.3.2 SCHOOLING AND GROUPING

Chemical communication is important for creating social cohesion and spatial patterns in fish groups at all ages (Liley 1982). The primary source of conspecific attractants seems to occur in the skin mucus. Immature charr (*Salvelinus alpinus*), herring (*Clupea harengus*) larvae and juveniles, minnows (*Phoxinus phoxinus*), even if blinded, and cichlid fry (*Hemichromis bimaculatus*) will all aggregate in the vicinity of conspecifics and avoid heterospecifics. The lateral line system, with neurons in cranial nerves VII (facial) and IX (glossopharyngeal), is also important in controlling spatial distribution of individual fish within a group.

2.3.3 LONG-DISTANCE MIGRATION AND HOMING

In some species migration occurs annually or biannually in response to environmental cues such as photoperiod. Migration may also occur with distinctive daily rhythmicity, e.g., when birds that are normally diurnal nonetheless migrate exclusively at night. Thus, there may be some common mechanisms underlying migration and other kinds of annual or daily rhythms (see Section 2.2). Because some migratory species migrate in one direction at one time of year and in the reverse direction at another time of year, understanding mechanisms of migration includes answering two questions: (1) what causes sustained migratory movement? (2) what causes the reversal in direction as a function of time of year? Progress in discovering neural and hormonal mechanisms has been slow for many taxa (including birds and most invertebrates) and close to nonexistent for others (reptiles, mammals) (Rankin 1991, Berthold 1993, Wingfield and Farner 1993, Gwinner *et al.* 1997, Owens 1997).

Many insects normally migrate when they are not in a reproductive state, whereas others (e.g., colonising insects) can arrive ready to reproduce (Nijhout 1994). Therefore, the causal relationship between migration and reproductive hormones such as the gonadotrophic juvenile hormones will likely depend on which of these two patterns occurs. There is evidence that juvenile hormones stimulate migratory behaviour in several species of insects (Rankin 1991). Little is known about the neurochemistry of insect migration, but octopamine, which is known to stimulate flight motor patterns, may be involved (Rankin 1991, Orchard *et al.* 1993).

In diadromous fishes, osmoregulatory functions of cortisol may be important in both reproductive activity and seasonal migrations. Meier (1984) proposed a neuroendocrine model which has received some support from work with the stickleback *Gasterosteus aculeatus* (Audet *et al.* 1986) in which cortisol plays a critical role in salinity preferences and reproductive indices. Age-related changes in motivational state for migration of European grayling (*Thymallus thymallus*) appear to be associated with the functional state of the catecholaminergic system (Pavlov *et al.* 1994).

The pioneering work by Hasler (1966) demonstrated that salmon are capable of returning to the headwaters of their origin with great precision by using their olfactory system. Such events occur during hormonally-linked sensitive periods. In addition, olfactory receptor neurons have a unique ability to turn over during a fish's life. Selection for neurons specific to certain odours may occur, facilitating sensitivity to home stream odours. This sensitivity and the concomitant increase in olfactory learning ability is associated with elevated thyroid activity and may be triiodothyronine/thyroxine induced (Hoar 1953, Baggerman 1962, Fontaine 1975, Morin and Døving 1992, Nevitt and Dittman 1998). This hypothesis builds upon Harden Jones' (1968) concept of experiential learning (storing and recall of odour information) to facilitate the age-specific temporal distribution of migrating smolts. Behavioural comparisons between wild and hatchery reared salmon support this concept (Jonsson *et al.* 1994).

Morin and Døving (1992) suggested three possible nonmutually exclusive environmental odour cues: (1) pheromones released by migrating smolts are sensed by adults as they ascend the river, (2) imprinted odours from soils, etc. in the headwaters are sensed by ascending adults, and (3) classical imprinting in which fish learn the physical characteristics of their conspecifics and use that information while migrating. Similar physiological processes may be invoked to explain each hypothesis.

Mechanosensory stimulation may be important in migratory orientation. Direction of stream movement is critical to determining orientation (rheotaxis). Either ablation of superficial neuromasts but not canal neuromasts or pharmacological blocks of the lateral line system induced a significant increase in the velocity threshold for rheotactic behaviour in three species of fishes (Montgomery *et al.* 1997).

Some birds fly relatively long distances in order to find food and water and then return later that day to the home site. Neural mechanisms underlying the ability to find the way home after such a displacement have been studied in only one species, the domestic pigeon. Hippocampal lesions cause deficits in the homing performance of pigeons, suggesting a role for this telencephalic structure in homing (Bingman *et al.* 1998).

2.3.4 JUVENILE DISPERSAL BY BIRDS AND MAMMALS

Once they are capable of living on their own, the young of many species leave the natal area and disperse to a new region and home range. Several features of juvenile dispersal suggest an involvement of hormones (Smale *et al.* 1997). First, timing or probability of dispersal in some cases is predictable from body fat or condition, suggesting links to chemical regulators of food intake and energy use such as leptin or glucocorticoids (Belthoff and Dufty 1998). Secondly, dispersal is often sexually dimorphic; one sex disperses a greater distance or with a greater probability than the other (Pusey 1987). The only experiments to date on hormonal mechanisms of juvenile dispersal in free-ranging animals suggest that organisational hormone effects might be responsible for greater male dispersal in ground squirrels (*Spermophilus beldingi*) (Smale *et al.* 1997).

2.4 FORAGING

A number of factors are known to affect foraging behaviour, including motivation, satiety controls in the CNS and PNS, morphology (e.g., mouth and gills in fishes), sensory processes (e.g., visual imaging) prey density and distribution, social facilitation and social learning, and circadian and seasonal rhythms. Although there is a vast literature on the neurochemistry and neuroendocrinology of food intake and nutrient preferences in laboratory animals (a few fishes, birds, rodents and primates) (Stricker 1990), relatively little is known about neural and hormonal mechanisms of foraging behaviour in wild species or in free-ranging animals.

2.4.1 REGULATION OF FEEDING AND FOOD INTAKE

Stimulation and inhibition of feeding behaviour is controlled by internal mechanisms that respond to circulating levels of specific compounds, including hormones and peptides (LeBail and Boeuf 1997). In the medicinal leech (*Hirudo medicinalis*) serotonin induces all motor components of the feeding response (salivation, biting, pharynx contraction, body wall distension and mucus secretion) (Lent 1985). While injections of serotonin to sated leeches did not induce additional feeding, it did increase the rate of biting (West *et al.* 1991); depletion of serotonin resulted in lower biting frequencies (Lent 1985, O'Gara *et al.* 1991). In the carnivorous leech *Haemopis marmorata* reserpine decreased the amount of food ingested (Goldburt *et al.* 1994), although it did not eliminate feeding behaviour as with the medicinal leech. This may be due to the feeding pattern of the two leeches. Bloodsucking species feed infrequently during the year and excretion of reserpine is, therefore, low. Carnivorous forms feed several times per day and the increased metabolism and excretion of reserpine may reduce its effectiveness. There may exist additional, reserpine-resistant neural pathways that regulate feeding in *Haemopis* spp. (Goldburt *et al.* 1994). In another haemophagic species, the blood fluke larva, eicosanoids (oxygenated metabolites of C-20 polyunsaturated fatty acids) are important in assisting skin penetration behaviour (reviewed in Stanley-Samuelson 1994).

The feeding behaviour of hydra (*Hydra vulgaris*) involves a suite of biologically active chemicals that control mouth opening and closure (Pierobon *et al.* 1997), suggesting that relatively simple feeding movements require a complex array of signals for proper coordination and timing. Levels of arachidonic acid, acting as a second messenger after a chemical stimulus induces the feeding response, are increased through the hydrozoan neuropeptide antho-RFamide.

The extensive arborisation of the cerebral neuron C-PR in cerebral and pedal ganglia may play an important role in the appetitive phase of feeding behaviour in *Aplysia* (Teyke *et al.* 1997). Feeding behaviour in *Aplysia fasciata* is affected by pheromones released by conspecifics and detected by chemosensory rhinophores during mating (Blumberg and Susswein 1998). Specific peptides from the atrial gland and bag cells induced an increase in consumption. Opioids increase food consumption in the pond snail by activating cells located near the buccal ganglia where the central pattern generator for feeding is found (D'Yakonova *et al.* 1992). Opioids found in amoebas and molluscs, and cholecystokinin (CCK) found in molluscs, are important in inducing feeding behaviour and satiety, respectively (Morley 1995).

Serotonin increases the number of searching behaviours in the pond snail *Lymnaea stagnalis* but not in *Helix pomatia* (Teyke 1996). Opioid antagonists were found to inhibit the feeding response but induce defensive and escape behaviours (D'Yakonova and Sakharov 1994). Dopamine apparently is also needed to stimulate the feeding response (Kemenes 1997). Inhibition of feeding behaviour is also induced in land snails (*Helix lucorum*) by injections of glutamic acid or N-methyl-DL-aspartic acid (Stepanov *et al.* 1993). Stimulation of the hunting behaviour of the pteropod mollusc *Clione limacina* can be elicited by inhibiting acetylcholinesterase activity. Activation of the feeding rhythm generator and motoneurons innervating the protractor tentacle muscles is accomplished by application of GABA to the isolated CNS (Panchin *et al.*

1994). Postprandial elevated levels of glucose in the haemolymph of sand fiddler crabs (*Uca pugilator*) stimulate a feeding inhibitory factor from the sinus gland (Sears and Rittschof 1991), allowing them to return rapidly to the burrow without stopping for additional food. Teyke (1996) observed that nitric oxide is involved in memory acquisition and food finding ability in the snail *Helix pomatia*; memory recall and olfactory orientation were not NO-dependent.

In goldfish, injections of bombesin into the third brain ventricle (which borders hypothalamic areas associated with satiety controls — Morley 1995) decreased food intake, induced spitting-out of ingested food pellets, and increased growth hormone (Himick and Peter 1994). Evidence also implicates CRH, serotonin and CCK in the regulation of food intake in fishes (DePedro *et al.* 1995a,b, 1998; Peyon *et al.* 1998).

Reproduction and migration take energy, and therefore many animals increase food intake prior to or during these activities. There are close links between reproductive hormones and food intake and weight gain (Buntin *et al.* 1999, Wade and Schneider 1992). Seasonal changes in food intake and body fat in wild migratory bird species usually disappear if the birds are gonadectomised, but castrated male European common quail (*Coturnix coturnix* — a migratory species) continue to show twice a year fattening (Boswell *et al.* 1993). In the migratory dark-eyed junco (*Junco hyemalis*), both testosterone and opioids regulate food intake (Deviche 1992).

In rats, insulin from the pancreas and leptin from fat cells are key hormonal signals telling the brain whether the animal is in positive or negative energy balance. There are receptors for these in the lateral and ventromedial hypothalamic nuclei, which are known to be involved in food intake and weight regulation. Direct administration of insulin or leptin into these nuclei alters feeding behaviour through mechanisms involving biogenic amines, e.g., norepinephrine, and peptides, e.g., CCK, NPY and CRH (Woods *et al.* 1998). Direct administration of NPY into the brain of domestic chickens increases food intake, whereas CCK decreases it, effects similar to those seen in domestic rats and suggesting conserved mechanisms regulating food intake (Kuenzel *et al.* 1999). On the other hand, brain administration of the opioid β-endorphin stimulates food intake in chickens, whereas opioids nearly always depress food intake in mammals, suggesting some differences between birds and mammals that are not yet understood (Denbow 1999).

Mechanisms of sex-specific differences in feeding behaviour have been examined. Toguyeni *et al.* (1997) observed that food utilisation efficiency and nutrient retention were higher in male than female tilapia (*Oreochromis niloticus*) and was correlated with level of social interaction, triiodothyronine levels and sex steroids. In the fiddler crabs, *Uca pugilator* and *U. pugnax*, chemoreceptor neurons in the claws and legs of females have lower threshold levels and greater sensitivity than males (Weisburg and Derby 1995). Females, therefore, can feed at food levels too low for males to detect and can orient to more dilute stimuli.

2.4.2 FORAGING STRATEGIES

Honeybee workers perform different tasks at different times in their adult life. This developmental progression involves morphological as well as major neural and behavioural changes, occurs at a variable time depending on colony need, and is regulated by juvenile hormones (Withers *et al.* 1993, Fahrbach 1997).

In fishes, ecological factors, e.g., prey density and circadian and seasonal rhythms, are highly correlated with visual imaging. Different foraging styles demand different neural processing systems. Whereas both ambush and cruise foragers are assumed to search for prey only at the outer perimeter of their search space, saltatory foragers, such as many zooplanktivores, search for prey items throughout their search space but only for the brief time they are stationary (reviewed by O'Brien *et al.* 1990). Once a visual search image is created, the ability to remember that image is critical to efficient foraging. In goldfish substance P enhances learning and memory of a foraging task, an effect possibly mediated by the dopamine system (Mattioli *et al.* 1997). The mollusc *Clione limacina* is able to detect prey because the tactile and chemical stimuli produced by the prey activate specific cerebral neurons that initiate the feeding motor programme in the buccal ganglia as well as swimming and tail flexion (Norekian 1995).

Visual cues of feeding may play a role in that a feeding stimulus in an adjacent aquarium stimulates fish to feed (Davis and Bardach 1965). Such observations may explain the mechanism of socially facilitated behaviour in minnows (Morgan and Colgan 1988) and the circadian rhythms entrained by such behaviour (Magurran 1984, Kavaliers 1989).

There are trade-offs between social and reproductive behaviour and foraging. Facing serious energetic problems, especially in cold weather, due to the energy costs of flight, the limits on body fat stores imposed by flight, and the high body surface to volume ratio, small birds starve to death quickly once output exceeds input. Territorial defence may be energetically costly, and like incubation is incompatible with sustained foraging. In field studies of male white-crowned sparrows (*Zonotrichia leucophrys*), spring snowstorms during breeding caused a rise in corticosterone and also a shift from reproductive to foraging activities; experimental administration of corticosterone during decent weather produced the same behavioural shift (Wingfield and Farner 1983, Wingfield *et al.* 1998). In the red-sided garter snake (*Thamnophis sirtalis parietalis*) intracerebroventricular administration of NPY decreased courtship and increased feeding, as if it might be a mechanism for shifting between these two incompatible behavioural priorities (Morris and Crews 1990).

Some species of birds, e.g., black-capped chickadees (*Parus atricapillus*), marsh tits (*Parus palustris*) and Clark's nutcrackers (*Nucifraga columbiana*), store numerous small food objects such as seeds or nuts in scattered locations in response to temporary food abundance and then later find and eat them. Lesion experiments implicate the hippocampus as a key brain region underlying this spatial ability (Balda and Kamil 1998, Sherry 1998). Hippocampal volumes are larger in the autumn, the season of greatest food storing (Clayton and Lee 1998), raising the possibility of hormonal regulation of seasonal food storing via the hippocampus.

Omnivorous birds and mammals, both wild and domestic, avoid novel foods that have made them ill, a phenomenon known as taste aversion learning. In rats the neuroanatomical structures that seem to be involved in this highly adaptive process include the basolateral amygdala, a telencephalic region (Morris *et al.* 1999).

2.5 PREDATOR/ANTIPREDATOR BEHAVIOUR

Escape reflex behaviour has been observed in many species of invertebrates. A pelagic marine copepod, *Euchaeta rimana*, uses antennal mechanoreceptors to detect prey-created disturbances that are separate from the turbulences created by its own swimming and feeding (Yen and Strickler 1996). Avoidance behaviour in the snail *Megalobulimus sanctipauli* is modulated by opioids by increasing the latency of the avoidance reaction (Romero *et al.* 1994). A small cardioactive peptide and serotonin are capable of increasing the amplitude of withdrawal reactions in snails and increasing the habituation rate of synaptic responses to the rhythmic stimulation in the command neurons for this reaction (Bravarenko 1994). FMRFamide has the opposite effects. Serotonin also facilitated retraction and searching behaviours in the snail *Cepaea nemoralis*, whereas leu- and met-enkephalins attenuated these behaviours (D'Yakonova *et al.* 1995). In the snail *Helix lucorum* serotonergic cells of the pedal ganglia are responsible for modulating withdrawal behaviour (Zakharov *et al.* 1995).

Nonspecific information is obtained about the presence of predators by *Daphnia magna* through chemical cues released by crushed conspecifics (Pijanowska 1997). When such cues or alarm pheromones are provided, *D. magna* respond by migrating to the bottom of the aquarium where they form aggregations. The freshwater amphipod *Gammarus lacustris* decreases its activities when exposed to crushed conspecifics or to chemical stimuli from either dragonfly larvae or northern pike (*Esox lucius*). These substances may be species specific in that the amphipod response to northern pike stimuli is to move to the bottom of an aquarium, a response that does not occur to dragonfly larvae stimuli (Wudkevich *et al.* 1997). In addition to chemical cues from the predators, changes in light intensity, e.g., shadows from predators, plus chemical cues from crushed conspecifics induced a dramatic decrease in daphnid swimming speed and increased motivation to display the somersault escape reflex (Pijanowska and Kowalczewski 1997). Eicosanoids are important in predator avoidance behaviours displayed by octocorals and nudibranchs (reviewed by Stanley-Samuelson 1994). The nerve cells for escape reflexes are present at hatching and are sensitive to shadow and tactile stimuli (Drewes and Brinkhurst 1990). The ontogeny of the behaviour was observed in several species of starfish and was correlated with the distribution and development of catecholamine-containing cells (Dautov and Nezlin 1990). *Daphnia galeata* is able to adjust its life history traits to differential levels of fish kairomones and, therefore, fish densities (Reede 1995).

Escape responses in fishes range from the classical C-start reflex of goldfish (Eaton *et al.* 1991) to the burying behaviour of eastern sand darters (*Ammocrypta pellucida*) (Simon 1991). Similar behaviours exist in larval amphibians.

Sensory stimuli create action potentials in specific sensory organs (octavolateralis system: anterior and posterior lateral lines, inner ear; eye, including the ability to see polarised light to detect prey or predators — Hawryshyn 1992) which are processed by higher brain centres. Each component of this system operates by similar mechanisms (Platt *et al.* 1989) and provides information to the animal about its immediate environment. Information from the octavolateralis system is relayed to the Mauthner cells (M-cell), a set of two large neurons in the dorsal medulla. The structure of the M-cell and its afferent and efferent neurons has been well described (Roberts 1992, Figure 2.10). Evidence suggests that the neurotransmitters GABA, L-glutamate and possibly glycine may be involved in creating the message and directing the predator escape response (Diamond 1968, Korn *et al.* 1993, Legendre 1997). With additional integration in the cerebellum, the message is relayed to specific supraspinal systems in the brain stem, e.g., the M-cell. Interneurons from the M-cell activate both white and red muscle to create the specific movement, e.g., swim, bury in sand. Additional visual, sound, etc. inputs will modulate the intensity and duration of the response and appear to be processed separately by higher brain centres (Wubbels *et al.* 1993), e.g., cerebellum (Mori 1993), torus semicircularis and telencephalon (Nicol and Lamming 1993). Important to survival is the ability of the M-cell to override ongoing locomotor activity. Svoboda and Fetcho (1996) observed that during swimming, an induced single action potential from the M-cell was sufficient to change the swimming pattern to an escape response.

Larval American eel (*Anguilla anguilla*) do not display the typical C-start reflex. Rather, they show a withdrawal response. Meyers *et al.* (1998) correlated this behaviour with the absence of inhibitory networks impinging upon the M-cell. Withdrawal occurs rather than C-start because both M-cells are able to fire simultaneously, and therefore induce a bilateral contraction of both sets of lateral trunk muscles vs. the contralateral trunk muscle contraction typical of most fishes. This structure-function relationship may be typical of many anguilliform fishes.

Experience or learning, i.e., activities of higher brain centres, are part of the feedback mechanism of this behaviour. Dill (1974) described the role of experience of zebrafish with predators in rate of escape response. Xu and coworkers (Xu *et al.* 1998) examined the effect of NMDA antagonists, e.g., MK-801, on conditioning of escape responses. This behavioural paradigm has been

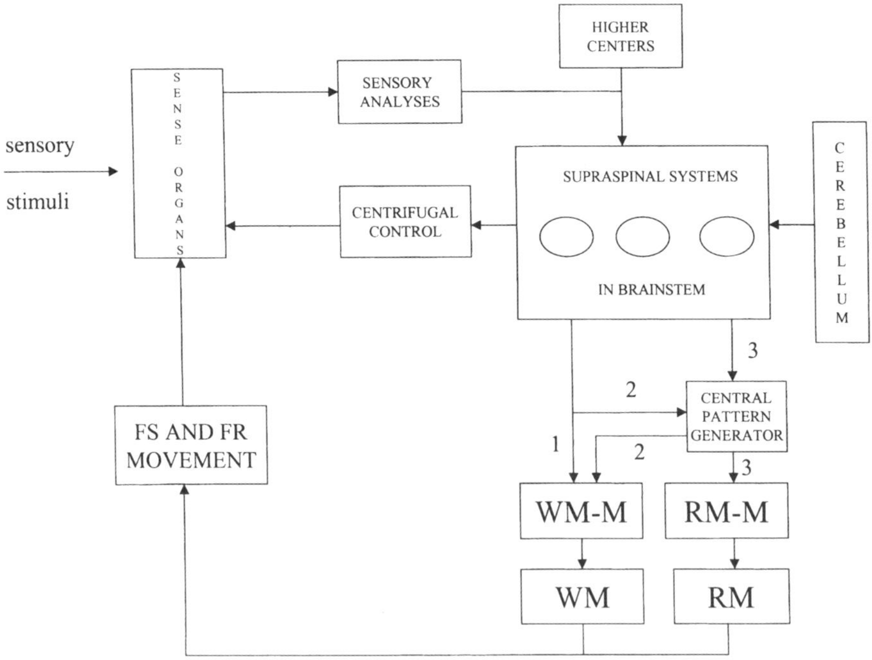

Figure 2.10 The organisation of the motor control system for escape behaviour in fishes. FS, fast start movement; FR, fast rhythmic movement; RM, red muscle; RM-M, red muscle motoneuron; WM, white muscle; WM-M, white muscle motoneuron. Redrawn from original. Original source: Figure 7 in Roberts (1992) *Reviews in Fish Biology and Fisheries* (vol. **2**, pp. 243–266), © Kluwer Academic Publishers, reprinted with kind permission from Kluwer Academic Publishers.

used to model the lateral line-Mauthner cell response (Janssen and Corcoran 1993).

Terrestrial vertebrates (with the possible exception of some birds) also recognise odours of predators and avoid areas with those odours (Weldon 1990). Those species that are active hunters, such as some snakes, or that must find prey located at a distance, such as some seabirds, use odours to help find and recognise prey (Halpern 1992, Wenzel 1986). In most cases the specific chemicals responsible for these responses have not been identified, however.

2.6 SOCIAL BEHAVIOUR

Social organisation is the context in which individual behaviour occurs, and therefore understanding mechanisms of behaviour requires understanding how these function in the normal social environment. This is why caution is appropriate in extrapolating from laboratory studies in which animals are tested singly or with one stranger to behaviour in free-ranging animals. When animals in a social context (free-ranging or semi-free-ranging) are experimentally manipulated or otherwise undergo a change in a mechanism, the consequences will extend to the physiology and behaviour of other individuals in the population as well. For example, when a female macaque monkey comes into oestrus or is ovariectomised, the dynamics of the entire group change (Wallen and Tannenbaum 1997).

2.6.1 AGGRESSION

One of the key individual behaviours underlying group level phenomena is aggression. It has long been recognised that there is some relationship between gonadal sex steroids and aggressive behaviour in a diverse array of vertebrates (Svare 1983). This is particularly the case for types of male aggressive behaviour that coincide with the breeding season, seem to have a function related to reproduction (e.g., access to mates), and occur more frequently or more intensely in males than in females. Experiments with a wide variety of wild and free-ranging animals confirm that testicular androgens or their brain-produced oestrogenic and androgenic metabolites are the most important class of aggression related hormones not only for overt fighting by males, but also for ritualised threat displays and other forms of aggressive communication (Svare 1983, Wingfield and Farner 1993, Houck and Woodley 1995). Gonadal hormones are involved in fish aggression as well (Hoar 1962, Baggerman 1966, Smith and Hoar 1967). As with any general statement about mechanisms of social behaviour, however, there are always exceptions. Furthermore, much less is known about the hormonal basis of female aggression even in species where females are quite aggressive (see Section 2.9.3 for maternal aggression in mammals).

2.6.1.1 Territoriality

There are now a number of experiments in which males have been castrated (or given an antiandrogen) or treated with testosterone and were later found to have smaller or larger territories, respectively (Moore and Lindzey 1992, Wingfield and Farner 1993). In the most dramatic cases, male birds with reduced androgen action lost their territories and males with enhanced testosterone not only expanded their territories but also acquired additional female mates.

In some species of fishes and reptiles, alternative male types (morphs) differ in territorial behaviour. In several different fish species with two male types, territorial (type I) male fish have higher 11-ketotestosterone levels than nonterritorial (type II) males (Brantley *et al.* 1993), e.g., stoplight parrotfish (*Sparisoma viride*), which also show social modification of androgen levels (Cardwell and Liley 1991). Male tree lizards (*Urosaurus ornatus*) can be either aggressively territorial and site faithful or less aggressive and less site faithful or even nomadic. This behavioural polymorphism is associated with morphological polymorphism (size and dewlap colour) and with different hormonal profiles in response to dry conditions. Manipulations, e.g., castration or administration of testosterone or progesterone at hatching, alter the ratio of the two male morphological types, raising the interesting possibility that the behavioural and hormonal components of the two alternative phenotypes might be hormonally organised as well (Moore *et al.* 1998).

2.6.1.2 Dominance

The relationship between hormones and dominance is complex. One reason for this complexity is species differences in whether hormone-dependent physical characteristics, e.g., size in fishes or antlers in ungulates, play a major role in determining dominance rank (Lincoln *et al.* 1972, Metcalfe *et al.* 1992). Another reason is that hormone levels are labile, responding rapidly to engagement in an aggressive interaction (Sapolsky 1992, Wingfield and Marler 1994). Hormone levels are more likely to be related to rank during periods of instability (change in ranks), when much overt aggression is occurring, than during periods of stability, when fewer aggressive encounters are occurring (Sapolsky 1992, Wingfield and Farner 1993). Furthermore, dominance itself is a relationship between individuals that know each other, not an inherent property of an individual that it carries with it throughout life. It has been known for many years that experimental manipulations of testosterone change the dominance status of male chickens only if the males are placed into new groups. No change in rank occurs following testosterone treatment if the animals remain in their existing group where they already know each other as individuals, have a past history together, and social inertia prevents changes in rank (Harding 1983). This phenomenon is widespread, occurring in insects (Trumbo 1996), birds (Wiley *et al.* 1999), ungulates (Bouissou 1983) and primates (Zumpe and Michael 1996). In short, testosterone seems to be permissive (perhaps even necessary) for males to achieve high rank, but is by no means sufficient. Sensitivity to social context and history is likely to be a general feature of the relationship between hormones and aggression, and perhaps other social behaviours as well (Wallen and Schneider 2000).

In male cichlids (*Haplochromis burtoni*) hypothalamic GnRH-containing neurons are significantly larger in dominant males; they become smaller when position within the hierarchy drops (Soma *et al.* 1996). Social interactions were also important in regulating urinary concentrations of testosterone, 11-ketotestosterone, 17α,20α-dihydroxy-4-pregnen-3-one-20-sulphate (17,20-αP), and 17α,20β-dihydroxy-4-pregnen-3-one-20-sulphate in male cichlids. Whereas there was a positive correlation between social status and urinary sex steroid concentrations, the addition of ovulating females resulted in a return to levels similar to isolated males, except for 17,20-αP which increased (Oliveira *et al.* 1996).

The relationship between adrenal corticosteroids and aggression and dominance is also complex. While these steroids facilitate aggression in animals such as teleost fishes and domestic mice (Leshner 1983, Munro and Pitcher 1983), they decrease it in lizards and sparrows (DeNardo and Licht 1993, Wingfield *et al.* 1998). Depending on the social system, either low or high rank can itself be stressful and, like other stressors, capable of inducing changes in energy balance (Creel *et al.* 1996, Fox *et al.* 1997). While rainbow trout α-males expend little energy to hold a breeding territory, subordinate males must expend higher amounts of energy obtaining equal amounts of food in less desirable sections of the stream. Conversely, pupfish α-males must expend more energy defending a nest site from egg-eating and satellite male conspecifics (Feldmeth 1983).

2.6.1.3 Other hormones; other kinds of aggression

Several other hormones have been implicated in aggression as well. ACTH stimulates aggression in tilapia, mice and hamsters (Leshner 1983, Munro and Pitcher 1983). Thyroxine stimulation of locomotor activity indirectly increases aggression in that higher levels of activity increase the chances for social contact. This may also be why growth hormone affects the intensity of aggressive encounters between juvenile rainbow trout (Jonsson *et al.* 1998). Conversely, pineal secretion of melatonin may decrease aggression because of its sedative effects.

Not all aggressive behaviour would be expected to require a special hormonal regulatory mechanism. Most animals, regardless of age, sex, or reproductive status, will defend themselves against threats and attacks from conspecifics or predators, or (in the case of predatory species) will hunt and kill for food.

2.6.1.4 Neurochemistry of aggression

Investigations of the neurochemistry of aggressive behaviour have thus far relied heavily on laboratory rodent models. All of the well-studied neurotransmitters have been found to be related to aggression, based on experiments using pharmacological manipulations (Albert *et al.* 1993, Haller *et al.* 1998, Parmigiani *et al.* 1998). These relationships are not always specific to aggression or to one particular kind of aggression, and the same manipulation will sometimes increase aggression and sometimes decrease it, depending on the species and the type of aggression. Some of this inconsistency may be due to drugs that interact with multiple neurotransmitter receptor types that have different anatomical distributions, or to time of day. The development of more receptor specific drugs and their administration to selected brain regions implicated in aggression, e.g., the lateral septum, amygdala, certain hypothalamic nuclei, and the periaqueductal grey area, may help clarify the picture (Albert *et al.* 1993, Weiger 1997, Haller *et al.* 1998, Simon *et al.* 1998).

A particular focus of recent attention has been the indoleamine serotonin. Manipulating serotonin levels by systemic drug treatment affects aggressive behaviour in both invertebrates, e.g., lobsters and crayfish, and vertebrates, e.g., teleost fishes, domestic cats, laboratory rodents and monkeys (Weiger 1997). Differences occur in brain serotonin between salmonids of differing ranks (Winberg and Nilsson 1993).

Dopamine and the invertebrate compound octopamine can also act as a modulators for aggression along with serotonin. When injected into squat lobsters (*Munida quadripina*) both serotonin and octopamine induce postures similar to subordinate individuals (Antonsen and Paul 1997). That this response is different in other crustacea indicates that species differences exist for the role of these substances in motor output and behavioural control. In snails serotonin also facilitates defensive behaviours, while dopamine, acting presynaptically on defensive behaviour command neurons to increase their threshold level for stimulation, suppresses this activity (Chistyakova 1989). The balance between these two neurotransmitters' actions results in a behavioural 'decision'.

Peptides, e.g., AVP and its nonmammalian homologue AVT, may act as neuromodulators of aggression. In birds, AVT sometimes increases male aggression and sometimes decreases it. Which effect occurs is predictable from the social organisation of the species, with colonial species such as zebra finches (*Taeniopygia guttata*) that tolerate other males in relatively close proximity showing an increase and territorial species such as field sparrows (*Spizella pusilla*) and violet-eared waxbills (*Uraeginthus granatina*) that are highly intolerant of nearby males showing a decrease (Goodson 1998). In male hamsters, AVP facilitates aggression (including flank marking, an aggressive scent marking behaviour) while serotonin may antagonise the AVP system (Ferris *et al.* 1997).

2.6.2 MATING SYSTEMS AND MATE CHOICE

Comparative studies of amphibians, birds, and rodents have found different circulating hormone levels or brain neuromodulators in species with different mating systems (Wingfield and Farner 1993, Houck and Woodley 1995, Emerson and Hess 1996, Young *et al.* 1998). Experiments with fishes, passerine birds and voles lend support to the idea that hormones and neuromodulators may actually be mechanisms producing some of the striking diversity in mating systems seen in animals.

In cichlids (*Aequidens pulcher*) dopamine system activation reduces male–female pair formation while norepinephrine activation facilitates such behaviour (Nechaev 1993). Male song sparrows (*Melospiza melodia*) and white-crowned sparrows (*Zonotrichia leucophrys*) given testosterone implants acquired second female mates (were polygynously mated), unlike controls who were monogamously paired (Wingfield and Farner 1993). This is interesting in light of correlational studies showing a longer period of elevated testosterone during the breeding season in males of polygynous than monogamous species (Beletsky *et al.* 1995). Male prairie voles (*Microtus ochrogaster*) are socially monogamous and form pairbonds, whereas meadow voles (*Microtus pennsylvanicus*) are polygynous and do not form pairs. In both voles and deermice (*Peromyscus*) species differences in monogamy/biparental care vs. polygyny/maternal care are correlated with species differences in AVP-immunoreactive fibre staining in the lateral septum; pairing stimulates an increase in the density of these fibres in male prairie voles (Young *et al.* 1998, Bester-Meredith *et al.* 1999).

Sex steroids are known to influence mate choice (Adkins-Regan 1998). Animals with the right hormone levels are preferred as mating partners because they convey signals (odours, appearance or behaviour) that are attractive (e.g., Enstrom *et al.* 1997). Activational effects of sex steroids are known to be an important determinant of the strength of mating preferences. They are also responsible for seasonal changes in social preferences in female mammals which prefer the company of females in the nonbreeding season and of males in the breeding season (Ferkin and Zucker 1991, Michael and Zumpe 1993). Organisational effects of sex steroids have been shown to be an important part of the developmental processes leading to adult sexual partner preference (preference for males or females as mating partners) in both birds and mammals (Adkins-Regan 1998).

2.6.3 COMMUNICATION

Behaviour that communicates is a fundamental individual level process creating group level social organisation. Animals have many ways to communicate with each other. Both production and reception of communicative signals have been studied in some detail in several animal groups, leading to insights about both hormonal and neurochemical mechanisms involved.

2.6.3.1 Communication by chemosignals

The use of olfactory, gustatory and other chemical senses to derive information from conspecific chemical cues is widespread with the possible exception of birds (Mason 1992, Brown and MacDonald 1995, Houck and Woodley 1995, Sorensen *et al.* 1998). Some of these chemical cues are specific molecules that are released or deposited via distinctive behaviours and that reliably elicit relatively specific behavioural responses in receivers. These are often referred to as pheromones. In some cases these pheromones are similar chemically to internal chemical messengers such as steroids or neuropeptides, e.g., the sulphated sex steroid metabolites of fishes. Not all chemosignals fit neatly into conventional definitions of a pheromone, however (Johnston 1998).

Regardless of whether the chemosignal is a pheromone, the production, dissemination and reaction to chemosignals are often hormone dependent processes in animals, especially in the case of chemosignals with direct or indirect reproductive functions. Gonadal sex steroids are key stimulators of scent marking behaviour in those mammals that have been studied, and animals in a hormonal state of reproduction spend more time sniffing scent marks (Gandelman 1983, Yahr 1983, Takahashi 1990).

Fishes, like other aquatic animals, make extensive use of chemical signalling. Chemical alarm pheromones (*Schreckstoff*, possibly hypoxanthine-3(N)-oxide; Pfeiffer *et al.* 1985) may be released by mechanical damage to epidermal club cells to warn other ostariophysan fishes about the presence of a predator. Experiential learning by young brook stickleback (*Culea inconstans*) seems to be important in deriving information from conspecific skin extracts, e.g., training naive fish to associate a neutral visual stimulus, a predatory species (pike), or a nonpredatory species (goldfish) with skin extract, a conditioned response that remains for up to two months. Interestingly, finescale dace (*Chrosomus neogaeus*) and fathead minnow were found to benefit from the stickleback skin extract, indicating interspecific communication (Chivers and Smith 1994a, b).

In adult male Atlantic salmon, but not juveniles or adult females, testosterone is a potent odorant (Moore and Scott 1991). In rainbow trout levels of urinary 17α,20β-dihydroxy-4-pregnen-3-one-20-sulphate (17,20-P) were highest in unovulated females, testosterone glucuronide (T-G) was highest in ovulating females, and oestradiol glucuronide (E-G) was highest in ovulated females, regardless of male presence. However, male excretion patterns were affected by the presence of a female; 17,20-P and T-G increased significantly when males were paired with an ovulated/nesting female (Scott and Liley 1994). This is important because exposure to 17,20-P induces male spawning behaviour and sperm motility (Defraipont and Sorensen 1993).

In goldfish, female urine contains chemicals produced in the kidney under the influence of oestradiol that induce male prespawning behaviour (Yamazaki 1990). Pheromones may also be used by goldfish to prevent sexually active males from courting nonovulatory females. Spermiated male goldfish appear to avoid the 17,20-P released by preovulatory females (Bjerselius *et al.* 1995). The olfactory system (cranial nerve 1) rather than the terminal nerve (cranial nerve 0) mediates the chemosensory response to pheromones in goldfish (Fujita *et al.* 1991). The parts of the goldfish olfactory system that are sensitive to pheromones (the medial olfactory tracts) are separate from those used for general odour detection (the lateral olfactory tracts) (Dulka 1993). These regions apparently innervate distinct regions in the brain. Odours associated with reproductive activity (sex steroids and prostaglandins) are processed by the medial olfactory system and feeding associated odours (amino acids, organic acids, nucleotides and bile salts/acids) by the lateral system (Hara 1994).

Male terrestrial salamanders (e.g., *Plethodon* and *Desmognathus*) produce proteinaceous courtship phermones from specialised glands that develop during the breeding season; these pheromones increase female receptivity and keep her near the male long enough for him to deposit a spermatophore (Rollmann *et al.* 1999). There also may be territorial chemosignals in terrestrial salamanders. Male aquatic newts (*Cynops pyrrhogaster*) release a sex attractant from the cloaca, a decapeptide (sodefrin) produced by the abdominal gland (Kikuyama *et al.* 1997). Female rough-skinned newts (*Taricha granulosa*) produce a molecule that elicits clasping by males (Moore 1999).

Mammals, snakes, lizards and amphibians have been known for some time to have two anatomically distinct olfactory systems, a main olfactory system originating in the main olfactory epithelium, and an accessory olfactory system originating in the vomeronasal organ epithelium (Wysocki 1979, Halpern 1992, Dulka 1993). The accessory system mediates important behavioural responses to communicative, especially sexual, odours in snakes and lizards, hamsters and house mice (Wysocki 1979, Halpern 1992, Halem *et al.* 1999), but does not have this function in all vertebrates or all mammals (Dorries *et al.* 1997, Johnston 1998).

2.6.3.2 Communication by acoustic signals

The courtship vocalisation of midshipmen fish (*Porichthys notatus*) is modulated by vasotocin (AVT) and may be facilitated by 11-ketotestosterone (Brantley *et al.* 1993, Goodson and Bass 2000). Male frogs and toads are well known for their distinctive calls to which females are strongly attracted. Steroids, e.g., testosterone or 5α-dihydrotestosterone, and the peptide AVT stimulate calling by males in several species of frogs and toads (Boyd 1997). Female phonotaxis (approach to a speaker playing male calls) in *Xenopus*, American toads (*Bufo americanus*) and bullfrogs (*Rana catesbeiana*) is stimulated by AVT and by hormones such as human chorionic gonadotropin, prostaglandins or progesterone (Picker 1983, Schmidt 1985, Boyd 1997). Male and unreceptive female frogs emit release calls when clasped by a male attempting to mate. In the bullfrog AVT decreases release calling in females while increasing it in males (Boyd 1997); both effects are increases in sex-appropriate mating behaviour. In the frog *Xenopus* there are steroid receptors throughout the vocal control pathway (Kelley 1997). In *Rana* AVT is present in the vocal control areas, sexually dimorphic and steroid regulated (Boyd 1997). Taken together, these behavioural and neurochemical studies suggest that testosterone stimulates calling via AVT.

Birds communicate extensively by songs and other vocalisations. Where these vocalisations occur mainly during the breeding season or are emitted only by one sex (e.g., male song in many temperate zone species), they are often dependent to some degree on gonadal sex steroids (Brenowitz and Kroodsma 1996). Most male galliform (chicken-like) birds emit loud long-distance signals (crowing). These calls completely disappear following castration, and are restored by administration of testosterone (Balthazart 1983), indicating strict dependence of signalling on testicular hormones. Both androgenic and oestrogenic metabolites of testosterone are required for normal male-typical singing and song display in redwings (*Agelaius phoeniceus*) and zebra finches (Harding 1992). In galliforms, 5α-dihydrotestosterone is more potent than oestradiol, but testosterone is more potent still, again suggesting that combinations of metabolites are normally involved (Balthazart 1989).

Major progress has been made identifying some of the brain regions that are involved in the production, learning and perception of songs by oscine passerine birds (the largest subgroup of the order Passeriformes) (Figure 2.11) (Arnold *et al.* 1987, Nottebohm 1991, Brenowitz and Kroodsma 1996). Forebrain song system nuclei contain many of the major neurotransmitters and some of the peptides in Table 1.1 (Harding 1992, Barclay *et al.* 1996). Except for Area X, they also contain androgen receptors, and nuclei HVC and RA are located in a part of the brain that is rich in aromatase (Schlinger and Arnold 1995, Balthazart and Ball 1998, Metzdorf *et al.* 1999). Sex steroids influence not only the overall size of most of these nuclei but also (depending on the specific nucleus) the size and dendritic arborisation of individual neurons and the levels and turnover of norepinephrine and dopamine (DeVoogd 1986, Barclay and Harding 1990).

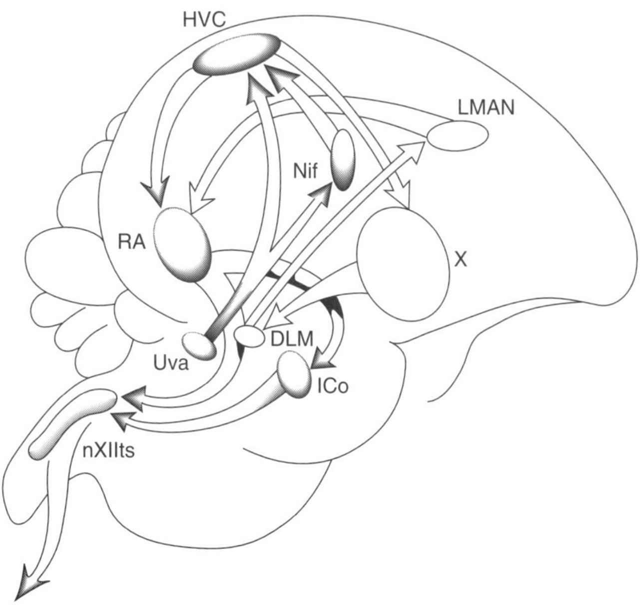

Figure 2.11 The neural song system of passerine birds. The output is to the syrinx (the vocal organ). HVC, high vocal centre; RA, robustus archistriatalis; IMAN lateral part of the nucleus magnocellularis of the anterior neostriatum; X, Area X; DLM, medial part of the dorsolateral nucleus of the anterior thalamus; ICo, nucleus intercollicularis; nXIIts, tracheosyringeal portion of the hypoglossal (cranial nerve XII) nucleus. Original source: Figure 1 in Casto and Ball (1996) *Hormones and Behaviour* (vol. **30**, pp. 387–406), © 1996 by Academic Press, reproduced by permission of the publisher. All rights of reproduction in any form reserved.

Thus, the song system has many notable characteristics that are correlated with important features of song behaviour, such as sexual dimorphism and steroid responsiveness.

Neural mechanisms of vocal behaviour have also been examined in a few nonpasserine birds. A small parakeet, the budgerigar (*Melopsittacus undulatus*), has a vocal control system which, by analogy to that of passerines, also includes sizeable telencephalic nuclei containing some of the same neurotransmitters (Ball 1994, Striedter 1994, Cookson *et al.* 1996). In the ring dove (*Streptopelia* sp.) there is evidence that preoptic - hypothalamic norepinephrine is an inhibitory regulator of the bow - coo and nest - coo displays (Barclay and Cheng 1992). Several lines of evidence point to the intercollicular nuclei in the midbrain (Figure 2.11) as an important steroid sensitive vocalisation area in ring doves and galliform birds (Balthazart and Ball 1993, Cheng 1993, Beani *et al.* 1995, Yazaki *et al.* 1999).

2.7 SEXUAL BEHAVIOUR

Sexual behaviour includes copulation and other mating behaviours that serve to bring gametes together, plus those courtship activities that are closely tied temporally to mating behaviour, especially courtship displays immediately preceding copulation. Many communicative and social behaviours discussed in the previous section may have direct or indirect sexual functions as well. Chemical regulators of sexual behaviour include chemosignals as well as hormones and neurochemicals.

2.7.1 HORMONAL MECHANISMS

In contrast to vertebrates, the hormones that have been implicated in insect sexual behaviour do not come from the gonads and are not steroids (Nijhout 1994). Rising levels of juvenile hormones stimulate sexual receptivity in females of species in several different orders; a similar hormonal requirement is seen in some males as well. In molluscs, eicosanoids exert species-specific effects on reproductive behaviour, e.g., barnacle hatching and bivalve spawning (Stanley-Samuelson 1994).

Most insects use pheromones to locate and attract copulatory partners (Raina 1993, Sorensen *et al.* 1998). Sex pheromone production by moths and butterflies is regulated by juvenile hormones or by PBANs (pheromone biosynthesis activating neuropeptides — a family of similar peptides with 15 - 33 amino acids) from the retrocerebral complex (Raina 1993, Nijhout 1994). Juvenile hormones increase sensitivity to the female produced sex pheromone in the male moth *Agrotis ipsilon* (Anton and Gadenne 1999).

Vertebrate gonads, unlike those of many invertebrates, produce both gametes and hormones regulating the behaviour that will bring gametes together. The same gonadal sex steroids that are conducive to gamete maturation and release (spermatogenesis, follicular development and ovulation) also activate mating behaviour, perhaps because mating behaviour has some costs that necessitate ensuring that it will coincide with availability of mature gametes. In a wide array of vertebrate species, both wild and domestic, removal of the testes or ovaries of reproductively active animals reduces or eliminates courtship and mating behaviour. Sex steroid replacement (testosterone in males, oestradiol or oestradiol plus progesterone in females) restores the behaviour (Balthazart 1983, Gans and Crews 1992, Nelson 2000). Although castration studies in fishes are difficult because incomplete castration allows the remaining cells to excrete sufficient hormones to induce reproductive behaviour (Stacey 1983), those that have been successful demonstrate the role of gonadal hormones in regulating male sexual behaviour. These behaviours are restored by administering the synthetic androgen methyltestosterone (Hoar 1962, Baggerman 1966, Smith and Hoar 1967).

In some cases mating behaviour is completely dependent on gonadal hormones; gonadectomised animals never again court or copulate. In others, gonadal hormones are permissive, allowing mating to occur if other conditions are appropriate, or are facilitatory but not absolutely essential. As with aggressive behaviour, social context and prior relationships affect the hormone - behaviour relationship (Wallen and Schneider 2000). Even when these are taken into account, however, there seem to be some real differences among species in the degree of coupling between mating and gonadal hormones, although most of these differences are not well understood. One possible explanation when gonadectomy does not eliminate mating behaviour is that sex steroids produced elsewhere (including the brain) are supporting the behaviour. Another possibility is that the behaviour is partially uncoupled from the physiology of gamete maturation because it serves functions in addition to fertilisation (e.g., mate assessment or pair formation) and overlaps with but is not temporally limited to the period of gamete maturation. The most extreme uncoupling of mating from gamete maturation occurs in species with dissociated reproduction (see Section 2.2.2).

In males of many species that have been studied (including newts, lizards, birds and mammals) both oestrogenic and androgenic metabolites of testosterone acting via oestrogen and androgen receptors activate courtship and mating (Meisel and Sachs 1994, Nelson 2000). Each individual component (such as a particular display or copulatory motor act) may depend more on one metabolite (e.g., 5α-dihydrotestosterone or oestradiol), but a combination of metabolites is required for the full male behaviour repertoire to be expressed.

In male teleost fishes 11-ketotestosterone is more effective than testosterone in stimulating secondary sexual characters, reproductive behaviour and spermatogenesis. 11-ketotestosterone is not aromatizable, and its effectiveness for behaviour is therefore surprising given the importance of androgen aromatisation in other vertebrates. Interestingly, in many teleost fishes significant differences exist in the plasma levels of males and females only for 11-ketotestosterone, not testosterone. However, receptor-like binding in specific organs has been identified only for testosterone and not yet for 11-ketotestosterone (Borg 1994).

Female sexual behaviour and courtship (proceptivity) are often activated or at least facilitated by ovarian oestrogens (Nelson 2000). Some, but not all, species (e.g., lizards, *Anolis carolinensis* and some female mammals) require progesterone as well, either oestrogen followed by progesterone (e.g., rats and hamsters) or progesterone followed by oestrogen (e.g., sheep) (Whittier and Tokarz 1992). Where females have relatively high levels of circulating testosterone, it may contribute to receptivity via aromatisation (Rissman 1991, Moore *et al.* 1994).

Small implants of sex steroids placed in the brains of gonadectomised male and female lizards, quail and rodents can activate sexual behaviour with a latency of several days (Moore and Lindzey 1992, Meisel and Sachs 1994, Panzica *et al.* 1996). The medial preoptic area (MPOA) is often the most effective site for male sexual behaviour, whereas the ventromedial hypothalamus is often the most effective site for female sexual behaviour. This is evidence for the causal role of brain steroid receptors in these regions in hormonally stimulated sexual behaviour. The behaviour is usually not as complete, vigorous or frequent as in animals receiving systemic sex steroid treatment, both because peripheral actions of the steroids are missing and because normally multiple hormone sensitive brain regions would be involved.

The hormone sensitive circuitry for sexual behaviour is best understood for the female rat and the male hamster (Figure 2.12). Male hamsters require both testosterone and female odours to mate; either castration or interruption of the accessory (vomeronasal) olfactory system prevents mating. These two circuits (hormonal and chemosensory) are anatomically somewhat separate and operate

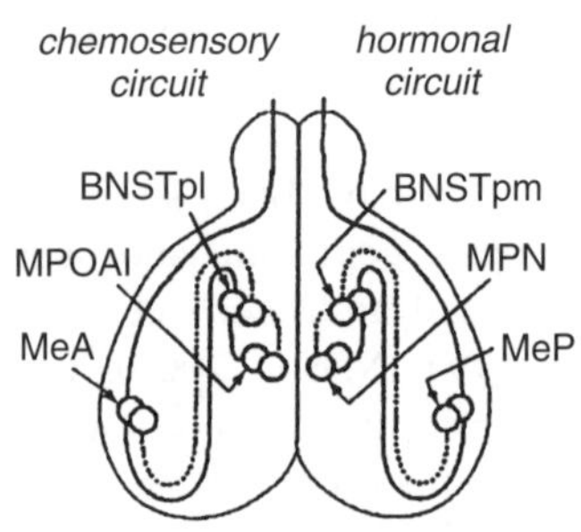

Figure 2.12 Forebrain circuitry for the sexual behaviour of the male Syrian hamster. Anterior is at the top. BNST, bed nucleus of the stria terminalis (BNSTpi, posterointermediate subdivision; BNSTpm, posteromedial subdivision); MPOAl, medial preoptic area (lateral subdivision); MPN, medial preoptic nucleus; Me, medial amygdaloid nucleus (MeA, anterior subdivision; MeP, posterior subdivision). Original source: Figure 1 in Wood and Newman (1995b) *Journal of Neuroscience* (vol. **15**, pp. 7261–7269), © 1995 Society for Neuroscience, reprinted by permission.

in parallel, but must be able to interact anatomically in order for mating to occur (Wood and Newman 1995a,b).

Prostaglandins (PGs) are also involved in sexual behaviour. Ovulation induces spawning behaviour in fishes with external fertilisation, e.g., goldfish, an effect apparently mediated by PGs (Stacey and Goetz 1982, Stacey 1983). PGs also facilitate receptivity in some rodents. However, they inhibit receptivity in crickets (*Teleogryllus commodus*), green anoles (*Anolis carolinensis*), garter snakes (*Thamnophis sirtalis parietalis*) and female guinea pigs (*Cavia porcellus*), and may be a mechanism for the postmating termination of receptivity that occurs in these species (Whittier and Tokarz 1992).

2.7.2 NEUROCHEMICAL MECHANISMS

In marine bivalves, sexual behaviour is relegated to broadcast release of eggs and sperm. Serotonin induced spawning in males and females of two scallop species, *Chlamys farreri nipponensis* and *Patinopecten yessoensis*, with PGs modulating that effect, at least *in vitro* (Matsutani 1990). Males were more sensitive to serotonin than females and, correlating with the ability of norepinephrine also to induce spermiation, only males displayed an increase in gonadal norepinephrine. Serotonin had an indirect effect on sexual activity and gonadal maturation of male and female sand fiddler crabs and red swamp crayfish (*Procambrus clarkii*) by first stimulating the release of gonadal-stimulating hormone (GSH) from brain and thoracic ganglia. In males serotonin also stimulates androgenic gland development. In both sexes dopamine inhibits sexual development. In females methionine enkephalin slows ovarian growth, while red pigment-concentrating hormone, using calcium as a second messenger, stimulates ovarian maturation without affecting the ovary directly (Fingerman 1997). The courtship display of blue crabs (*Callinectes sapidus*) requires the interaction of dopaminergic, serotonergic, octopaminergic and proctolinergic systems. Dopamine is also involved in female sexual receptivity in *Drosophila* (Neckameyer 1998). In the amphipod *Gammarus lacustris* serotonin stimulates the clinging behaviour involved in sexual activity while octopamine suppresses it (Helluy and Holmes 1990).

Several factors, both environmental and neural, affect seasonal egg-laying behaviours in nudibranchs by altering the function of the neuroendocrine α-bag cell. At colder temperatures, a specific neuropeptide hyperpolarised the bag cells, shortened the length of synaptically evoked bag cell discharges and decreased the number of action potentials for each discharge. The opposite effect occurred at warmer temperatures (Redman and Berry 1991). This autoinhibitory–autoexcitatory system allows the animal to respond to changes in ocean temperatures and time its egg laying accordingly (see Begnoche *et al.* 1996 for additional information about the peptidergic system). During egg laying *Aplysia* feeding behaviour is suppressed (Ter Maat and Ferguson

1996). Homologues to the *Aplysia* bag cell neuropeptide control of egg-laying behaviour are seen in the intercerebral white cells of the nudibranch *Archidoris montereyensis* and the caudodorsal cells of the pond snail *Lymnaea stagnalis* (Wiens and Brownell 1994).

Pheromones are important in the reproductive behaviour of several marine and freshwater invertebrates. Newly oviposited *Aplysia* eggs are a source of waterborne chemicals that induce breeding aggregations for this normally solitary animal and facilitate male–female coordination of sexual activities within the group (Painter 1992). The albumen gland, the site for packaging the egg clusters, is the source of this peptide pheromone (Fan *et al.* 1997). A related product, a carboxyl terminal arginine peptide, expressed in the bag cells acts as a neurohormone and nonsynaptic neurotransmitter to initiate egg laying and coordinate reproductive behaviours, e.g., body movements associated with egg laying (Bernheim and Mayeri 1995) and synchronised larval release (Rittschof *et al.* 1990). Sex pheromones released in the urine of female lobsters (*Homarus americanus*) and crabs (*Carcinus maenas*) stimulate receptors in the male antennules and also initiate the female's own sexual activity (Cowan 1991, Bamber and Naylor 1996).

Precopulatory activity in simultaneous hermaphrodites, e.g., the pond snail, requires a complex series of behaviours involving the interplay of peptidergic CNS neurons that inhibit penile retractor muscle contractions and serotonergic neurons that facilitate the behaviour (Li *et al.* 1992). Immunocytochemical analysis of the male copulatory apparatus show at least ten separate messenger molecules in the neurons that innervate it (De Boer *et al.* 1997). A positive correlation exists between the level of male activity and size of the prostate gland (De Boer *et al.* 1997).

All the major neurotransmitters and many of the neuromodulators in Table 2.1 have been shown in those few species that have been studied to be involved in mating behaviour and its hormonal regulation (Ottinger and Balthazart 1987, Balthazart and Ball 1992, Meisel and Sachs 1994, Micevych and Hammer 1995, Weiger 1997). Some tend to be predominantly facilitatory (e.g., dopamine) while others are predominantly inhibitory (e.g., serotonin, opioids). Direct evidence comes from pharmacological manipulations (systemic or intracranial drug administration) that increase sexual behaviour in castrated animals or decrease it in castrates receiving sex steroid replacement, thus ruling out the possibility that the drug has worked by altering gonadal sex steroid production. Indirect evidence comes from biochemical studies of sex steroid regulation of neurochemicals and their receptors and anatomical studies finding that the neurochemicals and their receptors are located in the same brain regions or neurons as sex steroids and their receptors (see Section 2.1.1.3). Much current research with laboratory animals is directed at determining which receptor subtypes in which specific brain regions are involved in copulation and in developing systems-level models relating the neurochemistry of different brain regions to the behavioural processes involved in copulation (e.g., arousal, motivation and reward) (Hull 1995).

There has been increasing interest in the role of both central and peripheral nitric oxide in sexual behaviour. Its importance in the peripheral vascular mechanisms of penile erection in primates is well established, but its central actions extend to nonmammalian species and even insects (Elphick *et al.* 1994, Hull 1995, Zerani and Gobbetti 1996).

Peptides, e.g., AVT, oxytocin, and GnRH, are part of the set of mechanisms responsible for activation of sexual behaviour in several vertebrates that have been studied thus far (Dornan and Malsbury 1989, Carter 1992). GnRH stimulates clasping in sexually inactive male great plains toads (*Bufo cognatus*) (Propper and Dixon 1997), AVT stimulates egg-laying behaviour in female rough-skinned newts (*Taricha granulosa*) (Moore *et al.* 1994), and AVT in combination with T stimulate mating behaviour (clasping) in male newts, whereas corticosterone, CRH, ACTH, or $GABA_A$ inhibit clasping, turning it off when danger threatens (Figure 2.13) (Moore and Orchinik 1991, Moore *et al.* 1994). This rapid inhibitory effect of corticosterone occurs via membrane receptors. GnRH facilitates receptivity in steroid primed female lizards (*Anolis carolinensis* and *Iguana iguana*) (Whittier and Tokarz 1992), and centrally administered chicken

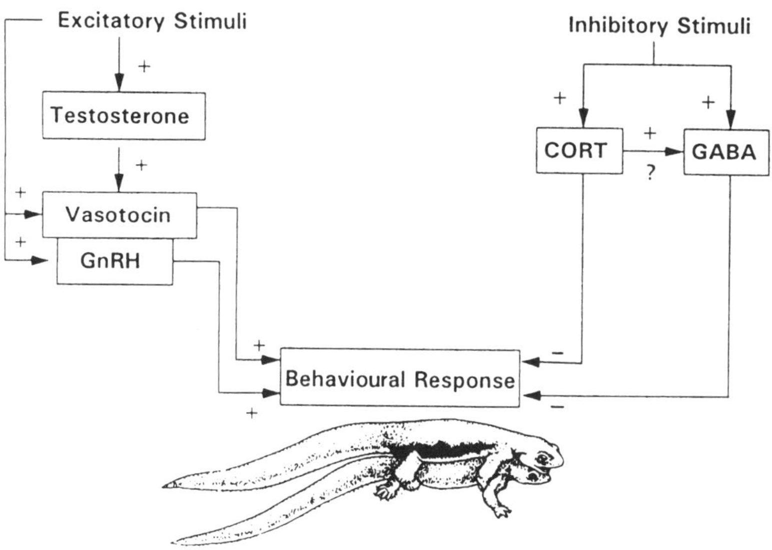

Figure 2.13 Hormonal and neurochemical regulation of sexual behaviour in the male rough-skinned newt (*Taricha granulosa*). In response to inhibitory stimuli such as environmental stressors, pathways involving corticosterone and GABA are activated that antagonise the excitatory pathways regulated by vasotocin and GnRH. See Table 2.1 for abbreviations. Original source: Figure 2 in Moore and Orchinik (1991) *Seminars in the Neurosciences* (vol. 3, pp. 489–496), © 1996 by Academic Press, reproduced by permission of the publisher. All rights of reproduction in any form reserved.

GnRH-II stimulates copulatory solicitation in female white-crowned sparrows (*Zonotrichia leucophrys*) (Maney *et al.* 1997).

2.8 SEX REVERSAL

Mating behaviour, like many of the other behaviours discussed in the previous sections, is often sexually dimorphic to some degree, as are some of its neural substrates (Panzica *et al.* 1996, Cooke *et al.* 1998). As discussed in Section 2.1.6, some of these sex differences originate through organisational actions of sex steroids. Oestrogens coming from the ovaries or from brain aromatisation of circulating androgens are involved in the process that creates sex differences in mating and other dimorphic behaviour in both birds and mammals (Gerall *et al.* 1992, Balthazart and Ball 1995, Cooke *et al.* 1998). Therefore, exposure of embryos (eggs) or very young animals to abnormal levels of steroids, especially oestrogens, or to chemicals affecting steroid action, especially oestrogen action, has the potential partially or completely to sex reverse the nervous system and behaviour, an effect which will be permanent (Christian and Gillies 1999).

2.8.1 SEX REVERSAL UNDER EXPERIMENTAL CONDITIONS

Injection of as little as 1 μg of oestradiol benzoate into Japanese quail eggs will result in males that all fail to mount as adults (Adkins 1979). Spraying bird eggs with oestrogens has the same effect, one that would clearly be devastating in a wild population (Adkins-Regan 1981). Much less is known about effects of early hormone exposure on adult behaviour of other nonmammalian vertebrates. The most systematic work has examined contributions of androgens to the development of sex differences in vocal behaviour in the frog *Xenopus* (Kelley 1997). In amphibians and fishes, experimentally produced gonadal sex reversal generally seems to be accompanied by behavioural sex reversal as well, but without further research one cannot know if this sex reversed behaviour is resulting from organisational actions of the treatment or from the sex reversed adult activational hormonal environment that is secondary to gonadal sex reversal (Adkins-Regan 1987). There is some evidence for organisational effects of gonadal hormones on the adult behaviour of leopard geckos (*Eublepharis macularius*) (Rhen and Crews 2000) and some evidence suggesting that in the garter snake (*Thamnophis sirtalis*) neonatal testosterone may organise the nervous system to respond in adulthood to the activational effect of rising temperature on courtship (Moore and Lindzey 1992).

2.8.2 NATURALLY OCCURRING SEX REVERSAL

Sequential hermaphroditism, or adult sex reversal, is common among teleost fishes. Although most species that show sequential hermaphroditism are protogynous, i.e., female first then male, with large, dominant males controlling access to the females, some protandrous species do exist. In the marine teleost *Pseudanthias squamipinnis*, removal of the dominant male induces the dominant female to change to a male; removal of the dominant male and female induces the second-ranking female to change to a male (Shapiro 1988). Importantly, a minimum number of females in the population is required for the process to proceed once the male is removed.

Because the gonads themselves change sex, factors influencing the gonads are critical in the process of sex change. Most researchers have, in fact, found that the HPG axis is involved. Grober and Bass (1991) implicated GnRH in the bluehead wrasse, *Thalassoma bifasciatum*. Sex reversal was directly related to the number or biosynthetic activity of GnRH-positive neuronal somata, especially in the preoptic area which is involved with sexual development, and their fibre projections. Furthermore, Grober *et al.* (1991) identified 11-ketotestosterone induction of increased GnRH cell numbers during the initial phases of sex change but not the terminal phase.

In the protandrous anemonefish (*Amphiprion melanopus*), a change in the plasma profile of cortisol and gonadal steroids occurred after the dominant female was removed (Godwin and Peter 1993). Androgen levels of sex-changing males initially decreased but eventually increased relative to control males. Oestrogen and cortisol levels did not change until nearly 3 weeks after female removal. Oestrogen levels did not increase until after oogonial proliferation and therefore were not related to the onset of female function. Although Guiguen *et al.* (1993) also found low androgen levels in transitional males of the protandrous sea bass (*Lates calcarifer*), oestradiol was high suggesting an important role in sex inversion for this species.

Although sex change is associated with changes in circulating steroid levels, gonadal sex steroids are not necessarily required in order for behavioural sex reversal to occur (Godwin *et al.* 1996). Other factors may play a role in sex reversal, e.g., changes in concentrations of specific pheromones due to the removal of a dominant male or female, prostaglandins and neuropeptide Y (Shapiro 1988, Kramer and Imbriano 1997).

2.9 PARENTAL BEHAVIOUR

Parental behaviour, like sexual behaviour, is often regulated by hormones, but there are important differences between these two behaviours and their relationships to hormones. Whereas many invertebrates and all vertebrates mate, so that one expects homologies at both behavioural and mechanistic levels, parental behaviour has a more disjunct distribution in the animal kingdom. The sex of the parent varies as well; mechanisms for parental behaviour might be different in male parents from in female parents, especially in animals such as mammals with gestation and lactation by females only (Brown 1985, De Vries 1995). Parenting may involve specialised external structures for producing infant food, e.g., epidermal mucus glands of discus fish (*Symphysodon discus*), pigeon and dove crop sacs, or mammary glands; the physiological regulators of these structures, such as prolactin, might also regulate parental behaviour (Blüm 1972, Hildeman 1975). Parental behaviour is not a single behaviour, but a coordinated sequence of rather different behaviours (e.g., compare incubation vs. posthatching care, or care of young vs. aggressive defence of the young). In species that do not mate and parent simultaneously, hormonal profiles are different during the two reproductive phases. In species with slow development, parenting may go on for weeks, months, or years. The objects of parenting (the young) grow, wean, fledge, etc. (i.e., change) in ways that require an adjustment of parental behaviour.

This need for flexible responsiveness to the needs of changing young helps explain two important outcomes of research on hormones and parental behaviour (Rosenblatt and Snowdon 1996). First, the behaviour is seldom strictly dependent on peripheral hormones. Instead, stimuli from the young, especially if repeated or extended in time, may be sufficient to elicit parental behaviour regardless of the age, sex, or hormonal status of the animals thus exposed. Second, this is particularly likely to happen when animals have had prior parental experience. These research conclusions have important implications for hypotheses about the mechanisms underlying alloparenting (helpers at the nest), in which individuals other than the parents, such as older offspring, assist with parental care (Schoech *et al.* 1996, Brown and Vleck 1998).

2.9.1 INSECTS, FISHES, AMPHIBIANS AND REPTILES

Little experimental work has been done on hormonal mechanisms of parental behaviour in amphibians or in parental reptiles such as crocodilians (Townsend *et al.* 1991, Houck and Woodley 1995). In fishes, prolactin is an important hormone in inducing egg fanning behaviour (Blüm and Fiedler 1965, de Ruiter *et al.* 1986). In cichlids prolactin induces behaviours directed to a distinct reference point, e.g., eggs or nest, and suppresses agonistic displays. FSH and LH inhibit the effects induced by prolactin; LH causes increased levels of aggression. In three-spined sticklebacks, implanted pituitary prolactin lobes induces fanning behaviour. Evoked potentials in the area dorsalis pars dorsalis of the telencephalon of sunfish and cichlids increased when prolactin was introduced (Fiedler and Blüm 1972). Because of the importance of the telencephalon in formation of associative connections (Karamyan *et al.* 1967), prolactin-sensitive neurons in this region may be important in coordinating parental care activities. Prolactin effects on parental care activities in male bluegills are mediated by dopamine (Kindler *et al.* 1991).

In parental insects, as in vertebrates, there are interesting correlations between changes in reproductive hormone levels and the occurrence of parental behaviour that deserve experimental investigation to see if the hormonal changes are causing the behaviour to appear, and, if so, whether this is an effect on the nervous system or on the periphery (Trumbo 1996). In earwigs (Dermaptera) levels of juvenile hormones and ecdysteroids are low during the early to mid-parental phase, whereas burying beetles *Nicrophorus orbicollis* have a surge of juvenile hormones at the onset of parental behaviour. Stimuli from the young both influence hormone levels and prolong parental behaviour, another notable parallel with vertebrates (Trumbo 1996).

2.9.2 BIRDS

The circulating hormonal profiles that coincide with different phases of parental behaviour (nest building, incubation, chick care) have been determined in a number of species of wild, mainly north temperate zone, birds (Wingfield and Farner 1993). A common pattern in females and in males of species with paternal care is for sex steroids to be relatively high during the nest building phase (when copulation and egg laying are also occurring) and relatively low during incubation (Figure 2.14), and for prolactin to rise as sex steroids fall (Figure 2.15). In birds with precocial young, e.g., galliforms and waterfowl, prolactin tends to be elevated during all of incubation and then fall when the chicks hatch (Figure 2.15). In songbirds, which usually have altricial young requiring direct feeding after hatching, prolactin may remain elevated after hatching. Doves, which have altricial young that are fed with prolactin stimulated crop milk, show a third pattern, in which prolactin rises toward the end of incubation and remains elevated during the chick feeding phase. In large penguins that spend long periods of time at sea before returning to care for young, prolactin remains elevated during the lengthy absences (Lormée *et al.* 1999). These correlations between sex steroids and nest building, between prolactin and incubation, and between prolactin and posthatching chick care make it very tempting to assume that sex steroids stimulate nest building and prolactin stimulates incubation and chick care.

The experiments necessary to test these ideas have been carried out in only a few species, however. Oestradiol or oestradiol plus progesterone stimulate nest building or other nest-orientated behaviour in female chickens, canaries and ring doves, especially if the social and environmental setting is appropriate (Balthazart 1983). Treatments that elevate testosterone in wild male birds reduce posthatching chick feeding, an effect which is consistent with the low levels of testosterone normally present during this phase (Oring *et al.* 1989, Ketterson *et al.* 1992, Wingfield and Farner 1993). In contrast, high testosterone does not interfere with paternal care by male prairie voles, but instead facilitates it (De Vries 1995). In chickens and turkeys oestradiol, progesterone and prolactin are candidates for stimulation of incubation (El Halawani *et al.* 1986, Buntin 1996). Oestradiol plus progesterone treatment produces a brief period of nest sitting which is sustained if prolactin is added; suppression of prolactin with bromocryptine interrupts already occurring incubation. In female ring doves, the

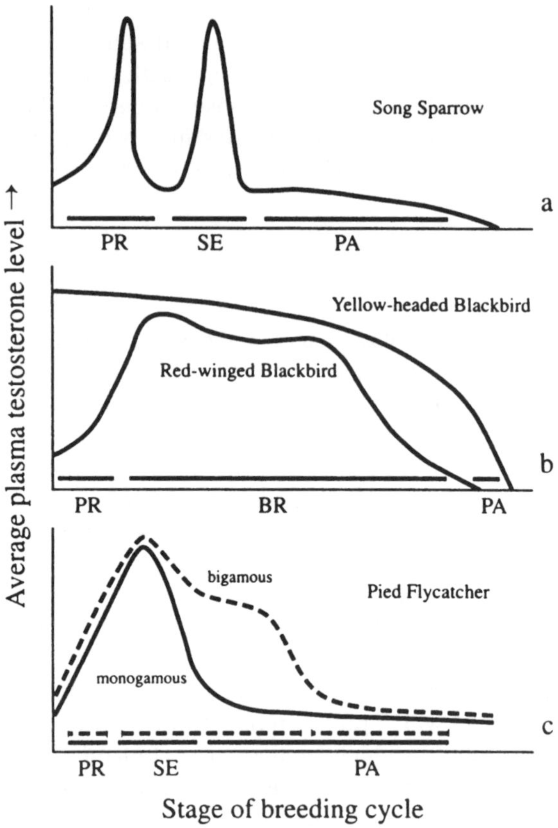

Figure 2.14 Testosterone profiles during the breeding season in free-living birds. (a) Song sparrows (*Melospiza melodia*); (b) redwings (*Agelaius phoeniceus*) and yellow-headed blackbirds (*Xanthocephalus xanthocephalus*); (c) monogamous and bigamous pied flycatchers (*Ficedula hypoleuca*). PR, prebreeding; SE, sexual; PA, parental; BR, breeding. Original source: Figure 1 in Beletsky *et al.* (1995) *Current Ornithology* (vol. **12**), © 1995 by Kluwer Academic/Plenum Publishers, reproduced by permission.

most effective hormone for stimulating the onset of incubation is progesterone (Buntin 1996). The posthatching chick feeding behaviour of both male and female ring doves can be stimulated by administration of ovine prolactin. A central action alone is sufficient to increase chick feeding behaviour, as shown by the effectiveness of intracranial administration of ovine prolactin (Buntin 1996). Ovine prolactin stimulates risky nest defence in female willow ptarmigan (*Lagopus lagopus*) (Pedersen 1989).

Stimuli from nests, eggs and chicks alter the hormone levels of birds, a mechanism for appropriate coordination of parental behaviour with offspring developmental stage. Exposure to chicks causes oestradiol to fall in chicken hens and testosterone to fall in male bobwhite quail (*Colinus virginianus*)

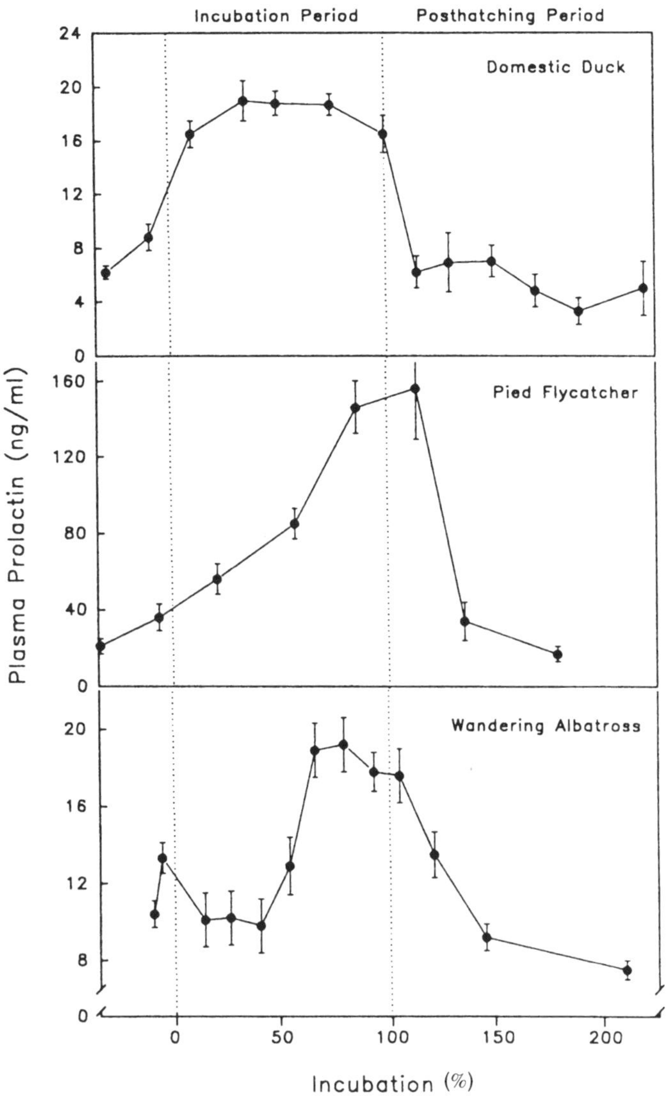

Figure 2.15 Prolactin in relation to incubation in domestic ducks, pied flycatchers (*Ficedula hypoleuca*) and wandering albatrosses (*Diomedea exulans*). Original source: Figure 2 in Buntin (1996) in Rosenblatt and Snowdon (eds) *Parental Care*, © 1996 by Academic Press, reproduced by permission of the publisher. All rights of reproduction in any form reserved. Data of Hall and Goldsmith (1983) Silverin and Goldsmith (1983) and Hector and Goldsmith (1985).

(Richard-Yris *et al.* 1983, Vleck and Dobrott 1993). Removing the nest and eggs is followed fairly quickly by a fall in prolactin and a rise in pituitary gonadotropins in several species, including polyandrous shorebirds where the male is the incubating parent (Wingfield and Farner 1993).

2.9.3 MAMMALS

In the Norway rat (*Rattus norvegicus*), the best studied species, hormones are critical primarily for inducing maternal behaviour immediately after the first litter of pups is born (Numan 1994, Bridges 1996). Oestradiol (which increases during gestation) and prolactin (which increases shortly before birth) have been shown to facilitate rapid onset maternal behaviour even when administered intracranially. Prior oestradiol priming is necessary for prolactin to have this effect. Progesterone, which decreases markedly shortly before parturition, is inhibitory. Once the female has been exposed to the pups for several days, the behaviour no longer requires gonadal or pituitary hormones. Furthermore, responses to subsequent litters will occur rapidly regardless of the female's hormonal condition, as if the nervous system has been permanently sensitised (Bridges 1996).

Experiments using manipulations, e.g., lesions, point to the MPOA, the bed nucleus of the stria terminalis (BNST), and the nucleus accumbens as some of the key brain areas underlying maternal behaviour in the rat (Numan 1994, Bridges 1996). Oestradiol and prolactin implants placed in the MPOA and the ventral BNST facilitate maternal behaviour. Neurons in these areas have both prolactin and oxytocin receptors. Several neurotransmitters and peptides are known to regulate maternal behaviour in rats (Figure 2.16).

Among the other mammals that have been studied experimentally are domestic rabbits, voles, domestic sheep and, to a very limited extent, macaque monkeys (see Maestripieri and Pryce 1996 and Zehr 1998 for macaques). Pregnant rabbits dig a nest burrow and line the nest with hair pulled from the ventrum. Oestradiol plus progesterone treatment of ovariectomised females stimulates digging, and bromocryptine treatment (which suppresses prolactin production) of intact females blocks hair pulling (González-Mariscal and Rosenblatt 1996). In the prairie vole (*Microtus ochrogaster*, a species with paternal care) AVP administration to the lateral septum increases paternal behaviour; an AVP antagonist decreases the behaviour (Wang and Insel 1996). The maternal behaviour of sheep appears to be the most tightly regulated by hormones and other physiological changes surrounding birth of any mammal yet studied (Lévy *et al.* 1996). Furthermore, the behaviour itself (acceptance of the lamb as indicated by sniffing, licking and vocalising) shows a pronounced imprinting phenomenon. Under normal circumstances the female must give birth and must be exposed to a lamb during the first 12 h after birth for her to accept it and behave maternally toward it. Recognition and acceptance of the lamb are based on chemosensory cues. The physiological milieu that has been shown experimentally to produce this critical period for imprinting includes recent steroid priming (increases in oestradiol and progesterone in late gestation), preoptic/hypothalamic oxytocin released when parturition mechanically stimulates the vaginocervical region, and main olfactory bulb norepinephrine (Lévy *et al.* 1996).

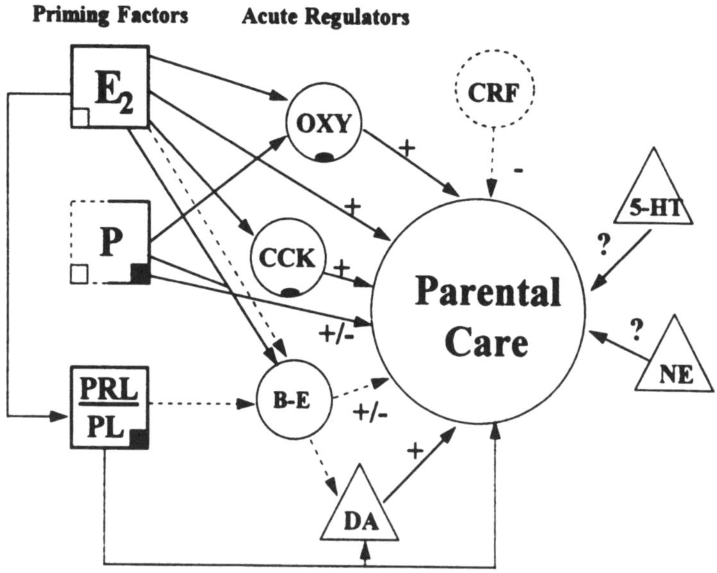

Figure 2.16 Hormonal and neurochemical regulation of maternal behaviour in the female rat. Solid lines represent stimulatory actions, while dashed lines represent inhibitory effects. Squares are hormones, triangles are peptides, and circles are biogenic amines. A filled-in square or oval within a symbol denotes that the action of that factor is oestrogen dependent, while an empty square within a symbol denotes that the action of the factor is PRL or PL dependent. PL, placental lactogen; other abbreviations are in Table 2.1. Original source: Figure 7 in Bridges (1996) in Rosenblatt and Snowdon (eds) *Parental Care*, © 1996 by Academic Press, reproduced by permission of the publisher. All rights of reproduction in any form reserved.

Mammals and some other animals are well known for their highly aggressive defence of the young. In domestic mice and rats, the hormones of pregnancy and lactation plus suckling stimuli from the young have been shown to facilitate this behaviour. In deermice (*Peromyscus*) and hamsters, prolactin appears to be involved (Svare 1983, Numan 1994).

2.10 CONCLUSIONS

A number of mechanisms of behaviour are shared between distantly related species to a degree that is remarkable but no longer surprising in the light of modern molecular neurobiology. Chemical mechanisms of behaviour (molecules) are limited in number, but each is versatile in its actions and consequences. They work together in integrated, feedback regulated systems that amplify signals coming in yet are buffered against the kinds of perturbations likely to occur in a normal environment. Chemical disrupters of these systems are likely to affect a wide array of species and behaviours in ways that could negatively impact the fitness of individuals.

ACKNOWLEDGEMENTS

The authors' research was supported by NSF grant IBN-9514088 (EAR) and NIEHS grant ES04184 (DNW) during the preparation of this chapter.

REFERENCES

Adamo SA and Chase R (1991) Central arousal and sexual responsiveness in the snail, *Helix aspersa. Behavioral and Neural Biology*, **55**, 194-213.

Adkins EK (1979) Effect of embryonic treatment with estradiol or testosterone on sexual differentiation of the quail brain: Critical period and dose-response relationships. *Neuroendocrinology*, **29**, 178-185.

Adkins-Regan E (1987) Hormones and sexual differentiation. In *Hormones and Reproduction in Fishes, Amphibians, and Reptiles*, Norris DO and Jones RE (eds), Plenum, New York.

Adkins-Regan E (1998) Hormonal mechanisms of mate choice. *American Zoologist*, **38**, 166-178.

Adkins-Regan EK (1981) Early organizational effects of hormones: an evolutionary perspective. In *Neuroendocrinology of Reproduction: Physiology and Behavior*, Adler NT (ed.), Plenum Press, New York, pp. 159-228.

Adler N, Pfaff D and Goy RW (eds) (1985) *Handbook of Behavioral Neurobiology*, vol. 7: *Reproduction*, Plenum, New York.

Aguilera G (1998) Corticotropin releasing hormone, receptor regulation and the stress response. *Trends in Endocrinology and Metabolism*, **9**, 329-336.

Albert DJ, Walsh ML and Jonik RH (1993) Aggression in humans: what is its biological foundation? *Neuroscience and Biobehavioral Reviews*, **17**, 405-425.

Angers A, Storozhuk M, Duchaine T and Castellucci VF (1998) Cloning and functional expression of an *Aplysia* 5-HT receptor negatively coupled to adenylate cyclase. *Journal of Neuroscience*, **18**, 5586-5593.

Anton S and Gadenne C (1999) Effect of juvenile hormone on the central nervous processing of sex pheromone in an insect. *Proceedings of the National Academy of Sciences USA*, **96**, 5764-5767.

Antonsen BL and Paul DH (1997) Serotonin and octopamine elicit stereotypical agonistic behaviors in the squat lobster *Munida quadraspina* (Anomura, Galatheidae). *Journal of Comparative Physiology* A, **181**, 501-510.

Arnold AP and Breedlove SM (1985) Organizational and activational effects of sex steroids on brain and behavior: A reanalysis. *Hormones and Behavior*, **19**, 469-498.

Arnold AP, Bottjer SW, Nordeen EJ, Nordeen KW and Sengelaub DR (1987) Hormones and critical periods in behavioral and neural development. In *Imprinting and Cortical Plasticity Comparative Aspects of Sensitive Periods*, Rauschecker JP and Marler P (eds), Wiley, New York, pp. 55-97.

Aschoff J (ed.) (1981) *Handbook of Behavioral Neurobiology*, vol. 4: *Biological Rhythms*, Plenum Press, New York.

Audet C, FitzGerald GJ and Guderley H (1986) Photoperiod effects on plasma cortisol levels in *Gasterosteus aculeatus*. *General and Comparative Endocrinology*, **61**, 76-81.

Baggerman B (1962) Some endocrine aspects of fish migration. *General and Comparative Endocrinology*, **1** (Suppl. 1), 188-205.

Baggerman B (1966) On the endocrine control of reproductive behaviour in the male three-spined stickleback (*Gasterosteus aculeatus* L.). *Symposia of the Society for Experimental Biology*, **20**, 427-456.

Balda RP and Kamil AC (1998) The ecology and evolution of spatial memory in corvids of the southwestern USA: the perplexing pinyon jay. In *Animal Cognition in Nature: The Convergence of Psychology and Biology in Laboratory and Field*, Balda RP, Pepperberg IM and Kamil AC (eds), Academic Press, San Diego, pp. 29-64.

Ball GF (1993) The neural integration of environmental information by seasonally breeding birds. *American Zoologist*, **33**, 185-199.

Ball GF (1994) Neurochemical specializations associated with vocal learning and production in songbirds and budgerigars. *Brain, Behavior and Evolution*, **44**, 234-246.

Balthazart J (1983) Hormonal correlates of behavior. In *Avian Biology*, vol. 8, Farner DS, King JR and Parkes KC (eds), Academic Press, New York, pp. 221-366.

Balthazart J (1989) Steroid metabolism and the activation of social behavior. In *Advances in Comparative and Environmental Physiology*, vol. 3, Balthazart J (ed.), Springer-Verlag, Berlin, pp. 105-159.

Balthazart J and Ball GF (1992) Is dopamine interacting with aromatase to control sexual behavior in male quail? *Poultry Science Reviews*, **4**, 217-233.

Balthazart J and Ball GF (1993) Neurochemical differences in two steroid-sensitive areas mediating reproductive behaviors. In *Advances in Comparative and Environmental Physiology*, vol. 15, Springer-Verlag, Berlin, pp. 133-161.

Balthazart J and Ball GF (1995) Sexual differentiation of brain and behavior in birds. *Trends in Endocrinology and Metabolism*, **6**, 21-29.

Balthazart J and Ball GF (1998) New insights into the regulation and function of brain estrogen synthase (aromatase). *Trends in Neurosciences*, **21**, 243-247.

Bamber SD and Naylor E (1996) Mating behavior of male *Carcinus maenas* in relation to a putative pheromone: behavioural changes in response to antennule restriction. *Marine Biology*, **125**, 483-488.

Barclay SR and Cheng M-F (1992) Role of catecholamines in the courtship behavior of male ring doves. *Pharmacology Biochemistry and Behavior*, **41**, 739-747.

Barclay SR and Harding CF (1990) Differential modulation of monoamine levels and turnover rates by estrogen and/or androgen in hypothalamic and vocal control nuclei of male zebra finches. *Brain Research*, **523**, 251-262.

Barclay SR, Harding CF and Waterman SA (1996) Central DSP-4 treatment decreases norepinephrine levels and courtship behavior in male zebra finches. *Pharmacology Biochemistry and Behavior*, **53**, 213-220.

Beani L, Panzica G, Briganti F, Persichella P and Dessi-Fulgheri F (1995) Testosterone-induced changes of call structure, midbrain and syrinx anatomy in partridges. *Physiology and Behavior*, **58**, 1149-1157.

Begnoche VL, Moore SK, Blum N, van Gils C and Mayeri E (1996) Sign stimulus activates a peptidergic neural system controlling reproductive behavior in *Aplysia*. *Journal of Neurophysiology*, **75**, 2161-2166.

Beletsky LD, Gori DF, Freeman S and Wingfield JC (1995) Testosterone and polygyny in birds. In *Current Ornithology*, vol. 12, Power DM (ed.), Plenum, New York, pp. 1-42.

Belthoff JR and Dufty AM (1998) Corticosterone, body condition and locomotor activity: a model for dispersal in screech-owls. *Animal Behaviour*, **55**, 405-415.

Bernheim SM and Mayeri E (1995) Complex behavior induced by egg-laying hormone in *Aplysia*. *Journal of Comparative Physiology* A, **176**, 131-136.

Berthold P (1993) *Bird Migration: A General Survey*, Oxford University Press, Oxford.

Bester-Meredith JK, Young LJ and Marler C (1999) Species differences in paternal behavior and aggression in *Peromyscus* and their associations with vasopressin immunoreactivity and receptors. *Hormones and Behavior*, **36**, 25-38.

Bicker G and Menzel R (1989) Chemical codes for the control of behaviour in arthropods. *Nature*, **337**, 33-39.

Bingman V, Riters LV, Strasser R and Gagliardo A (1998) Neuroethology of avian navigation. In *Animal Cognition in Nature: The Convergence of Psychology and Biology in*

Laboratory and Field, Balda RP, Pepperberg IM and Kamil AC (eds), Academic Press, San Diego, pp. 201-226.

Bjerselius R, Olsen KH and Zheng W (1995) Behavioral and endocrinological responses of mature male goldfish to the sex pheromone 17-alpha, 20 beta-dihydroxy-4-pregnen-3-one in the water. *Journal of Experimental Biology*, **198**, 747-754.

Block GD, Hudson DJ and Lickey ME (1974) Extraocular photoreceptors can entrain the circadian oscillator in the eye of *Aplysia*. *Journal of Comparative Physiology*, **89**, 237-249.

Blüm V (1972) Experimente mit Teleosteer-Prolaktin. *Zool JG (Allg. Zool.)*, **77**, 335.

Blüm V and Fiedler K (1965) Hormonal control of reproductive behavior in some cichlid fish. *General and Comparative Endocrinology*, **5**, 186-196.

Blumberg S and Susswein AJ (1998) Consummatory feeding movements in *Aplysia fasciata* are facilitated by conspecifics with access to mates, by reproductive tract homogenates and by bag cell peptides. *Journal of Comparative Physiology* A, **182**, 175-181.

Borg B (1994) Androgens in teleost fishes. *Comparative Biochemistry and Physiology* C, **109**, 219-245.

Boswell T, Hall MR and Goldsmith AR (1993) Annual cycles of migratory fattening, reproduction and moult in European quail (*Coturnix coturnix*). *Journal of Zoology*, **231**, 627-644.

Bouissou M-F (1983) Hormonal influences on aggressive behavior in ungulates. In *Hormones and Aggressive Behavior*, Svare BB (ed.), Plenum, New York, pp. 507-534.

Boyd S (1997) Brain vasotocin pathways and the control of male sexual behavior in the bullfrog. *Brain Research Bulletin*, **44**, 345-350.

Brantley RK, Wingfield JC and Bass AH (1993) Sex steroid levels in *Porichthys notatus*, a fish with alternative reproductive tactics, and a review of the hormonal bases for male dimorphism among teleost fishes. *Hormones and Behavior*, **27**, 332-347.

Brauth SE, Hall WS and Dooling RJ (eds) (1991) *Plasticity of Development*, MIT Press: Cambridge, MA.

Bravarenko NI (1994) Participation of cardioactive peptides in habituation and sensitization of synaptic responses in command neurons for avoidance behaviour of snail. *Zhurnal Vysshei Nervnoi Deyatel'Nosti Imeni I P Pavlova*, **44**, 301-306.

Brenowitz EA and Kroodsma DE (1996) The neuroethology of birdsong. In *Ecology and Evolution of Acoustic Communication in Birds*, Kroodsma DE and Miller EH (eds), Comstock Publishing, Ithaca, pp. 285-304.

Bridges RS (1996) Biochemical basis of parental behavior in the rat. In *Parental Care: Evolution, Mechanisms, and Adaptive Significance (Advances in the Study of Behavior,* vol. 25*)*, Rosenblatt JS and Snowdon CT (eds), Academic Press, San Diego, pp. 215-242.

Brown JL and Vleck CM (1998) Prolactin and helping in birds: has natural selection strengthened helping behavior? *Behavioral Ecology*, **9**, 541-545.

Brown RE (1985) Hormones and paternal behavior in vertebrates. *American Zoologist*, **25**, 895-910.

Brown RE (1994) *An Introduction to Neuroendocrinology*, Cambridge University Press, Cambridge.

Brown RE and MacDonald DW (eds) (1985) *Social Odours in Mammals*, Clarendon Press, Oxford.

Buntin JD (1996) Neural and hormonal control of parental behavior in birds. In *Parental Care: Evolution, Mechanisms, and Adaptive Significance (Advances in the Study of Behavior,* vol. 25*)*, Rosenblatt JS and Snowdon CT (eds), Academic Press, San Diego, pp. 161-214.

Buntin JD, Hnasko RM and Zuzick PH (1999) Role of the ventromedial hypothalamus in prolactin-induced hyperphagia in ring doves. *Physiology and Behavior*, **66**, 255-261.

Butler AB and Hodos W (1996) *Comparative Vertebrate Neuroanatomy*, Wiley-Liss, New York.

Capaldi EA, Robinson GE and Fahrbach SE (1999) Neuroethology of spatial learning: the birds and the bees. *Annual Review of Psychology*, **50**, 651-682.

Cardwell JR and Liley R (1991) Androgen control of social status in males of a wild population of stoplight parrotfish, *Sparisoma viride* (Scaridae). *Hormones and Behavior*, **25**, 1-18.

Cardwell JR, Stacey NE, Tan ESP, McAdam DSO and Lang SLC (1995) Androgen increases olfactory receptor response to a vertebrate sex pheromone. *Journal of Comparative Physiology* A, **176**, 55-61.

Carter CS (1992) Oxytocin and sexual behavior. *Neuroscience and Biobehavioral Reviews*, **16**, 131-144.

Cash WB and Holberton RL (1999) Effects of exogenous corticosterone on locomotor activity in the red-eared slider turtle, *Trachemys scripta elegans*. *Journal of Experimental Zoology*, **284**, 637-644.

Cassone VM (1990) Effects of melatonin on vertebrate circadian systems. *Trends in Neurosciences*, **13**, 457-463.

Casto JM and Ball GF (1996) Early administration of 17β-estradiol partially masculinizes song control regions and α_2-adrenergic receptor distribution in European starlings (*Sturnus vulgaris*). *Hormones and Behavior*, **30**, 387-406.

Celotti F, Negri-Cesi P and Poletti A (1997) Steroid metabolism in the mammalian brain: 5alpha-reduction and aromatization. *Brain Research Bulletin*, **44**, 365-376.

Cheng M-F (1993) Vocal, auditory and endocrine systems: three-way connectivities and implications. *Poultry Science Reviews*, **5**, 37-47.

Cheng M-F and Zuo M (1994) Proposed pathways for vocal self-stimulation: met-enkephalinergic projections linking the midbrain vocal nucleus, auditory-responsive thalamic regions and neurosecretory hypothalamus. *Journal of Neurobiology*, **25**, 361-379.

Chistyakova AMV (1989) Role of dopamine and serotonin in the modulation of snail defensive behavior. *Zhurnal Vysshei Nervnoi Deyatel'Nosti Imeni I P Pavlova*, **39**, 941-948.

Chivers DP and Smith RJF (1994a) Fathead minnows, *Pimephales promelas*, acquire predator recognition when alarm substance is associated with the sight of unfamiliar fish. *Animal Behaviour*, **48**, 597-605.

Chivers DP and Smith RJF (1994b) Intra- and interspecific avoidance of areas marked with skin extract from brook sticklebacks (*Culea inconstans*) in a natural habitat. *Journal of Chemical Ecology*, **20**, 1517-1524.

Christian M and Gillies G (1999) Developing hypothalamic dopaminergic neurones as potential targets for environmental estrogens. *Journal of Endocrinology*, **160**, R1-R6.

Clare AS, Freet RK and McClary M Jr (1994) On the antennular secretion of the cyprid of *Balanus amphitrite amphitrite*, and its role as a settlement pheromone. *Journal of the Marine Biological Association UK*, **74**, 243-250.

Clayton NC and Lee DW (1998) Memory and the hippocampus in food-storing birds. In *Animal Cognition in Nature: The Convergence of Psychology and Biology in Laboratory and Field*, Balda RP, Pepperberg IM and Kamil AC (eds), Academic Press, San Diego, pp. 99-118.

Coast GM and Webster SG (eds) (1998) *Recent Advances in Arthropod Endocrinology*, Cambridge University Press, Cambridge.

Cooke B, Hegstrom CD, Villeneuve LS and Breedlove SM (1998) Sexual differentiation of the vertebrate brain: principles and mechanisms. *Frontiers in Neuroendocrinology*, **19**, 323-362.

Cookson KK, Hall WS, Heaton JT and Brauth SE (1996) Distribution of choline acetyltransferase and acetylcholinesterase in vocal control nuclei of the budgerigar (*Melopsittacus undulatus*). *Journal of Comparative Neurology*, **369**, 220-235.

Corrent G and Eskin A (1982) Transmitterlike action of serotonin in phase shifting a rhythm from the *Aplysia* eye. *American Journal of Physiology*, **242**, R333-R337.

Corrent G, Eskin A and Kay I (1982) Entrainment of the circadian rhythm from the eye of *Aplysia*: Role of serotonin. *American Journal of Physiology*, **242**, R326-R332.

Cowan DF (1991) The role of olfaction in courtship behavior of the American lobster *Homarus americanus*. *Biological Bulletin (Woods Hole)*, **181**, 402-407.

Creel S, Creel NM and Monfort SL (1996) Social stress and dominance. *Nature*, **379**, 212.

Crews D (1984) Gamete production, sex hormone secretion, and mating behavior uncoupled. *Hormones and Behavior*, **18**, 22-28.

Crews D (1993) The organizational concept and vertebrates without sex chromosomes. *Brain, Behavior and Evolution*, **42**, 202-214.

D'Yakonova VE and Sakharov DA (1994) Neurotransmitter basis of behavior in a mollusc: control of choice between searching and defensive responses to a novel object. *Zhurnal Vysshei Nervnoi Deyatel'Nosti Imeni I P Pavlova*, **44**, 526-531.

D'Yakonova VE, D'Yakonova TL and Sakharov DA (1992) Endogenous opioids of model gastropods: coordination of motor programs for feeding and defensive behavior in the pond snail *Lymnaea stagnalis*. *Biologicheski Membrany Moscow*, **9**, 1074-1076.

D'Yakonova VE, Elofsson R, Carblerg M and Sakharow DA (1995) Complex avoidance behavior and its neurochemical regulation in the land snail *Cepaea nemoralis*. *General Pharmacology*, **26**, 773-777.

Darlison MG and Richter D (1999) Multiple genes for neuropeptides and their receptors: co-evolution and physiology. *Trends in Neurosciences*, **22**, 81-88.

Dautov SSH and Nezlin LP (1990) Nervous system and behavior of starfish larvae during ontogenesis. *Ontogenez*, **21**, 167-176.

Davis RE and Bardach JE (1965) Time-coordinated prefeeding activity in fish. *Animal Behaviour*, **13**, 154-162.

De Boer PACM, TerMaat A, Pieneman AW, Croll RP, Kurokawa M and Jansen RF (1997) Functional role of peptidergic anterior lobe neurons in male sexual behavior of the snail *Lymnaea stagnalis*. *Journal of Neurophysiology*, **78**, 2823-2833.

Defraipont M and Sorensen PW (1993) Exposure to the pheromone 17α,20β-dihydroxy-4-pregnen-3-one enhances the behavioural spawning success, sperm production and sperm motility of male goldfish. *Animal Behaviour*, **46**, 245-256.

Demski LS (1984) The evolution of neuroanatomical substrates of reproductive behavior: sex steroid and LHRH-specific pathways including the terminal nerve. *American Zoologist*, **24**, 809-830.

DeNardo DF and Licht P (1993) Effects of corticosterone on social behavior of male lizards. *Hormones and Behavior*, **27**, 184-199.

Denbow DM (1999) Food intake regulation in birds. *Journal of Experimental Zoology*, **283**, 333-338.

Denny JS (1987) Guidelines for the culture of fathead minnows *Pimephales promelas* for use in toxicity tests. *EPA/600/3-87/001*. Environmental Protection Agency, Duluth, MN, 42 pp.

DePedro N, Gancedo B, Alonso GAL, Delgado MJ and Alonso BM (1995a) Alterations in food intake and thyroid tissue content by corticotropin-releasing factor in *Tinca tinca*. *Revista Española de Fisiología*, **51**, 71-75.

DePedro N, Gancedo B, Alonso GAL, Delgado MJ and Alonso BM (1995b) CRF effect on thyroid function is not mediated by feeding behavior in goldfish. *Pharmacology Biochemistry and Behavior*, **51**, 885-890.

DePedro N, Pinillos ML, Valenciano AI, Alonso BM and Delgado MJ (1998) Inhibitory effect on serotonin on feeding behavior in goldfish: Involvement of CRF. *Peptides*, **19**, 505-511.

de Ruiter AJH, Wendelaar SE, Slijkhuis H and Baggerman B (1986) The effect of prolactin on fanning behavior in the male three-spined stickleback, *Gasterosteus aculeatus* L. *General and Comparative Endocrinology*, **64**, 273-283.

Deviche P (1992) Testosterone and opioids interact to regulate feeding in a male migratory songbird. *Hormones and Behavior*, **26**, 394-405.

DeVoogd TJ (1986) Steroid interactions with structure and function of avian song control regions. *Journal of Neurobiology*, **17**, 177-201.

De Vries GJ (1995) Studying neurotransmitter systems to understand the development and function of sex differences in the brain: the case of vasopressin. In *Neurobiological Effects of Sex Steroid Hormones*, Micevych PE and Hammer RP (eds), Cambridge University Press, Cambridge, pp. 254-280.

Diamond J (1968) The activation and distribution of GABA and L-glutamate receptors on goldfish Mauthner neurones: An analysis of dendritic remote inhibition. *Journal of Physiology*, **194**, 669-723.

Dill LM (1974) The escape response of the zebra danio (*Brachydanio rerio*). II. The effect of experience. *Animal Behaviour*, **22**, 723-730.

Dornan WA and Malsbury CW (1989) Neuropeptides and male sexual behavior. *Neuroscience and Biobehavioral Reviews*, **13**, 1-15.

Dorries KM, Adkins-Regan E and Halpern BP (1997) Sensitivity and behavioral responses to the pheromone androstenone are not mediated by the vomeronasal organ in domestic pigs. *Brain, Behavior and Evolution*, **49**, 53-62.

Drewes CD and Brinkhurst RO (1990) Giant nerve fibers and rapid escape reflexes in newly hatched aquatic oligochaetes, *Lumbriculus variegatus* (Family Lumbriculidae). *Invertebrate Reproduction and Development*, **17**, 91-96.

Dulka JG (1993) Sex pheromone systems in goldfish: comparisons to vomeronasal systems in tetrapods. *Brain, Behavior and Evolution*, **42**, 265-280.

Eaton RC, Di Domenico R and Nissanov J (1991) Mauthner cell in sensorimotor integration by the brainstem escape network. *Brain, Behavior and Evolution*, **37**, 272-285.

El Halawani ME, Silsby JL, Betinke EJ and Fehrer SC (1986) Hormonal induction of incubation behavior in ovariectomized female turkeys (*Meleagris gallopavo*). *Biology of Reproduction*, **35**, 59-67.

Elphick MR, Green IC and O'Shea M (1994) Nitric oxide signalling in the insect nervous system. In *Insect Neurochemistry and Neurophysiology*, Borkovec AB and Loeb M J (eds), CRC Press, Boca Raton, pp. 129-132.

Emerson SB and Hess DL (1996) The role of androgens in opportunistic breeding, tropical frogs. *General and Comparative Endocrinology*, **103**, 220-230.

Enstrom DA, Ketterson ED and Nolan V (1997) Testosterone and mate choice in the dark-eyed junco. *Animal Behaviour*, **54**, 1135-1146.

Eskin A (1979) Circadian system of the *Aplysia* eye: properties of the pacemaker and mechanisms of its entrainment. *Federation Proceedings*, **38**, 2573-2579.

Ewer J, Gammie SC and Truman JW (1997) Control of insect ecdysis by a positive-feedback endocrine system: roles of eclosion hormone and ecdysis triggering hormone. *Journal of Experimental Biology*, **200**, 869-881.

Fahrbach SE (1997) Regulation of age polyethism in bees and wasps by juvenile hormone. In *Advances in the Study of Behavior*, vol. 26, Slater PJB, Rosenblatt JS, Snowdon CT and Milinski M (eds), Academic Press, San Diego, pp. 285-316.

Falcon J (1999) Cellular circadian clocks in the pineal. *Progress in Neurobiology*, **58**, 121-162.

Fan X, Wu B, Nagle GT and Painter SD (1997) Molecular cloning of a cDNA encoding a potential water-borne pheromonal attractant released during *Aplysia* egg laying. *Molecular Brain Research*, **48**, 167-170.

Feldmeth CR (1983) Costs of aggression in trout and pupfish. In *Behavioural Energetics: The Cost of Survival in Vertebrates*, Asprey WP and Lustick SI (eds), Ohio State University Press, Columbus, OH, pp. 117-138.

Ferkin MH and Zucker I (1991) Seasonal control of odour preferences of meadow voles (*Microtus pennsylvanicus*) by photoperiod and ovarian hormones. *Journal of Reproduction and Fertility*, **92**, 433-441.

Fernald RD and White RB (1999) Gonadotropin-releasing hormone genes: phylogeny, structure, and functions. *Frontiers in Neuroendocrinology*, **20**, 224-240.

Fernandez de Miguel F and Arechiga H (1994) Circadian locomotor activity and its entrainment by food in the crayfish *Procambarus clarkii*. *Journal of Experimental Biology*, **190**, 9-21.

Ferris CF, Melloni RH, Koppel G, Perry KW, Fuller RW and Delville Y (1997) Vasopressin/serotonin interactions in the anterior hypothalamus control aggressive behavior in golden hamsters. *Journal of Neuroscience*, **17**, 4331-4340.

Fiedler K and Blüm V (1972) Prolactin-sensitive neurons in the forebrain of some fishes. *Experientia*, **28**, 1384-1385.

Fingerman M (1997) Roles of neurotransmitters in regulating reproductive hormone release and gonadal maturation in decapod crustaceans. *Invertebrate Reproduction and Development*, **31**, 47-54.

Fontaine M (1975) Physiological mechanisms in the migration of marine and amphihaline fish. *Advances in Marine Biology*, **13**, 241-355.

Fox HE, White SA, Kao MHF and Fernald RD (1997) Stress and dominance in a social fish. *Journal of Neuroscience*, **17**, 6463-6469.

Francis RC, Soma K and Fernald RD (1993) Social regulation of the brain-pituitary-gonadal axis. *Proceedings of the National Academy of Sciences USA*, **90**, 7794-7798.

Fraser NHC, Metcalfe NB and Thorpe JE (1993) Temperature-dependent switch between diurnal and nocturnal foraging in salmon. *Proceedings of the Royal Society London B*, **252**, 135-139.

Fujita I, Sorensen PW, Stacey NE and Hara TJ (1991) The olfactory system, not the terminal nerve, functions as the primary chemosensory pathway mediating responses to sex pheromones in male goldfish. *Brain, Behavior and Evolution*, **38**, 313-321.

Gäde G, Hoffmann K-H and Spring JH (1997) Hormonal regulation in insects: facts, gaps, and future directions. *Physiological Reviews*, **77**, 963-1032.

Gandelman R (1983) Gonadal hormones and sensory function. *Neuroscience and Biobehavioral Reviews*, **7**, 1-17.

Gans C and Crews D (eds) (1992) *Biology of the Reptilia*. Volume 18, Physiology E: *Hormones, Brain, and Behavior*, University of Chicago Press, Chicago.

Gelinas D and Callard GV (1993) Immunocytochemical and biochemical evidence for aromatase in neurons of the retina, optic tectum and retinotectal pathways in goldfish. *Journal of Neuroendocrinology*, **5**, 635-641.

Gerall AA, Moltz H and Ward IL (eds) (1992) *Handbook of Behavioral Neurobiology*, vol. 11: *Sexual Differentiation*, Plenum, New York.

Godwin J, Crews D and Warner RR (1996) Behavioural sex change in the absence of gonads in a coral reef fish. *Proceedings of the Royal Society London* B, **263**, 1683-1688.

Godwin JR and Peter T (1993) Sex change and steroid profiles in the protandrous anemonefish *Amphipiron melanopus* (Pomacentridae, Teleostei). *General and Comparative Endocrinology*, **91**, 144-157.

Goldberg JI, Koehncke NK, Christopher KJ, Neumann C and Diefenbach TJ (1994) Pharmacological characterization of a serotonin receptor invoked in an early embryonic behavior of *Helisoma trivolvis*. *Journal of Neurobiology*, **25**, 1545-1557.

Goldburt V, Sabban BA and Kleinhaus AL (1994) Serotonin depletion inhibits feeding in carnivorous leeches (*Haemopis*). *Behavioral and Neural Biology*, **61**, 47-53.

Goldman BD and Nelson RJ (1993) Melatonin and seasonality in mammals. In *Melatonin: Biosynthesis, Physiological Effects, and Clinical Applications*, Yu H-S and Reiter RJ (eds), CRC Press, Boca Raton, FL, pp. 225-252.

González-Mariscal G and Rosenblatt JS (1996) Maternal behavior in rabbits: a historical and multidisciplinary perspective. In *Parental Care: Evolution, Mechanisms, and Adaptive Significance* (*Advances in the Study of Behavior*, vol. 25), Rosenblatt JS and Snowdon CT (eds), Academic Press, San Diego, pp. 333-360.

Goodson JL (1998) Vasotocin and vasoactive intestinal polypeptide modulate aggression in a territorial songbird, the violet-eared waxbill (Estrildidae: *Uraeginthus granatina*). *General and Comparative Endocrinology*, **111**, 233-244.

Goodson JL and Bass A (2000) Forebrain peptides modulate sexually polymorphic vocal circuitry. *Nature*, **403**, 769-772.

Gould E, Reeves AJ, Fallah M, Tanapat P, Gross CG and Fuchs E (1999) Hippocampal neurogenesis in adult Old World primates. *Proceedings of the National Academy of Sciences USA*, **96**, 5263-5267.

Gould JL (1996) Specializations in honey bee learning. In *Neuroethological Studies of Cognitive and Perceptual Processes*, Moss CF and Shettleworth SJ (eds), Westview Press, Boulder, CO, pp. 11-30.

Grober MS and Bass AH (1991) Neuronal correlates of sex-change in labrid fishes: LHRH-like immunoreactivity. *Brain, Behavior and Evolution*, **38**, 302-312.

Grober MS, Jackson IMD and Bass AH (1991) Gonadal steroids affect LHRH preoptic cell number in a sex/role changing fish. *Journal of Neurobiology*, **22**, 734-741.

Guiguen Y, Jalabert B, Thouard E and Fostier A (1993) Changes in plasma and gonadal steroid hormones in relation to the reproductive cycle and the sex inversion process in the protandrous seabass, *Lates calcarifer*. *General and Comparative Endocrinology*, **92**, 327-338.

Gwinner E, Hau M and Heigl S (1997) Melatonin: generation and modulation of avian circadian rhythms. *Brain Research Bulletin*, **44**, 439-444.

Halem HA, Cherry JA and Baum MJ (1999) Vomeronasal neuroepithelium and forebrain Fos responses to male pheromones in male and female mice. *Journal of Neurobiology*, **39**, 249-263.

Hall MR and Goldsmith AR (1983) Factors affecting prolactin secretion during breeding and incubation in the domestic duck (*Anas platyrhynchos*). *General and Comparative Endocrinology*, **49**, 270-276.

Haller J, Makara GB and Kruk MR (1998) Catecholaminergic involvement in the control of aggression: hormones, the peripheral sympathetic, and central noradrenergic systems. *Neuroscience and Biobehavioral Reviews*, **22**, 85-97.

Halpern M (1992) Nasal chemical senses in reptiles: structure and function. In *Biology of the Reptilia*, Volume 18, Physiology E: *Hormones, Brain, and Behavior*, Gans C and Crews D (eds), University of Chicago Press, Chicago, pp. 423-524.

Hara TJ (1994) Olfaction and gustation in fish: An overview. *Acta Physiologica Scandinavica*, **152**, 207-217.

Harden Jones FR (1968) *Fish Migration*, Edward Arnold Press, London.

Harding CF (1983) Hormonal influences on avian aggressive behavior. In *Hormones and Aggressive Behavior*, Svare BB (ed.), Plenum, New York, pp. 435-469.

Harding CF (1992) Hormonal modulation of neurotransmitter function and behavior in male songbirds. *Poultry Science Reviews*, **4**, 261-273.

Hasler AD (1966) *Underwater Guideposts: Homing of Salmon*, University of Wisconsin Press, Madison.

Hawryshyn CW (1992) Polarization vision in fish. *American Scientist*, **80**, 164-175.

Hector JAL and Goldsmith AR (1985) The role of prolactin during incubation: comparative studies of three *Diomedea* albatrosses. *General and Comparative Endocrinology*, **60**, 236-243.

Helluy S and Holmes JC (1990) Serotonin, octopamine, and the clinging behavior induced by the parasite *Polymorphus paradoxus* (Acanthocephala) in *Gammarus lacustris* (Crustacea). *Canadian Journal of Zoology*, **68**, 1214-1220.

Henson SA and Warner RR (1997) Male and female alternative reproductive behaviors in fishes: a new approach using intersexual dynamics. *Annual Review of Ecology and Systematics*, **28**, 571-592.

Hermann PM and Bulloch AGM (1998) Developmental plasticity of respiratory behavior in *Lymnaea*. *Behavioral Neuroscience*, **112**, 656-667.

Hildeman WH (1975) A cichlid fish, *Symphosodon discus*, with unique nutritive habits. *Cichlidae* (London), **2**, 32.

Himick BA and Peter RE (1994) Bombesin acts to suppress feeding behavior and alter serum growth hormone in goldfish. *Physiology and Behavior*, **55**, 65-72.

Hoar WS (1953) Control and timing of fish migration. *Biological Reviews*, **28**, 437-452.

Hoar WS (1962) Reproductive behavior of fish. *General and Comparative Endocrinology*, **1** (Suppl.), 206-216.

Houck LD and Woodley SK (1995) Field studies of steroid hormones and male reproductive behaviour in amphibians. In *Amphibian Biology*, vol. 2: *Social Behaviour*, Heatwole H and Sullivan BK (eds), Surrey Beatty & Sons, New South Wales, Australia, pp. 677-703.

Huang ZY, Plettner E and Robinson GE (1998) Effects of social environment and worker mandibular glands on endocrine-mediated behavioral development in honey bees. *Journal of Comparative Physiology* A, **183**, 143-152.

Hull EM (1995) Dopaminergic influences on male rat sexual behavior. In *Neurobiological Effects of Sex Steroid Hormones*, Micevych PE and Hammer RP (eds), Cambridge University Press, Cambridge, pp. 234-253.

Ito M (1989) Long-term depression. *Annual Review of Neuroscience*, **12**, 85-102.

Jacklet JW (1974) The effect of constant light and light pulses on the circadian rhythm in the eye of *Aplysia*. *Journal of Comparative Physiology*, **90**, 33-45.

Jacobs LF (1996) Sexual selection and the brain. *Trends in Ecology and Evolution*, **11**, 82-86.

Janssen J and Corcoran J (1993) Lateral line stimuli can override vision to determine sunfish strike trajectory. *Journal of Experimental Biology*, **176**, 299-305.

Johnston RE (1998) Pheromones, the vomeronasal system, and communication: from hormonal responses to individual recognition. In *Olfaction and Taste XII. An International Symposium (Annals of the New York Academy of Sciences*, **855**, 333-348), Murphy C (ed.), New York Academy of Sciences, New York.

Jonsson E, Johnsson J and Bjornsson BT (1998) Growth hormone increases aggressive behavior in juvenile rainbow trout. *Hormones and Behavior*, **33**, 9-15.

Jonsson N, Hansen LP and Jonsson B (1994) Juvenile experience influences timing of adult river ascent in Atlantic salmon. *Animal Behaviour*, **48**, 740-742.

Jordan CL (1999) Glia as mediators of steroid hormone action on the nervous system: an overview. *Journal of Neurobiology*, **40**, 434-445.

Karamyan AI, Malyukova IV and Sergeev BF (1967) Participation of the telencephalon of bony fish in the accomplishment of complex conditioned — reflex and general — behavior reactions. Behavior and reception in fish. *Akad. Nauk SSSR*, Minesterstvo Rybnogo Khozyaustva SSSR, Ikhtiologicheskaya Komissiya. Publ. Nauka, Moscow, pp. 109-114 (translated from the Russian).

Katz PS and Harris-Warrick RM (1999) The evolution of neuronal circuits underlying species-specific behavior. *Current Opinion in Neurobiology*, **9**, 628-633.

Kavaliers M (1989) Day-night rhythms of shoaling behavior in goldfish: opioid and pineal involvement. *Physiology and Behavior*, **46**, 167-172.

Kelley DB (1997) Generating sexually differentiated songs. *Current Opinion in Neurobiology*, **7**, 839-843.

Kemenes G (1997) In vivo neuropharmacological and in vitro laser ablation techniques as tools in the analysis of neuronal circuits underlying behavior in a molluscan model system. *General Pharmacology* **29**, 7-15.

Ketterson ED, Nolan V, Wolf L and Ziegenfus C (1992) Testosterone and avian life histories: effects of experimentally elevated testosterone on behavior and correlates of fitness in the dark-eyed junco (*Junco hyemalis*). *American Naturalist*, **140**, 980-999.

Kikuyama S, Toyoda F, Yamamoto K, Tanaka S and Hayashi H (1997) Female-attracting pheromone in newt cloacal glands. *Brain Research Bulletin*, **44**, 415-422.

Kindler PM, Bahr JM, Gross MR and Philipp DP (1991) Hormonal regulation of parental care behavior in nesting male bluegills: do the effects of bromocriptine suggest a role for prolactin? *Physiological Zoology*, **64**, 310-322.

Korn H, Bausela F, Charpier S and Faber DS (1993) Synaptic noise and multiquantal release at dendritic synapses. *Journal of Neurophysiology*, **70**, 1249-1254.

Kramer CR and Imbriano MA (1997) Neuropeptide Y (NPY) induces gonad reversal in the protogynous bluehead wrasse, *Thalassoma bifasciatum* (Teleostei: Labridae). *Journal of Experimental Zoology*, **279**, 133-144.

Kuenzel WJ, Beck MM and Teruyama R (1999) Neural sites and pathways regulating food intake in birds: a comparative analysis to mammalian systems. *Journal of Experimental Zoology*, **283**, 348-364.

Lam TJ (1983) Environmental influences on gonadal activity in fish. In *Fish Physiology*, vol. 9B, Hoar WS, Randall DJ and Donaldson EM (eds), Academic Press, New York, pp. 65-105.

Laming PR, Kimelberg H, Robinson S, Salm A, Hawrylak N, Muller C, Roots B and Ng K (2000) Neuronal-glial interactions and behavior. *Neuroscience and Biobehavioral Reviews*, **24**, 295-340.

LeBail PY and Boeuf G (1997) What hormones may regulate food intake in fish. *Aquatic Living Resources*, **10**, 371-379.

Legendre P (1997) Pharmacological evidence for two types of postsynaptic glycinergic receptors on the Mauthner cell of 52-h-old zebrafish larvae. *Journal of Neurophysiology*, **77**, 2400-2415.

Lent CM (1985) Serotonergic modulation of the feeding behavior of the medicinal leech. *Brain Research Bulletin*, **14**, 643-655.

Leshner AI (1983) Pituitary-adrenocortical effects on intermale agonistic behavior. In *Hormones and Aggressive Behavior*, Svare BB (ed.), Plenum, New York, pp. 27-38.

Lévy F, Kendrick KM, Keverne EB, Porter RH and Romeyer A (1996) Physiological, sensory, and experiential factors of parental care in sheep. In *Parental Care: Evolution, Mechanisms, and Adaptive Significance (Advances in the Study of Behavior,* vol. 25*)*, Rosenblatt JS and Snowdon CT (eds), Academic Press, San Diego, pp. 385-422.

Li KW, Smit AB and Geraerts WPM (1992) Structural and functional characterization of neuropeptides involved in the control of male mating behavior of *Lymnaea stagnalis*. *Peptides*, **13**, 633-638.

Liley NR (1982) Chemical communication in fish. *Canadian Journal of Fisheries and Aquatic Sciences*, **39**, 22-35.

Lincoln GA, Guiness F and Short RV (1972) The way in which testosterone controls the social and sexual behavior of the red deer stag (*Cervus elaphus*). *Hormones and Behavior*, **3**, 375-396.

Lormée H, Jouventin P, Chastel O and Mauget R (1999) Endocrine correlates of parental care in an antarctic winter breeding seabird, the emperor penguin, *Aptenodytes forsteri*. *Hormones and Behavior*, **35**, 9-17.

Luine V, Nottebohm F, Harding C and McEwen BS (1980) Androgen affects cholinergic enzymes in syringeal motor neurons and muscle. *Brain Research*, **192**, 89-107.

Lunt GG and Olsen RW (eds) (1988) *Comparative Invertebrate Neurochemistry*, Cornell University Press, Ithaca.

Maestripieri D and Zehr JL (1998) Maternal responsiveness increases during pregnancy and after estrogen treatment in macaques. *Hormones and Behavior*, **34**, 223-230.

Magurran A (1984) Gregarious goldfish. *New Scientist*, **103**, 32-33.

Maney DL, Richardson RD and Wingfield JC (1997) Central administration of chicken gonadotropin-releasing-hormone-II enhances courtship behavior in a female sparrow. *Hormones and Behavior*, **32**, 11-18.

Maney DL, Schoech SJ, Sharp PJ and Wingfield JC (1999) Effects of vasoactive intestinal polypeptide on plasma prolactin in passerines. *General and Comparative Endocrinology*, **113**, 323-330.

Marler P (1991) The instinct for vocal learning: songbirds. In *Plasticity of Development*, Brauth SE, Hall WS and Dooling RJ (eds), MIT Press, Cambridge, MA, pp. 107-126.

Martinoli MG, Williams LM, Kah O, Titchener LT and Pelletier G (1991) Distribution of central melatonin binding sites in the goldfish (*Carassius auratus*). *Molecular and Cellular Neuroscience*, **2**, 78-85.

Mason RT (1992) Reptilian pheromones. In *Biology of the Reptilia*, volume 18, Physiology E: *Hormones, Brain, and Behavior*, Gans C and Crews D (eds), Chicago University Press, Chicago, pp. 114-228.

Matsutani T (1990) Endogenous factors controlling spawning in marine bivalves. In *Advances in Invertebrate Reproduction*, vol. 5, Hoshi M and Yamashita O (eds), Elsevier Science Publishers BV (Biomedical Division), Amsterdam, pp. 231-237.

Mattioli R, Santangelo EM, Costa ACC and Vasconcelos L (1997) Substance P facilitates memory in goldfish in an appetitively motivated learning task. *Behavioral Brain Research*, **85**, 117-120.

McAllister AK, Katz LC and Lo DC (1999) Neurotrophins and synaptic plasticity. *Annual Review of Neuroscience*, **22**, 295-318.

Meier AH (1984) Temporal synergism of circadian neuroendocrine oscillations regulates seasonal conditions in the Gulf killifish. *American Fisheries Society*, **113**, 422-431.

Meisel RL and Sachs BD (1994) The physiology of male sexual behavior. In *The Physiology of Reproduction* (2nd edn), vol. 2, Knobil E and Neill JD (eds), Raven, New York.

Mendonça MT (1987) Timing of reproductive behaviour in male musk turtles, *Sternotherus odoratus*: effects of photoperiod, temperature, and testosterone. *Animal Behaviour*, **35**, 1002-1014.

Metcalfe NB, Huntingford FA and Thorpe JE (1992) Social effects on appetite and development in Atlantic salmon. In *The Importance of Feeding Behaviour for the Efficient Culture of Salmonid Fishes*, Thorpe JE and Huntingford FA (eds), World Aquaculture Series, Baton Rouge, LA, pp. 29-40.

Metzdorf R, Gahr M and Fusani L (1999) Distribution of aromatase, estrogen receptor, and androgen receptor mRNA in the forebrain of songbirds and nonsongbirds. *Journal of Comparative Neurology*, **407**, 115-129.

Meyers JR, Copanas EH and Zottoli SJ (1998) Comparison of fast startle responses between two elongate bony fish with anguilliform type of locomotion and the implications for the underlying neuronal basis of escape behavior. *Brain, Behavior and Evolution*, **52**, 7-22.

Micevych PE and Hammer RP (eds) (1995) *Neurobiological Effects of Sex Steroid Hormones*, Cambridge University Press, Cambridge.

Michael RP and Zumpe D (1993) A review of hormonal factors influencing the sexual and aggressive behavior of macaques. *American Journal of Primatology*, **30**, 213-241.

Montgomery JC, Baker CF and Carton AG (1997) The lateral line can mediate rheotaxis in fish. *Nature*, **389**, 960-963.

Moore A and Scott AP (1991) Testosterone is a potent odorant in precocious male Atlantic salmon (*Salmo salar* L.) parr. *Philosophical Transactions of the Royal Society of London B Biological Sciences*, **332**, 241-244.

Moore FL, Lowry CA and Rose JD (1994) Steroid-neuropeptide interactions that control reproductive behaviors in an amphibian. *Psychoneuroendocrinology*, **19**, 581-592.

Moore FL (1999) In *Advances in Chemical Communication in Vertebrates*, Johnston RE, Müller-Schwarze D and Sorensen P (eds), Plenum, New York.

Moore FL and Lowry CA (1998) Comparative neuroanatomy of vasotocin and vasopressin in amphibians and other vertebrates. *Comparative Biochemistry and Physiology* C, **119**, 251-260.

Moore FL and Orchinik M (1991) Multiple molecular actions for steroids in the regulation of reproductive behaviors. *Seminars in the Neurosciences*, **3**, 489-496.

Moore MC and Lindzey J (1992) In *Biology of the Reptilia*, volume 18, Physiology E: *Hormones, Brain, and Behavior*, University of Chicago Press, Chicago, pp. 70-113.

Moore MC and Marler CA (1988) Hormones, behavior, and the environment: an evolutionary perspective. In *Processing of Environmental Information in Vertebrates*, Stetson MH (ed.), Springer-Verlag, New York, pp. 71-84.

Moore MC, Hews DK and Knapp R (1998) Hormonal control and evolution of alternative male phenotypes: generalizations of models for sexual differentiation. *American Zoologist*, **38**, 133-151.

Morgan MJ and Colgan PM (1988) The role of experience in foraging shoals of bluntnose minnows (*Pimephales notatus*). *Behavioural Processes*, **16**, 87-93.

Mori S (1993) Localization of extratectally evoked visual response in the corpus and valvula cerebelli in carp, and cerebellar contribution to 'dorsal light reaction' behavior. *Behavioral Brain Research*, **59**, 33-40.

Morin LP (1994) The circadian visual system *Brain Research Reviews*, **67**, 102-127.

Morin P-P and Døving KB (1992) Changes in the olfactory function of Atlantic salmon, *Salmo salar*, in the course of smoltification. *Canadian Journal of Fisheries and Aquatic Sciences*, **49**, 1704-1713.

Morita M (1990) Photoperiod and melatonin control of planarian asexual reproduction. In *Advances in Invertebrate Reproduction*, vol. 5, Hoshi M and Yamashita O (eds), Elsevier Science Publishers BV (Biomedical Division).

Morley JE (1995) The role of peptides in appetite regulation across species. *American Zoologist*, **35**, 437-445.

Morris R, Frey S, Kasambira T and Petrides M (1999) Ibotenic acid lesions of the basolateral, but not the central, amygdala interfere with conditioned taste aversion: evidence from a combined behavioral and anatomical tract-tracing investigation. *Behavioral Neuroscience*, **113**, 291-302.

Morris YA and Crews D (1990) The effects of exogenous neuropeptide Y on feeding and sexual behavior in the red-sided garter snake (*Thamnophis sirtalis parietalis*). *Brain Research*, 1990, **530**, 339-41.

Munro AD and Pitcher TJ (1983) Hormones and agonistic behaviour in teleosts. In *Control Processes in Fish Physiology*, Rankin JC, Pitcher TJ and Duggan RT (eds), Wiley Interscience, New York, pp. 155-175.

Nechaev IV (1993) Role of dynamics of the catecholamines in mating pair formation and offspring care in the fish *Aequidens pulcher* (Cichlidae): Neuro-pharmacological correlation of male behaviour in the presence of female. *Zhurnal Obshchei Biologii*, **54**, 691-705.

Neckameyer WS (1998) Dopamine and mushroom bodies in *Drosophila*: experience -dependent and -independent aspects of sexual behavior. *Learning and Memory*, **5**, 157-165.

Nelson RJ (2000) *An Introduction to Behavioral Endocrinology* (2nd edn), Sinauer, Sunderland, MA.

Nevitt G and Dittman A (1998) A new model for olfactory imprinting in salmon. *Integrative Biology*, **1**, 215-223.

Nevitt GA, Pentcheff ND, Lohmann KS and Zimmer-Faust RK (1995) Evidence for hydrodynamic orientation by spiny lobsters in a patch environment. *Journal of Experimental Biology*, **198**, 2049-2054.

Nicol AU and Lamming PR (1993) Sustained potential shifts, alterations in acoustic evoked potential amplitude and brachycardiac responses to the onset of illumination in the goldfish (*Carassius auratus*). *Journal of Comparative Physiology* A, **173**, 353-362.

Nijhout HF (1994) *Insect Hormones*, Princeton University Press, Princeton.

Norekian TP (1995) Prey capture phase of feeding behavior in the pteropod mollusc, *Clione limacina*: Neuronal mechanisms. *Journal of Comparative Physiology* A, **177**, 41-53.

Norgren RB and Silver R (1989) Retinohypothalamic projections and the suprachiasmatic nucleus in birds. *Brain, Behavior and Evolution*, **34**, 73-83.

Norris D (1996) *Vertebrate Endocrinology* (3rd edn), Academic Press, San Diego.

Nottebohm F (1991) Reassessing the mechanisms and origins of vocal learning in birds. *Trends in Neurosciences*, **14**, 206-211.

Nottebohm F and Arnold AP (1976) Sexual dimorphism in vocal control areas of the songbird brain. *Science*, **194**, 211-213.

Numan M (1994) Maternal behavior. In *The Physiology of Reproduction* (2nd edn), vol. 2, Knobil E and Neill JD (eds), Raven, New York, pp.1569-1645.

O'Brien WJ, Browman HI and Evans BI (1990) The search behavior of foraging animals. *American Scientist*, **78**, 152-160.

O'Gara BA, Chae H, Latham LB and Friessen WO (1991) Modification of leech behaviour patterns by reserpine-induced amine depletion. *Journal of Neuroscience*, **11**, 96-110.

Okimoto DK and Stetson MH (1999) Presence of an intrapineal circadian oscillator in the teleostean family poeciliidae. *General and Comparative Endocrinology*, **114**, 304-312.

Oliveira RF, Almada VC and Canario AVM (1996) Social modulation of sex steroid concentrations in the urine of male cichlid fish *Oreochromis mossambicus*. *Hormones and Behavior*, **30**, 2-12.

Orchard I, Ramirez J-M and Lange AB (1993) A multifunctional role for octopamine in locust flight. *Annual Review of Entomology*, **38**, 227-249.

Oring LW, Fivizzani AJ and El Halawani ME (1989) Testosterone-induced inhibition of incubation in the spotted sandpiper (*Actitis mecularia*). *Hormones and Behavior*, **23**, 412-423.

Orlov DV (1997) The role of larval settling behaviour in determination of the specific habitat of the hydrozoan *Dynamena pumila* (L.). Larval settlement in *Dynamena pumila* (L.). *Journal of Experimental Marine Biology and Ecology*, **208**, 73-85.

Ottinger MA and Balthazart J (1987) Brain monoamines and sexual behavior in Japanese quail: effects of castration and steroid replacement therapy. *Behavioural Processes*, **14**, 197-216.

Owens DW (1997) Hormones in the life history of sea turtles. In *The Biology of Sea Turtles*, Lutz P and Musick J (eds), CRC Press, Boca Raton, FL, pp. 315-341.

Painter SD (1992) Coordination of reproductive activity in *Aplysia*: peptide neurohormones, neurotransmitters, and pheromones encoded by the egg-laying hormone family of genes. *Biological Bulletin*, **183**, 165-172.

Panchin YV, Gamkrelidze GN, Popova LB and Deliagina TG (1994) Neuronal basis of hunting and feeding behaviour in the pteropod mollusk *Clione limacina*. *Netherlands Journal of Zoology*, **44**, 170-183.

Panzica GC, Viglietti-Panzica C and Balthazart J (1996) The sexually dimorphic medial preoptic nucleus of quail: a key brain area mediating steroid action on male sexual behavior. *Frontiers in Neuroendocrinology*, **17**, 51-125.

Parmigiani S, Ferrari PF and Palanza P (1998) An evolutionary approach to behavioral pharmacology: using drugs to understand proximate and ultimate mechanisms of different forms of aggression in mice. *Neuroscience and Biobehavioral Reviews*, **23**, 143-153.

Pasmanik M and Callard GV (1988) Changes in brain aromatase and 5α-reductase activities correlate significantly with seasonal reproductive cycles in goldfish (*Carassius auratus*). *Endocrinology*, **122**, 1349-1356.

Pavlov DS, Legkii BP, Nechaev IV and Popova NK (1994) Age-related changes of behaviour and state of the catecholamine system in the European grayling fry. *Izvestiya Akademii Nauk Seriya Biologicheskaya (Moscow)*, **1**, 71-80.

Pedersen HC (1989) Effects of exogenous prolactin on parental behaviour in free-living female willow ptarmigan *Lagopus l. lagopus*. *Animal Behaviour*, **38**, 926-934.

Peyon P, Lin XW, Himick BA and Peter RE (1998) Molecular cloning and expression of cDNA encoding brain preprocholecystokinin in goldfish. *Peptides*, **19**, 199-210.

Pfeiffer W, Riegelbauer G, Meir G and Scheibler B (1985) Effects of hypoxanthine-3(N)-oxide and hypoxanthine-1(N)-oxide on central nervous excitation of the black tetra *Gymnocorymbus ternetzi* (Characidae, Ostariophysi, Pisces) indicated by dorsal light response. *Journal of Chemical Ecology*, **11**, 507-524.

Picker MD (1983) Hormonal induction of the aquatic phonotactic response of *Xenopus*. *Behaviour*, **84**, 74-90.

Pierobon P, DePetrocellis L, Minei R and DiMarzo V (1997) Arachidonic acid as an endogenous signal for the glutathione-induced feeding response in *Hydra*. *Cell Molecular Life Science*, **53**, 61-68.

Pijanowska J (1997) Alarm signals in *Daphnia*? *Oecologia*, **112**, 12-16.

Pijanowska J and Kowalczewski A (1997) Predators can induce swarming behaviour and locomotory response in *Daphnia*. *Freshwater Biology*, **37**, 649-656.

Platt C, Popper AN and Fay RR (1989) The ear as part of the octavolateralis system. In *The Mechanosensory Lateral Line: Neurobiology and Evolution*, Coombs S, Görner P and Münz H (eds), Springer-Verlag, New York, pp. 633-651.

Popek W (1983) Seasonal variations in circadian rhythm of hypothalamic catecholamine content in the eel (*Anguilla anguilla* L.). *Comparative Biochemistry and Physiology* C, **75**, 193-198.

Propper CR and Dixon TB (1997) Differential effects of arginine vasotocin and gonadotropin-releasing hormone on sexual behaviors in an anuran amphibian. *Hormones and Behavior*, **32**, 99-104.

Prosada Rao PD and Sharma SC (1982) Retinofugal pathways in juvenile and adult channel catfish, *Ictalurus (Ameiurus) punctatus*: an HRP and autoradiographic study. *Journal of Comparative Neurology*, **219**, 37-48.

Pryce CR (1996) Socialization, hormones, and the regulation of maternal behavior in nonhuman simian primates. In *Parental Care: Evolution, Mechanisms, and Adaptive Significance (Advances in the Study of Behavior*, vol. 25*)*, Rosenblatt JS and Snowdon CT (eds), Academic Press, San Diego, pp. 423-476.

Purves D, Augustine GJ, Fitzpatrick D, Katz LC, LaMantia A-S and McNamara JO (eds) (1997) *Neuroscience*, Sinauer, Sunderland, MA.

Pusey AE (1987) Sex-biased dispersal and inbreeding avoidance in birds and mammals. *Trends in Ecology and Evolution* **2**, 295-299.

Raina AK (1993) Neuroendocrine control of sex pheromone biosynthesis in lepidoptera. *Annual Review of Entomology*, **38**, 329–350.

Raisman G (1997) An urge to explain the incomprehensible: Geoffrey Harris and the discovery of the neural control of the pituitary gland. *Annual Review of Neuroscience*, **20**, 533–566.

Rand MN and Breedlove SM (1992) Androgen locally regulates rat bulbocavernosus and levator ani size. *Journal of Neurobiology*, **23**, 17–30.

Rankin MA (1991) Endocrine effects on migration. *American Zoologist*, **31**, 217–230.

Redman RS and Berry RW (1991) Temperature-dependent peptidergic feedback: potential role in seasonal egg laying in *Aplysia*. *Journal of Neuroscience*, **11**, 1780–1785.

Reebs SG and Colgan PW (1991) Nocturnal care of eggs and circadian rhythms of fanning activity in two normally diurnal cichlid fishes, *Cichlasoma nigrofasciatum* and *Herotilapia multispinosa*. *Animal Behaviour*, **41**, 303–311.

Reede T (1995) Life history shifts in response to different levels of fish kairomones in *Daphnia*. *Journal of Plankton Research*, **17**, 1661–1667.

Rhen T and Crews D (2000) Organization and activation of sexual and agonistic behavior in the leopard gecko, *Eublepharis macularius*. *Neuroendocrinology*, **71**, 252–261.

Richard-Yris MA, Garnier DH and Leboucher G (1983) Induction of maternal behavior and some hormonal and physiological correlates in the domestic hen. *Hormones and Behavior*, **17**, 345–355.

Ringelberg J (1969) Spatial orientation of planktonic crustaceans 2. The swimming in the vertical plane. *Verhandlungen Internationale Vereinigung für theoretische und angewandte Limnologie*, **17**, 841–847.

Rissman EF (1991) Evidence that neural aromatization of androgen regulates the expression of sexual behaviour in female musk shrews. *Journal of Neuroendocrinology*, **3**, 441–448.

Rissman EF, Li X, King JA and Millar RP (1997) Behavioral regulation of gonadotropin-releasing hormone production. *Brain Research Bulletin*, **44**, 459–464.

Rittschof D, Forward Jr RB and Erickson BW (1990) Larval release in brachyuran crustaceans: Functional similarity of peptide pheromone receptor and catalytic site of trypsin. *Journal of Chemical Ecology*, **16**, 1359–1370.

Robel P and Baulieu E-E (1995) Neurosteroids and neuroactive steroids. In *Neurobiological Effects of Sex Steroid Hormones*, Micevych PE and Hammer RP (eds), Cambridge University Press, Cambridge, pp. 281–296.

Roberts BL (1992) Neural mechanisms underlying escape behaviour in fishes. *Reviews in Fish Biology and Fisheries*, **2**, 243–266.

Rollmann SM, Houck LD and Feldhoff RC (1999) Proteinaceous pheromone affecting female receptivity in a terrestrial salamander. *Science*, **285**, 1907–1909.

Romero SMB, Hoffmann A and Menescal de Oliveira L (1994) Is there an opiate receptor in the snail *Magalobulimus sanctipauli*? Action of morphine and naloxone. *Comparative Biochemistry and Physiology* C, **107**, 37–40.

Roselli CE, Abdelgadir SE and Resko JA (1997) Regulation of aromatase gene expression in the adult rat brain. *Brain Research Bulletin*, **44**, 351–358.

Rosenblatt JS and Snowdon CT (eds) (1996) *Parental Care: Evolution, Mechanisms, and Adaptive Significance* (*Advances in the Study of Behavior*, vol. 25), Academic Press, San Diego.

Saldanha CJ, Leak RK and Silver R (1994) Detection and transduction of daylength in birds. *Psychoneuroendocrinology*, **19**, 641–656.

Salter MW and De Koninck Y (1999) An ambiguous fast synapse: a new twist in the tale of two transmitters. *Nature Neuroscience*, **2**, 199–200.

Sapolsky RM (1992) Neuroendocrinology of the stress-response. In *Behavioral Endocrinology*, Becker JB, Breedlove SM and Crews D (eds), MIT Press, Cambridge, MA, pp. 287–324.

Scharrer E and Scharrer B (1963) *Neuroendocrinology*, Columbia University Press, New York.

Schlinger BA and Arnold AP (1995) Estrogen synthesis and secretion by the songbird brain. In *Neurobiological Effects of Sex Steroid Hormones*, Micevych PE and Hammer RP (eds), Cambridge University Press, Cambridge, pp. 297-323.

Schlinger BA, Slotow RH and Arnold AP (1992) Plasma estrogens and brain aromatase in winter White-crowned sparrows. *Ornis Scandinavica*, **23**, 292-297.

Schmidt RS (1985) Prostaglandin-induced mating call phonotaxis in female American toad: facilitation by progesterone and arginine vasotocin. *Journal of Comparative Physiology* A, **156**, 823-829.

Schmidt BM, Gerdes D, Feuring M, Fulkenstein E, Christ M and Wehling M (2000) Rapid, nongenomic steroid actions: a new age? *Frontiers in Neuroendocrinology*, **21**, 57-94.

Schoech SJ, Mumme RL and Wingfield JC (1996) Prolactin and helping behaviour in the cooperatively breeding Florida scrub-jay, *Aphelocoma coerulescens*. *Animal Behaviour*, **52**, 445-456.

Schreck CB (1981) Stress and compensation in teleostean fishes: response to social and physical factors. In *Stress and Fish*, Pickering AD (ed.), Academic Press, London, pp. 295-321.

Schultz LM and Clark GA (1997) GABA-induced synaptic facilitation at type B to A photoreceptor connections in *Hermissenda*. *Brain Research Bulletin*, **42**, 377-383.

Scott AP and Liley NR (1994) Dynamics of excretion of $17\alpha,20\beta$-dihydroxy-4-pregnen-3-one 20-sulphate, and the glucuronides of testosterone and 17β-oestradiol, by urine of reproductively mature male and female rainbow trout (*Oncorhynchus mykiss*). *Journal of Fisheries Biology*, **44**, 117-129.

Sears MA and Rittschof D (1991) Control of chemically stimulated feeding behavior in sand fiddler crabs *Uca pugilator*: evidence for hemolymph feeding inhibitory factor. *Journal of Chemical Ecology*, **17**, 2337-2346.

Shapiro DY (1988) Behavioral influences on gene structure and other new ideas concerning sex change in fishes. *Environmental Biology of Fishes*, **23**, 283-297.

Sherry DF (1998) The ecology and neurobiology of spatial memory. In *Cognitive Ecology: The Evolutionary Ecology of Information Processing and Decision Making*, Dukas R (ed.), University of Chicago Press, Chicago, pp. 261-296.

Short RV and Balaban E (eds) (1994) *The Differences between the Sexes*, Cambridge University Press, Cambridge.

Silverin B and Goldsmith AR (1983) Reproductive endocrinology of free-living pied flycatchers (*Ficedula hypoleuca*): prolactin and FSH secretion in relation to incubation and clutch size. *Journal of Zoology (London)*, **200**, 119-130.

Silverin B, Baillien M, Foidart A and Balthazart J (2000) Distribution of aromatase activity in the brain and peripheral tissues of passerine and nonpasserine avian species. *General and Comparative Endocrinology* **117**, 34-53.

Simon NG, Cologer-Clifford A, Lu SF, McKenna SE and Hu S (1998) Testosterone and its metabolites modulate 5HT(1A) and 5HT(1B) agonist effects on intermale aggression. *Neuroscience and Biobehavioral Reviews*, **23**, 325-336.

Simon TP (1991) Startle response and causes of burying behavior in captive eastern sand darters, *Ammocrypta pellucida* (Putnam). *Proceedings of the Indiana Academy of Sciences*, **100**, 155-160.

Smale L, Nunes S and Holekamp KE (1997) Sexually dimorphic dispersal in mammals: patterns, causes, and consequences. *Advances in the Study of Behavior*, vol. 26. Academic Press, San Diego.

Smith RJF and Hoar WS (1967) The effects of prolactin and testosterone on the parental behaviour of the male stickleback *Gasterosteus aculeatus*. *Animal Behaviour*, **15**, 342-352.

Soma KK, Francis RC, Wingfield JC and Fernald RD (1996) Androgen regulation of hypothalamic neurons containing gonadotropin-releasing hormone in a cichlid fish: integration with social cues. *Hormones and Behavior*, **30**, 216-226.

Sorensen PW, Christensen TA and Stacey NE (1998) Discrimination of pheromonal cues in fish: emerging parallels with insects. *Current Opinion in Neurobiology*, **8**, 458-467.

Spieler RE and Clougherty JJ (1989) Free-running locomotor rhythms of feeding-entrained goldfish. *Zoological Science*, **6**, 813-816.

Spieler RE and Noeske TA (1984) Effects of photoperiod and feeding schedule on diel variations of locomotor activity, cortisol and thyroxine in goldfish. *Transactions of the American Fisheries Society*, **113**, 528-539.

Spieler RE, Noeske-Hallin TA, DeRosier TA and Poston HA (1987) Some dietary amino acids and meal-feeding phase shifts of locomotor activity. *Medical Science Research*, **15**, 921-922.

Stacey NE (1983) Hormones and reproductive behaviour in teleosts. In *Control Processes in Fish Physiology*, Rankin JC, Pitcher TJ and Duggan RT (eds), Wiley Interscience, New York, pp. 117-129.

Stacey NE and Goetz FW (1982) Role of prostaglandins in fish reproduction. *Canadian Journal of Fisheries and Aquatic Sciences*, **39**, 92-98.

Stanley-Samuelson DW (1994) The biological significance of prostaglandins and related eicosanoids in invertebrates. *American Zoologist*, **34**, 589-598.

Staub NL and DeBeer M (1997) The role of androgens in female vertebrates. *General and Comparative Endocrinology*, **108**, 1-24.

Stepanov II, Smirnova AG and Sapronov NS (1993) Effects of glutamic acid on defensive and feeding behaviour of the land snail *Helix lucorum*. *Zhurnal Vysshei Nervnoi Deyatel'Nosti Imeni I P Pavlova*, **43**, 915-924.

Strand FL (1999) *Neuropeptides: Regulators of Physiological Processes*, MIT Press, Cambridge, MA.

Stricker E (ed.) (1990) *Handbook of Behavioral Neurobiology*, vol. 10, *Neurobiology of Food and Fluid Intake*, Plenum, New York.

Striedter GF (1994) The vocal control pathways in budgerigars differ from those in songbirds. *Journal of Comparative Neurology*, **343**, 35-56.

Striedter GF (1998) Progress in the study of brain evolution: from speculative theories to testable hypotheses. *Anatomical Record*, **253**, 105-112.

Sukhedo M (1992) The behavior of parasitic flatworms in vivo: what is the role of the brain? *Journal of Parasitology* **78**, 231-242.

Svane I and Dolmer P (1995) Perception of light at settlement: a comparative study of two invertebrate larvae, a scyphozoan planula and a simple ascidian tadpole. *Journal of Experimental Marine Biology and Ecology*, **187**, 51-61.

Svare BB (1983) *Hormones and Aggressive Behavior*, Plenum, New York.

Svoboda KR and Fetcho JR (1996) Interactions between the neural networks for escape and swimming in goldfish. *Journal of Neuroscience*, **16**, 843-852.

Syed NI and Winlow W (1991) Coordination of locomotor and cardiorespiratory networks of *Lymnaea stagnalis* by a pair of identified interneurons. *Journal of Experimental Biology*, **158**, 37-62.

Takahashi LK (1990) Hormonal regulation of sociosexual behavior in female mammals. *Neuroscience and Biobehavioral Reviews*, **14**, 403-413.

Takase M, Ukena K, Yamazaki T, Kominami S and Tsutsui K (1999) Pregnenolone, pregnenolone sulfate, and cytochrome P450 side-chain cleavage enzyme in the amphibian brain and their seasonal changes. *Endocrinology*, **140**, 1936-1944.

Ter Maat A and Ferguson GP (1996) Neuronal input contributes to sequence of *Aplysia* egg laying behaviors. *Journal of Comparative Physiology* A, **179**, 775-783.

Teyke T (1996) Nitric oxide, but not serotonin, is involved in acquisition of food-attraction conditioning in the snail *Helix pomatia*. *Neuroscience Letters*, **206**, 29-32.

Teyke T, Xin Y, Weiss KR and Kupfermann I (1997) Ganglionic distribution of inputs and outputs of C-PR, a neuron involved in the generation of a food-induced arousal state in *Aplysia*. *Invertebrate Neuroscience*, **2**, 235-244.

Tinbergen N (1963) On aims and methods of ethology. *Zeitschrift für Tierpsychologie* **20**, 410-433.

Toguyeni A, Fauconneau B, Boujard T, Fostier A, Kuhn ER, Mol KA and Baroiller JF (1997) Feeding behavior and food utilization in tilapia, *Oreochromis niloticus*: effect of sex ratio and relationship with endocrine status. *Physiology and Behavior*, **62**, 273-279.

Townsend DS, Palmer B and Guillette LJ (1991) The lack of influence of exogenous testosterone on male parental behavior in a neotropical frog (*Eleutherodactylus*): a field experiment. *Hormones and Behavior*, **25**, 313-322.

Tramontin AD and Brenowitz EA (2000) Seasonal plasticity in the adult brain. *Trends in Neurosciences*, **23**, 251-258.

Truman JW (1992) Hormonal regulation of behavior: insights from invertebrate systems. In *Behavioral Endocrinology*, Becker JB, Breedlove SM and Crews D (eds), MIT Press, Cambridge, MA, pp. 423-451.

Trumbo ST (1996) Parental care in invertebrates. In *Parental Care: Evolution, Mechanisms, and Adaptive Significance* (*Advances in the Study of Behavior*, vol. 25), Rosenblatt JS and Snowdon CT (eds), Academic Press, San Diego, pp. 3-52.

Underwood H (1990) The pineal and melatonin: regulators of circadian function in lower vertebrates. *Experientia*, **46**, 120-128.

Underwood H (1992) Endogenous rhythms. In *Biology of the Reptilia*, volume 18, Physiology E: *Hormones, Brain, and Behavior*, Gans C and Crews D (eds), University of Chicago Press, Chicago, pp. 229-297.

Underwood H, Barrett RK and Siopes T (1990) The quail's eye: a biological clock. *Journal of Biological Rhythms*, **5**, 257-265.

Vleck CM and Dobrott SJ (1993) Testosterone, antiandrogen, and alloparental behavior in bobwhite quail foster fathers. *Hormones and Behavior*, **27**, 92-107.

Von Elert E and Loose C (1996) Predator-induced diel vertical migration in *Daphnia*: Enrichment and preliminary chemical characterization of a kairomone exuded by fish. *Journal of Chemical Ecology*, **22**, 885-895.

Wada M (1984) Effects of ventricularly implanted sex steroids on calling and locomotor activity in castrated male Japanese quail. *Hormones and Behavior*, **18**, 130-139.

Wade GN (1976) Sex hormones, regulatory behaviors and body weight. In *Advances in the Study of Behavior*, vol. 6, Rosenblatt JS, Hinde RA, Shaw E and Beer C (eds), Academic Press, New York, pp. 201-279.

Wade GN and Schneider JE (1992) Metabolic fuels and reproduction in female mammals. *Neuroscience and Biobehavioral Reviews*, **16**, 235-272.

Wallen K and Schneider JE (2000) *Reproduction in Context: Social and Environmental Influences on Reproductive Physiology and Behavior*, MIT Press, Cambridge, MA.

Wallen K and Tannenbaum PL (1997) Hormonal modulation of sexual behavior and affiliation in rhesus monkeys. In *The Integrative Neurobiology of Affiliation*, Carter CS, Lederhendler II and Kirkpatrick B (eds), New York Academy of Sciences, New York, pp. 185-202.

Wang Z and Insel TR (1996) Parental behavior in voles. In *Parental Care: Evolution, Mechanisms, and Adaptive Significance* (*Advances in the Study of Behavior*, vol. 25), Rosenblatt JS and Snowdon CT (eds), Academic Press, San Diego, pp. 361-384.

Weber DN (1993) Exposure to sublethal levels of waterborne lead alters reproductive behavior patterns in fathead minnows (*Pimephales promelas*). *Neurotoxicology*, **14**, 347-358.

Weber DN and Spieler RE (1987) Effects of the light-dark cycle and scheduled feeding on behavioral and reproductive rhythms of the cyprinodont fish, Medaka, *Oryzias latipes*. *Experientia*, **43**, 621-624.

Weber DN and Spieler RE (1994) Behavioral mechanisms of metal toxicity in fishes. In *Aquatic Toxicology: Molecular, Biochemical, and Cellular Perspectives*, Malins DC and Ostrander GK (eds), Lewis Publ., Boca Raton, FL, pp. 421-467.

Weeks JC and Levine RB (1995) Steroid hormone effects on neurons subserving behavior. *Current Opinion in Neurobiology*, **5**, 809-815.

Weeks JC, Jacobs GA and Miles CI (1989) Hormonally mediated modifications of neuronal structure, synaptic connectivity, and behavior during metamorphosis of the tobacco hornworm, *Manduca sexta*. *American Zoologist*, **29**, 1331-1344.

Weiger WA (1997) Serotonergic modulation of behaviour: a phylogenetic overview. *Biological Reviews*, **72**, 61-95.

Weisburg MJ and Derby CD (1995) Regulation of sex-specific feeding behavior in fiddler crabs: physiological properties of chemoreceptor neurons in claws and legs of males and females. *Journal of Comparative Physiology A*, **176**, 513-526.

Weldon PJ (1990) Responses by vertebrates to chemicals from predators. In *Chemical Signals in Vertebrates*, vol. 5, MacDonald DW, Müller-Schwarze D and Natynczuk SE (eds), Oxford University Press, Oxford, pp. 500-521.

Wendelaar-Bonga SE (1997) The stress response in fish. *Physiological Reviews*, **77**, 591-625.

Wenzel BM (1986) The ecological and evolutionary challenges of procellariiform olfaction. In *Chemical Signals in Vertebrates*, vol. 4, Duvall D, Müller-Schwarze D and Silverstein RM (eds), Plenum Press, New York.

West BR, Nichter LS and Halpern D (1991) Leech therapy: when once is not enough. *Blood Coagulation, Fibrinolysis and Kinin*, **2**, 197-200.

Whittier JM and Tokarz RR (1992) In *Biology of the Reptilia*, vol. 18, Physiology E: *Hormones, Brain, and Behavior*, Gans C and Crews D (1992), University of Chicago Press, Chicago, pp. 24-69.

Wiens BL and Brownell PH (1994) Neuroendocrine control of egg-laying behavior in the nudibranch, *Archidoris montereyensis*. *Journal of Comparative Neurology*, **344**, 619-625.

Wilczynski W, Allison JD and Marler CA (1993) Sensory pathways linking social and environmental cues to endocrine control regions of amphibian forebrains. *Brain, Behavior and Evolution*, **42**, 252-264.

Wiley RH, Steadman L, Chadwick L and Wollerman L (1999) Social inertia in white-throated sparrows results from recognition of opponents. *Animal Behaviour*, **57**, 453-463.

Willard AL (1981) Effects of serotonin on the generation of the motor program for swimming by the medicinal leech. *Journal of Neuroscience*, **1**, 936-944.

Winberg S and Nilsson GE (1993) Time course of changes in brain serotonergic activity and brain tryptophan levels in dominant and subordinate juvenile arctic charr. *Journal of Experimental Biology*, **179**, 181-195.

Wingfield JC and Farner DS (1993) Endocrinology of reproduction in wild species. In *Avian Biology*, vol. IX, Farner DS, King JR and Parkes KC (eds), Academic Press, London, pp. 164-328.

Wingfield JC and Marler P (1994) Communication in vertebrate reproduction: the role of hormones. In *The Physiology of Reproduction* (2nd edn), vol. 2, Knobil E and Neill JD (eds), Raven, New York, pp. 1647-1677.

Wingfield JC, Maney DL, Breuner CW, Jacobs JD, Lynn S, Ramenofsky M and Richardson RD (1998) Ecological bases of hormone-behavior interactions: the emergency life history stage. *American Zoologist*, **38**, 191-206.

Withers GS, Fahrbach SE and Robinson GE (1993) Selective neuroanatomical plasticity and division of labour in the honeybee. *Nature*, **364**, 238 - 240.

Wood DE, Gleason RA and Derby CD (1995) Modulation of behavior by biogenic amines and peptides in the blue crab, *Callinectes sapidus*. *Journal of Comparative Physiology* A, **177**, 321 - 333.

Wood RI and Newman SW (1995a) Hormonal influences on neurons of the mating behavior pathway in male hamsters. In *Neurobiological Effects of Sex Steroid Hormones*, Micevych PE and Hammer RP (eds), Cambridge University Press, Cambridge, pp. 3 - 39.

Wood RI and Newman SW (1995b) Integration of chemosensory and hormonal cues is essential for mating in the male Syrian hamster. *Journal of Neuroscience*, **15**, 7261 - 7269.

Woods SC, Seeley RJ, Porte D and Schwartz MW (1998) Signals that regulate food intake and energy homeostasis. *Science*, **280**, 1378 - 1383.

Woolley CS (1999) Electrophysiological and cellular effects of estrogen on neuronal function. *Critical Reviews in Neurobiology*, **13**, 1 - 20.

Wubbels RJ, Kroese ABA and Schellart NA (1993) Response properties of lateral line and auditory units in the medulla oblongata of the rainbow trout (*Oncorhynchus mykiss*). *Journal of Experimental Biology*, **179**, 77 - 92.

Wudkevich K, Wisenden BD, Chivers DP and Smith RJF J (1997) Reactions of *Gammarus lacustris* to chemical stimuli from natural predators and injured conspecifics. *Journal of Chemical Ecology*, **23**, 1163 - 1173.

Wysocki CJ (1979) Neurobehavioral evidence for the involvement of the vomeronasal system in mammalian reproduction. *Neuroscience and Biobehavioral Reviews*, **3**, 301 - 341.

Xu X, Boshoven W, Lombardo B and Spranger J (1998) Comparison of the amnestic effects of NMDA receptor antagonist MK-801 and nitric oxide synthase inhibitors: L-NAME and L-NOARG in goldfish. *Behavioral Neuroscience*, **112**, 892 - 899.

Yahr P (1983) Hormonal influences on territorial marking behavior. In *Hormones and Aggressive Behavior*, Svare BB (ed.), Plenum, New York, pp. 145 - 176.

Yamazaki F (1990) The role of urine in sex discrimination in the goldfish *Carassius auratus*. *Bulletin of the Faculty of Fisheries Hokkaido University*, **41**, 155 - 161.

Yazaki Y, Matsushima T and Aoki K (1999) Testosterone modulates stimulation-induced calling behavior in Japanese quails. *Journal of Comparative Physiology A*, **184**, 13 - 19.

Yen J and Strickler JR (1996) Advertisement and concealment in the plankton: what makes a copepod hydrodynamically conspicuous? *Invertebrate Biology*, **115**, 191 - 205.

Young LJ and Crews D (1995) Comparative neuroendocrinology of steroid receptor gene expression and regulation: relationship to physiology and behavior. *Trends in Endocrinology and Metabolism*, **6**, 317 - 323.

Young LJ, Wang Z and Insel TR (1998) Neuroendocrine bases of monogamy. *Trends in Neurosciences*, **21**, 71 - 75.

Zakharov IS, Ierushalimsky VN and Balaban PM (1995) Pedal sertonergic neurons modulate the synaptic input of withdrawal interneurons of *Helix*. *Invertebrate Neuroscience*, **1**, 41 - 52.

Zakon HH (1996) Hormonal modulation of communication signals in electric fish. *Developmental Neuroscience*, **18**, 115 - 123.

Zakon HH (1998) The effects of steroid hormones on electrical activity of excitable cells. *Trends in Neurosciences*, **21**, 202 - 207.

Zerani M and Gobbetti A (1996) NO sexual behaviour in newts. *Nature*, **382**, 31.

Zielinski WJ, vom Saal FS and Vandenbergh JG (1992) The effect of intrauterine position on the survival, reproduction and home range size of female house mice (*Mus musculus*). *Behavioral Ecology and Sociobiology*, **30**, 185–191.

Zumpe D and Michael RP (1996) Social factors modulate the effects of hormones on the sexual and aggressive behavior of macaques. *American Journal of Primatology*, **38**, 233–261.

Zupanc GKH (1999) Neurogenesis, cell death and regeneration in the adult gymnotiform brain. *Journal of Experimental Biology*, **202**, 1435–1446.

3

Environmental Contaminants Altering Behaviour

MACE G. BARRON

P.E.A.K. Research, Longmont, CO, USA

3.1 INTRODUCTION

The number and type of environmental chemicals that can alter behaviour is large. For example, a partial list of neurotoxicants (National Research Council [NRC] 1992) includes 88 chemicals ranging from anaesthetic gases and solvents, to metals and pesticides. Specific behavioural effects are also caused by multiple chemicals and chemical mixtures. The NRC (1992) shows 183 chemicals causing convulsions, 135 chemicals causing equilibrium changes, and 125 chemicals inducing narcosis. In addition, many chemicals and chemical classes cause the same neurological syndromes (Niesink *et al.* 1999). For example, seizures can be induced by solvents, metals, carbamate and organophosphorus (OP) insecticides (acetylcholinesterase [AChE] inhibitors), and organochlorine pesticides.

Drummond and Russom (1990) categorised over 300 organic chemicals from 49 chemical classes according to three acute behavioural toxicity syndromes in fish: hypoactivity (depressed locomotor activity, loss of startle response), hyperactivity (accelerated locomotor activity, overreaction to stimuli, increased ventilatory activity), and physical deformity (convulsions, tetany). Multiple syndromes were observed for most of the chemical classes. The hypoactivity syndrome was caused by narcotic chemicals including alcohols, ketones, esters, phthalates, amides, ethers, secondary and tertiary amines, pyridines, and barbitals. The hyperactivity syndrome was observed after exposure to chemicals that disrupt metabolic activity or function (e.g., uncouplers of oxidative phosphorylation), including diols, aldehydes, primary and secondary amines, and halogenated and nitro-substituted phenols. The physical deformity syndrome was induced by chemicals that disrupt or damage the nervous system (e.g., AChE inhibitors, CNS seizure agents) including pesticides and halogenated benzenes. Later, Russom *et al.* (1997) categorised these organic chemicals into

Behavioural Ecotoxicology. Edited by Giacomo Dell'Omo.

five groups based on mode of action: narcotics, oxidative phosphorylation uncouplers, AChE and CNS seizure agents, respiratory inhibitors, and chemicals with electrophile reactivity mechanisms. Two or all three of the behavioural toxicity syndromes of Drummond and Russom (1990) were observed for each of these mode-of-action categories.

Behavioural toxicants can act through different modes of action after acute or chronic exposures. Acute chemical exposure may cause general behavioural syndromes such as hypoactivity, hyperactivity and seizures through reversible epithelial damage and disruption of biochemical processes (Russom *et al.* 1997). In contrast, chronic exposure to behavioural toxicants may occur through irreversible damage to nerve tissues resulting in encephalopathy, cerebellar dysfunction, as well as peripheral and central neuropathy (NRC 1992, Niesink *et al.* 1999). Thus, categorising chemicals into discrete groupings and developing a classification scheme is inevitably imprecise and based on generalisations.

3.1.1 CATEGORISING BEHAVIOURAL TOXICANTS

This chapter generalises the behavioural toxicity of a broad range of chemicals to a diversity of animal species to arrive at a classification scheme for environmental contaminants. Our classification scheme, summarised in Table 3.1, is based on general grouping of environmental contaminants based on their chemical similarities and their predominant mode of action. In addition, Table 3.1 generalises the predominant type of behavioural impairments caused by the pollutants, but does not indicate the likelihood of behavioural effects occurring in the environment. We have focused on chemicals occurring in terrestrial (soil, plant) and aquatic (water, sediment) environments and that accumulate in organisms; widespread atmospheric contaminants such as SO_2, O_3, or NO can also affect behaviour (Weiss 1998) are not discussed here.

Our classification scheme includes the following eight categories:

1. narcotic chemicals: low molecular weight solvents causing narcosis (e.g., alkanes, alcohols, and ketones),
2. excitatory agents: halo and nitro substituted phenols and anilines causing hyperactivity through uncoupling of oxidative phosphorylation,
3. metals causing tissue damage or biochemical impairments,
4. organometals causing nerve damage,
5. ChE inhibitors: organophosphate and carbamate pesticides,
6. reactive chemical: with an electrophilic mechanism of action (e.g., aldehydes, alkenes, alkynes, and alcohols),
7. CNS seizure agents: organochlorine (OC) pesticides and pyrethroids, and
8. endocrine disrupters: polycyclic aromatic compounds (PAHs) and other xenobiotics causing modulation of hormone or receptor activity.

We recognise that classifications based on chemical classes can be problematic. For example, chemical classes inducing narcosis also include reactive

Table 3.1 Classification Scheme for Behavioural Toxicants (adapted from NRC 1992, Russom *et al*. 1997 and Niesink *et al*. 1999)

Categories	Representative chemicals	Sources	Relative exposure[a] in environment aquatic/terrestrial	Exposure/mode of action	Major behavioural effects
Narcotics (e.g., low molecular weight alkanes, alcohols, ketones)	Pentanol, dichlorpropane, naphthol, phenol, octanone	Industrial use, petroleum	+/−	Acute/narcosis	Hypoactivity
Excitatory agents (e.g., polyhalo- and nitro-anilines, phenols, pyridines)	Dinitrophenols, pentachlorophenol, tetrachloroanilines	Industrial use	+/−	Acute/oxidative phosphorylation Uncoupling	Hyperactivity, hyperreactivity
Metals	Lead, cadmium, copper, mercury,	Agricultural use, combustion, mining	+++/+	Acute and chronic/membrane damage, metabolic interaction	Feeding activity, learning, reproduction and parental care
Organometals	Methylmercury	Biocide use	+/+	Chronic/nerve tissue damage	Reproduction, parental care and learning
Che inhibitors	OP and carbamate insecticides, malathion, dimethoate chlorpyrifos, carbaryl	Agricultural use	+/++	Acute/ChE inhibition	Hypoactivity, behavioural depression
Reactive chemicals	Acrolein, benzaldehyde	Industrial use	+/−	Acute/electrophilic reaction with cell macromolecules	Incoordination, hyporeactivity
CNS seizure agents	OC pesticides, pyrethroids, DDT, dieldrin, lindane fenvalerate, permetrin	Agricultural use	+/++	Acute/central nervous system interaction	Incoordination, seizures, hyperactivity, ataxia, convulsions, learning deficits
Endocrine disrupters	PAHs, PCBs, TCDD	Industrial use, petroleum, combustion	+/++	Chronic/endocrine disruption	Reproductive and social behaviour

[a]Relative risk level of behavioural effects: − minimal; + low; ++ moderate; +++ high

compounds acting through an electrophilic-based mode of action (Russom *et al.* 1997). Thus, our classification scheme should be viewed as a generalisation of the behavioural effects of environmental contaminants, and is not intended to substitute chemical-specific and species-specific evaluations of behavioural alterations.

Section 3.2 describes the source, use, environmental fates, mode of action, and main behavioural effects of the chemicals in each category. Section 3.3 provides conclusions and summarises the taxonomy for behaviour toxicants; in addition, it discusses the relative ecological risks of environmental contaminants altering behaviour.

3.2 SOURCE, FATE, MODE OF ACTION AND BEHAVIOURAL EFFECTS OF BEHAVIOURAL TOXICANTS

3.2.1 NARCOTIC CHEMICALS

Narcotics are a structurally diverse group of low molecular weight industrial organic chemicals that cause hypoactivity and depression of sensory functions. They are the largest group of behavioural toxicants and include alkanes, benzenes, simple alcohols and phenols, ketones, esters, and phthalates. Narcotic chemicals are used as solvents (e.g., for metal cleaning) and chemical intermediates, and in paints, adhesives and protective materials. Narcotics enter the environment during manufacturing processes, and alkane and benzene compounds are also important components of petroleum (Albers 1995). They are not reactive, and generally have moderate water solubility, high soil mobility, high volatility, exhibit rapid equilibration between exposure water and tissue, low bioaccumulation and partition primarily in the vapour phase. Their mode of action is through a nonspecific mechanism causing anaesthetic effects that may involve changes in membrane fluidity and altered membrane protein functions secondary to interaction with membrane lipids or direct binding to proteins (van Wezel *et al.* 1996).

3.2.1.1 Behavioural effects

Narcotic chemicals cause a syndrome of hypoactivity and reduced responsiveness to outside stimuli. The anaesthetic effects of narcotic chemicals result from a nonspecific partitioning of the chemical in nerve tissue. Narcotics have been extensively studied in fish because these chemicals are water soluble and are rapidly absorbed.

Few studies have evaluated the effects of sublethal or chronic exposure to narcotic chemicals. In one study, medaka (*Orizias latipes*) appeared to be more susceptible to predation after exposure to sublethal concentrations of phenol, 1-octanol, and nitrophenol (Carlson *et al.* 1998). Behavioural effects of acute exposure of birds and mammals to narcotics include slowness, hind limb weakness, paralysis, drowsiness, vertigo, giddiness, rapid and shallow breathing, loss of coordination, and tremors (Molnar *et al.* 1986). Thus, narcotics seem to affect motor control, rather than cognitive capacity (Bushnell 1989). Narcotics can also cause initial hyperexcitability and convulsions, and cause degeneration of the spinal cord and spinal root (Windus-Podehl *et al.* 1983). Chronic exposure of mammals to narcotics may result in axonal neuropathy, ataxia (altered muscle coordination and movement), and encephalopathy (Niesink *et al.* 1999).

3.2.2 EXCITATORY AGENTS

The excitatory agents are a small group of phenolic chemicals with a single aromatic ring (e.g., phenols, anilines, pyridines) containing multiple electron withdrawing substituents (e.g., halogens and nitro groups). Chemicals include chlorophenols and nitrophenols in this category which are used as solvents, chemical intermediates, and biocides. Chlorophenols, particularly pentachlorophenol (PCP), have been widely used as pesticides and wood preservatives. The excitatory agents generally have lower water solubility, higher affinity for sediments and soil, low volatility, and are generally more persistent in the environment than narcotic agents (Rice and O'Keefe 1995). The mode of action of the excitatory agents involves an uncoupling of oxidative phosphorylation in mitochondria, resulting in dissociation of the electron transport process from the generation of ATP (Terada 1990).

3.2.2.1 Behavioural effects

The excitatory agents cause hyperactivity and overreaction to physical and environmental stimuli in fish, birds, and mammals. General behavioural responses to acute exposure to these chemicals include irritant responses, increased respiratory rate, tremors, convulsions, and loss of righting reflex. Gomez *et al.* (1998) studied the effect of PCP on the predator–prey relationships of two rotifers. PCP decreased the number of preys captured, the number of prey ingested, and increased the time to first ingestion by the predator rotifer. In another study, medaka exposed to 2,4-dinitrophenol were more susceptible to predation (Carlson *et al.* 1998). In mallards and pheasants, acute exposure to PCP caused wing shivers and twitching, jerks and shakes, ataxia, imbalance, tremors, and spasms (Hudson *et al.* 1984).

3.2.3 METALS

Metals are inorganic elements distributed in the earth's crust; natural sources of exposure include soil/dust, volcanoes, decaying vegetation, forest fires, and sea spray. Metals are used in electroplating, ammunitions, alloys, semiconductors, fuel additives and electronic components. Anthropogenic activities that introduce metals into the environment include smelting, iron foundries, combustion sources (e.g., incinerators, car exhausts), mining operations and tailings, sewage sludge, hunting, and agricultural uses (e.g., copper herbicides). The environmental chemistry of metals is complex because each metal (1) exhibits chemical speciation (changes in chemical form in response to changes in physical/chemical conditions), (2) has diverse affinities for organic ligands, and (3) has an array of biological effects (Hammond and Beliles 1980). In general, metals have limited volatility, moderate to high water solubility (pH-dependent), strong soil adsorption, and low bioaccumulation.

Metals causing behavioural toxicity include lead, mercury, copper, selenium, aluminium, tin, cobalt, cadmium, chromium, manganese, and zinc. Metals cause behavioural toxicity by competing or interfering with normal biochemical processes, or by damaging nerve or epithelial tissues. For example, exposure to lead can damage the hippocampal cholinergic system (Bielarczyk *et al.* 1996), and the neuromuscular junction (Atchison and Narahashi 1984). Copper impairs sensory mechanisms of fish through damage to the olfactory epithelium, resulting in impairments of orientation, behavioural avoidance, and responsiveness to external food stimuli (Hansen *et al.* 1999).

3.2.3.1 Behavioural effects

Although there are relatively few metals compared with the number of organic chemicals in each behavioural category of toxicant effects, generalisations on metal effects on behaviour is difficult because of the diversity of responses observed in animals. Nearly all metals have been reported to elicit behavioural responses in aquatic organisms, and often the most sensitive indicators of heavy metal toxicity in fish are behavioural responses. The behavioural effects of metals have been extensively studied in fish and include behavioural avoidance, abnormal respiratory responses, impaired feeding behaviour, and altered orientation (Atchison *et al.* 1987). Fish exhibit behavioural avoidance to metals at sublethal concentrations but may not exhibit this response at higher concentrations (Giattina and Garton 1983, Hartwell *et al.* 1989). Exposure of fish to metals, including copper, zinc, and cadmium, also elicits cough responses and altered ventilatory behaviour, possibly in response to irritation of the gills or the buccal cavity (Diamond *et al.* 1990). The feeding behaviour of fish is also affected by metal exposure, including changes in the quantity of food intake, feeding time, and prey localisation (Sullivan *et al.* 1978, Sandheinrich and Atchison 1989). Other studies suggest that metals affect the swimming performance or alter the downstream migration of anadromous fish (Lorz *et al.* 1978, Waiwood and Beamish 1978). Behaviours that are guided by olfaction are also impaired by metal exposure, possibly through damage of the olfactory mucosa. Metal exposure also causes changes in activity levels and frequency and patterns of movements. Ellgaard and Guillot (1988) observed a 33% decrease in activity in bluegill sunfish, *Lepomis macrochirus*, following exposure to 40 μg copper/l and higher, while Drummond *et al.* (1973) observed an increase in activity in the brook trout, *Salvelinus fontinalis*, for 6–8 h after exposure to 6 μg copper/l and higher. Maintenance behaviours have also been shown to increase following exposure to copper. For instance, bluegills exposed to copper showed an increase in the frequency of yawns, jerks, fin flicks, threats, nudges and nips (Henry and Atchison 1986). Activity levels have been shown to increase following exposure to zinc and cadmium (Bengtsson 1974). In addition, metal exposure resulted in impaired learning, reduced aggressive behaviour in subordinate fish, and increased in dominant fish (Atchison *et al.* 1987).

The majority of studies reporting on the behavioural effects of metals on terrestrial wildlife have focused on lead (Rice 1993). In birds, lead causes neurotoxicity and behavioural impairments, including hyperactivity, distractibility, disruption of parental behaviour, impaired sibling recognition, impaired predator avoidance, and decreased learning ability (a more detailed description of the effects of lead on birds is given in Chapter 9). Acutely toxic doses of lead may cause loss of coordination and paralysis. For example, Douglas-Stroebel (1997) observed decreased swimming activity in mallard ducklings ingesting lead-contaminated sediments compared with ducklings that ingested reference area sediments. Eisler (1988) concluded that lead causes neurobehavioural deficits such as learning impairment at dose levels that produce no obvious signs of toxicity and at very low blood levels. In rats, lead exposure alters development, affects specific motor skills, and can result in long-term cognitive deficits (Kuhlmann *et al.* 1997, Mello *et al.* 1998). Laboratory studies with mammalian species have demonstrated that lead impairs learning and memory functions by disrupting hippocampal functions at nearly all life stages of animals (Rice 1993). Sublethal lead exposure during prenatal and early development of animals and humans can produce behavioural changes that persist beyond the period of exposure (Needleman and Gatsonis 1990, Riess and Needleman 1992).

3.2.4 ORGANOMETALS

Organometals are metals (e.g., tin, lead, mercury) that are covalently bonded to carbon groups (e.g., methyl, ethyl). In contrast to inorganic forms, organometals have low water solubility, and can extensively bioaccumulate. Sources of organometals include biocides (e.g., trialkyl tins), chemical catalysts and stabilisers (e.g., mono- and dialkyl tins), chemical additives (e.g., tetraethyl lead), and environmental alkylation of inorganic metals (e.g., methylmercury). Organometals have high affinity for soil, sediment and particulates, and are translocated via atmospheric deposition and runoff (Zillioux *et al.* 1993). Organotin compounds appear to be inhibitors of oxidative phosphorylation (WHO

1980). Acute exposure to organometals such as methylmercury causes deregulation of the dopaminergic system in the CNS (Atchison and Hare 1994). Chronic exposure to most organometals can cause peripheral neuropathy and impaired learning and memory (Cohen *et al.* 1987).

3.2.4.1 Behavioural effects

An example of an organometal with significant behavioural effects is methylmercury (MeHg), which is formed by microbial methylation of inorganic mercury. MeHg induced hyperactivity or hypoactivity, depressed swimming performance, and decreased avoidance of predator crabs by larval mummichogs (*Fundulus heteroclitus*; Zhou and Weis 1998). Heinz (1979) evaluated the behavioural effects of MeHg on birds by feeding the compound to three generations of mallard ducks *Anas platyrhynchos*. Ducklings from parents fed MeHg were less responsive than controls to tape-recorded maternal calls, but were hyperresponsive to a frightening stimulus in avoidance tests; there were no significant differences in locomotor activity in an open-field test. The author concluded that MeHg had a specific effect on approach behaviour to maternal calls and avoidance behaviour, and not a generalised effect on activity level or exploratory behaviour. In nestling great egrets chronically exposed to dietary MeHg, Bouton *et al.* (1999) observed reduced activity, increased tendency to seek shade, and reduced fishing activity. Humans can be exposed to this neurotoxic agent by consuming predatory fishes (Weiss 1995).

3.2.5 CHOLINESTERASE INHIBITORS

ChE inhibitors are organophosphate and carbamate pesticides which are intentionally applied in agricultural and residential uses, and may also be introduced into the environment through accidental runoff, overspray, or aerial drift. Examples of OPs include diazinon, chlorpyrifos, parathion, and dimethoate; representative carbamates are carbofuran and carbaryl. They are formulated as esters, amides or thiol derivatives which affect their physical/chemical properties and environmental fate (WHO 1986a,b). In general, they have low solubility in water, high soil binding and low volatility. Compared with the chlorinated pesticides (e.g., DDT, dieldrin), OPs and carbamates are relatively nonbioaccumulative and nonpersistent, with half-lives ranging from days to weeks. Carbamate insecticides are generally even less persistent in the environment and less hydrophobic than OPs.

OP and carbamate pesticides cause toxicity through inhibition of cholinesterase enzymes (ChEs) and particularly acetylcholinesterase (AChE), the enzyme that hydrolyses acetylcholine (Hill 1995). Excessive accumulation of this primary neurotransmitter results in the blockage of synaptic transmission in the cholinergic system. Death likely results when neurotransmissions are blocked in the brain's respiratory centre or in the myoneural junctions of respiratory muscles when brain AChE activity is inhibited by more than 70–80%. Determining exposure to ChE inhibitors based on chemical residues in tissues is problematic because they are rapidly metabolised and excreted from animals. Thus, determination of brain or blood cholinesterase activity has become an accepted indicator of exposure to these chemicals both in vertebrates and invertebrates (Dell'Omo *et al.* 1996c).

3.2.5.1 Behavioural effects

In general, acute exposures to ChE-inhibiting compounds cause intensified conditioned motor reflexes at low doses and a general behavioural impairment at higher doses (Clark 1971, Peakall 1985). Chronic low level exposure to some OPs may cause peripheral neuropathy, resulting in tremors and paralysis (Hill 1995). The larvae of an aquatic insect (caddisfly) exposed to the OP diazinon exhibited an increase in duration and variability of ventilation activity (van der Geest *et al.* 1999). Rainbow trouts, *Oncorhynchus mykiss*, acutely exposed to chlorpyrifos, malathion, or carbaryl did not exhibit behavioural signs of intoxication (Bradbury *et al.* 1991), whereas Japanese medaka, *Orizias latipes*, exhibited hypoactivity and loss of equilibrium following acute exposure to chlorpyrifos (Rice *et al.* 1997). Medaka also appeared to be more susceptible to predation after exposure to sublethal concentrations of chlorpyrifos and carbaryl (Carlson *et al.* 1998).

Inhibition of ChEs by OPs and carbamates produces a general behavioural depression in wildlife that is similar across different species and chemicals (Dell'Omo *et al.*[2] 2002; Grue[1] *et al.* 1997). In birds, exposure to the OP acephate and to the carbamate aldicarb caused similar effects, consisting, with increasing severity, of ataxia, imbalance, hopping and falling, mild spasms in the legs and feet, wing-drop and wing spread, hyporeactivity, convulsions, tetany, and immobility (Hudson *et al.* 1984). Kenaga (1974) reported that house sparrows (*Passer domesticus*), common grackles (*Quiscalus quiscala*) and red-winged blackbirds (*Agelaius phoeniceus*) were repelled by seeds containing 1% chlorpyrifos, while ring-necked pheasants (*Phasianus colchicus*) were not. Mallard ducklings exposed to toxic levels of methyl-parathion preened and loafed on land, and emitted distress calls, while control ducklings primarily fed and swam in open water (Fairbrother *et al.* 1988). Exposed mallards also exhibited changes in incubation behaviour, including nest abandonment and reduced nest attentiveness (Bennett *et al.* 1991). Sublethal parathion exposure to laughing gulls (*Larus atricilla*) also caused a transient reduction in nest attentiveness (White et al. 1983). A more exhaustive description of the effects of these toxicants on bird behaviour is given in Chapter 9.

In mammals exposure to the OP dimethoate altered open field exploration, digging, grooming and sniffing behaviours in the common shrew (*Sorex araneus*), and affected the open field behaviour of wood mice (*Apodemus *sylvaticus*; Dell'Omo and Shore 1996a; Dell'Omo *et al.* 1997). Exposure to the same sublethal doses in the wild caused a transient reduction in nocturnal activity in wood mice but no effects on retrappability of bank voles (*Clethrionomys glareolus*, Dell'Omo and Shore 1996b, Shore and Dell'Omo 1998).

3.2.6 REACTIVE CHEMICALS

Reactive chemicals are low-molecular weight compounds that can react with DNA, enzymes, and other proteins, causing irritation or damage to mucous membranes and nerve tissues (Verhaar *et al.* 1996). Reactive chemicals include aldehydes, unsaturated aliphatics (e.g., alkenes, alkynes), and alcohols (e.g., propargylics). These chemicals contain structural features that allow electrophilic or proelectrophilic (i.e., metabolic activation to an electrophilic compound) reaction with the nucleophilic moieties (e.g., amino, thiol groups) of cellular macromolecules. For example, hexachlorobutadiene is metabolically transformed to a reactive sulphur conjugate that damages renal tissue (WHO 1994). Sources of reactive chemicals include biocides (e.g., acrolein), chemical intermediates (such as those for polymer production), products of organic pyrolysis (e.g., automobile emissions, manufacturing, pulp and paper effluents), solvents, and industrial fluids. Because of their various chemical structures, reactive chemicals can have either low (e.g., chlorinated alkenes) or high (e.g., acrolein) water solubility, moderate volatility, and variable soil binding.

3.2.6.1 Behavioural effects

Reactive chemicals can alter the functionality of cellular components resulting in a diversity of behavioural and neurotoxic responses, including narcosis, hyperactivity and convulsions (Russom *et al.* 1997). Acute acrylamide exposure causes CNS effects, including ataxic tremors, severe tonic-clonic convulsions, and other signs of diffuse central excitation whereas chronic exposures to acrylamide causes peripheral neuropathy, tremors, incoordination, ataxia, and muscular weakness. Loss of reflexes and blunting of tactile sensitivity, such as loss of sensitivity to vibrations, has also been observed in mammals exposed to acrylamide (Crofton *et al.* 1996, Ko *et al.* 1999). Acrolein and benzaldehyde irritated the mucous membranes in rainbow trout, and caused an increase in cough rate through a direct effect at the respiratory surface (McKim *et al.* 1987). Mallards exposed to acrolein exhibited slow responses, incoordination, vocalisations reduced, wing tremors, running and falling, withdrawing, and reluctance to leave the swimming area (Hudson *et al.* 1984).

3.2.7 CNS SEIZURE AGENTS: ORGANOCHLORINE PESTICIDES AND PYRETHROIDS

The CNS seizure agents include two general classes of chemicals: the organochlorine (OCs) and the pyrethroid insecticides. The OCs are a group broad spectrum of chlorinated insecticides produced commercially beginning in the 1940s for agricultural and residential use (Blus 1995). Most uses of OCs have been banned worldwide because of extreme environmental persistence (half-lives range from months to years), and extensive bioaccumulation and magnification in the food web. The OCs include ethanes (e.g., DDT), cyclodienes (e.g., chlordane, dieldrin), hexachlorocyclohexane (lindane), chlorinated camphenes (toxaphene), mirex, and chlordecone (kepone). OCs have high lipid and low water solubility, and high affinity for sediments and soils. They are translocated through soil runoff and airborne dusts, and suspended sediments (WHO 1989). Pyrethroids include the synthetic insecticides cypermethrin, permethrin, and fenvalerate. Pyrethroids are substantially less persistent in the environment than the OCs, and are less hydrophobic. Acute exposure to pyrethroids may damage CNS tissues or affect ion permeability (Norton 1980).

3.2.7.1 Behavioural effects

Acute exposure to OCs and pyrethroids increases spontaneous locomotor activity with a high incidence of convulsions, spasms and tetany (Murphy 1980). Effects also include paralysis, abnormal coordination, and decreased memory retention (Blus 1995). Chronic or low-dose exposure to OCs may alter reproductive behaviour (e.g., courtship, parenting) through an endocrine disruptive mechanism (e.g., hormone, receptor, or enzyme modulation; Uphouse and Williams 1989).

Decreased foraging activity of honey bees was observed for a few hours following spraying of the pyrethroid alpha-cypermethrin (Shires 1984). Goodrich and Lech (1990) reported that exposure to the OC lindane altered the spatial orientation of *Daphnia magna* and increased the randomness of vertical migration. Fish exposed to the OCs endosulfan and endrin, and the pyrethroids cypermethrin and fenvalerate, exhibited increased cough frequency, recurrent tremors and tetany of pectoral fins, convulsions, rapid pectoral movements, and tonic paralysis (Bradbury *et al.* 1991). Fish exposed to DDT exhibited hyperactivity, and changes in feeding, in salinity or temperature preference, and orientation (WHO 1989). The pyrethroids cypermethrin and fenvalerate caused seizures in the rainbow trout (Bradbury *et al.* 1991), and permethrin caused both hyper- and hypoactivity, loss of equilibrium and excessive lateral flexure in medaka (Rice *et al.* 1997). Medaka appeared also to be more susceptible to predation after exposure to sublethal concentrations of fenvalerate (Carlson *et al.* 1998). Exposure of amphibians to DDT during limb development and tail resorption caused uncoordinated hyperactivity and loss of equilibrium (Cooke 1972).

In birds OCs cause behavioural effects at approximately ten times lower tissue concentrations than lethal levels, and at similar tissue concentrations causing biochemical and physiological impairments (Peakall 1985). Effects consisted of decreased aggression, impaired avoidance, reduced nest defence and reduced nest attentiveness. In mallards, pheasants and other granivorous birds acute exposure to aldrin and chlordane caused ataxia, hyper- and hypoactivity, circling, excessive drinking, tremors, increased vocalisations, violent wing-beating convulsions and seizures (Hudson *et al.* 1984). Exposure of birds to the DDT degradate DDE caused ducklings to be hyper-responsive to maternal calls (Heinz 1976), and decreased the courtship activity of ring doves (*Streptotelia risoria*) (WHO 1989). Bats (*Tadarida brasiliensis*) environmentally exposed to DDE showed hyperactivity, intermittent audiogenic seizures (produced by sound), and violent contractions of chest muscles (Geluso *et al.* 1976). Rats exposed to lindane were less active, exhibited decreased sexual receptivity, and had higher error rates in maze performance (WHO 1991).

3.2.8 ENDOCRINE DISRUPTERS

An increasing number of environmental contaminants, encompassing many behavioural toxicant categories, are believed to interfere with the endocrine functions of aquatic and terrestrial organisms. Among them polycyclic aromatic compounds play an important role as Endocrime disrupters. These are a diverse group of chemicals that contain two or more benzene or substituted benzene rings, and include polycyclic aromatic hydrocarbons (PAHs) (e.g., anthracene, benzo(a)pyrene) and heterocycles (e.g., acridine, dibenzothiophenes), and poli halogenated aromatic hydrocarbons, including polychlorinated biphenyls (PCBs), dibenzo-p-dioxins (PCDDs) and dibenzofurans (PCDFs). PAHs primarily enter the environment from either petrogenic (i.e., petroleum combustion and discharges) or pyrogenic (thermal decomposition and pyrolysis) sources (Albers 1995). PCBs are used as lubricants in electric transformers and in manufacturing of plastic and paints. They were introduced into the environment as commercial mixtures of over a hundred different congeners, with the precise congener composition dependent on the manufacturing process (EPA 1980). They are known to cross the placenta and intoxicate the foetus and can accumulate in wildlife and human tissues. PCDDs and PCDFs are produced by chemical (e.g., kraft process pulp and paper mills) and combustion (e.g., waste incineration) processes (Rice and O'Keefe 1995). They also include hundreds of individual congeners, differing in the number and pattern of chlorine substitution. In general, polycyclic aromatic compounds have low water solubility, low volatility, and high affinity for sediments and soil. Increasing the number of halogens on a polycyclic aromatic compound generally increases its environmental persistence and bioaccumulative properties.

These chemicals can modulate enzyme activity, or alter hormone or vitamin binding to receptors or carrier proteins. Many of the toxic effects of polycyclic aromatic compounds are attributed to binding to the aryl hydrocarbon (Ah) receptor (Barron *et al.* 1995). The toxicity of polycyclic aromatic compounds is determined by the molecular conformation and chemical structure. For example, the most toxic PCBs, PCDDs and PCDFs have outward (meta, para) halogen substitutions rather than inward (ortho) substitutions, which increases binding affinity for the Ah receptor. In addition, TCDD and coplanar PCBs may act as 'mixed' thyroid hormone agonist/antagonist, causing hypothyroid-like effects in some tissues and hyperthyroid-like effects in other tissues in rats (Schantz *et al.* 1997). Ortho-chlorine substituted PCBs induce changes in neuromuscular activity and decreases in dopamine levels (Safe 1994, Choksi *et al.* 1997), associated with behavioural changes and learning deficits (Fischer *et al.* 1998). Noncoplanar PCBs appear to alter nerve signal transduction pathways and interrupt intracellular calcium homeostasis (Barron *et al.* 1995, Fischer *et al.* 1998).

3.2.8.1 Behavioural effects

Acutely toxic doses of PCBs or TCDD can cause ataxia, tremors, paralysis and cessation of feeding in both birds and mammals (Eisler 1986). Chronically, these agents affect social interactions, such as mating and rearing of young. Prenatal exposure in rodents, monkeys, and humans resulted in delayed ontogeny of reflexes, altered activity level, and impaired learning (Tilson *et al.* 1990). Juvenile rainbow trout chronically exposed to TCDD exhibited lethargic swimming, feeding inhibition, and lack of response to external stimuli (Mehrle *et al.* 1988). PCB-induced behavioural effects in birds include decreased parental attentiveness during incubation in ring doves (Peakall and Peakall 1973), altered courtship behaviour in mourning doves (*Zenaida macroura*) (Tori and Peterle 1983), and impaired avoidance of visual cliff in pheasants (Dahlgren and Linder 1971). McCarty and Secord (1999) observed abnormal nest building behaviour and lowered nest quality in PCB-contaminated tree swallows. Similarly Woodford *et al.* (1998) reported altered parental nest attentiveness and food provisioning in osprey environmentally exposed to TCDD. PCBs (Aroclor 1254) caused a transient reduction in motor activity in rats tested in a maze (Goldey and Crofton 1998). Exposure to PCBs also affects learning in mammals: female rats exposed to individual PCB congeners during gestation and lactation exhibited spatial learning deficits (Schantz *et al.* 1995), and perinatal exposure resulted in poor visual discrimination and increased activity (Holene *et al.* 1995). PCBs can also reduce sleeping times in animals that normally hibernate (Eisler 1986).

3.3 CONCLUSIONS

Generalising toxicant-behaviour interactions for a large array of chemicals and animal species is inevitably problematic. Individual chemicals as well as classes of chemicals cause multiple behavioural effects, and the behavioural responses of animals will vary with the exposure regimen, dosage and environmental conditions. Despite these limitations, we have broadly categorised the majority of chemicals causing behavioural toxicity into eight general groups based on chemical structure, mode of action and general behavioural effects.

Quantitatively assessing the risks of behavioural impairments would require a comprehensive evaluation of exposure routes, environmental conditions, species assemblages, and comparison of contaminant concentrations causing behavioural impairments and other manifestations of toxicity. A general approach was used by Shore and Douben (1994) to identify lead, PCBs and selected OCs, and OPs and carbamates as having high environmental hazard based on their relative toxicity to small mammals and potential environmental exposures. Toxicity endpoints included survival, growth and reproductive parameters; however, behavioural effects were not specifically considered.

The complexity of assessing the behavioural effects of environmental contaminants is exemplified by spilled oil, which can cause behavioural impairments both from physical effects (oiling) or from hundreds of chemical constituents. The behavioural toxicity of just one component of oil, benzene, can be complex. For example 0.8 mg/l of benzene caused premature spawning of Pacific herrings, *Clupea harengus*, decreased survival rates of fry and eggs, loss of equilibrium and increased surfacing behaviour (Struhsaker 1977). Exposure of larval Pacific herring to 6.7 mg/l of benzene produced an initial erratic swimming activity with violent contractions, followed by lethargy and reduced feeding that continued for 1 to 2 days following exposure. At 10 to 35 mg/l of benzene, the respiration rate of herrings increased 25–45% (Struhsaker *et al.* 1974).

Despite the limitations of generalising potential environmental impacts, we have attempted broadly to summarise the relative risks of behavioural effects of contaminants in aquatic and terrestrial environments (see Table 3.1). For aquatic organisms, particularly fish, the greatest risks of behavioural effects appear to be from exposure to metals. Metals are ubiquitous in both sediments and surface waters and may cause behavioural toxicity at the sublethal concentrations present in many aquatic habitats. In terrestrial environments, the greatest risks of behavioural alterations appear to be from exposure to pesticides (OPs, OCs) and polycyclic aromatic compounds (e.g., PCBs, dioxins). Endocrine disrupting compounds represent a risk because they are persistent, bioaccumulative and ubiquitous, and can alter the behaviour of wildlife during chronic exposures. Table 3.1 is intended as a general guide of the behavioural toxicity of environmental contaminants, and the reader is referred to other chapters of this book and to the specific articles referenced for more detailed information.

ACKNOWLEDGEMENTS

We thank Jim Hansen for assistance.

REFERENCES

Albers PH (1995) Petroleum and individual polycyclic aromatic hydrocarbons. In *Handbook of Ecotoxicology*, Hoffman DJ, Rattner BA, Burton GA and Cairns J Jr (eds), Lewis Publishers, Boca Raton, pp. 330–355.

Atchison WD and Hare MF (1994) Mechanism of methylmercury-induced neurotoxicity. *FASEB Journal*, **8**, 622-629.

Atchison GJ, Henry MG and Sandheinrich MB (1987) Effects of metals on fish behaviour: a review. *Environmental Biology of Fishes*, **18**, 11-25.

Atchison WD and Narahashi T (1984) Mechanisms of action of lead on neuromuscular junctions. *NeuroToxicology*, **5**, 267-282.

Barron MG, Galbraith H and Beltman D (1995) Comparative reproductive and developmental toxicology of PCBs in birds. *Comparative Biochemistry and Physiology*, **112C**, 1-14.

Bengtsson BE (1974) The effect of zinc on the ability of the minnow, *Phoxinus phoxinus* L., to compensate for torque in a rotating water current. *Bulletin of Environmental Contamination and Toxicology*, **12**, 654-658.

Bennett RS, Williams BA, Schmedding DW and Bennett JK (1991) Effects of dietary exposure to methyl parathion on egg laying and incubation in mallards. *Environmental Toxicology and Chemistry*, **10**, 501-507.

Bielarczyk H, Tian X and Suszkiv JB (1996) Cholinergic denervation-like changes in rat hippocampus following developmental lead exposure. *Brain Research*, **708**, 108-115.

Blus LJ (1995) Organochlorine pesticides. In *Handbook of Ecotoxicology*, Chapter 13, Hoffman DJ, Rattner BA, Burton GA and Cairns J Jr (eds), Lewis Publishers, Boca Raton, FL, pp. 275-300.

Bouton SN, Frederick PC, Spalding MG and McGill H (1999) Effects of chronic, low concentrations of dietary methylmercury on the behaviour of juvenile great egrets. *Environmental Toxicology and Chemistry*, **18**, 1934-1939.

Bradbury SP, Carlson RW, Niemi GJ and Henry TR (1991) Use of respiratory-cardiovascular responses of rainbow trout (*Oncorhynchus mykiss*) in identifying acute toxicity syndromes in fish: part 4. Central nervous system seizure agents. *Environmental Toxicology and Chemistry*, **10**, 115-131.

Bushnell PJ (1989) Behavioural effects of acute p-xylene inhalation in rats: autoshaping, motor activity, and reversal learning. *Neurotoxicology and Teratology*, **10**, 569-577.

Carlson RW, Bradbury SP, Drummond RA and Hammermeister DE (1998) Neurological effects on startle response and escape from predation by medaka exposed to organic chemicals. *Aquatic Toxicology*, **43**, 51-68.

Choksi NY, Kodavanti PRS and Booth RG (1997) Effects of polychlorinated biphenyls (PCBs) on brain tyrosine hydroxylase activity and dopamine synthesis in rats. *Fundamental and Applied Toxicology*, **39**, 76-80.

Clark G (1971) Organophosphate insecticides and behaviour, a review. *Aerospace Medicine*, **42**, 735-740.

Cohen CA, Messing RB and Sparber SB (1987) Selective learning impairment of delayed reinforcement autoshaped behavior caused by low doses of trimethyltin. *Psychopharmacology*, **93**, 301-307.

Cooke AS (1972) The effects of DDT, dieldrin and 2,4-D on amphibian spawn and tadpoles. *Environmental Pollution*, **3**, 51-68.

Crofton KM, Padilla S, Tilson HA, Anthony DC, Raymer JH and MacPhail RC (1996) The impact of dose rate on the neurotoxicity of acrylamide: the interaction of administered dose, target tissue concentrations, tissue damage, and functional effects. *Toxicology and Applied Pharmacology*, **139**, 163-176.

Dahlgren RB and Linder RL (1971) Effects of polychlorinated biphenyls on pheasant reproduction, behaviour, and survival. *Journal of Wildlife Management*, **35**, 315-319.

Dell'Omo G, Bryenton R and Shore RF (1997) Effects of exposure to an organophosphate pesticide on behaviour and acetylcholinesterase activity in the common shrew, *Sorex araneus*. *Environmental Toxicology and Chemistry*, **16**, 272-276.

Dell'Omo G, Pleskacheva M, Wolfer D and Shore RF (2002) Comparative effects of exposure to an organophosphate pesticide on locomotor activity of laboratory mice and five species of wild rodents. *Bulletin of Environmental Contamination and Toxicology*, in press.

Dell'Omo G and Shore RF (1996a) Behavioral and physiological effects of acute sublethal exposure to dimethoate on wood mice, *Apodemus sylvaticus*. I-Laboratory studies. *Archives of Environmental Contamination and Toxicology*, **31**, 202-207.

Dell'Omo G and Shore RF (1996b) Behavioral effects of acute sublethal exposure to dimethoate on wood mice, *Apodemus sylvaticus*. II-Field studies on radio-tagged mice in a cereal ecosystem. *Archives of Environmental Contamination and Toxicology*, **31**, 538-542.

Dell'Omo G, Shore RF and Fishwick SK (1996c) The relationships between brain, serum, and whole blood ChE activity in the wood mouse (*Apodemus sylvaticus*) and the common shrew (*Sorex araneus*) after acute sublethal exposure to dimethoate. *Biomarkers*, **1**, 202-207.

Diamond JM, Parson MJ and Gruber D (1990) Rapid detection of sublethal toxicity using fish ventilatory behaviour. *Environmental Toxicology and Chemistry*, **9**, 3-11.

Douglas-Stroebel E (1997) *Effects of Coeur d'Alene River Basin Sediment on Mallard Duckling Brain Biochemistry, Behavior, and Growth*. Masters Thesis. Frostburg State University, Frostburg, MA, 116 pp.

Drummond RA and Russom CL (1990) Behavioural toxicity syndromes: a promising tool for assessing toxicity mechanisms in juvenile fathead minnows. *Environmental Toxicology and Chemistry*, **9**, 37-46.

Drummond RA, Spoor WA and Olson GF (1973) Some short-term indicators of sublethal effects of copper on brook trout, *Salvelinus fontinalis*. *Journal of the Fisheries Research Board of Canada*, **30**, 698-701.

Eisler R (1986) Polychlorinated biphenyl hazards to fish, wildlife, and invertebrates: a synoptic review. US Fish and Wildlife Service, US Department of Interior. *Biological Report 85 (1.7), Contaminant Hazards Reviews Report No. 7*.

Eisler R (1988) Lead hazards to fish, wildlife, and invertebrates: a synoptic review. US Fish and Wildlife Service, US Department of Interior. *Biological Report 85 (1.14), Contaminant Hazards Reviews Report No. 14*.

EPA (1980) Ambient water quality criteria for polychlorinated biphenyls. US Environmental Protection Agency, Office of Water, Washington, DC, EPA 440/5-80-068.

Ellgaard EG and Guillot JL (1988) Kinetic analysis of the swimming behaviour of bluegill sunfish, *Lepomis macrochirus rafinesque*, exposed to copper: hypoactivity induced by sublethal concentrations. *Journal of Fish Biology*, **33**, 601-608.

Fairbrother A, Meyers SM and Bennett RS (1988) Changes in mallard hen and brood behaviours in response to methyl parathion-induced illness of ducklings. *Environmental Toxicology and Chemistry*, **7**, 499-503.

Fischer LJ, Seegal RF, Ganey PE, Pessah IN and Kodavanti PRS (1998) Symposium overview: toxicity of non-coplanar PCBs. *Toxicological Sciences*, **41**, 49-61.

Geluso KN, Altenbach JS and Wilson DE (1976) Bat mortality: pesticide poisoning and migratory stress. *Science*, **194**, 184-186.

Giattina JD and Garton RR (1983) A review of the preference-avoidance responses of fishes to aquatic contaminants. *Residue Reviews*, **87**, 43-90.

Goldey ES and Crofton KM (1998) Thyroxine replacement attenuates hypothyroxinemia, hearing loss, and motor deficits following developmental exposure to Aroclor 1254 in rats. *Toxicological Sciences*, **45**, 94-105.

Gomez A, Cecchine G and Snell TW (1998) Effect of pentachlorophenol on predator-prey interaction of two rotifers. *Aquatic Toxicology*, **37**, 271-282.

Goodrich MS and Lech JJ (1990) A behavioural screening assay for *Daphnia magna*: a method to assess the effects of xenobiotics on spatial orientation. *Environmental Toxicology and Chemistry*, **9**, 21-30.

Grue CE, Gibert PL and Seeley ME (1997) Neurophysiological and behavioural changes in non-target wildlife exposed to organophosphate and carbamate pesticides: thermoregulation, food consumption, and reproduction. *American Zoologist*, **37**, 369-388.

Hammond PB and Beliles RP (1980) Metals. In *Casarett and Doull's Toxicology* (2nd edn), Doull J, Klaassen CD and Amdur MO (eds), Macmillan Publishers Co., Inc., New York, pp. 409-467.

Hansen JA, Rose JD, Jenkins RA, Gerow KG and Bergman HL (1999) Chinook salmon (*Oncorhynchus tshawytscha*) and rainbow trout (*Oncorhynchus mykiss*) exposed to copper: neurophysiological and histological effects on the olfactory system. *Environmental Toxicology and Chemistry*, **18**, 1979-1991.

Hartwell SI, Jin JH, Cherry DS and Cairns J Jr (1989) Toxicity versus avoidance response of golden shiner, *Notemigonus crysoleucus*, to five metals. *Journal of Fish Biology*, **35**, 447-456.

Heinz GH (1976) Behaviour of mallard ducklings from parents fed 3 ppm DDE. *Bulletin Environmental Contamination and Toxicology*, **16**, 640-645.

Heinz GH (1979) Methylmercury: reproductive and behavioural effects on three generations of mallard ducks. *Journal of Wildlife Management*, **43**, 394-401.

Heinz GH, Hill EF and Contrera JF (1980) Dopamine and norepinephrine depletion in ring doves fed DDE, dieldrin, and Aroclor 1254. *Toxicology and Applied Pharmacology*, **53**, 75-82.

Henry MG and Atchison GJ (1986) Behavioural changes in social groups of bluegills exposed to copper. *Transactions of the American Fisheries Society*, **115**, 590-595.

Hill EF (1995) Organophosphorous and carbamate pesticides. In *Handbook of Ecotoxicology*, Hoffman DJ, Rattner BA, Burton GA and Cairns J Jr (eds), Lewis Publishers, Boca Raton, FL, pp. 243-274.

Holene E, Nafstad I, Skaare JU, Bernhoft A, Engen P and Sagvolden T (1995) Behavioural effects of pre- and postnatal exposure to individual polychlorinated biphenyl congeners in rats. *Environmental Toxicology and Chemistry*, **14**, 967-976.

Hudson RH, Tucker RK and Haegele MA (1984) *Handbook of Toxicity of Pesticides to Wildlife*. US Fish and Wildlife Service Resource Publication 153. 90 pp.

Kenaga EE (1974) Evaluation of the safety of chlorpyrifos to birds in areas treated for insect control. *Residue Review*, **50**, 1-43.

Ko MH, Chen WP, Lin-Shiau SY and Hsieh ST (1999) Age-dependent acrylamide neurotoxicity in mice: morphology, physiology, and function. *Experimental Neurology*, **158**, 37-46.

Kuhlmann AC, McGlothan JL and Guilarte TR (1997) Developmental lead exposure causes spatial learning deficits in adult rats. *Neuroscience Letters*, **233**, 101-104.

Lorz HW, Williams RH and Fustish CA (1978) *Effects of Several Metals on Smolting of Coho Salmon*. US Environmental Protection Agency, Corvallis, Oregon.

Lyon RC, McComb JA, Schreurs J and Goldstein DB (1981) A relationship between alcohol intoxication and the disordering of brain membranes by a series of short-chain alcohols. *Journal of Pharmacology and Experimental Therapeutics*, **218**, 669-675.

McCarty JP and Secord AL (1999) Nest-building behaviour in PCB-contaminated tree swallows. *The Auk*, **116**, 55-63.

McKim JM, Schmieder PK, Niemi GJ, Carlson RW and Henry TR (1987) Use of respiratory-cardiovascular responses of rainbow trout (*Salmo gairdneri*) in identifying acute toxicity syndromes in fish: Part 2. Malathion, carbaryl, acrolein, and benzaldehyde. *Environmental Toxicology and Chemistry*, **6**, 313-328.

Mehrle PM, Buckler DR, Little EE, Smith LM, Petty JD, Peterman PH, Stalling DL, DeGraeve GM, Coyle JJ and Adams WJ (1988) Toxicity and bioconcentration of 2,3,7,8-tetrachlorodibenzo-p-dioxin and 2,3,7,8-tetrachlorodibenzofuran in rainbow trout. *Environmental Toxicology and Chemistry*, **7**, 47-62.

Mello CF, Kraemer CK, Filippin A, Morsch VM, Rodrigues ALS, Martins AF and Rubin MA (1998) Effect of lead acetate on neurobehavioural development of rats. *Brazilian Journal of Medical and Biological Research*, **31**, 943–950.

Molnar J, Paksy KA and Naray M (1986) Changes in the rat's motor behaviour during 4-hr inhalation exposure to prenarcotic concentrations of benzene and its derivatives. *Acta Physiologica Hungarica*, **67**, 349–354.

Murphy S (1980) Pesticides. In *Casarett and Doull's Toxicology* (2nd edn) Doull J, Klaassen CD and Amdur MO (eds), Macmillan Publishers Co., Inc., New York, pp. 357–408.

Needleman HL and Gatsonis CA (1990) Low-level lead exposure and the IQ of children. *Journal of the American Medical Association*, **263**, 673–678.

Niesink RJM, Jaspers RMA, Kornet LMW, van Ree JM and Tilson HA (1999) *Introduction to Neurobehavioural Toxicology*, CRC Press, Boca Raton.

Norton S (1980) Toxic responses of the central nervous system. In *Casarett and Doull's Toxicology*, 2nd edn, Doull J, Klaassen CD and Amdur MO (eds), Macmillan Publishers Co., Inc., New York, pp. 179–205.

NRC (1992) *Environmental Neurotoxicology*. National Research Council, Committee on Neurotoxicology and Models for Assessing Risks, Washington, DC.

Peakall DB and Peakall ML (1973) Effects of a polychlorinated biphenyl on the reproduction of artificially- and naturally-incubated dove eggs. *Journal of Applied Ecology*, **10**, 863–868.

Peakall DB (1985) Behavioural responses of birds to pesticides and other contaminants. *Residue Reviews*, **96**, 46–77.

Rice DC (1993) Lead-induced changes in learning: evidence for behavioural mechanisms from experimental animal studies. *NeuroToxicology*, **14**, 167–178.

Rice CP and O'Keefe P (1995) Sources, pathways, and effects of PCBs, dioxins, and dibenzofurans. In *Handbook of Ecotoxicology*, Hoffman DJ, Rattner BA, Burton GA and Cairns J Jr (eds), Lewis Publishers, Boca Raton, pp. 424–468.

Rice PJ, Drewes CD, Klubertanz TM, Bradbury SP and Coats JR (1997) Acute toxicity and behavioural effects of chlorpyrifos, permethrin, phenol, strychnine, and 2,4-dichlorophenol to 30-day-old Japanese medaka (*Oryzias latipes*). *Environmental Toxicology and Chemistry*, **16**, 696–704.

Riess JA and Needleman HL (1992) Cognitive, neural, behavioural effects of low-level lead exposure. In *The Vulnerable Brain and Environmental Risk*, Vol. 2, *Toxins in food*, Isaacson RL and Jensen KF (eds), Plenum Press, New York, pp. 111–126.

Russom CL, Bradbury SP, Broderius SJ, Hammermeister DE and Drummond RA (1997) Predicting modes of toxic action from chemical structure: acute toxicity in the fathead minnow (*Pimephales promelas*). *Environmental Toxicology and Chemistry*, **16**, 948–967.

Safe S (1994) Polychlorinated biphenyls (PCBs): environmental impact, biochemical and toxic responses, and implications for risk assessment. *Critical Reviews in Toxicology*, **24**, 87–149.

Sandheinrich MB and Atchison GJ (1989) Sublethal copper effects on bluegill, *Lepomis macrochirus*, foraging behaviour. *Canadian Journal of Fisheries and Aquatic Sciences*, **46**, 1977–1985.

Schantz SL, Moshtaghian J and Ness DK (1995) Spatial learning deficits in adult rats exposed to ortho-substituted PCB congeners during gestation and lactation. *Fundamental and Applied Toxicology*, **26**, 117–126.

Schantz SL, Seo B-W, Moshtaghian J and Amin S (1997) Developmental exposure to polychlorinated biphenyls or dioxins: do changes in thyroid function mediate effects on spatial learning? *American Zoologist*, **37**, 399–408.

Shires SW (1984) The effects of a new pyrethroid insecticide WL85871 on foraging honey bees (*Apis mellifera* L.). *Pesticide Science*, **15**, 491–499.

Shore RF and Dell'Omo G (1998) Does sub-lethal exposure to organophosphate pesticide affect capture rates in free-living rodents? *Bulletin of Environmental Contamination and Toxicology*, **61**, 440-447.

Shore RF and Douben PET (1994) Predicting ecotoxicological impacts of environmental contaminants on terrestrial small mammals. *Reviews of Environmental Contamination and Toxicology*, **134**, 49-89.

Struhsaker JW, Eldridge MB and Echeverria T (1974) Effects of benzene on eggs and larvae of Pacific herring and northern anchovy. In *Pollution and Physiology of Marine Organisms*, Venbery FJ and Vernbeg WB (eds), Academic Press, New York, pp. 253-284.

Struhsaker JW (1977) Effects of benzene on spawning Pacific herring, *Clupea harengus pallasi. Fishery Bulletin*, **75**, 43-49.

Sullivan JF, Atchison GJ, Kolar DJ and Mcintosh AW (1978) Changes in the predator-prey behaviour of fathead minnows (*Pimephales promelas*) and largemouth bass (*Micropterus salmoides*) caused by cadmium. *Journal of the Fisheries Research Board of Canada*, **35**, 446-451.

Terada H (1990) Uncouplers of oxidative phosphorylation. *Environmental Health Perspectives*, **87**, 213-218.

Tilson HA, Jacobson JL and Rogan WJ (1990) Polychlorinated biphenyls and the developing nervous system: cross species comparisons. *Neurotoxicology and Teratology*, **12**, 239-248.

Tori GM and Peterle TJ (1983) Effects of PCBs on mourning dove courtship behaviour. *Bulletin of Environmental Contamination and Toxicology*, **30**, 44-49.

Uphouse L and Williams J (1989) Sexual behaviour of intact female rats after treatment with o,p,DDT or p,p,DDT. *Reproductive Toxicology*, **3**, 33-42.

van der Geest HG, Greve GD, de Haas EM, Scheper BB, Kraak MHS, Stuijfzand SC, Augustijn KH and Admiraal W (1999) Survival and behavioural responses of larvae of the caddisfly *Hydropsyche angustipennis* to copper and diazinon. *Environmental Toxicology and Chemistry*, **18**, 1965-1971.

van Wezel AP, Cornelissen G, van Miltenburg JK and Opperhuizen A (1996) Membrane burdens of chlorinated benzenes lower the main phase transition temperature in dipalmitoyl-phosphatydylcholine vesicles: implications for toxicity by narcotic chemicals. *Environmental Toxicology and Chemistry*, **15**, 203-212.

Verhaar HJM, Rorije E, Borkent H, Seinen W and Hermens JLM (1996) Modeling the nucleophilic reactivity of small organochlorine electrophiles: a mechanistically based quantitative structure-activity relationship. *Environmental Toxicology and Chemistry*, **15**, 1011-1018.

Waiwood KG and Beamish FWH (1978) Effects of copper, pH and hardness on the critical swimming performance of rainbow trout (*Salmo gairdneri richardson*). *Water Research*, **12**, 611-619.

Weiss B (1988) Behavior as an endpoint for inhaled toxicants. In: McNelland & Henderson (eds) *Concepts in Inhalation Toxicology*. Hemisphere, New York, pp. 493-513.

Weiss B (1995) Perspectives on methylmercury as a global health hazard. *NeuroToxicology*, **16**, 577-578.

White DH, Mitchell CA and Hill EF (1983) Parathion alters incubation behaviour of laughing gulls. *Bulletin of Environmental Contamination and Toxicology*, **31**, 93-97.

WHO (1980) *Tin and organotin compounds. Environmental Health Criteria 15*. World Health Organization, Geneva. 109 pp.

WHO (1986a) *Organophosphorous insecticides: A general introduction. Environmental Health Criteria 63*. World Health Organization, Geneva. 181 pp.

WHO (1986b) *Carbamate pesticides: A general introduction. Environmental Health Criteria 64*. World Health Organization, Geneva. 137 pp.

WHO (1989) *DDT and its derivatives — Environmental aspects. Environmental Health Criteria 83*. World Health Organization, Geneva. 98 pp.

WHO (1991) *Lindane. Environmental Health Criteria 124*. World Health Organization, Geneva. 208 pp.

Windus-Podehl G, Lyftog C, Zieve L and Brunner G (1983) Encephalopathic effect in rats. *Journal of Laboratory and Clinical Medicine*, **101**, 586-592.

Woodford JE, Karasov WH, Meyer MW and Chambers L (1998) Impact of 2,3,7,8-TCDD exposure on survival, growth, and behaviour of ospreys breeding in Wisconsin, USA. *Environmental Toxicology and Chemistry*, **17**, 1323-1331.

Zhou T and Weis JS (1998) Swimming behaviour and predator avoidance in three populations of *Fundulus heteroclitus* larvae after embryonic and/or larval exposure to methylmercury. *Aquatic Toxicology*, **43**, 131-148.

Zillioux EJ, Porcella DB and Benoit JM (1993) Mercury cycling and effects in freshwater wetland ecosystems. *Environmental Toxicology and Chemistry*, **12**, 2245-2264.

4

Relationship Between Behaviour and the Biochemical/Physiological Biomarkers of Exposure to Environmental Pollutants

DAVID B. PEAKALL[1], HELEN THOMPSON[2] AND ERIK BAATRUP[3]

[1]*Deceased*

[2]*Central Science Laboratory, Department for Environment, Food and Rural Affairs, Sand Hutton, York, UK*

[3]*Institute of Biological Sciences, University of Aarhus, Denmark*

4.1 INTRODUCTION

At the landmark conference 'Pesticides in the Environment and their Effects on Wildlife' held at the Institute for Terrestrial Ecology in 1965 the advantages of using behavioural parameters as an index of sublethal toxicity were considered (Warner *et al.* 1966). These were as follows:

1. The behaviour (or activities) of an organism represents the final integrated result of a diversity of biochemical and physiological processes. Thus, a single behavioural parameter is generally more comprehensive than a physiological or biochemical parameter.
2. Behavioural patterns are known to be highly sensitive to changes in the steady state of an organism. This sensitivity is one of the key values for its use in exploring sublethal toxicity.
3. Behavioural measurements can usually be made without direct physical harm to the organism. Later, Little (1990) concluded that observations of behaviour provide a unique toxicological perspective, one that links the biochemical and ecological consequences of environmental contamination. Thus, on the one hand, chemically-induced changes of behaviour

Behavioural Ecotoxicology. Edited by Giacomo Dell'Omo.

are intrinsically linked to impaired biochemical, physiological, or general metabolic processes within the animal. In the other direction, the ecological importance of an altered behaviour in population maintenance is intuitively obvious since behaviour can be considered as the functional interface between the individual and the population.

In this chapter we examine the relationship between behavioural changes and biochemical and physiological biomarkers in order to assess the first two statements of Warner and coworkers.

4.2 TYPES OF BEHAVIOURAL STUDIES

A wide variety of tests have been used ranging from the operant tests which are readily automated so that large numbers of animals can be tested under reproducible conditions to field studies where the variables are hard to control. Operant and open field laboratory tests allow specific and complex behaviours to be analysed in detail under controlled conditions and are therefore particularly suitable for studying mechanistic relationships with other biomarker responses. However, the relevance of such laboratory studies to the real world is difficult to gauge, although it is likely that a provoked change in behaviour or decreased ability to perform a specific task is likely to be disadvantageous to the organism. In contrast, field tests have a high ecological relevance but suffer from the plethora of unknown variables and even the actual effect on survival or breeding success is difficult to demonstrate. The tests that are available are discussed in detail in Chapters 5 and 6.

4.3 SPECIFIC BIOMARKERS

The relationship of specific biomarkers to behavioural changes will now be considered. By far the largest sets of data compare behavioural changes with the inhibition of esterases and the levels of neurotransmitters, where the biomarkers in both cases are considered to be directly related to behaviour. There is very little data relating the relative sensitivity of other biomarkers, even those as widely studied as mixed function oxidases (MFOs), aminolaevulinic acid dehydrase (ALAD) and metallothionein (MT) to behavioural changes since these are generally regarded as biomarkers of exposure rather than effect. Here the approach used is to look at the dose dependence of biomarkers and behavioural changes to a given pollutant. Ideally, the biochemical and physiological biomarker studies should be made as part of the behavioural study although this limits the literature available considerably. Some attempt has been made to pair studies but this is fraught with difficulties. Matching the dose and duration of exposure exactly is rarely possible without allowing other possible variables. The few studies that we have been able to match well enough are given in Table 4.1.

Table 4.1 Paired studies of biomarkers and behavioural changes

Species	Dosage and duration of exposure	Main findings	Reference
American kestrel, *Falco sparvierus*	DDE, 6 ppm, 40 days	No effect on prey-capturing ability.	Rudolph *et al.* (1983)
	DDE, 35 ppm, 2 weeks	No effect on prey-capturing ability.	Rudolph *et al.* (1984)
	DDE, 0.3, 3, 6 and 10 ppm. Several weeks*	Dose-dependent eggshell thinning over range of 3–10 ppm DDE.	Lincer (1975)
	DDE, 10 ppm, 27 weeks	Increase of ethylmorphine demethylase activity.	Gillett *et al.* (1970)
Mourning dove, *Zenaida macroura*	PCB (Aroclor 1254) 10 and 40 ppm, 42 days	Increased length of courting period and onset of nest initiation at lower dose. General behavioural scores decreased at higher dose.	Tori and Peterle (1983)
Ring dove, *Streptotelia risoria*	PCB (Aroclor 1254) 10 ppm, 56 days	Decrease of dopamine and norepinephrine. No effect on AAT or plasma electrolytes.	Heinz *et al.* (1980)
	DDE, 10 and 50 ppm 31–35 and 59–63 days	Decrease in courtship activity at 10 ppm at longer time and at both time periods at 50 ppm.	Haegele and Hudson (1977)
	DDE, 2, 20, and 200 ppm, 56 days	Decrease of dopamine and norepinephrine at 20 and 200 ppm. Increase of AAT at highest dose.	Heinz *et al.* (1980)
	DDT, 10 ppm, 21 days.	Threefold increase in MFO activity. Decrease in oestradiol level in blood.	Peakall (1970)

*Dosing started in mid-March, dates of egg-laying not available, probably April/May.

4.3.1 CHOLINESTERASES

Cholinesterases (ChEs) are present in the brain as acetylcholinesterase (AChE) and in blood and plasma as butyrylcholinesterase (BChE). The relationship between AChE inhibition and behavioural changes in mammals and/or birds has attracted much attention and has been reviewed several times (Clark 1971,

Bignami *et al.* 1975, Grue *et al.* 1983, 1991, Peakall 1985). Here we focus on the more recent studies of ecotoxicological significance for mammals and birds and have broadened the data set by including information on fish and invertebrates. These studies are listed in Table 4.2.

A considerable amount of data exists for the effects of organophosphorus compounds (OPs) on the operant behaviour of laboratory animals which has been tabulated by Grue *et al.* (1991). Since this has little ecotoxicological

Table 4.2 Comparison of AChE inhibition with behavioural alterations

Species	Chemical	Decrease of AChE[1]	Main findings	Reference
Carabid beetle, *Pterostichus cupreus*	Dimethoate	39–73% whole body	Dose dependent effects on locomotor behaviour.	Jensen *et al.* (1997)
Freshwater fish, *Channa punctatus*	Endosulfan	38–78%	Surfacing activity and distance travelled affected. Serotin and dopamine levels also measured.	Gopal *et al.* (1985)
Mallard, *Anas platyrhynchos*	Carbofuran	53–84%	No effect on behaviour of ducklings.	Martin and Forsyth (1993)
Starling, *Sturnus vulgaris*	Chloro-fenvinphos	57–88%	Posture, feeding, flying, singing and foraging were affected in a dose-dependent manner.	Hart (1993)
Black-capped chickadee, *Parus montanus*	Fenitro-thion	50–60%	No effect on memory for cache-sites locations.	Mineau *et al.* (1994)
White-throated sparrow, *Zonotrichia albicollis*	Fenitro-thion	41–58%	Increased time foraging, decreased singing and time spent hopping.	Forsyth and Martin (1993)
House sparrow, *Passer domesticus*	Chloro-fenvinphos	15–52%	No consistent changes in numbers of seeds dropped.	Fryday *et al.* (1994)
Wood mouse, *Apodemus sylvaticus*	Dimethoate	65–75%	Open field experiments in laboratory showed effects for the first 6 h.	Dell'Omo and Shore (1996a)
			Field experiments showed decreased locomotor activity for first 6 h.	Dell'Omo and Shore (1996b)
Common shrew, *Sorex araneus*		70%	Reduction of rearing, exploring and locomotor activity. Some other parameters not affected.	Dell'Omo *et al.* (1997)

significance, suffice it to say that in virtually all cases the degree of inhibition of AChE was greater than 50%.

One of the most detailed studies was made on wood mice (*Apodemus sylvaticus*) by Dell'Omo and Shore (1996a,b). In one experiment they exposed the mice to three doses of the organophosphorous pesticide dimethoate (5, 15 and 50 mg/kg). They found significant changes in sniffing, rearing, grooming and overall activity at the highest dose but no consistent changes at the low dosages. These changes were noted only during the first 6 h, with no significant alterations from control data at 12 and 24 h indicating fast recovery. The activity of both serum and brain AChE was significantly reduced by the two higher dosages. At the highest dose the AChE activity remained significantly decreased at both 12 and 24 h. In a second experiment free-living mice were radio-tagged and the experimental group was subsequently dosed with 50 mg/kg dimethoate. Exposure significantly reduced locomotor activity during the first night but no effects were seen on the subsequent night. Thus the laboratory and field studies were in good agreement.

Wills (1972) reviewed data on the effects of diisopropyl-fluorophosphate (DFP) administration on plasma ChE, brain AChE activity and behaviour in male rats. Total (100%) inhibition of plasma ChE was associated with 93% brain AChE inhibition and spontaneous trembling and hyperactivity, 40% plasma ChE inhibition was associated with 80% brain AChE inhibition and intermittent trembling and motor resistance and 10% plasma ChE inhibition was associated with 65% brain AChE inhibition and with transient fasciculations and slight motor restlessness.

Shultz *et al.* (1995) investigated the effects of life-long exposure of rats to dichlorvos on behavioural competence as adults. Dose related increases in running time and horizontal activity with decreased vertical activity and defecation were observed in an open field test and incorrect choices in a maze were observed. All these alterations in behaviour were associated with blood and brain AChE activity inhibited by 35–60%.

Buerger *et al.* (1991) investigated the effects of methyl parathion on Northern bobwhite (*Colinus virginianus*) survivability comparing laboratory and field data. Administration of 6 mg/kg methyl parathion in the laboratory resulted in 40% depression of brain AChE activity 4 hours post treatment. Bobwhites administered the same level of methyl parathion in the field had a lower survival rate primarily due to increased predation and separation from their coveys.

The extensive use of organophosphorous and carbamate pesticides and their ultimate release into aquatic environments have naturally prompted research on the sublethal effects of these environmental contaminants on fish. A few of these studies are included here. The swimming performance of rainbow trout (*Oncorhynchus myciss*), brook trout (*Salvelinus fontinalis*) and coho salmon (*Oncorhynchus kitusch*) was studied following 25%, 65% and 75% brain AChE reduction produced by sublethal exposure to malathion in the water (Post and Leasure 1974). The concentrations of malathion necessary to reduce brain AChE to desired levels were significantly different in the three salmonid species. For instance, a 65% AChE reduction required 90 μg/l malathion in rainbow trout and 200 μg/l in coho salmon. The data indicate that 20-30% AChE can be lost in the fish brains without noticeable influence on the ability to perform physical work, whereas the performance is diminished to about one-third in fish with AChE levels reduced by approximately 75%. In contrast to the wood mice, trout and salmon need 25-42 days to recover normal brain AChE levels. Unfortunately, AChE inhibition in the peripheral nervous system was not included in this study. In a similar study, Da Silva and coworkers (1993) injected *Callichthys callichthys* with 270 mg/kg of methyl parathion and found that the initial 90% reduction in blood plasma cholinesterase was accompanied by increased resting, decreased swimming speed, swimming distance and surfacing, and abolished food search behaviour. Recovery of plasma cholinesterase began 8 days after the methyl parathion injection and reached 90% enzyme activity after about 40 days.

Feeding behaviour and brain AChE activity were compared in groups of bream (*Abramis brama*) affected by 1.87 mg/l of the OP insecticide DDVP (dichlorvos) (Pavlov *et al.* 1992). After 4 days of exposure, AChE activity was inhibited about 68% and the consumption of food reduced about 33%. When the toxicant delivery was stopped neither feeding efficiency nor brain AChE activity attained preexposure levels for at least 48 h. The authors noted that such simple mechanistic comparison of a changed behaviour with brain AChE activity is insufficient evidence of direct links between the two responses. Following DDVP exposure, fish were therefore treated with OP antidotes. Injections with atropine (m-acetylcholine receptor blocker) and the AChE-reactivator TMP-4 (dipiroxim, 1,1′-trimethylene-bis (4-formilpiridine bromide) dioxim), which are drugs specifically directed upon the cholinergic nervous system, resulted in recovery of fish feeding behaviour within 12 hours. These findings suggest that the cholinergic system contribute to the control of fish feeding behaviour and that the inhibition of AChE, the necessary component of the cholinergic nerve transmission, by DDVP is responsible for the DDVP-induced impairment of the fish feeding behaviour.

The literature on AChE-behaviour relationships in invertebrates is scanty despite the fact that terrestrial species are particularly liable to direct spraying or unintended wind dispersal of insecticides. Kasturi Bai and Reddy (1977) found a good correlation between thoracic AChE inhibition and symptoms of poisoning by OPs and carbamates in the honeybee (*Apis cerania indica*) with symptoms starting at 25% thoracic AChE inhibition, hyperactivity at 40%, knockdown at about 63% and death above 90%.

Impaired burrowing activity and AChE inhibition in the central nervous tissues were measured in earthworms *(Pheretima posthuma)* following 5-80 min exposure to solutions containing 1-8 μg/l of the carbamate insecticide carbaryl (Gupta and Sundararaman 1991). It was found that the ability of worms to burrow in soil decreased proportionally with the dose and the time of exposure to carbaryl. The reduction was evident at the lowest test concentration (1 μg/l) and the shortest exposure (5 min) and gradually increased to complete loss of burrowing after 40 min of exposure to 4 and 8 μg/l. Likewise, AChE inhibition in the nerve ring and the nerve cord was correlated to the dose of carbaryl but with a temporal delay when compared with the behavioural response. Thus, a significant AChE reduction in the central nervous tissues first appeared after 20 min of exposure to the lowest dose. This apparently slower biochemical response merely reflected different response times of the two measured biomarkers. Presumably, carbaryl quickly penetrated the moist surface of the earthworm and affected the peripheral motor nerves and hereby the burrowing ability before reaching the central nervous tissue which was used for the AChE analysis. This emphasises the importance of experimental designs which ensure that the measured biomarker responses are in equilibrium. Proper relationships between different levels of biological complexity can only be obtained if the higher level effect quantitatively expresses the underlying biochemical disturbances. Another problem with the earthworm experiment and other studies on small invertebrates is the limited amount of available tissue. While the behavioural response is measured on the individual animal, it is often necessary to pool tissue samples (or whole animals) for the chemical analyses, which restrain the identification of causal relations.

A study that directly compared the activity of AChE and locomotor behaviour in the individual animal is that of Jensen *et al.* (1997) on the carabid beetle (*Pterostichus cupreus*). The beetles were exposed to dimethoate at three doses (0.64, 1.35, and 2.32 μg/g) which resulted in 27%, 54%, and 61% inhibition of AChE for male beetles and 29%, 35%, and 59% for females 24 h after the application. In the case of males significant changes in all locomotory indices (path length, active time, average velocity, stops per walked metre and turning rate) were found even at the lowest dosage. With females none of these indices were significantly altered at the lowest dosage, suggesting a sex-dependent behavioural sensitivity to reduced AChE activity. The strong correlation between AChE activity and locomotor behaviour suggests a mechanistic relationship between the two biomarkers.

A study of the impact of the widespread forest spraying with fenitrothion to control spruce budworm (*Choristoneura fumiferana*) on the breeding behaviour of white-throated sparrows (*Zonotrichia albicollis*) was carried out by Busby *et al.* (1990). The insecticide was applied aerially twice, once at 420 g/ha and 8 days later at 210 g/ha. An earlier study (Busby *et al.* 1983) had shown that this spray regimen resulted in inhibition of brain AChE by 42% and 30% respectively. These workers followed the fate of the ringed sparrows and found that the adult population was reduced by a third in the sprayed area due to mortality and territory abandonment. Other behavioural responses that were affected were strength of territorial defence and disruption of normal incubation patterns. Overall the reproductive success was only a quarter of that in control areas.

While the measurement of AChE is a clearer and more reliable indicator of exposure to OPs and carbamates than behavioural changes it is important to realise that alteration of behaviour can affect the use of this assay in the field. Mineau and Peakall (1987) concluded that behavioural changes caused collection biases as there is a tendency for heavily exposed birds to seek shelter and thus be less likely to be collected. This has been confirmed by experimental studies of the time that starlings spent under cover when exposed to a near-lethal dose of chlorfenvinphos (Fryday *et al.* 1996). Mineau and Peakall (1987) found that a plot of mean AChE inhibition against the spray dosage of fenitrothion did not give a good dose-response curve and that the plot of the percentage of birds with more than 20% inhibition against dosage was more meaningful. A 5-year study of the effects of fenitrothion spraying of forests by Millikin and Smith (1990) concluded that although fenitrothion affected the behaviour of forest songbirds — such as moving out of the canopy — they did not have a long-term effect on populations. The difficulty with studies of this type are, of course, the large number of confounding variables.

Brewer *et al.* (1988) investigated the effects of a field application of methyl parathion on radio-tagged mallard (*Anas platyrhynchos*), blue-winged (*A. discors*), cinnamon teal (*A. cyanoptera*) and wood (*Aix sponsa*) ducks. Brood abandonment and nesting hen mortality was observed only on the methyl parathion treated field postspray. Brain AChE activity was depressed in two of three nesting hens. Of 37 ducklings in the control field 58% survived until day 22 whereas only 16% of 24 ducklings in the treated field survived to day 22. The average daily rate of loss of ducklings was also greater in the treated than the control field. Baker (1985) investigated in the laboratory and field the effects of malathion on salamanders. Brain AChE inhibition was significant following exposure to malathion on an artificial substrate in the laboratory. Behaviour experiments showed no effects on feeding, endurance or coordination but a possible decrease in digestive efficiency at high dosages. No effects were observed on ChE activity or behaviour in the field even after ten applications of malathion.

Galindo *et al.* (1985) undertook an experiment to investigate the susceptibility of quail treated with methyl parathion to predation by a cat. Birds received 2, 4 or 8 mg/kg methyl parathion and the time spent in various activities (still, walking, running or flying) was measured before and after the predator was introduced. The only significant change was an increase in time spent still in quails at the highest dosage. This change was associated with brain AChE activity 42.8% of normal.

Sheets *et al.* (1997) investigated the relation between cholinesterase inhibition and behaviour in rats using a functional observational battery, figure of eight maze, and detailed clinical observations. Six organophosphates were administered in the diet. The ChE measures were consistently the most sensitive indices of exposure with 20% inhibition of plasma, erythrocyte and brain ChE occurring before any neurobehavioural symptoms.

Moulton *et al.* (1996) investigated the effects of aldicarb and acephate on the freshwater mussel. Both compounds reduced shell closure responsiveness of mussels at 5 mg/l. ChE activity of the adductor muscle was significantly inhibited in a dose-related manner at all levels from 0.1 mg aldicarb/l and 1.3 mg acephate/l.

Padilla (1995) discussed the lack of data relating ChE inhibition to behavioural effects and quoted the US EPA 1990 'Given the lengthy history of anti-cholinesterase agents, the lack of information on the correlation between central and peripheral nervous system effects and cholinesterase inhibition is surprising'. Padilla went on to review the data of Roney *et al.* (1986) which correlated plasma and brain ChE inhibition in rats with taste aversion by three organophosphates. The correlation between brain AChE inhibition and taste aversion was linear for dichlorvos but sigmoidal for parathion and DFP. However, a linear relationship between taste aversion and plasma ChE inhibition was observed for all three organophosphates suggesting the inhibition of the peripheral and not central cholinesterase activity was mediating the taste aversion in the treated rats. Padilla also reviewed data from a functional observational battery and motor activity in rats after administration of paraoxon and its correlation with brain AChE activity (Padilla *et al.* 1992). Gait and body temperature appeared to exhibit a threshold pattern with brain AChE inhibition. Motor activity levels appeared not to have a threshold with 26% inhibition of brain AChE associated with a slight but significant decrease in motor activity and higher levels of depression associated with larger decreases. Padilla's review concludes that the correlation of the degree of ChE inhibition with clinical or behavioural endpoints depends on both the endpoint and the cholinesterase inhibitor.

The studies above suggest that two important conclusions can be drawn from the combination of the AChE and behavioural assays. First, there is hardly any doubt that the behavioural modifications were caused by a reduced AChE activity and that proper behavioural tests may provide biological meaningful measures of OP and carbamate poisoning. Secondly, the quantitative connection between the two biomarker responses has extended the biological relevance of the AChE assay because of its documented linkage to higher level effects.

These considerations invite a more general discussion of the sensitivity, chemical specificity and biological relevance of the AChE and behavioural biomarkers. If one wants to know whether or not an organism has been exposed to an OP or carbamate insecticide the assay for AChE is the only available approach with sufficient sensitivity and specificity. In fact, it can be argued that chemical specific biomarkers can only be obtained at the molecular level because all responses at higher levels of organisation integrate multiple biochemical and physiological actions. At present, only a few chemical specific biomarkers have been identified. In addition to AChE inhibition, vitellogenin induction in male vertebrates exposed to oestrogenic compounds (Sumpter and Jobling 1995) is a good example. Also, the AChE assay is the most sensitive because higher level reactions, including behavioural changes, in most cases are dimmed by compensatory countermeasures of the organism at weak intoxications. However, except for demonstrating particular groups of chemicals in the environment this high sensitivity of the AChE assay and other biochemical biomarkers is of purely academic interest if not related to consequences at higher biological levels.

Therefore, if we instead ask the question whether a specific chemical or mixture of chemicals affect the organism as such, behavioural tests are obviously more meaningful because the responses in most cases can be related to important life functions or organism health. The sensitivity of behaviour assays depends on the employed method, animal species, the specific behaviour looked at, and especially, it depends on which disturbed mechanisms are responsible for the observed behavioural change. For instance, an animal may react to traces of neurotoxins or repellent substances but behave normally despite strongly suppressed metabolic pathways. Further, technical advances such as computer-aided monitoring and various transmitters have improved the sensitivity of behaviour assays by increasing the possibility to detect and quantify simple and complex behavioural patterns. Concerning specificity, behavioural changes may be induced by innumerable substances interfering with neural, endocrine, or metabolic processes. So, unless the chemical exposure is well defined, behavioural responses usually lack chemical specificity and should be considered as effect biomarkers of general chemical stress when applied to field or semifield conditions. An exception to this rule is the disturbances in sexual behaviour which can be ascribed to endocrine disrupters (Bayley *et al.* 1999). Also, behavioural modifications have been successfully used to assess polluted waste sites (Bayley *et al.* 1997, Sorensen *et al.* 1997). This issue is discussed more fully in Section 4.5.

Neurotoxic esterases are a specific class of esterases the inhibition of which is related to a delayed neuropathy. Larsen *et al.* (1986) investigated the effects of feeding a single oral dose of triortholyl phosphate a neurotoxin, on turkeys. At 24 h after dosing, neurotoxic esterase activity was inhibited in a dose-related manner (59% at 125 mg/kg, 47% at 250 mg/kg, and 33% at 500 mg/kg of the control level) and clinical signs were observed 2 weeks later with histological examination showing only mild lesions.

4.3.2 NEUROTRANSMITTERS

Another biomarker that has been extensively studied in conjunction with behavioural studies are the neurotransmitters. Ali *et al.* (1980) exposed rats to dichlorvos for 10 days and then studied open field behaviour and locomotor activity for a further 10 days. Ambulation and preening were decreased day

5 through day 10 with the lowest activity at day 7. Total movements were significantly altered day 3 through day 9 with the greatest deviation (about 50%) in the middle of the period. Both dopamine and norepinephrine were decreased in the cerebral hemisphere on days 7 and 10, changes in other parts of the brain were less marked.

Farage-Elawar and Blaker (1992) examined the effects of two carbamate pesticides on neurochemical parameters and behaviour of chicks who had be exposed *in ovo* at day 15. The experiments have little ecotoxicological significance so suffice it to say that the behavioural experiments showed more clear-cut effects than the measurements of neurochemicals. In the paired experiments of the effects of PCBs on doves (Table 4.1) significant effects on both dopamine and norepinephrine were seen at doses similar to those causing effects on nesting behaviour.

Weber *et al.* (1991) examined the effect of lead (0.5 and 1 mg/l) on the feeding abilities of fathead minnows (*Pimephales promelas*) and their levels of neurotransmitters. Time spent feeding on daphnids and number of miscues were significantly higher than in control fish, but reaction distance showed little effect. The levels of both serotonin and norepinephrine were increased but levels of dopamine were not affected.

Gopal *et al.* (1985) examined the effect of two dosages of endosulfan on the surfacing activity and distance travel per unit time of the freshwater fish *Channa punctatus* and measured the levels of neurotransmitters and AChE activity over a 96-h period. The dosages used were 0.006 and 0.01 ppm, the higher dose causing lethality. The AChE activity was markedly reduced and claims of increases of dopamine and decreases of serotonin are made. In view of the small changes and the small sample size it is hard to take the claims of statistical significance seriously. Increases of both behavioural parameters were found at 4, 24 and 48 h but not after 96 h.

Studies by Sharma *et al.* (1976) on serotonin are given in Table 4.3 (Section 4.3.2.4) and discussed in Chapter 10.

Table 4.3 Effect of dieldrin on neurotransmitters and behaviour of mallard

Dietary dieldrin (ppm)	4 ppm	10 ppm	30 ppm
% change in EPN activity	+64	+92	+550
% change in serotonin levels	−14	−58	−78
% change in encounter scores	—	—	—
Adults	28 —	43 —	45 —
Young	34	31	33

All changes, except effect of the lowest dose on serotonin were statistically significant. Effects on O-demethylation of p-nitroanisol, norepinephrine and dopamine were significant only at the highest dosage.
EPN O-ethyl-O-(4-nitrophenyl)phenylphosphonothioate
After Sharma *et al.* (1976).

4.3.3 PLASMA ENZYMES

The effect of oil on the plasma enzymes (alanine aminotransferase AAT, ornithine carbamyl transferase and hydroxybutyric acid dehydrogenase) and behaviour of mallard ducklings was studied by Szaro *et al.* (1978). A wide range of dosage were used (0.025, 0.25, 2.5, and 5% oil in diet), the higher three doses had effects on avoidance but not on open field behaviour. Elevation of plasma enzymes was noted but was small and was not dose dependent. In a subsequent study (Szaro *et al.* 1981) the effects of 0.5% and 5% oil in the diet were studied on older ducklings. Avoidance behaviour and open field behaviour was affected, the latter markedly, at the high dose. The enzyme activity was not affected at 6 and 12 weeks at either dosage, but after 18 weeks some small but significant effects were seen.

In general, plasma enzymes have not proved to be a particularly sensitive biomarker. The studies on ring doves (*Streptopelia risoria*) exposed to DDE and PCBs (Table 4.1) showed little effect on AAT levels. While there is a considerable earlier work (Dieter 1974, 1975, Christensen *et al.* 1982) on such enzymes as the transaminases, lactate dehydrogenase and the phosphatases there has been little recent research work in this field.

Inhibition of plasma ALAD has been used as a sensitive biomarker of lead exposure in a wide range of avian species particularly in relation to the environmental implications of the use of lead shot. Work and Smith (1996) showed a significant negative correlation between blood lead levels and ALAD activity in Laysan albatross (*Diomedea immutabilis*) chicks and a significant association between elevated blood lead and droopwing in chicks.

4.3.4 MIXED FUNCTION OXIDASES

Although the MFOs are one of the most widely studied of biomarkers there are remarkably few studies linked to behaviour changes. It is a very complex system with new gene families and new genes being discovered all the time. Walker (1998a) gives the current count as 74 gene families and 481 genes. Originally evolved to handle naturally occurring toxic compounds they now play an important role in the detoxification of man-made chemicals. The activity of MFOs is affected by a wide variety of compounds of environmental interest, the organochlorines (OCs), OPs, pyrethroids and polynuclear aromatic compounds. Although not highly specific, induction of MFOs can be a very sensitive biomarker. For example, a study of the effect of dioxin on flounder (*Platichthys flesus*) found a four-fold increase in total cytochrome and 27-fold increase in 7-ethoxyresorefin-O-deethylase without obvious changes to body weight, condition of the fish or other biomarkers — retinols, thyroid hormones (Besselink *et al.* 1997).

Nagler and Cyr (1997) investigated hatching success and Cyp 1A1 levels in plaice following exposure to contaminated sediments. They showed a significant negative correlation between male Cyp 1A1 levels, which were increased 2–3-fold, and hatch success (35–48% decreased) with increased levels of Cyp1A1 associated with decreased hatch success.

Murk *et al.* (1996) reported the effects of polychlorinated aromatic hydrocarbons on common tern (*Sterna hirundo*) reproduction. They showed higher hepatic EROD activity and ratio of plasma retinol over yolk sac retinoids correlated with later egg laying, prolonged incubation period and smaller eggs and chicks. In contrast, Janz and Bellward (1996) showed that a 13–43-fold induction of hepatic EROD activity in chickens through exposure to TCDD had no effect on hatchability, body growth or plasma thyroid levels. Exposure of pigeons resulted in 6–15-fold induction of EROD with decreased hatchability and decreased growth.

Hoffman *et al.* (1993) reported the responses of common tern eggs collected from industrialised vs. non-industrialised sites. Day-old hatchlings from industrialised areas exhibited reduced femur length to body weight ratio, developmental abnormalities and elevated hepatic aryl hydrocarbon hydroxylase (AHH) activity. Femur length to body weight ratios was inversely related to AHH activity.

Sharma *et al.* (1976) studied both MFOs and neurotransmitters while carrying out behavioural tests on mallard ducklings exposed to dieldrin. A summary of this study is given in Table 4.2. While all parameters were affected, the largest change was with the MFOs.

The reason for the lack of studies relating MFOs to behavioural change would appear to be the lack of any direct relationship between the induction of MFOs and behaviour. On the other hand, comparative studies would still be worthwhile considering that the MFO exposure biomarker and the behaviour effect biomarker both cover a broad spectrum of compounds and therefore would be suitable for studies of chemically complex pollutions.

4.3.5 HORMONES

McArthur *et al.* (1983) exposed ring doves to two dosages of organochlorine mixtures based on levels found in the Great Lakes. Hormonal levels were measured at various times during the reproductive cycle. The following statistically significant changes were found: androgen was decreased by both doses in the isolation period and by the high dose 4 days postpairing; oestrogen was decreased by both doses 4 days postpairing and thyroxin was increased by both doses mid-incubation. No statistically significant behavioural effects were seen in the low-dose group, but percentage of time spent brooding was decreased and the percent of time feeding young was halved at the higher dose. Both doses affected the phenology of the breeding cycle.

There has been increasing use of plasma vitellogenin levels in detection of exposure of fish to endocrine disrupting chemicals. The impact of 4-nonylphenol on smoltification and vitellogenin levels in salmon were reported by Madsen *et al.* (1997). 4-Nitrophenol significantly inhibited the onset of smoltification together with activation of the vitellogenic system.

Vitellogenin levels and egg production in fathead minnows were investigated following exposure to 17-beta-oestradiol by Kramer *et al.* (1998). Egg production, expressed as eggs laid per female, was significantly correlated with plasma vitellogenin in both males and females. Spawning was inhibited in a way that indicated that the cycle of recrudescence was reset to the beginning of the oogenic cycle.

4.4 USE OF BEHAVIOURAL TESTS IN HAZARD ASSESSMENT

In 1989 Heinz wrote an editorial for Environmental Toxicology and Chemistry entitled 'How lethal are sublethal effects?' He considered that two requirements must be fulfilled before we have proof that contaminant-induced sublethal effect measured in the laboratory leads to a lethal effect in nature. He concluded: 'First, we must show that a contaminant will also cause the sublethal effect in the field. Second, we need proof that it was the sublethal effect that lead to the death of adults or reproductive failure in the field'. Hart (1993), in his paper which is a combination of presenting original work and a review, cites Heinz's discussion of sublethal effects and considers that while the first criterion had been meet by a number of studies — and there now several including the study by Dell'Omo and Shore (1996b) discussed in Section 4.2.1 — the only study linking AChE inhibition to behavioural changes that he considers meets both criteria is that of Busby *et al.* (1990) on white-throated sparrows already cited. The Boxworth study on tree sparrows (*Passer montanus*; Hart *et al.* 1992) proved a negative — just as valuable as positive — that although AChE inhibition occurred reproduction was not significantly affected.

Heinz cites DDE-induced eggshell thinning as one of the few examples where both of his criterion are met. Although a behavioural component was suggested by Ratcliffe (1973) in the decline of the peregrine (*Falco peregrinus*), time-lapse studies by Enderson *et al.* (1972) which recorded 4200 h of the activities at seven eyries failed to provide any evidence of abnormal behaviour. A study based on 300 h of observer watching of 12 eyries from hides (Nelson 1976) also failed to show any abnormal behaviour.

We would like to reiterate Heinz's point of that proof of effects in the field are important. We do not need more information on whether or not behaviour is altered by cholinesterase-inhibiting compounds. Any such studies should be taken to the next step to find the consequences of these changes or they should not be undertaken. One quote that is widely used when considering that the behavioural changes observed are likely to increase the risk of the individual is 'few failures in the life of an organism are as unforgiving as the failure to avoid a predator' (Lima and Dill 1990). Baatrup and Bayley (1998) refer to a study where groups of the predatory mite *Hypoaspis aculeifer* and one of its prey, the collembolan *Folsomia fimetaria*, were exposed to a low concentration of the OP dimethoate in the soil. Subsequent analyses of predator-prey interactions in an experimental set-up showed that the weak intoxication reduced the probability of the collembola surviving the 30 min experimental period from 50% in controls to 17% in the exposed group. While this may be critical for the individual it still may have little impact on the population as the total number of individuals removed by the predator is unlikely to be altered — just a different selection is made.

In general, molecular biomarkers are a more sensitive and reliable means of demonstrating exposure of organisms to specific classes of pollutants than are behavioural changes. Behavioural changes cannot be related to exposure to specific chemicals or even specific classes of pollutants. Thus functions such as locomotor activity and open field behaviour are affected by a wide range of contaminants. This lack of specificity can be turned to advantage. Automated behavioural tests such as the earthworm avoidance test (Yeardley *et al.* 1996) and the altered locomotory behaviour of woodlice (*Oniscus asellus*; Sorensen *et al.* 1997, Bayley *et al.* 1997) are valuable to show that organisms are exposed to pollutants at a biological significant level. So that in the case of waste sites such changes could be used to trigger further investigations.

Sorensen *et al.* (1997) compared the locomotor behaviour together with glycogen and protein levels in woodlice collected from a 2500 m^2 well-defined pollution of pyrolysed plastic severely contaminated with heavy metals and a complex mixture of organic compounds with woodlice collected from an unpolluted reference site 100 m from the edge of the pollution. The behaviour measurements, employing computer-aided video tracking, showed that the woodlice collected from the litter layer overlying the pyrolysed plastic were less than half as active and had a more disrupted movement pattern than the animals from the nearby reference area. This aberrant behaviour coincided with significantly lowered glycogen levels and a tendency towards reduced protein levels in the polluted woodlice. Although it seems reasonable to assume a certain relationship between the reduced activity and diminished glycogen and protein levels, a correlation study was precluded, because the two measurements were not performed on the same individuals. Also, it is uncertain whether the low nutritional state of the polluted animals was caused by disrupted metabolic pathways, an increased energy demand for detoxification processes or was simply the result of reduced food consumption. One would expect a relationship between locomotor activity and available metabolic energy reserves, but recent studies on woodlice and carabid beetles with inhibited glycolysis and cytochrome electron transport chain have indicated that this is not the case. Even with very low AEC-levels (adenylate energy charge) animals maintained almost normal movement activity until shortly before death.

The applicability of altered woodlouse locomotor behaviour in the assessment of contaminated sites was also addressed in a study (Bayley *et al.* 1997) where woodlice were collected from a closed iron foundry. The locomotor behaviour of the animals was automatically recorded in the laboratory and compared with woodlice from four unpolluted sites separated by more than 100 km. Subsequently, linear discriminant analysis was used to identify those elements of the locomotor behaviour which are the best indicators of chemical stress. The data demonstrated that the foundry woodlice moved significantly differently from the control populations which mutually were remarkably similar in their locomotor behaviour. In a subsequent unpublished study woodlouse behaviour was compared between three sites with different types of pollution and four clean reference sites (see also Chapter 5, 5.6.5). The contaminated sites included a former rubbish dump polluted by zinc, cadmium and lead, a shut-down coal gas works with soil heavily contaminated with cyanide and various organic pollutants and a partially renovated tar asphalt works with polycyclic aromatic hydrocarbons (PAHs) in the soil. The locomotor behaviour of the woodlice from the former rubbish dump and coal gas works was again significantly different from all controls, whereas the behaviour of the animals from the PAH-contaminated soil was not. However, it was not possible to discriminate the three different pollution types from each other. Possibly such a distinction can be achieved by including other biomarkers in the discriminant model. Hopefully, future combinations of biochemical and behavioural biomarkers will provide us with such biologically meaningful screening tools in environmental management, perhaps even with predictive capabilities.

Warner *et al.* (1966) refers to behavioural changes as a culmination of many systems and thus considers behavioural change as more comprehensive than a physiological or biochemical parameter. In fact, this may be the reason for the comparative lack of sensitivity of behaviour changes, the organism being able to use this complexity to compensate. A behaviour such as prey capturing is clearly vital to raptors but studies on the American kestrel (*Falco sparverius*) (Rudolph *et al.* 1983, 1984) have shown that this ability is unaffected by both an OC or an OP or indeed both together. Even the level of DDE which would cause eggshell thinning to an extent that would inhibit reproduction and of acephate that caused 43% reduction in AChE activity did not effect the kestrel's ability to capture prey.

Warner and coworkers concern for nondestructive tests was ahead of its time, but it is now a major concern. A recent conference on alternative methodologies was held under the *Scientific Group on Methodologies for the Safety Evaluation of Chemicals* (Stokes *et al.* 1998) to consider the validity of nondestructive testing to evaluate the safety of chemicals. A conference on nondestructive biomarkers was held earlier (Fossi and Leonzio 1994). In neither volume are behavioural tests considered in any detail. In a report produced for the *European Centre for the Validation of Alternative Methods* (Walker 1998b) it is stated 'there are many ways in which behavioural changes caused by chemicals can be measured, but the number of critical studies has been disappointingly small. What must be emphasised here is that behavioural assays have considerable potential for detecting the early stages of neurotoxicity. However, since they lack specificity, they need to be used in combination with more specific biomarker assays'.

Besides ethical considerations the nondestructive use of biomarkers allows serial measurement on the same organism and use an animal as its own control thereby reducing the number of animals needed and the degree of biological variation. There are many examples where the use of nondestructive biomarkers could not only replace destructive residue tests but give more meaningful information. For example, the use of ALAD inhibition in the studies by Burger (1990) of the effects of lead on the behaviour of herring gulls (*Larus argentatus*). These behavioural studies (discussed in detail in Chapter 10) were backed up by residue work (Burger and Gochfeld, 1990) but nondestructive measurement of ALAD activity in blood on the chicks during the behavioural studies as they went along would have been most valuable.

In conclusion, the nonspecific nature of behavioural changes caused by chemicals can be used to detect if any of a wide range of chemicals is present in sufficient concentrations to have an effect. Behavioural changes also move us along the path towards determining if chemicals are causing an adverse effect although we agree with Heinz (1989) of the importance of proof that behavioural changes are actually harmful to the organism. A better linkage of behavioural studies to biomarkers on the one hand and to changes in reproduction and other population effects on the other would be most valuable to all concerned.

REFERENCES

Ali SF, Chandra O and Hasan M (1980) Effects of an organophosphate (dichlorvos) on open field behavior and locomotor activity: correlation with regional brain monoamine levels. *Psychopharmacology*, **68**, 37-42.

Baatrup E and Bayley M (1998) Animal locomotor behavior as a health biomarker of chemical stress. In *Diversification in Toxicology — Man and Environment*, Seiler JP, Autrup JL and Autrup H (eds), *Archives of Toxicology* (Suppl. 20), 163-178.

Baker KN (1985) Laboratory and field experiments on the responses by two species of woodland salamander to malathion treated substrates. *Archives of Environmental Contamination and Toxicology*, **14**, 685-692.

Bayley M, Baatrup E and Bjerregaard P (1997) Woodlouse locomotor behavior in the assessment of clean and contaminated field sites. *Environmental Toxicology and Chemistry*, **16**, 2309-2314.

Bayley M, Nielsen JR and Baatrup E (1999) Guppy sexual behaviour as an effect biomarker of estrogen mimics. *Ecotoxicology and Environmental Safety*, **43**, 68-73.

Besselink HT, van Santen E, Vorstman W, Vethaak AD, Koeman JH and Brouwer A (1997) High induction of cytochrome P4501A activity without changes in retinoid and thyroid hormone levels in flounder (*Platichthys flesus*) exposed to 2,3,7,8-tetrachlorodibenzo-p-dioxin (TCDD). *Environmental Toxicology and Chemistry*, **16**, 816-823.

Bignami G, Rosic N, Michalek H, Milosevic H and Gatti GL (1975) Behavioral toxicity of antocholinesterase agents: methodological, neurochemical and neuropsychological aspects. In *Behavioral Toxicology*, Weiss B and Laties VG (eds), Plenum Press, New York, pp. 155-210.

Brewer LW, Driver C, Kendall RJ, Zenier C and Lacher TE (1988) Effects of methyl parathion in ducks and duck broods. *Environmental Toxicology and Chemistry*, **7**, 375-380.

Burger J (1990) Behavioral effects of early postnatal lead exposure in herring gull (*Larus argentatus*) chicks. *Pharmacology, Biochemistry and Behavior*, **35**, 7-13.

Burger J and Gochfeld M (1990) Tissue levels of lead in experimentally exposed herring gull (*Larus argentatus*) chicks. *Journal of Toxicology and Environmental Health*, **29**, 219-234.

Buerger TT, Kendall RJ, Mueller BS, Devos T and Williams BA (1991) Effects of methyl parathion on Northern bobwhite survivability. *Environmental Toxicology and Chemistry*, **10**, 527-532.

Busby DG, Pearce PA, Garrity NR and Reynolds LM (1983) Effect of an organophosphorus insecticide on brain cholinesterase activity in white-throated sparrows exposed to aerial forest spraying. *Journal of Applied Ecology*, **20**, 255-263.

Busby DG, White LM and Pearce PA (1990) Effects of aerial spraying of fenitrothion on breeding of white-throated sparrows. *Journal of Applied Ecology*, **27**, 743-755.

Christensen GM, Olson D and Riedel B (1982) Chemical effects on the activity of eight enzymes: a review and a discussion relevant to environmental monitoring. *Environmental Research*, **29**, 247-255.

Clark G (1971) Organophosphate insecticides and behavior, a review. *Aerospace Medicine*, **42**, 735-740.

Da Silva HC, Medina HSG, Fanta E and Bacila M (1993) Sub-lethal effects of the organophosphate Folidol 600 (methyl parathion) on *Callichthys callichthys* (Pisces: Teleostei). *Comparative Biochemistry and Physiology*, **105C**, 197-201.

Dell'Omo G, Bryenton R and Shore RF (1997) Effects of exposure to an organophosphate pesticide on behavior and acetylcholinesterase activity in the common shrew *Sotex araneus*. *Environmental Toxicology and Chemistry*, **16**, 272-276.

Dell'Omo G and Shore RF (1996a) Behavioral and physiological effects of acute sublethal exposure to dimethoate on wood mice, *Apodemus sylvaticus* [I — Laboratory studies]. *Archives of Environmental Contamination and Toxicology*, **31**, 91-97.

Dell'Omo G and Shore RF (1996b) Behavioral and physiological effects of acute sublethal exposure to dimethoate on wood mice, *Apodemus sylvaticus* [II — Field studies on radio-tagged mice in a cereal ecosystem]. *Archives of Environmental Contamination and Toxicology*, **31**, 538-542.

Dieter MP (1974) Plasma enzyme activities in coturnix quail fed graded doses of DDE, polychlorinated biphenyl, malathion, and mercuric chloride. *Toxicology and Applied Pharmacology*, **27**, 86-98.

Dieter MP (1975) Further studies on the use of enzyme profiles to monitor residue accumulation in wildlife: plasma enzymes in starlings fed graded concentrations of morsodren, DDE, Arochlor 1254, and malathion. *Archives of Environmental Contamination and Toxicology*, **3**, 142-150.

Enderson JH, Temple SA and Swartz LG (1972) Time-lapse photographic records of nesting peregrine falcons. *Living Bird*, **11**, 113-128.

Farage-Elawar M and Blaker WD (1992) Chick embryo exposure to carbamates alters neurochemical parameters and behavior. *Journal of Applied Toxicology*, **12**, 421-426.

Forsyth DJ and Martin PA (1993) Effects of fenitrothion on survival, behavior, and brain cholinesterase activity of white-throated sparrows (*Zonotrichia albicollis*). *Environmental Toxicology and Chemistry*, **12**, 91-103.

Fossi MC and Leonzio C (eds) (1994) *Nondestructive Biomarkers in Vertebrates*, Lewis Publishers, Boca Raton, pp. 345.

Fryday SL, Hart ADM and Denis NJ (1994) Effects of exposure to an organophosphate on the seed-handling efficiency of the house sparrow. *Bulletin of Environmental Contamination and Toxicology*, **53**, 869-876.

Fryday SL, Hart ADM and Langton SD (1996) Effects of exposure to an organophosphorus pesticide on the behavior and use of cover by captive starlings. *Environmental Toxicology and Chemistry*, **15**, 1590-1596.

Galindo JC, Kendall RJ, Driver CJ and Lacher TE (1985) The effect of methyl parathion on susceptibility of bobwhite quail *Colinus virginianus* to domestic cat predation. *Behaviour and Neural Biology*, **43**, 21-36.

Gillett JW, Porter RD, Wiemeyer SN, Gram TH, Schroeder DH and Gillette JR (1970) Induction liver microsomal activities in Japanese quail with supplemental information on the induction of microsomal ethylmorphine demethylase activity in sparrowhawk livers by diets containing DDT and dieldrin or DDE. In *Biological Impact of Pesticides in the Environment*, Gillett JW (ed.), Oregon State University, Corvallis, pp. 59-64.

Gopal K, Anand M, Mehrotra S and Ray PK (1985) Neurobehavioural changes in fresh water fish *Channa punctatus* exposed to endosulfan. *Journal of Advances in Zoology*, **6**, 74-80.

Grue CE, Fleming WJ, Busby DG and Hill EF (1983) Assessing hazards of organophosphate pesticides to wildlife. *Transactions of the North American Wildlife and Natural Resources Conference*, **48**, 200-220.

Grue CE, Hart ADM and Mineau P (1991) Biological consequences of depressed cholinesterase activity in wildlife. In *Cholinesterase-inhibiting insecticides — their impact on wildlife and the environment*, Mineau P (ed.), Elsevier, Amsterdam, pp. 151-209.

Gupta SK and Sundararaman V (1991) Correlation between burrowing capability and AChE activity in the earthworm, *Pheretima posthuma*, on exposure to carbaryl. *Bulletin of Environmental Contamination and Toxicology*, **46**, 859-865.

Haegele MA and Hudson RH (1977) Reduction of courtship behaviour induced by DDE in male ringed turtle doves. *Wilson Bulletin*, **89**, 593-601.

Hart ADM (1993) Relationships between behavior and the inhibition of acetylcholinesterase in birds exposed to organophosphorus pesticides. *Environmental Toxicology and Chemistry*, **12**, 321-336.

Hart ADM, Thompson HM, Fletcher MR, Greig-Smith PW, Hardy AR and Langton SD (1992) Effects of summer aphicides on tree sparrows. In *Pesticides, Cereal Farming and the Environment*, Greig-Smith P, Frampton G and Hardy T (eds), The Boxworth Project, HSMO, London, pp. 175-193.

Heinz GH (1989) How lethal are sublethal effects? *Environmental Toxicology and Chemistry*, **8**, 463-464.

Heinz GH, Hill EF and Contrera JF (1980) Dopamine and norepinephrine depletion in ring doves fed DDE, dieldrin, and Aroclor 1254. *Toxicology and Applied Pharmacology*, **53**, 75-82.

Hoffman DJ, Smith GJ and Rattner BA (1993) Biomarkers of contaminant exposure in common terns and black-crowned night herons in the Great Lakes. *Environmental Toxicology and Chemistry*, **12**, 1095-1103.

Janz DM and Bellward GD (1996) In ovo 2,3,7,8-tetrachlorodibenzo-p-dioxin exposure in three avian species. 1. Effects on thyroid hormones and growth during the perinatal period. *Toxicology and Applied Pharmacology*, **139**, 281-291.

Jensen CS, Garsdal L and Baatrup E (1997) Acetylcholinesterase inhibition and altered locomotor behavior in the carabid beetle *Pterostichus cupreus*. A linkage between biomarkers at two levels of biological complexity. *Environmental Toxicology and Chemistry*, **16**, 1727-1732.

Kasturi Bai AR and Reddy CC (1977) Inhibition of acetylcholinesterase as a criterion to determine the degree of insecticide poisoning in *Apis cerana indica*. *Journal of Apicultural Research*, **16**, 112-114.

Kramer VJ, Miles-Richardson S, Pierens SL and Giesy JP (1998) Reproductive impairment and induction of alkaline-labile phosphate, a biomarker of estrogen exposure, in fathead minnows (*Pimephales promelas*) exposed to waterborne 17-beta-estradiol. *Aquatic Toxicology*, **40**, 335-360.

Larsen C, Jortner BS and Ehrich M (1986) Effect of neurotoxic organophosphorus compounds in turkeys. *Journal of Toxicology and Environmental Health*, **17**, 365-374.

Lima SL and Dill LM (1990) Behavioral decisions made under the risk of predation; a review and prospectus. *Canadian Journal of Zoology*, **68**, 619-640.

Lincer JL (1975) DDE-induced eggshell thinning in the American kestrel: a comparison of the field situation and laboratory results. *Journal of Applied Ecology*, **12**, 781-793.

Little EE (1990) Behavioral toxicology: stimulating challenges for a growing discipline. *Environmental Toxicology and Chemistry*, **9**, 1-2.

Martin PA and Forsyth DJ (1993) Survival and behaviour of captive mallard broods exposed to carbofuran-sprayed vegetation: a field experiment. *Ecotoxicology*, **2**, 79-92.

Madsen SS, Mathiesen AB and Korsgaard B (1997) Effects of 17-beta-estradiol and 4-nonylphenol on smoltification and vitellogenesis in Atlantic salmon (*Salmo salar*). *Fish Physiology and Biochemistry*, **17**, 303-312.

McArthur MLB, Fox GA, Peakall DB and Philogene BJR (1983) Ecological significance of behavioral and hormonal abnormalities in breeding ring doves fed an organochlorine chemical mixture. *Archives of Environmental Contamination and Toxicology*, **12**, 343-353.

Millikin RL and Smith JNM (1990) Sublethal effects of fenithrothion on forest passerines. *Journal of Applied Ecology*, **27**, 983-1000.

Mineau P, Boag PT and Beninger RJ (1994) Effects of fenitrothion on memory for cache-site locations in black-capped chickadees. *Environmental Toxicology and Chemistry*, **13**, 281-290.

Mineau P and Peakall DB (1987) An evaluation of avian impact assessment techniques following broad-scale forest insecticide sprays. *Environmental Toxicology and Chemistry*, **6**, 781-791.

Moulton CA, Fleming WJ and Purnell CE (1996) Effects of two cholinesterase inhibiting pesticides on freshwater mussels. *Environmental Toxicology and Chemistry*, **15**, 131-137.

Murk AJ, Boudewijn TJ, Meininger PL, Bosveld ATC, Rossaert G, Ysebaert T, Meire P and Dirksen S (1996) Effects of polyhalogenated aromatic hydrocarbons and related contaminants on common tern reproduction: integration of biological, biochemical and chemical data. *Archives of Environmental Contamination and Toxicology*, **31**, 128-140.

Nagler JJ and Cyr DG (1997) Exposure of male American plaice (*Hippoglossoides platessoides*) to contaminated marine sediments decreases the hatching success of their progeny. *Environmental Toxicology and Chemistry*, **16**, 1733-1738.

Nelson RW (1976) Behavioral aspects of egg breakage in peregrine falcons. *Canadian Field-Naturalist*, **90**, 320-329.

Padilla S (1995) Regulatory and research issues related to cholinesterase inhibition. *Toxicology*, **102**, 215-220.

Padilla S, Moser VC, Pope CN and Brimijoin WS (1992) Parathion toxicity is not potentiated by prior reduction in blood acetylcholinesterase. *Toxicology and Applied Pharmacology*, **117**, 110-115.

Pavlov DD, Chuiko GM, Gerassimov YV and Tonkopiy VD (1992) Feeding behavior and brain acetylcholinesterase activity in bream (*Abramis brama* L.) as affected by DDVP, an organophosphorus insecticide. *Comparative Biochemistry and Physiology*, **103C**, 563-568.

Peakall DB (1970) p,p′-DDT: effect on calcium metabolism and concentration of estradiol in the blood. *Science*, **168**, 592-594.

Peakall DB (1985) Behavioral responses of birds to pesticides and other contaminants. *Residue Reviews*, **96**, 45-77.

Post G and Leasure RA (1974) Sublethal effect of malathion to three salmonid species. *Bulletin of Environmental Contamination and Toxicology*, **12**, 312-319.

Ratcliffe DA (1973) Studies of the recent breeding success of the peregrine, *Falco peregrinus*. *Journal of Reproduction and Fertility* (Suppl.), **19**, 377-389.

Roney PL, Costa LG and Murphy SD (1986) Conditioned taste aversion induced by organophosphate compounds in rats. *Pharmacology and Biochemical Behavior*, **24**, 737-742.

Rudolph SG, Anderson DW and Risebrough RW (1983) Kestrel predatory behaviour under chronic low level exposure to DDE. *Environmental Pollution* (Series A), **32**, 121-136.

Rudolph SG, Zinkl JG, Anderson DW and Shea PJ (1984) Prey-capturing ability of American kestrels fed DDE and acephate or acephate alone. *Archives of Environmental Contamination and Toxicology*, **13**, 367-372.

Sharma RP, Winn DS and Low JB (1976) Toxic, neurochemical and behaviorial effects of dieldrin exposure in mallard ducks. *Archives of Environmental Contamination and Toxicology*, **5**, 43-53.

Sheets LP, Hamilton BF, Sangha GK and Thyssen JH (1997) Subchronic neurotoxicity screening studies with six organophosphate insecticides: an assessment of behavior and morphology relative to cholinesterase inhibition. *Fundamental and Applied Toxicology*, **35**, 101-119.

Shultz H, Nagymajtenyi L and Desi I (1995) Lifetime exposure to dichlorvos affects behaviour of mature rats. *Human and Experimental Toxicology*, **14**, 721-726.

Sorensen FF, Weeks JM and Batrup E (1997) Altered locomotory behavior in woodlice (*Oniscus asellus* L.) collected at a polluted site. *Environmental Toxicology and Chemistry*, **16**, 685-690.

Stokes W, Marafante E, Peakall DB and Goldstein B (eds) (1998) Alternative testing methodologies. *Environmental Health Perspectives*, **106** (Suppl. 2), 403-620.

Sumpter JP and Jobling S (1995) Vitellogenesis as a biomarker for estrogenic contamination of the aquatic environment. *Environmental Health Perspectives*, **103**, 173-178.

Szaro RC, Dieter MP, Heinz GH and Ferrell JF (1978) Effects of chronic ingestion of South Louisiana crude oil on mallard ducklings. *Environmental Research*, **17**, 426-436.

Szaro RC, Hensler G and Heinz GH (1981) Effects of chronic ingestion of No. 2 fuel oil on mallard ducklings. *Journal of Toxicological and Environmental Health*, **7**, 789-799.

Tori GM and Peterle TJ (1983) Effects of PCBs on mourning dove courtship behavior. *Bulletin of Environmental Contamination and Toxicology*, **30**, 44-49.

Walker CH (1998a) Avian forms of cytochrome P450. *Comparative Biochemistry and Physiology* (Part C), **121**, 65-72.

Walker CH (1998b) Alterative approaches and tests in ecotoxicology: a review of the present position and the prospects for change, taking into account ECVAM's duties, topic selection and test criteria. *ATLA*, **26**, 649-677.

Warner RE, Peterson KK and Borgman L (1966) Behavioural pathology in fish: a quantitative study of sublethal pesticide toxication. *Journal of Applied Ecology*, **3** (Suppl.), 223-247.

Weber DN, Russo A, Seale DB and Spieler RE (1991) Waterborne lead affects feeding abilities and neurotransmitter levels of juvenile fathead minnows (*Pimephales promelas*). *Aquatic Toxicology*, **21**, 71-80.

Wills JH (1972) The measurement and significance of changes in the cholinesterase activities of erythrocytes and plasma in man and animals. *CRC Critical Reviews of Toxicology*, **1**, 153-202.

Work TM and Smith MR (1996) Lead exposure in Laysan albatross adults and chicks in Hawaii: prevalence, risk factors and biochemical effects. *Archives of Environmental Contamination and Toxicology*, **31**, 115-119.

Yeardley RB Jr, Lazorchak JM and Gast LC (1996) The potential of an earthworm avoidance test for evaluation of hazardous waste sites. *Environmental Toxicology and Chemistry*, **15**, 1532-1537.

Measuring Behaviour

5

Basic Behaviour: The use of Animal Locomotion in Behavioural Ecotoxicology

MARK BAYLEY
Department of Zoology, Institute of Biological Sciences, University of Aarhus, Denmark

5.1 INTRODUCTION

The use of behaviour in ecotoxicological research faces two major problems. First, the complexity of the behavioural repertoire increases with the level of neural organisation in animals. Many behavioural events are highly specific to a single species or even developmental stage, and difficult to elicit in the laboratory. Experiments including such behavioural patterns are therefore cumbersome and, although sensitive, are unsuitable for routine experiments (Døving 1991). Only basic elements of the behavioural repertoire such as motor activity are common to all taxa. The major component of motor activity in the vast majority of animals is locomotor behaviour (see Box 5.1). Thus, locomotor behaviour presents perhaps the best opportunity for developing broad generic test protocols in behavioural ecotoxicology.

The second is the issue of the difficulties involved in quantifying behavioural endpoints. As any observer of animal behaviour knows, defining and identifying the beginning and end of a behavioural pattern is often complex and requires great rigour on the part of the observer. Many of the more complex behavioural patterns, directly related to the Darwinian fitness of the animal, are therefore not easily quantified automatically. In this context locomotor behaviour has the major advantage over other behavioural events in that it is mathematically defined and therefore lends itself particularly well to automatic quantification.

It is in this context where locomotor behaviour is an automatically quantifiable endpoint present in all taxa that I would like to consider its use in ecotoxicology, in particular the use of quantitative analysis of spontaneous locomotor behaviour as a general biomarker of toxic stress.

Box 5.1

There is an implicit assumption in the use of the term 'motor behaviour' that all movement can be grouped into a single behavioural category. However, depending on the recording technique and experimental animal, many different motor movements may be included in this category (Geyer 1990). Each of these movements can be defined as a behaviour and studied in its own right (Skinner 1933). Thus as stated by Geyer (1990) '...attempts to detect indiscriminately any movement of the animal and to combine it into one category called motor activity are of questionable validity . . . such considerations have led most researchers to rely on measures of locomotor rather than motor behaviour. Locomotor activity can be operationally defined as movement from place to place and is virtually always one of the behavioural responses that provides a major contribution to any measure of general motor activity'.

5.2 LOCOMOTOR BEHAVIOUR IN ECOTOXICOLOGY

Animals use their freedom to move as one of the most important ways in which they adapt themselves to the conditions in which they live (Martin

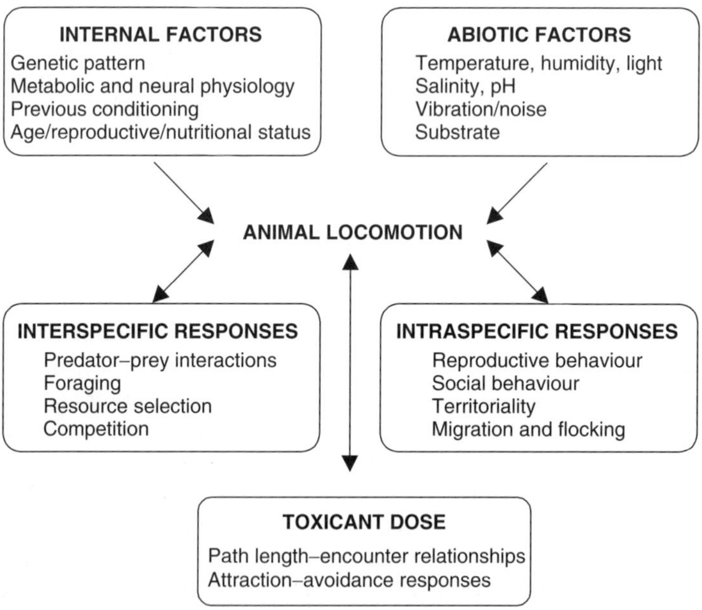

Figure 5.1 The integrative nature of animal locomotion. Animal locomotion is influenced by a variety of internal and abiotic factors. In most animals it is central in the animal's inter- and intraspecific relations. From an ecotoxicological perspective, locomotor behaviour is a major determinant factor in the amount of chemical encountered by the animal.

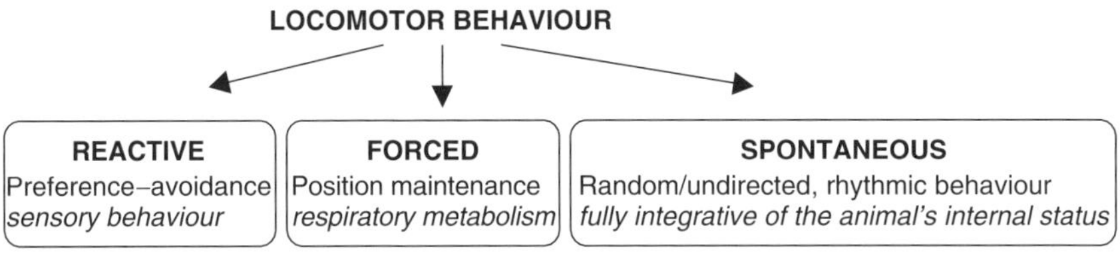

Figure 5.2 It is useful to make three subdivisions of locomotor behaviour in ecotoxicology. Reactive locomotion involves the animal's response to input from its sensory apparatus. Forced locomotor behaviour is, of the three, most directly related to the animal's respiratory physiology and can be used to measure stamina. Spontaneous locomotion, best known in uncovering circadian rhythms, is a measure of the animal's overall 'well-being' and is a highly sensitive indicator of general toxic stress.

and Bateson 1993). Locomotor activity is linked to, and expresses, a synthesis of the animal's physiological, metabolic, and neurological processes and its anatomical condition, whilst at the same time being central in many aspects of the more complex inter- and intraspecific interactions between animals, treated elsewhere in this book (Figure 5.1). Within the context of behavioural ecotoxicology as a whole, locomotion plays a uniquely important role in the amount of contaminant encountered by the animal and, therefore, the dose of chemical contracted. Locomotor behaviour can reasonably be divided into three sub-groups (Figure 5.2) drawing on different neural networks (Scherer 1992). First, most animals express *reactive locomotor responses* to both abiotic and biotic stimuli in the environment. Secondly, *forced locomotion*, or stamina, is closely related to the physiological fitness of the animal, and has great survival value for animals that need to maintain position in moving media as for example many species of fish (Beemish 1978, Rand 1985). This locomotor response against a current is termed rheotaxis. Finally, many animals exhibit *spontaneous locomotion* (SL) when placed in a uniform test arena (the prefixes free, random, undirected, general and spontaneous are used in the literature concerning this type of locomotion). All of these three subgroups of animal locomotor behaviour have been used in behavioural toxicology and ecotoxicology and I will treat them separately below.

5.3 REACTIVE LOCOMOTOR RESPONSES

Reactive locomotor responses are the most complex in the repertoire of locomotor behaviour, calling as they do on the animal's full sensory apparatus. As stated by Rand (1985) orientation to an environmental stimulus, be it abiotic, from conspecifics, from food or a predator, requires the fulfilment of at least three conditions: first, the animal must have a sensory apparatus allowing it to perceive the stimulus; secondly, the environment must contain direction vectors relative to the source of the stimulus and, finally, the animal must have the capability to perceive these vectors and use them in its reactive response.

In testing contaminants of the aquatic environment, reactive locomotor responses involved in testing for an animal's attraction (preference) to or avoidance (repellence) of the chemical contaminant are those most commonly applied in behavioural testing (Scherer 1992). There are a number of excellent reviews of this endpoint in aquatic organisms (Cherry and Cairns 1982, Beitinger and Freeman 1983, Giattina and Garton 1983, Rand 1985, Atchison *et al.* 1987). The use of locomotor endpoints in toxicity testing in the terrestrial environment is far less developed and widespread, but again the vast majority involve preference-avoidance studies and most of these concern the effects of pesticides on terrestrial invertebrates (Bartlett 1966, Danka and Collison 1987, Haynes 1988, Margolies and Kennedy 1988, Rieth and Levin 1988, Croft 1990, Hoy *et al.* 1990, El-Bassiouny 1991).

5.3.1 TWO TYPES OF PREFERENCE–AVOIDANCE BEHAVIOUR IN ECOTOXICOLOGY

Proof of an animal's attraction to, or avoidance of, a toxic chemical is certainly of considerable importance in assessing the hazard posed by an environmental contaminant and is probably the reason behind the large numbers of studies in this area. As pointed out by Rand (1985), the avoidance of toxic substances occurring naturally in the animal's habitat has adaptive significance. The response has evolved through several generations. However, avoidance responses are not likely to have evolved to the vast majority of anthropogenic contaminants that are such recent additions to the environment (Rand 1985, Døving 1991, Scherer 1992). Most of these chemicals are probably neither detected nor perceived as a danger by the animal. Further, there are several studies that have shown that pre-exposure of animals to the chemical used in the preference-avoidance reaction can increase attractance or reduce or even eliminate the avoidance response observed in 'novice' animals (Hartwell *et al.* 1987, Myllyvirta and Vuorinen 1989, McNicol and Scherer 1993). Despite these potential difficulties in translating the results of laboratory studies to expected effects in the field, avoidance responses to toxic chemicals have been validated under field conditions at least with copper (Atchison *et al.* 1987).

An alternative choice paradigm where the effects of environmental contaminants are tested on preference avoidance responses to natural gradients of, for example, humidity, temperature, pH, salinity or pheromones, has been proposed by various authors (Rand 1985, Scherer 1992). This approach, with its significant ecological importance, has been little used in either the aquatic (Scherer 1992, Scherer and McNicol 1998) or the terrestrial environment, where to my knowledge examples of this type of approach are sadly lacking. The use of this approach can further highlight the limitations of testing for a preference-avoidance response to the chemical alone. Even when an organism exhibits a clear avoidance response to a chemical under standardised laboratory conditions, the addition of a competing natural gradient can confound this response. Thus, Scherer and McNicol (1998) found a clear avoidance of low concentrations of Cu, Zn, and Pb under uniform lighting conditions in lake whitefish *Coregonus clupeaformis*. However, when given the choice between uncontaminated high light intensity and contaminated shade, the fish choose the contaminated side except at the highest concentration tested.

5.3.2 METHODS FOR DETECTING PREFERENCE–AVOIDANCE REACTIONS

The methods used for preference-avoidance studies are essentially similar both in the aquatic and terrestrial environments. There are two basic types. Where the contaminant can be perceived in the water or in the air surrounding the animal either the Y-maze or an open-gradient arena can be used. The complexity of the Y-maze can be increased to allow multiple choices and, particularly in the aquatic environment, the concentration gradient of the pollutant in the open arena can be softened to increase realism. For a more detailed description of the various arrangements available the reader is referred to the excellent reviews by Giattina and Garton (1983) and Rand (1985). The open arena is the preferred arrangement when the contaminant is present on the animal's substrate. The contaminated side of both the Y-maze and the open arena design should be alternated in the experiment to allow the exclusion of possible effects of unknown external influences such as variations in light, external vibration, and noise. The endpoints used to quantify the existence of preference-avoidance behaviour include the time distribution between contaminated and uncontaminated areas, the number of entries into the contaminated area, and the distribution of animals after a given time interval. The measurement of these endpoints can now be automated to a very high degree using many of the devices for tracking animal movements described in Section 5.5.1 below.

5.4 FORCED LOCOMOTION

Laboratory experiments involving forced locomotion are stress tests for determining the endurance or stamina and the respiratory metabolism of the animals. In the aquatic environment, this paradigm is measured by forcing the experimental subject to maintain the position against current. The ability to perform in this task is important for the survival in the wild (Rand 1985, Scherer 1992) and such tests have seen extensive use in aquatic ecotoxicology (Rand 1985, Little and Finger 1990). By contrast, this paradigm has not been used by terrestrial ecotoxicologists, despite the existence of suitable apparatuses for measuring respiratory parameters (Herreid 1981). According to Little and Finger (1990), the sensitivity of this endpoint in fish is variable, with some chemicals reducing the swimming capacity well below concentrations producing mortality and others failing to produce a response even at concentrations as high as 50% of the LC_{50}.

The methods used to elicit forced locomotion in the aquatic environment involve the use of swim tunnels or chambers, in which water current is produced by pumps and maintained at defined velocities and the ability of fish to maintain the position against the current can be monitored (Rand 1985). Alternatively, fish can be stimulated to swim to exhaustion by the addition of an electric screen at the downstream end of the chamber. The endpoints measured include the maximum sustainable locomotor velocity and the time to exhaustion at a constant current velocity.

5.5 SPONTANEOUS LOCOMOTION

When placed in a uniform environment where stimuli are deliberately kept at a minimum, few animals with motile capability will respond by remaining still. Many will become highly active, 'exploring' their new surroundings; afterwards, their general level of activity will fall but rarely disappear completely. This type of locomotor activity is frequently termed spontaneous locomotion. In many cases, SL assumes a rhythm in association with natural, ecologically-important rhythms such as, for example, the circadian rhythm determined by the light-dark cycle (Cloudsley-Thompson 1961). These rhythms in SL have been known for centuries (Nielsen 1984) and are to my knowledge the only aspect of the locomotor repertoire where the direct genetic basis has been documented (Jackson 1983, Bargiello *et al.* 1984, Costa *et al.* 1989, Jackson *et al.* 1989, Newby and Jackson 1991). There are a number of other factors that influence spontaneous locomotor behaviour, particularly in arthropods. The classic example is the effect of temperature (Shapely 1924, Josephson 1981). For example, Shapely (1924) found that the average time taken by 10 to 20 individual ants (*Liometopum apiculatum*) to run 30 cm could correctly predict the environmental temperature to within 1°C. Other factors influencing SL include: air humidity in terrestrial animals with high integumental permeability such as woodlice, *Oniscus asellus* (Den Boer 1961), food deprivation in insects (reviewed by Barton Browne 1975) and sex (Baatrup and Bayley 1993a, Bayley 1995). Minimisation of the variance in control group SL therefore requires great care in the regulation of abiotic factors in the laboratory. Further, they demand that the animals are acclimatised to the experimental conditions. Rand (1985) notes that laboratory studies on the behaviour of aquatic animals collected in the field, have adopted acclimatisation periods of hours to days in invertebrates and a week or longer in vertebrates such as fish. The best approach to this difficult problem is to standardise the length of the acclimatisation period as far as possible prior to experiments.

5.5.1 METHODS FOR QUANTIFYING SPONTANEOUS LOCOMOTION IN THE LABORATORY

Levels and patterns in spontaneous locomotor behaviour are species-specific and often highly variable over time (Rand 1985). In ecotoxicological studies, changes in spontaneous activity resulting from toxic stress are often quite conspicuous, but in the past, because of the traditional difficulties in quantifying locomotion, these changes have often been reported in a 'non-quantitative, anecdotal, or ancillary manner' (Scherer 1992).

The earliest methods for quantifying animal locomotion in much of the pioneering work on activity rhythms in terrestrial animals used the rockerbox actographs (Cloudsley-Thompson 1952). Actograph designs have been increased in flexibility with respect to choice of environment (aquatic and terrestrial), size of animal to be tested, and data collection, by replacing the mechanism used in early apparatuses with arrays of light beams or ultrasound

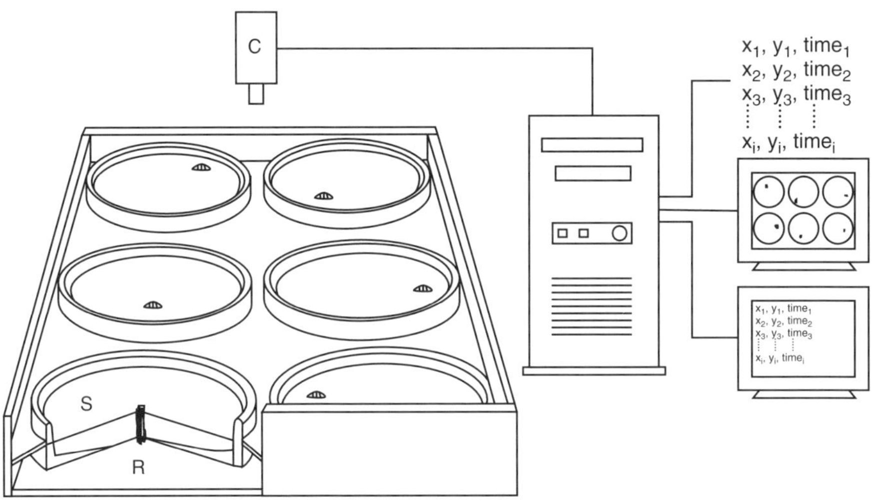

Figure 5.3 A diagrammatic representation of a set-up used to measure the spontaneous locomotor behaviour of woodlice moving on a soil surface with computer automated video tracking. The soil was maintained at saturation by connecting the soil to an underlying water reservoir. The image of the six arenas is captured using a black and white (sensitive to infrared for night tracking) or colour camera. The analogue signal is digitised using a frame grabber and image analysis software used to extract the position of the animals (*x*, *y* coordinates) up to 25 times per second. Such systems are virtually unlimited with respect to arena type (size, substrate, aquatic, terrestrial) and choice of animal, as long as sufficient contrast exists between the animal and its background (adapted from Bayley, 1995).

(Scherer 1992). Such systems can provide a good estimate of the duration of locomotor activity and have been used to considerable success in studying the temporal aspects of locomotor activity patterns and in automation of preference-avoidance studies. However, their spatial resolution is poor and a detailed analysis of parameters such as locomotor velocity and turning rate cannot be made. The automatic analysis of such detailed aspects of SL has been made possible with the advent of computer-automated video tracking systems (Lubinski *et al.* 1977, Sanberg *et al.* 1985, Lewis and Hurnik 1989, Baatrup and Bayley 1993a). An example of a modern computer-automated video tracking system is shown in Figure 5.3. With the rapid advances in computer technology of recent years, these systems have achieved high temporal and spatial resolution and are almost unlimited in the choice of animal and test conditions. Sometimes it is necessary to mark the animal with a colour contrasting that of the arena (Baatrup and Bayley 1993a,b). In addition, the possible effects due to the arena boundaries on animal's locomotion (Figure 5.4) have been circumvented, at least with terrestrial arthropods, by the use of a locomotor compensator, a sphere which can then be considered as an arena of infinite size (Kramer 1976), or similar methods (Preiss and Kramer 1986, Sorensen and Bell 1986).

5.6 SPONTANEOUS LOCOMOTION AS A GENERAL BIOMARKER OF TOXIC STRESS

The term 'biomarker' was originally used to describe biochemical, physiological, and histological indicators of either the exposure to, or the effects of, toxic substances in individual organisms. This concept was expanded by Depledge (1994) to include the use of other endpoints such as behaviour. In evaluating the utility of a potential biomarker it is useful to consider the criteria compiled by Huggett *et al.* (1992). These include its chemical specificity, relative sensitivity in relation to conventional endpoints, the temporal aspects of the response, field validation, and finally, its ability to predict effects at higher levels (population, community, etc.). In the following sections I will consider SL within this context.

5.6.1 CHEMICAL SPECIFICITY AND THE SENSITIVITY OF SPONTANEOUS LOCOMOTION

There is now little doubt that the prediction of Warner *et al.* (1966) that behaviour could be expected to be altered by a wide variety of xenobiotics is true, at least in the case of spontaneous locomotion. Under controlled laboratory conditions, SL has been affected by pesticides (Baatrup and Bayley 1993b, Bayley 1995, Sørensen *et al.* 1995), heavy metals (Bayley *et al.* 1995), and a variety of organic pollutants (Little and Finger 1990). Thus, whilst changes in locomotor behaviour are unlikely to be able to point towards exposure to a particular

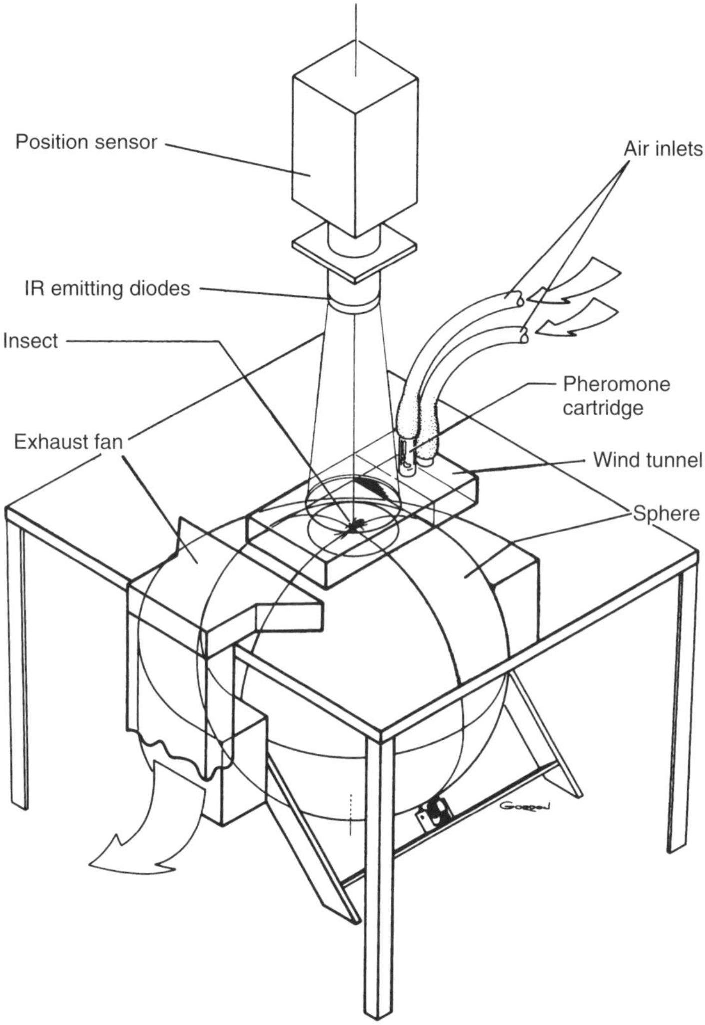

Figure 5.4 The locomotor compensator. An insect walks on a sphere of 50 cm in diameter, which is rotated around its two orthogonal axes by servo motors that compensate for the insect's locomotion. The motors respond to changes in the insect's position detected by the position sensor above the insect. Movement of the sphere, corresponding to the insect's locomotion is detected by two pulse generators at the equator of the sphere. Locomotor stimulants, such as pheromones, humidity gradients, or environmental contaminants with repellent properties, can be provided through the wind tunnel surrounding the insect. Although limited in terms of substrate and size of animal, this apparatus has the ingenious advantage of removing the effects of arena boundary on locomotion (Original source: Figure 17.1, p. 283, in Bell WJ, 1991. *Analysing search tracks* © 1991 by Kluwer Academic/Plenum Publishers, reproduced by permission).

chemical, they have value as general indicators of exposure and toxic stress (Huggett *et al.* 1992).

Spontaneous locomotion reveals toxic stress earlier than mortality. In a review of studies reporting both changes in spontaneous swimming behaviour and LC50 values, Little and Finger (1990) concluded that swimming behaviour was highly sensitive and consistently revealed changes at between 0.7% and 5% of the LC50 value. In the terrestrial environment there are very few studies allowing this type of comparison, but again SL appears to be a sensitive endpoint, at least with pesticides. Thus, Bayley (1995) showed long-term effects on woodlice (*Porcellio scaber*) locomotion after being exposed to soil contaminated with 10% of the 96-h LD20 application dose of the organophosphate (OP) pesticide dimethoate. Under similar conditions, Sørensen *et al.* (1995) showed alterations in SL in the collembolan *Folsomia candida* at 10% of the LD50 soil application dose of dimethoate.

Both hyperactivity and hypoactivity have been observed in animals exposed to toxic chemicals (Little and Finger 1990). In a study of the effect of exposure of woodlice to three soil application rates of dimethoate, Bayley and Baatrup (1996) reported hyperactivity at all three doses but the most pronounced response was upon the lowest dose. The authors argued that following an increase in exposure time or application dose, hypoactivity would undoubtedly precede paralysis and death. There is therefore, an inherent bimodality in these traditional measures of SL resulting in the possibility of no-effect concentrations occurring twice in the dose–response curve. However, a more detailed analysis of the locomotor behaviour of woodlice revealed increased number of stops in their locomotor pattern with increasing dose (Bayley and Baatrup 1996). This emphasises the value of selecting a measurement device enabling the analysis of SL in great detail.

5.6.2 TEMPORAL ASPECTS OF CHANGES IN SPONTANEOUS LOCOMOTION

The temporal aspects of a biomarker response can be divided into two areas. First, there is the question of the time necessary for the biomarker to respond after exposure of the animal to a toxin and, secondly, the question of how long the response will persist after cessation of exposure.

Spontaneous locomotor behaviour responds rapidly to exposure of the animal to xenobiotics. An examination of the review by Little and Finger (1990) of studies showing alterations in fish swimming behaviour, reveals that in 16 out of 18 cases, alterations in swimming activity were observed within 96 h of exposure. In studies of the effects of pesticides on terrestrial arthropod locomotion, the response time for changes in SL have been reported from within minutes upon pyrethroid (Baatrup and Bayley 1993b), to within 48 h upon OPs (Bayley 1995, Sørensen *et al.* 1995, Jensen *et al.* 1997).

It is not possible to make a sweeping conclusion about the recovery of an animal's SL to normal levels following the cessation of a sublethal exposure to a toxic chemical. The duration of the response depends on the chemical involved, the animal species, and the life stage. For example, Schricker and Stephen (1970) reported that the elevated flying speeds observed in honeybees (*Apis mellifera*) receiving an oral dose of the OP parathion, had returned to normal levels after 23 h from exposure. In contrast, Bayley (1995) showed that the hyperactivity measured in woodlice during a 48 h exposure to soil contaminated with dimethoate, did not decrease after 3 weeks of recovery on uncontaminated soil with access to food (Figure 5.5). In another study, larvae of the carabid beetle *Pterostichus cupreus* were raised on copper contaminated food and soil (Bayley *et al.* 1995). On emergence from their pupa, adults were sexed, weighed, and placed in containers with uncontaminated soil and food for 7 days prior to measurement of their spontaneous locomotor behaviour and their metal content. Copper-exposed animals were similar to controls with respect to their appearance, larval development time, and emergence weight, but showed a dramatically depressed locomotor behaviour. Residue analysis of adults revealed that exposed animals had similar or slightly elevated copper content than controls. It was concluded that the copper-induced alterations in locomotor behaviour were the result of internal structural damage during larval development or metamorphosis and were therefore an expression of a prolonged or permanent effect.

5.6.3 SPONTANEOUS LOCOMOTION AND CHEMICAL ENCOUNTER

The ways in which reactive locomotor responses can be of importance in determining the exposure of the animal to environmental contaminants and therefore their hazard, have been discussed in Section 5.3. In the terrestrial environment, the amount of xenobiotic encountered by an animal also depends on the extent of its locomotor activity. Harris (1971) suggested that one of the explanations for the frequently observed increases in the toxicity of pesticides to arthropods at elevated temperatures was simply due to the associated increase in the animals' locomotor activity. Salt and Ford (1984) developed a stochastic model of insecticide pickup from plant surfaces by insect larvae, in which the animals' locomotion played a key role. In a sensitivity analysis of this model, Jepson *et al.* (1990) showed that a simple hazard index based on the area of contact between the insect and its substrate, its tolerance, and its walking velocity, correctly predicted the ranked susceptibilities of several beetle species to residual deposits of pesticides. This relationship between spontaneous locomotor behaviour and pesticide uptake has been experimentally demonstrated for woodlice walking on OP-contaminated soil

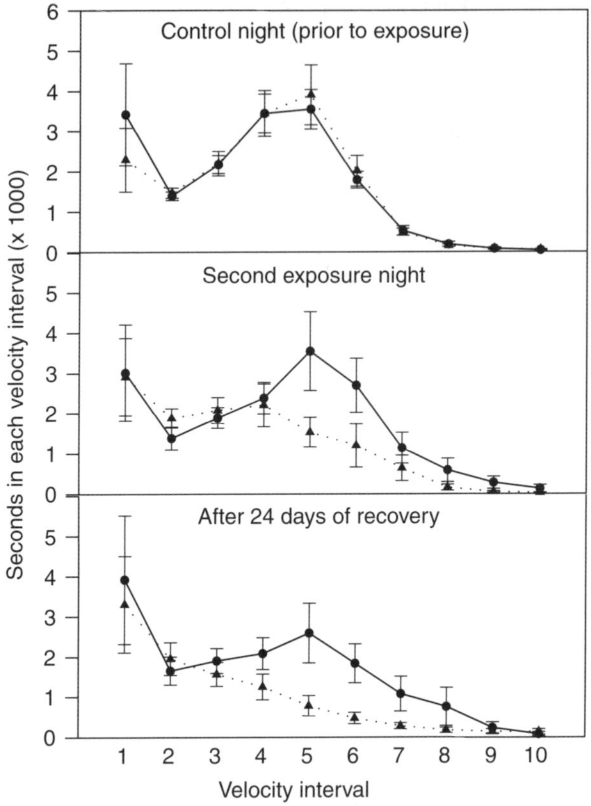

Figure 5.5 The mean velocity distributions of 20 exposed (• and solid line) and 20 control (▲ and dotted line) woodlice during a pre-exposure control night, after 2 days of exposure to dimethoate and after 21 days of recovery. Each velocity interval represents one-tenth of the individual animal's range from 1 mm s^{-1} to its maximum velocity recorded during the control night. When exposed to the unfamiliar surroundings for the first pre-exposure period all animals showed a high level of locomotor activity expressed in the middle peak of the velocity distribution curve. As control animals familiarised themselves with the arena, their movement speed decreased and by night 23 these animals preferred to move at low velocities. In contrast, dimethoate-exposed animals continued to exhibit the same stressed behaviour as evident during the pre-exposure night, despite being allowed to recover for 21 days with food on uncontaminated soil (adapted from Bayley, 1995).

surfaces (Bayley and Baatrup 1996) using ^{14}C-labelled dimethoate and computer automated video tracking. This study showed that uptake of this pesticide was strongly correlated to the animal's path length, their average velocity, and the duration of their activity (Figure 5.6).

These studies point toward the potential of using locomotor activity to explain interspecies susceptibilities to environmental contaminants and the observed differences in bioavailability from various substrates.

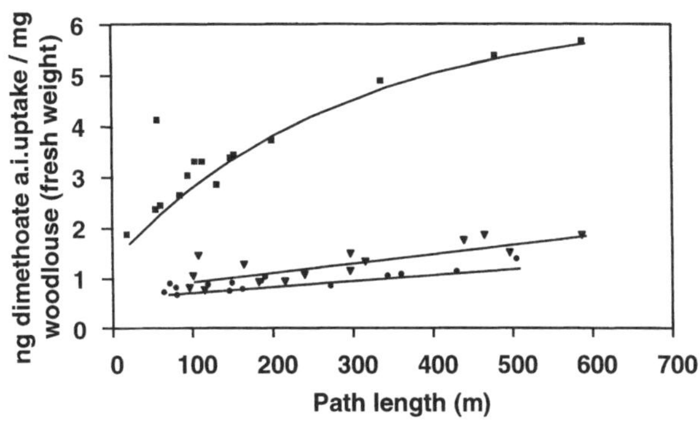

Figure 5.6 Residual uptake of ^{14}C-labelled dimethoate is strongly correlated to the path length of woodlice walking on a contaminated soil surface. Three application rates were tested: ■ 560 g ha^{-1}, ▼ 240 g ha^{-1}, and • 140 g ha^{-1} (adapted from Bayley and Baatrup, 1996).

5.6.4 FIELD VALIDATION AND LOCOMOTOR BEHAVIOUR IN ENVIRONMENTAL MONITORING

Numerous methods exist for tracking even small animals in the field, including radio-telemetry (Hayashi and Nakane 1989), radar tracking (Mascanzoni and Wallin 1986) and various marking techniques including radioactive labelling (Baars 1979) and colour tagging (Freilich 1989, Burris *et al.* 1990). These methods can provide valuable information on the gross morphology of animal movements in the field and can therefore be used to test laboratory generated preference-avoidance hypotheses. Direct and detailed quantitative analysis of the locomotor behaviour of small animals in their natural environment does not appear to be technologically possible at present, and full field validation of laboratory observed changes in spontaneous locomotor behaviour are therefore not possible.

However, partial validation is possible by collecting animals at polluted and clean field sites or along a pollution gradient and quantifying their locomotor behaviour in the laboratory after a suitable period of acclimation. In the terrestrial environment, woodlice are particularly suitable for such field studies because they are common, globally distributed in both urban and natural habitats and, as saprophages, live in and eat polluted plant litter (Hopkin *et al.* 1986).

Sørensen *et al.* (1997) collected woodlice along a short (200 m) and well-defined pollution gradient consisting of a variety of heavy metals and a complex mixture of organic compounds. These animals were brought to the laboratory and acclimatised for 2 days after which their spontaneous locomotion was measured using computer automated video tracking. The tracked animals were analysed for zinc, lead, cadmium and antimony. In addition, the glycogen and protein content was determined in woodlice collected in parallel. Animals collected from the most polluted site of the gradient were less than half as active and showed severely disrupted movement patterns when compared with animals from the clean part. This disrupted behaviour was associated with dramatically reduced glycogen reserves and elevated heavy metal burdens, in particular lead and cadmium.

In the terrestrial environment such clear and spatially-limited pollution gradients are extremely rare and a much broader knowledge of the variance among and within control populations is necessary if SL is to have broad utility in environmental monitoring. As stated by Underwood (1993), replicated disturbed areas are rarely available but no reason exists for not replicating controls areas (Underwood 1993).

These aspects were taken into account in a study on the effect of pollution at a disused iron foundry (lead 140–1500, zinc 400–2000, chromium 10–100, and nickel 11–40 μg metal mg^{-1} soil) on the spontaneous locomotor behaviour of the woodlouse (Bayley and Baatrup 1997). Woodlice were collected at the foundry and at four undisturbed woodland sites along a 300 km transect in Jutland, Denmark, acclimated in the laboratory for 2 days and, after quantification of their locomotor activity, analysed for metal content. The woodlice from the foundry were hyperactive in comparison to all control populations and suffered from significantly elevated body burdens of zinc and lead. Two important results emerged from this study: first, the remarkable consistency in locomotor behaviour in control populations indicated the potential of using the SL of this species in environmental monitoring; and second, complex chemical mixtures present at contaminated sites in the field altered woodlouse behaviour (Figure 5.7).

This experimental design has been repeated in a subsequent unpublished study on the island of Fynen, Denmark, where the woodlice can be considered genetically isolated from the animals on the Jutland peninsula. Here, the locomotor behaviour of woodlice collected at three sites with different types of pollution, was compared with the behaviour of woodlice from four clean reference sites separated by up to 80 km. The first polluted site was located at an industrial rubbish dump, closed since 1964 and where the soil was almost exclusively polluted by zinc (up to 2500 μg mg^{-1}), cadmium (up to 1.9 μg mg^{-1}) and lead (200–250 μg mg^{-1}). Woodlice were collected at the centre of this site, which is at the interface between a suburban garden and overgrown waste ground. The second site was a coal gas works abandoned in 1974 and later heavily overgrown with vegetation. The soil at this site was contaminated with cyanide (total 25 000 μg mg^{-1}, acid-washed 160 μg mg^{-1}), arsenic (11 μg mg^{-1}), tar (120 000 μg mg^{-1} total tar), and a variety of polycyclicaromatic hydrocarbons (PAH). Finally, woodlice were collected from a disused tar and asphalt factory, where the topsoil had been removed and replaced due to the extreme PAH pollution, but where spots of this pollution still remained. The results of this study confirmed the similarity of woodlouse behaviour from control sites. Further, the behaviour of woodlice

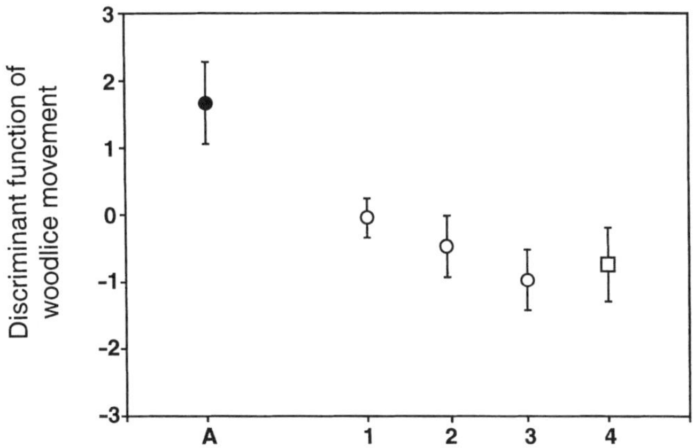

Figure 5.7 Post-hoc analysis of the discriminant function of woodlouse locomotion from the Jutland study. Discriminant analysis was used to combine the locomotor parameters (path length, movement velocity, turning rate, etc.,) into a single locomotor index best describing the differences between the polluted site (A) and control sites 1, 2 and 3. An increase in the index value indicates elevated locomotor behaviour in terms of path length and movement velocity. Control site 4 was used to validate the discriminant function. Eighty-nine percent of animals at this site were correctly identified as controls. Error bars denote 95% confidence intervals using Bonferroni correction for multiple comparisons (adapted from Bayley and Baatrup 1997).

from the dumpsite and the disused gas works were again disrupted when compared with controls but not to each other, whereas the behaviour of woodlice collected from the partially remediated asphalt factory could not be distinguished from that of control animals.

In the aquatic environment it has been recognised that chemical analysis alone is incapable of providing early warning of the thousands of potential pollutants in major waterways (Hendriks and Stouten 1993). Dynamic biological early-warning systems where the behaviour of aquatic organisms can be monitored *in situ*, analogous to the coal miner's canary (Scherer 1992), have a potentially useful role to play in the environmental monitoring of sensitive water bodies. The various types of monitors have been reviewed by Gruber *et al.* (1994) and include examples where the level of SL is used as the endpoint, in both invertebrates and fish.

5.6.5 CHANGES IN SPONTANEOUS LOCOMOTION AND HIGHER LEVEL EFFECTS

As stated by Henry and Atchison (1991), the ecological significance of a measured effect is best determined when a direct causal link can be made to the survival or reproductive success of the animal. Despite the intuitive links between an animal's locomotor behaviour and its success in interacting with its environment there is still little solid evidence that alterations in this parameter have any impact at the population level. Long-term alterations in spontaneous locomotor behaviour such as those reported by Bayley (1995) and Bayley *et al.* (1995) are strongly indicative of, but do not prove, a reduction in fitness.

One of the ways to bridge this gap is to study the influence of altered locomotor behaviour on interactions between a predator and its prey. The ability of a predator to succeed in killing its prey is an essential prerequisite to energy intake and therefore directly linked to growth and reproduction. From the prey's viewpoint, as expressed by Lima and Dill (1990), *few failures in the life of an organism are as unforgiving as the failure to avoid a predator.* Baatrup and Bayley (1998) reported a study of the predator–prey interactions between the predatory mite *Hypoaspis aculeifer* and one of its natural prey, the collembolan *Folsomia fimentaria*, using computer automated video tracking. In control animals, logistic analysis of interaction parameters revealed that the most important factors in determining capture were the average locomotor velocity of the mite and size of prey animal, whereas the average locomotor velocity of the prey animal and the hunger level of the mite had little influence. Exposure of mites and collembola to sublethal concentrations of dimethoate reduced the probability of the collembola surviving the 30-min interaction period from 50% in controls to only 17% and the data suggested that this change was partly explained by an associated impairment in the collembolan's locomotor escape response.

5.7 CONCLUSIONS

Methods for quantifying animal locomotor behaviour in the laboratory are now automated to a very high degree, allowing detailed analysis of this endpoint and facilitating its use in chemical screening studies. Knowledge of an animal's ability to sense and avoid an environmental contaminant (or be attracted to and over-exposed) can be determined in both the laboratory and the field and should be an important element in hazard analysis. Assessment of the effects of xenobiotics on an animal's ability to make choices in naturally occurring gradients, of importance to its survival in the field, can also be made using quantitative analysis of locomotor behaviour. Measures of the effect of xenobiotics on stamina and station-holding behaviour can also be assessed using analyses of animal locomotion. Finally, the use of SL as a sensitive biomarker of general toxic stress and indicative of the animal's 'well-being' has potential both in screening studies and in environmental monitoring.

REFERENCES

Atchison GJ, Henry MG and Sandheinrich MB (1987) Effects of metals on fish behavior: a review. *Environmental Biology of Fishes*, **18**, 11–25.

Baars MA (1979) Patterns of movement of radioactive carabid beetles. *Oecologia*, **44**, 125-140.

Baatrup E and Bayley M (1993a) Quantitative analysis of spider locomotion employing computer-automated video tracking. *Physiology and Behavior*, **54**, 83-90.

Baatrup E and Bayley M (1993b) Effects of the pyrethroid insecticide cypermethrin on the locomotor activity of the wolf spider *Pardosa amentata*: quantitative analysis employing computer-automated video tracking. *Ecotoxicology and Environmental Safety*, **26**, 138-152.

Baatrup E and Bayley M (1998) Animal locomotor behaviour as a health biomarker of chemical stress. *Archives of Toxicology* (Suppl. 20), 163-178.

Bargiello TA, Jackson TA and Young MW (1984) Restoration of circadian rhythms by gene transfer in *Drosophila*. *Nature*, **312**, 752-754.

Bartlett BR (1966) The repellant effect of some pesticides to hymenopterous and coccinellid predators. *Journal of Economic Entomology*, **58**, 294-296.

Barton Browne L (1975) Regulatory mechanisms in insect feeding. In *Advances in Insect Physiology Vol. 11*, Treherne JE, Berridge MJ and Wigglesworth VB (eds), Academic Press, London, pp.1-116.

Bayley M (1995) Prolonged effects of the insecticide dimethoate on locomotor behaviour in the woodlouse, *Porcellio scaber* Latr. (Isopoda). *Ecotoxicology*, **4**, 79-90.

Bayley M, Baatrup E, Heimbach U and Bjerregaard P (1995) Elevated copper levels during larval development cause altered locomotor behavior in the adult carabid beetle *Pterostichus cupreus* L. (Coleoptera: Carabidae). *Ecotoxicology and Environmental Safety*, **32**, 166-170.

Bayley M and Baatrup E (1996) Pesticide uptake and locomotor behaviour in the woodlouse: an experimental study employing video tracking and ^{14}C-labelling. *Ecotoxicology*, **5**, 35-45.

Bayley M and Baatrup E (1997) Woodlouse locomotor behavior in the assessment of clean and contaminated field sites. *Environmental Toxicology and Chemistry*, **16**, 2309-2314.

Beemish FWH (1978) Swimming capacity. In *Fish Physiology Vol. 7*, Hoar WJ and Randall DJ (eds), Academic Press, New York, pp. 101-187.

Beitinger TL and Freeman L (1983) Behavioural avoidance and selection responses of fish to chemicals. *Residue Reviews*, **90**, 35-55.

Bell WJ (1991) *Searching Behaviour. The Behavioural Ecology of Finding Resources*, Chapman and Hall, London, p. 283.

Burris JA, Bamford MS and Stewart AJ (1990) Behavioural responses of marked snails as indicators of water quality. *Environmental Toxicology and Chemistry*, **9**, 69-76.

Cherry DS and Cairns J Jr (1982) Biological monitoring, part V - Preference and avoidance studies. *Water Research*, **16**, 263-301.

Cloudsley-Thompson JL (1952) Studies in diurnal rhythms II. Changes in the physiological responses of the woodlouse *Oniscus asellus* to environmental stimuli. *Journal of Experimental Biology*, **29**, 295-303.

Cloudsley-Thompson JL (1961) *Rhythmic Activity in Animal Physiology and Behaviour*. Academic Press, New York.

Costa R, Zonta LA, Osti M and Jayakar SD (1989) Genetic analysis of adult locomotor activity in *Drosophila melanogaster*. *Heredity*, **63**, 107-117.

Croft BA (1990) *Arthropod Biological Control Agents and Pesticides*, John Wiley & Sons, New York, pp. 175-178.

Danka RG and Collison CH (1987) Laboratory evaluation of dimethoate repellance to honey bees. *Journal of Applied Entomology*, **104**, 211-214.

Den Boer PJ (1961) The ecological significance of activity patterns in the woodlouse *Porcellio scaber* Latr. (Isopoda). *Archives Néerlandaises de Zoologie*, **14**, 283-409.

Depledge MH (1994) The rational basis for the use of biomarkers as ecotoxicological tools. In *Nondestructive Biomarkers in Vertebrates*, Fossi MC and Leonzio C (eds), Lewis Publishers, Boca Raton, pp. 271-297.

Døving KB (1991) Assessment of animal behaviour as a method to indicate environmental toxicity. *Comparative Biochemistry and Physiology*, **100C**, 247-252.

El Bassiouny SA (1991) Changes in food-related behavioural patterns of some phytophagous insect species following exposures to an antifeedant. *Acta Phytopathologica et Entomologica Hungarica*, **26**, 483-496.

Freilich JE (1989) A method for tagging individual benthic macroinvertebrates. *Journal of the North American Benthological Society*, **8**, 351-354.

Geyer MA (1990) Approaches to the characterization of drug effects on locomotor activity in rodents. In *Modern Methods in Pharmacology, Vol. 6. Testing and Evaluation of Drugs of Abuse*, Adler MW and Cowan A (eds), Wiley-Liss, Inc., New York, pp. 81-99.

Giattina JD and Garton RR (1983) A review of the preference-avoidance responses of fishes to aquatic contaminants. *Residue Reviews*, **87**, 43-90.

Gruber D, Frago CH and Rasnake WJ (1994) Automated biomonitors-first line of defence. *Journal of Aquatic Ecosystem Health*, **3**, 87-92.

Harris CR (1971) Influence of temperature on the biological activity of insecticides in soil. *Journal of Economic Entomology*, **64**, 1044-1049.

Hartwell SI, Cherry DS and Cairns J Jr (1987) Field validation of avoidance of elevated metals by fathead minnows (*Pimephales promelas*) following *in situ* acclimation. *Environmental Toxicology and Chemistry*, **6**, 189-200.

Hayashi F and Nakane M (1989) Radio tracking and activity monitoring of the dobson fly larva, *Protohermes grandis* (Megaloptera: Corydalidae). *Oecologia*, **78**, 468-472.

Haynes KE (1988) Sublethal effects of neurotoxic insecticides on insect behaviour. *Annual Review of Entomology*, **33**, 149-168.

Hendriks AJ and Stouten MDA (1993) Monitoring the response of microcontaminants by dynamic *Daphnia magna* and *Leuciscus idus* assays in the Rhine delta: biological early warning as a useful supplement. *Ecotoxicology and Environmental Safety*, **26**, 265-279.

Henry MG and Atchison GJ (1991) Metal effects on fish behaviour-advances in determining the ecological significance of responses. In *Metal Ecotoxicology-Concepts and Applications*, Newman MC and McIntosh AW (eds), Lewis Publishers, Chelsea, MI, pp. 131-142.

Herreid CF II (1981) Energetics of pedestrian arthropods. In *Locomotion and Energetics in Arthropods*, Herreid CF II and Fourtner CR (eds) Plenum Press, New York, pp. 491-526.

Hopkin SP, Hardisty GN and Martin MH (1986) The woodlouse *Porcellio scaber* as a biological indicator of zinc, cadmium, lead and copper pollution. *Environmental Pollution (Ser B)*, **11**, 271-290.

Hoy CW, Adams AJ and Hall FR (1990) Behavioral response of *Plutella xylostella* (Lepidoptera: Plutellidae) populations to permethrin deposits. *Journal of Economic Entomology*, **83**, 1216-1221.

Huggett RJ, Kimerle RA, Mehrie PM, Bergman HL, Dickson KL, Fava JA, McCarthy JF, Parrish R, Dorn PB, McFarland V and Lahvis G (1992) Introduction. In *Biomarkers: Biochemical, Physiological and Histological Markers of Anthropogenic Stress*. Huggett RJ, Kimerle RA, Mehrie PM and Bergman HL (eds), Lewis Publishers, Boca Raton, pp. 1-3.

Jackson FR (1983) The isolation of biological rhythm mutations on the autosomes of *Drosophila melanogaster*. *Journal of Neurogenetics*, **1**, 3-15.

Jackson FR, Newby LM and DiBartolomeis SM (1989) *Drosophila dusky (dy)* mutations alter circadian rhythms. *Society of Neuroscience Abstracts*, **15**, 461-461.

Jensen CS, Garsdal L and Baatrup E (1997) Acetylcholinesterase inhibition and altered locomotor behavior in the carabid beetle *Pterostichus cupreus*. A linkage between biomarkers at two levels of biological complexity. *Environmental Toxicology and Chemistry*, **16**, 1727 - 1732.

Jepson PC, Chaudhry AG, Salt DW, Ford MG, Efe E and Chowdhury ABMNU (1990) A reductionist approach towards short-term hazard analysis for terrestrial invertebrates exposed to pesticides. *Functional Ecology*, **4**, 339 - 347.

Josephson RK (1981) Temperature and the mechanical performance of insect muscle. In *Insect Thermoregulation*, Heinrich B (ed.), Wiley, New York, pp. 19 - 44.

Kramer E (1976) The orientation of walking honeybees in odour fields with small concentration gradients. *Physiological Entomology* **1**, 27 - 37.

Lewis NJ and Hurnik JF (1989) Locomotion of broiler chickens in floor pens. *Poultry Science*, **69**, 1087 - 1093.

Little EE and Finger SE (1990) Swimming behavior as an indicator of sublethal toxicity in fish. *Environmental Toxicology and Chemistry*, **9**, 13 - 19.

Lima SL and Dill LM (1990) Behavioural decisions made under the risk of predation: a review and prospectus. *Canadian Journal of Zoology*, **68**, 619 - 640.

Lubinski KS, Dickson KL and Cairns J Jr (1977) Microprocessor-based interface converts video signals for object tracking. *Computer Design*, **Dec.**, 81 - 87.

Margolies DC and Kennedy GG (1988) Fenvalerate-induced aerial dispersal by the twospotted spider mite. *Entomologia Experimentalis et Applicata*, **46**, 233 - 240.

Martin P and Bateson P (1993) *Measuring Behaviour: an Introductory Guide*, 2nd edition, Cambridge University Press, Cambridge.

Mascanzoni D and Wallin H (1986) The harmonic radar: a new method of tracing insects in the field. *Ecological Entomology*, **11**, 387 - 390.

Myllyvirta TP and Vuorinen PJ (1989) Avoidance of bleached kraft mill effluent by pre-exposed *Coregonus albula* L. *Water Research*, **23**, 1219 - 1227.

McNicol RE and Scherer E (1993) Influence of cadmium pre-exposure on the preference - avoidance responses of lake whitefish (*Coregonus clupeaformis*) to cadmium. *Archives of Environmental Contamination and Toxicology*, **25**, 36 - 40.

Newby LM and Jackson FR (1991) *Drosophila ebony* mutants have altered circadian activity rhythms but normal eclosian rhythms. *Journal of Neurogenetics*, **7**, 85 - 101.

Nielsen ET (1984) Relation of behavioural activity rhythms to the changes of day and night. A revision of views. *Behaviour*, **89**, 147 - 173.

Preiss R and Kramer E (1986) Anemotactic orientation of gypsy moth males and its modification by the attractant pheromone (+)-disparlure during walking. *Physiological Entomology*, **11**, 185 - 198.

Rand GM (1985) Behavior. In *Fundamentals of Aquatic Toxicology*, Rand GM and Petrocelli SR (eds) Hemisphere Publishing Corporation, New York, pp. 221 - 263.

Rieth JP and Levin MD (1988) The repellant effect of two pyrethroid insecticides on the honeybee. *Physiological Entomology*, **13**, 213 - 218.

Salt DW and Ford MG (1984) The kinetics of insecticide action. Part III: the use of stochastic modelling to investigate the pick-up of insecticides from ULV-treated surfaces by larvae of *Spodoptera littoralis* Boisd. *Pesticide Science*, **15**, 382 - 410.

Sanberg PR, Hagenmeyer SH and Henault MA (1985) Automated measurements of multivariate locomotor behavior in rodents. *Neurobehavioral Toxicology and Teratology*, **7**, 87 - 94.

Scherer E (1992) Behavioral responses as indicators of environmental alterations - Approaches, results, developments. *Journal of Applied Ichthyology*, **8**, 122 - 131.

Scherer E and McNicol RE (1998) Preference-avoidance responses of lake whitefish (*Coregonus clupeaformis*) to competing gradients of light and copper, lead and zinc. *Water Research*, **32**, 924-929.

Schricker B and Stephen WP (1970) The effect of sublethal doses of parathion on honeybee behaviour. I. Oral administration and the communication dance. *Journal of Apiculture Research*, **9**, 141-153.

Shapely H (1924) Note on the thermokinetics of dolichoderine ants. *Proceedings of the National Academy of Sciences USA*, **10**, 436-439.

Skinner BF (1933) The measurement of spontaneous activity. *Journal of General Psychology*, **9**, 3-24.

Sorensen KA and Bell WJ (1986) Orientation responses of an isopod to temporal changes in relative humidity: simulation of a 'humid patch' in a 'dry habitat'. *Journal of Insect Physiology*, **32**, 51-57.

Sørensen FF, Bayley M and Baatrup E (1995) The effects of sublethal dimethoate exposure on the locomotor behaviour of the collembolan *Folsomia candida* (Isomotidae). *Environmental Toxicology and Chemistry*, **14**, 1587-1590.

Sørensen FF, Weeks JM and Baatrup E (1997) Altered locomotor behaviour in woodlouse (*Oniscus asellus* L.) collected at a polluted site. *Environmental Toxicology and Chemistry*, **16**, 685-690.

Underwood AJ (1993) The mechanics of spatially replicated sampling programmes to detect environmental impacts in a variable world. *Australian Journal of Ecology*, **18**, 99-116.

Warner RE, Peterson KK and Borgman L (1966) Behavioural pathology in fish: a quantitative study of sublethal pesticide toxication. *Journal of Applied Ecology*, **3**, 223-247.

6

Assessing Complex Behaviour in Ecotoxicology

HANS-PETER LIPP
Department of Neuroanatomy and Behaviour, Institute of Anatomy, University of Zürich, Switzerland

6.1 INTRODUCTION

Short- or long-lasting environmental contamination may have subtle effects on brain development and function which are neither noticed by naturalistic observations nor by standard neurobehavioural test batteries in the laboratory. Yet they may affect biological fitness in different ways, the net result being an effect at population level.

Complex behaviour is a uniquely sensitive endpoint of toxicological agents, as it is based on the interactions and orchestration of multiple brain systems. In the definition given here, it includes cognition, memory and organisation of behavioural sequences or strategies. However, quantitative assessment of this type of behavioural impairment in an ecotoxicological context faces several methodological challenges.

- Tests measuring cognition or the analysis of individual behavioural patterns often require careful and prolonged studies of individual animals. Thus, they are mostly used to test specific hypotheses addressing causal mechanisms and only rarely for screening.
- The behavioural test should be highly sensitive to minor pathological alterations in the brain but, at the same time, be relatively robust against context-dependent influences and variations of the test procedures.
- The test should provide some preliminary information about the brain systems possibly responsible for an observed impairment.
- Behavioural ecotoxicology should have procedures that permit to verify negative findings in the laboratory, because false-negative findings may entail unwarranted use of chemical compounds.
- Behavioural tests should permit correlative studies in the laboratory and the field.

- Finally, the same tests recognising impaired cognitive functions after exposure to environmental toxicants should also be applicable to species traditionally studied in ecology.

These problems concern essentially all species. Given the almost unlimited range of behavioural approaches, this chapter will necessarily focus on measuring complex behaviour of rodents as the classical laboratory animals in toxicology. Unlike the traditional pharmaco-toxicological approach centring on rats (Cory-Slechta 1989, Peele and Vincent 1989), it will present mainly paradigms used to analyse the behavioural effects of genetically engineered mutations in mice. There are valid methodological reasons for this approach as outlined below. An additional benefit is that behavioural ecotoxicology, by adapting protocols used in behavioural molecular biology, can profit from the rapidly growing test battery and know-how for behavioural analysis of knockout and transgenic mice (Crawley and Paylor 1997).

Thus, this chapter will first present the rationale for choosing the tests. This will be followed by a presentation of individual tests and their suitability for ecotoxicological behavioural analysis according to the criteria given above. General interpretation problems of behavioural analysis will be touched briefly, and the usefulness of hippocampal structures as markers for subtle toxicological effects will be discussed. Finally, new behavioural approaches will be presented which permit to combine both field and laboratory studies.

6.2 WHY USING PARADIGMS TO STUDY GENETICALLY MODIFIED MICE

Using behavioural tests for behavioural ecotoxicology that have been established for the analysis of transgenic mice appears to be odd at first glance. There appears to be a fundamental difference between the fields when one considers the mechanism of actions. In molecular neuroscience, the cause of the behavioural problem is precisely specified, the targeted gene, while in ecotoxicology the toxicants are ill defined and their mode of action mostly unknown. However, the field of behavioural and neural analysis of knockout mice is facing the problem that precise elimination of a gene is leading to a plethora of pleiotropic effects at the behavioural level, or even that many gene deletions remain without noticeable phenotype because of masking brain plasticity (Büeler *et al.* 1992). These represent largely the problems that are faced by behavioural ecotoxicology as well.

Approaching behavioural ecotoxicology from the angle of molecular neuroscience brings some additional practical and conceptual benefits. As the knockout technology must use mice, at least at present, many laboratories are adapting tests initially designed for rats. While the rat is a convenient animal for learning relatively complex tasks, mice are much more limited for a variety of reasons. Having adapted a test for mice makes it suitable for almost any other small mammalian species studied ecologically. Many new transgenic mouse models are specifically aimed at the molecular analysis of memory, learning and cognition (Campbell and Gold 1996, Chen and Tonegawa 1997, Silva *et al.* 1996, Lipp and Wolfer 1998), with a relatively minor interest in teratogenic effects or neurological disorders. This has led to an emphasis on behavioural tests revealing efficiently cognitive deficits. Likewise, phenotypic analysis of targeted mutations has a specific interest in pinpointing the brain systems involved. In order to exclude sensorimotor deficits as a cause of impairment, tests have focused on tasks thought to depend specifically on hippocampal or amygdalar functions.

Finally, molecular neuroscience is paying much more attention to the problem of individual genetic variability and genetic background, an issue often neglected in teratology and toxicology. Yet, such genetic variation may affect both physiological responsiveness to toxicants and individual or species-specific levels of behavioural abilities.

6.3 DESCRIPTIONS OF TESTS

The following presentation of tests and methods is highly selective. The selection is based on practical experience, economy in application (Bignami 1996), sensitivity to detect subtle alteration in specific brain mechanisms, suitability for computerised analysis (Kruk 1997), suitability for mass screening and, importantly, for its potential expansion to compare many different and wild-living species. Some procedures of ethological interest are thus also considered.

The main common denominator for many rodent species are tests assessing spatial abilities, as such brain mechanisms are of fundamental importance for survival and reproduction. Another class of general relevance for small species are tests assessing anxiety, flight and avoidance responses. On the other hand, potentially sensitive operant behavioural tests based on parametric control of behaviour by sensory stimuli or response schedules are largely omitted, as they remain confined to selected laboratory species such as rats, monkeys and pigeons.

It should be emphasised that the assessment of complex behaviour should always be preceded, or accompanied, by an evaluation of sensorimotor abilities as described in various test batteries for rats and mice (Cory-Slechta 1989, Crawley and Paylor 1997, Moser 1991; Tilson 1993).

6.3.1 TESTS ASSESSING SPATIAL BEHAVIOUR

6.3.1.1 High level path analysis of open field behaviour

The value of analysing locomotor activity has been discussed thoroughly in the previous chapter. The section here describes how to extract a maximum of information from an analysis of locomotion paths.

Open fields generally consist of a round or squared arena with a flat surface, exposed to illumination of variable intensity. Animals are placed for an observation period of several minutes (usually 10–20 min) in the arena. In order to assess habituation to this environment, the exposure can be repeated, usually on the next day. Because of the relatively short duration of the test, many animals can be tested during a day.

Rodents placed in an open field remain often motionless for a variable time. Afterwards, they start moving, first along the walls, and then increasingly more across the open arena. Locomotion is of variable speed and mostly interrupted by short stops during which the animals rear or groom (Golani *et al.* 1993). Simple measures of activity can be obtained by counting the number of times the animal crosses grids painted on the floor. More sophisticated analysis is possible by recording the path by means of a video camera connected to a computer equipped with tracking software (Figures 6.1 and 6.2a). Paths can then be stored digitally and can be re-analysed using special off-line programs (Wolfer and Lipp 1992, Wolfer *et al.* 2001). The numerous variables that can be extracted by digital analysis permit a detailed analysis of peculiarities and changes in open field behaviour (Table 6.1).

The main analytical problem is that many variables are correlated to some degree with the basal activity level of the subjects. By means of factor analysis of large number of animals, it is possible to recognise four factors that contribute

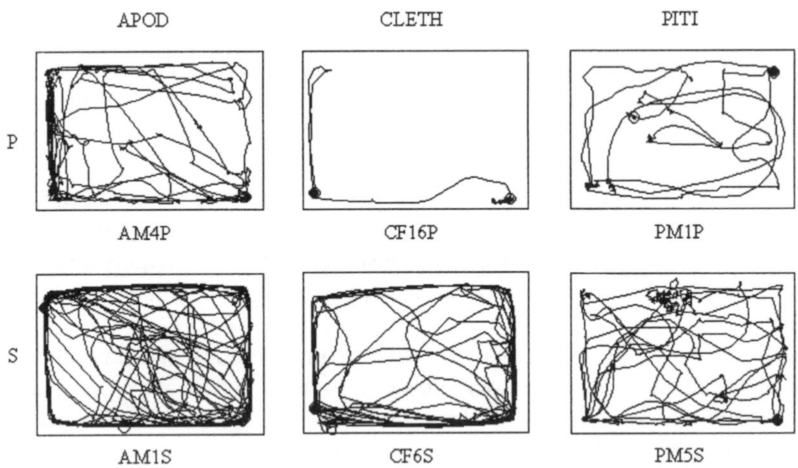

Figure 6.1 Computerised assessment of exposure to dimethoate (an organophosphate pesticide) on open field activity of three species of wild rodents. APOD = *Apodemus microps*, CLETH = *Clethrionomy glareolus*, PITI = *Pitimis terricola*; P = intraperitoneal treatment with 50 mg/kg dimethoate, S = saline injection (data from Dell'Omo *et al.* 2002). The example shows that even visual inspection of paths can discriminate easily between exposed and non-exposed animals. In addition, between-species differences are also evident in the activity of non-injected animals. Thus, such between-species variation must be accounted for by ecotoxicological studies.

Table 6.1 Open-field variables extracted by digital analysis of videotrackings

Latencies and counts	Description
BOL	number of fecal boli deposited (s)
LTM	latency to move out of start zone (s)
LTC	latency to centre field (s)
Trial variables	
PTH	path travelled (m), not considering noise during immobility
SPD	speed during activity (m/s)
ACT	% time active
RST	number of rests (x)
RTM	time spent resting (s)
RDU	average rest duration (s)
BUR	number of locomotory bursts (x)
BDU	average duration of locomotory bursts (s)
BPT	average path travelled per locomotory burst (m)
WDS	average distance to wall (m)
WCT	time in wall zone (s)
CVS	number of visits to centre field (x)
CTM	time spent in centre field (s)
CPT	path travelled in centre field (m)
STP	number of stops (x)
STC	number of stops in centre field (x)
LPI	path tortuosity (°/m). The path is divided into periods of floating, straight segments and curves with constant clockwise (negative) or counter-clockwise (positive) direction change. The absolute values of all direction changes are summed and divided by the total distance travelled.
APT	Cumulative circling bias (°).The path is divided into periods of immobility, straight segments and curves with constant clockwise (negative) or counter-clockwise (positive) direction change. All (signed) direction changes are summed. The absolute value of this sum is the cumulative circling bias.
API	Circling bias (°/m). APT divided by total distance travelled.

x = counts, m = meter, s = second, ° = degrees
Computed by WINTRACK (D.P. Wolfer) using XY-coordinates and sampling frequency

more or less independently to the overall variability of the scores (Table 6.2). One factor is always related to activity, discriminating active from passive animals, and includes variables such as path length, activity periods, speed of movement and others. A second factor usually refers to the agoraphobic tendencies of most small rodents (their fear of open spaces), and includes distance to the wall, time in centre of the field and other related variables that are generally interpreted as a covariate of anxiety. A third factor includes measures of within-session habituation (i.e., the decline of activity within a session, or lack thereof), and a fourth factor represents the temporal patterning

Table 6.2 Factor analysis of open-field behaviour of 1303 mice

Variables (averaged)	Factor 1	Factor 2	Factor 3	Factor 4
Path length	**0.91**	0.28	0.01	0.23
Moving speed	**0.93**	0.10	0.00	−0.04
Activity time	**0.74**	0.34	0.11	**0.47**
Distance to walls	0.18	**0.97**	0.05	0.10
Time in central part	0.22	**0.96**	0.04	0.10
Latency to move	**−0.42**	−0.03	**−0.66**	−0.24
Habituation within sessions	−0.13	0.07	**0.77**	−0.05
Habituation between sessions	0.03	−0.07	**0.49**	**−0.60**
Number of stops	0.26	0.12	0.21	**0.81**
% of variance explained	43	15	14	10
Tentative interpretation	'Activity'	'Anxiety'	'Adaptation'	'Exploration'

Bold print: factor scores above 0.40 (traditionally considered as relevant)
Varimax rotation with orthogonal solution

of activity, including stops, duration running bursts and related measures. It may be interpreted as a measure of exploration.

The factor analysis can also include group or treatment differences (e.g., exposed and nonexposed animals can be entered as dummy variables into the analysis). Further steps of analysis use the factor scores (usually tabulated by the computer programs) to run analysis of variance (ANOVA), thus permitting analysis of more independent or nested factors (Chiarotti *et al.* 1987, Puopolo *et al.* 1999).

The advantages of a detailed open field analysis are the remarkable sensitivity of this simple test to discover subtle changes in behaviour, and the ease of application to many species (Frynta 1994). Increased activity and differential patterning has been reported for adult rats after prenatally administered PCB (Hany *et al.* 1999). By contrast, acute exposure to dimethoate, an OP pesticide, resulted in decreased locomotor activity and a general behavioural impairment in wild rodent and insectivorous species (Dell'Omo and Shore 1996, Dell'Omo *et al.* 1997, 2002). The disadvantage of open field studies is that observed differences can be confounded by genetic factors (see Figure 6.1) and by many environmental factors (Peeler and Nowakowski 1987, Peeler 1990) which produce significantly different results across laboratories even after most rigorous standardisation (Crabbe *et al.* 1999). Thus, it is almost impossible to compare groups from different locations or time periods. Not much is known about the correlation between open field behaviour in the laboratory and under naturalistic conditions. However, one study has shown that subtle behavioural changes in the open field behaviour of mice, produced by prenatally administered oxazepam, can be recognised again by measuring the activity of the same animals in a large outdoor setting (Fiore *et al.* 1995). Lastly, changes in open field behaviour can be produced by alterations of many brain systems. This is probably the reason for its sensitivity, but obviously lowers the value of this test as a diagnostic tool for specifically impaired brain systems.

6.3.1.2 Water maze learning

The water maze test assessing navigational capacities of rodents has been popularised by Morris (1984). During the past 15 years, it became the most widely used testing procedure for spatial memory and learning in many fields of behavioural neuroscience, e.g., ageing (Fischer *et al.* 1992, Frick *et al.* 1995, Spencer *et al.* 1995), psychopharmacology (Arolfo and Brioni 1991, Riekkinen *et al.* 1991), neurodegenerative diseases (Müller *et al.* 1994, Nitta *et al.* 1994, Hsiao *et al.* 1996), lesion studies of extrahippocampal systems (Kolb and Whishaw 1991, Granon and Poucet 1995, Kraemer *et al.* 1996, Rogers and Hunter 1997) and of the hippocampal formation (Morris *et al.* 1990, Conrad and Roy 1993, McDonald and White 1994, Block and Schwarz 1997). More recently, it became a *de facto* standard method for the phenotypic analysis of genetically modified mice (Lipp and Wolfer 1998). Its popularity is based on several factors. It is one of the rare cognitive tasks which is rapidly learned, and can thus be applied to mice. Because of the many processes involved (see above) it is a sensitive indicator of minor behavioural changes that are barely recognised by other tests (Dell'Omo *et al.* 1995, Wolfer *et al.* 1998). Water maze learning is easily analysed by computer controlled video-tracking and requires little observational efforts. It is also thought to be insensitive to odour trails, but see Means *et al.* (1992). An important factor for its wide acceptance is the opportunity to dissociate learning performance from spatial memory and the relative sparing of visually guided (cued) escape learning by hippocampal lesions (Morris 1984). The set-up also permits easy implementation of different versions of assessing discrimination learning and working memory. However, these procedures take much longer than the standard task. Lastly, the Morris maze task is well learned by many species of small wild rodents used in ecological research (Schenk 1987, Kavaliers and Galea 1994, Galea *et al.* 1996, Pleskacheva *et al.* 2000), and has also been used to study perinatally administered ecotoxicants in rats, albeit with contradictory results (Provost *et al.* 1999, Seo *et al.* 1999).

The basic procedure is simple. Animals are released into a tank filled with water made opaque by the addition of white colourants or milk (Figure 6.2b). They have to search for a invisible platform submerged about 0.5 cm below the surface, usually during a maximal period of 60 - 120 s. Once the animal has reached the platform, it is removed and allowed to dry and to recover until the next trial. Variations of the learning procedure concern the intertrial interval, and the number of trials carried out per day (mostly 4 - 6). Learning can be made more difficult by permitting only one or two trials a day and by shortening the intertrial interval (Kogan *et al.* 1997, Gass *et al.* 1998). This can help in discovering subtle effects. Ordinarily, acquisition using a schedule of 4 - 6 trials

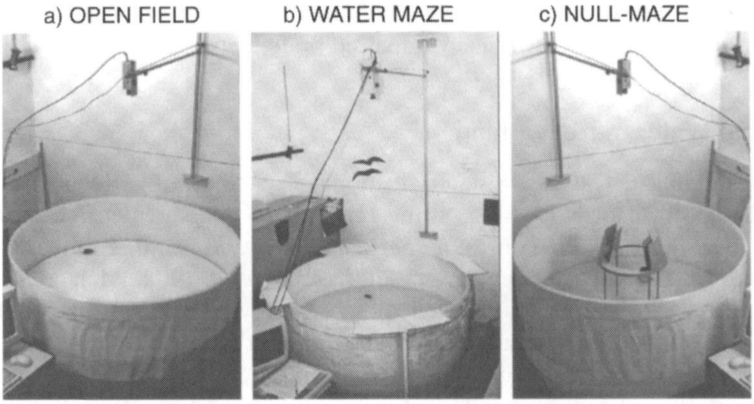

Figure 6.2 Computerised assessment of animal paths in three different tasks using the same or only slightly modified set-ups. Such arrangements greatly facilitate mass screening. a) Open field analysis permits to recognise basal activity levels and agoraphobic tendencies; b) the Morris water maze allows the analysis of spatial learning and memory and assess also impaired shifting of search strategies; c) the elevated null-maze measures, as the open field, basal activity levels and anxiety levels. In addition, it permits to recognise separately exploratory behaviour, albeit only after conducting factor analysis. These three apparatus alone have been shown to recognise many subtle anomalies in transgenic and knockout mice, but also prenatal drug exposures and environmental toxicants such as PCBs.

per day, with an intertrial interval of 30–45 min, is very fast (Figure 6.3). The acquisition of a spatial memory is tested by means of a probe trial, in which the platform has been removed or been relocated. Prolonged searching in the former target quadrant is taken as evidence for spatial map memory (Figure 6.3).

A dissection of the underlying factors is possible by means of a detailed analysis of the swim paths. Table 6.3 presents the variables that can be extracted off-line from stored digital data by means of WINTRACK. This set is extracted from every trial and tabulated for statistical analysis. The factor analysis presented in Table 6.4 has been obtained after running several factor analyses to recognise correlated variables using 3000 mice from eight strains and 32 lines of genetically modified mouse lines, including controls. Whatever the genetic background and the mutation, the behavioural variability in the Morris water maze can be reduced to three clusters of variables.

Factor 1 represents mainly variables related to the acquisition of the task, such as swim path and time, time in target quadrant and other correlated measures. Practically all measures of acquisition depend heavily on the propensity of mice to swim near the wall (wall hugging and thigmotaxis). This behaviour represents a natural response hierarchy of mice when facing this unusual and stressful situation. First they try to escape by swimming to the walls, then they swim along the walls. Depending on stress sensitivity and ability to inhibit this instinctive behaviour, they switch then to a more controlled search strategy which includes scanning of external cues and random scanning of the water

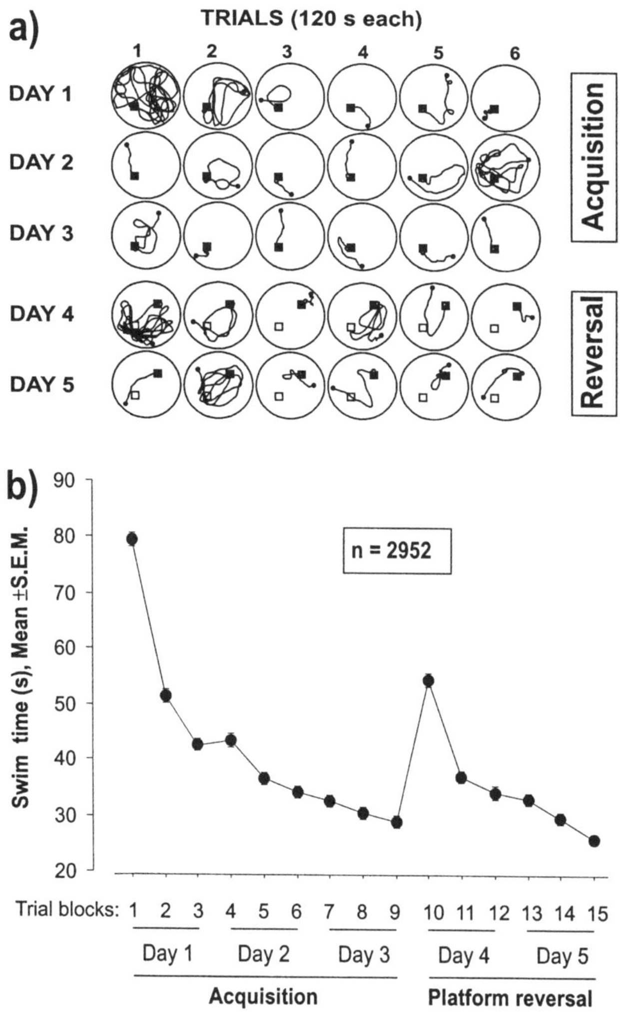

Figure 6.3 Qualitative and quantitative aspects of swimming navigation learning in mice. a) Documentation of typical learning by plotting swim paths. Note the rapid acquisition that is followed by variations in performance. Changing the platform position is followed by intensive searching over the old platform position. Paths plotted by WINTRACK (D.P. Wolfer). b) Average learning curve of a large number of mice, showing the generally rapid acquisition of this task during the first trials (day 1). The increase in swim time after platform reversal indicates the development of a spatial memory for the old target location. For examples of mice being unable to learn that task (see Figure 6.5).

surface by means of meandering swim paths until they find the platform. The next step involves recognising that platform finding permits escape from the water. As shown in Figures 6.3 and 6.6, many animals seem to learn the spatial location of the platform position in a single trial — a phenomenon known

Table 6.3 Swimming navigation: variables extracted by digital analysis of videotrackings

Variable name	Description
Acquisition/reversal	
ACI	airline corridor index: % path in airline corridor
ADB	average distance to release point (m)
AIR	airline distance between release point and goal (m)
API	Circling bias (°/m). APT divided by total distance swum
APT	Cumulative circling bias (°).The swim path is divided into periods of floating, straight segments and curves with constant clockwise (negative) or counter-clockwise (positive) direction change. All (signed) direction changes are summed. The absolute value of this sum is the cumulative circling bias
BEG	% time spent in start zone
CHI	number of goal chaining responses per distance swum (x/m)
CHN	number of goal chaining responses (x)
CSE	cumulative search error (m*s). Average distance between subject and goal is multiplied by the escape time. The time needed to reach the goal on a straight path is excluded from the calculation
DCG	average distance to current goal (m)
EFF	path efficiency: % path during which speed vector component toward goal is 75% or more
ERR	absolute direction error at 0.30 m from release point (°). Angle between vector pointing from release point to subject and vector pointing from release point to goal
FAI	% trials with max. swim time
FLO	time spent floating (s)
JMP	number of jumps off the platform (x)
LPI	path tortuosity (°/m). The swim path is divided into periods of floating, straight segments and curves with constant clockwise (negative) or counter-clockwise (positive) direction change. The absolute values of all direction changes are summed and divided by the total distance swum
NMS	number of near misses: approaches to platform by 0.05 m or less without hitting after swimming a distance at maximum 1.5 m longer than the airline distance between release point and goal
PBC	% path meeting the parallel to border swimming criterion and falling within the centre zone
PBD	parallel to border swimming: % path not meeting the efficiency criterion and having a radial speed vector component of 25% or less
PBW	% path meeting the parallel to border swimming criterion and falling within the wall zone
PTH	distance swum, not counting noise while floating (m)
QCG	% nonfloating time in current goal quadrant
RCG	average radial distance to goal (m). A circle is drawn through the goal platform with its centre in the centre of the pool. The subject's radial distance from this circle is averaged over time
SPD	average speed (m/s) excluding periods of floating and acute decelerations of 0.15 m/s or more
SSE	spatial search error (m). After the subject has swum the distance needed to reach the platform on a straight path, its distance to the goal is calculated every 0.15 m. The distance values are summed to obtain the spatial search error

TABLE 6.3 *(continued)*

Variable name	Description
TIM	escape time (s)
WAP	number of wall approaches (x)
WTI	% time spent in wall zone
Probe trial/reversal	
Q	% nonfloating time spent in quadrant (chance is 25%)
X	annulus crossings (x)
D	average linear distance (m)
P	average polar error (°). Average of absolute angle between vector pointing from pool centre to annulus and vector pointing from pool centre to subject
R	average radial distance (m). A circle is drawn through the annulus with its centre in the centre of the pool. The subjects radial distance from this circle is averaged over time
ERT	crossing preference (x/m). Crossings over former goal annulus are counted. The average number of crossings over adjacent control annuli is subtracted and the result is divided by the distance swum
Z	% nonfloating time in zone (chance is 12.5%)

x = counts, m = metre, s = second, ° = degrees
Computed by WINTRACK (D.P. Wolfer) using XY-coordinates and sampling frequency

as snapshot memory. However, they must learn to associate the position of the platform with escape from the water. Thus, variations in learning time reflect the influence of a multitude of factors, most of them unrelated to the formation of spatial memory. This process is generally reflected in a decrease of thigmotaxis that permits the adoption of more efficient search strategies. Apart from the first one or two trials, persistent thigmotaxis therefore represents the inability to switch from a primitive escape strategy to a better controlled search behaviour. A sudden increase in thigmotaxis is often observed when the animal faces a new situation, such as the reversal of the platform, or the switch to an another testing procedure in which the platform is marked with a visual cue (see also Figure 6.6a). Such regressions are usually of short duration when the animals are habituated to the water maze. In conclusion, prolonged thigmotaxis, entailing slow acquisition, is likely to be caused by problems in strategy switching. As such strategy switching depends on the interactions of many brain systems at different stages of the learning process, it is sensitive to a wide range of cerebral impairments beyond the hippocampal formation (Cain *et al.* 1996, Day and Schallert 1996). One may note that most of the mutant mice that have shown a behavioural phenotype in our set-up were characterised by mutation-dependent changes in thigmotactic behaviour.

Factor 2 reflects variations of swim speed and includes two processes. One is the control of the actual speed during movement, the other is the propensity of many mice (and other species) to float motionlessly. While the variation in

Table 6.4 Factor analysis of swimming navigation of 2952 mice

Variables (average of 30 trials)	Factor 1	Factor 2	Factor 3
Swim path length	**0.90**	0.11	−0.07
Time in target quadrant	**−0.73**	0.16	0.26
Thigmotaxis (time near wall)	**0.82**	−0.28	−0.06
Actual swimming speed	0.06	**0.86**	−0.07
Immobile floating	0.34	**−0.76**	−0.10
Spatial retention in probe trial	−0.18	−0.01	**0.97**
% of variance explained	42	21	14
Tentative interpretation	'Thigmotaxis'	'Speed control'	'Spatial memory'

Bold print: factor scores above 0.40 (traditionally considered as relevant)
Varimax rotation with orthogonal solution

actual swim speed is unrelated to any measure of success and performance, heavy floating can severely impair acquisition of the task (Wolfer *et al.* 1997). In milder forms, however, it can occur in many individuals during some trials without interfering with learning success. The origin and the significance of the behaviour is not known, nor its relation to genetic factors or brain systems.

Factor 3 reflects differential search time in the old target quadrant during the probe trial. The occurrence of this behaviour is a clear indicator that a spatial memory (a mental map) has been formed. However, it is not clear whether variations of this behaviour reflect the strength of spatial memory (the more popular interpretation) or whether it reflects the degree of perseverance by which the animals search for the disappeared platform. In the latter case, it would reflect a compound measure of spatial memory and strategy switching. It must be noted, however, that probe trial scores at chance level can occur in perfectly healthy wild voles learning the task well (Pleskacheva *et al.* 2000; see also swim paths in Figure 6.6), and that inability to learn the task is not necessarily indicative of a lack of spatial memory (see Section 6.3 below).

Impairments in water maze learning are considered as pathognomic for impaired hippocampal functions, particularly if they involve impaired probe trial scores, and if there is no significant difference between groups in a more simple version of the task in which the animals must swim to a visually marked platform. However, the considerations above show that such an exclusive localisation of brain damage is probably unlikely. Also, careful inspection of test with the cued platform indicate that there is often a more subtle impairment which is not significant because the task is easier which leads to flooring effects. Thus, lack of significant differences cannot be taken as a confirmation of the null hypothesis of no differences, at least not if the sample size is small and the probability for a type II error (the beta value) is high. Not much is known of the role of the hippocampus in mediating thigmotaxis. We have found that thigmotaxis may depend partially on hippocampal functions (Pleskacheva *et al.* 2000), and it would seem from early lesion studies that the inability to switch maze solving strategies is also characteristic of hippocampal lesions (Hostetter and Thomas 1967).

In summary, the Morris water maze is perhaps the best screening tool available for assessing disturbances in complex brain functions. Contrarily to a widely held belief, however, impairments in this task are in most cases no indicator for a specific malfunction of the hippocampus.

6.3.1.3 Radial maze learning

Radial maze learning is probably the learning task most frequently cited as being related to foraging habits, and appears thus particularly relevant for ecotoxicological studies of behaviour. The most popular device is the eight-arm radial maze as popularised 20 years ago by Olton and coworkers (1978a) (see also Figure 6.4a).

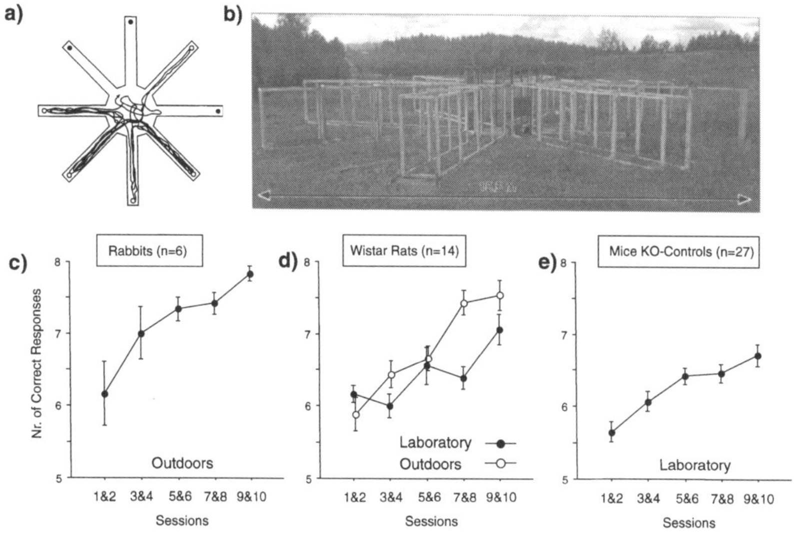

Figure 6.4 Radial maze learning of different species in the laboratory and in outdoor settings (Lipp *et al.* 2001). a) Diagram of radial maze. Correct performance is assumed if an animal enters a baited arm only once during a session. Multiple entries are denoted as errors; b) giant outdoor radial maze in a Russian field station, allowing the comparison of mammals and birds using the same metric scale; c) excellent radial maze learning of rabbits in the giant maze; d) Wistar rats learn the task both indoors and outdoors, the performance in the outdoor radial maze being slightly better; e) radial maze learning of mice in a laboratory set-up (Minichiello *et al.* 1999). Note that performance increases significantly over days, but that mice usually do not show a reliable final performance and continue to commit errors – a problem likely due to species-specific foraging strategies (Dell'Omo *et al.* 2000). Reprinted from Neuroscience and Biobehavioural Reviews **25**, Lipp HP, Pleskacheva MG, Gossweiler H, Ricceri L, Smirnova AA, Garin NN, Perepiolkina OP, Voronkov DN, Kupstov PA and Dell'Omo G. A large outdoor radial maze for comparative studies in birds and mammals, 83–99 (2001), with permission from Elsevier Science.

Typically, experimental subjects are placed on a central platform from which they have to collect invisible baits placed at the end of the arms. The most extensively studied species, the rat, adopts very quickly a win-shift foraging strategy, avoiding double entries. At least in rats, correct performance appears to depend primarily on memorising extramaze cues (Suzuki *et al.* 1980). Because of the ease by which the task is learnt by rats, the eight-arm radial maze became soon a standard apparatus for the analysis of normal and deficient spatial memory and learning, double entries being taken as evidence for impaired spatial working memory; for reviews see Roberts (1984) and Foreman and Ermakova (1998). The task may be run with all baited arms constantly open (the more easily learned version), or by closing the entries to the visited arms during a selected confinement period which challenges the short-term spatial memory to a greater extent. Modifications of the initial procedure by dividing arms in four never baited places and four places holding baits (Jarrard 1978) permit the distinction of spatial working memory errors (double entries into baited places) versus spatial reference memory errors (entering never baited arms). Learning by intramaze cues is possible by inserting, in random sequence, differentially textured floor plates (Jarrard 1983). Like swimming navigation, radial maze learning tasks can be adapted to many species. In mammals, radial maze learning was demonstrated and analysed in gerbils (Wilkie and Slobin 1983), hamsters (Etienne *et al.* 1994), opossums (Kimble and Whishaw 1994) and humans (Aadland *et al.* 1985, Overman *et al.* 1996). The use of radial maze for the evaluation of spatial memory in birds proved to be less easy as it usually requires changes of apparatus and procedures. Successful learning has been reported for ring doves (Wilkie *et al.* 1981), savannah sparrows (Moore and Osadchuk 1982) and, with differential results, for pigeons (Bond *et al.* 1981, Roberts and Van Veldhuizen 1985).

Adaptations of the radial maze tasks to versions without confined arms are known as radial maze analogues. These are often used for the study of spatial memory in bird species, mostly corvids (Spetch and Edwards 1986, Balda and Kamil 1988, Spetch and Honig 1988) and tits (Hilton and Krebs 1991). This applicability to an almost unlimited range of species is of particular value for behavioural ecotoxicology. The strengths and weaknesses of the approach are illustrated by the comparative analysis of laboratory data from mice and rats, and from learning of rabbits and rats in a giant radial maze permitting to study many species using the same metric scale and apparatus (Lipp *et al.* 2001; Figure 6.4). This comparison first shows that the size of the radial maze is not crucial, although the same rat strain shows better acquisition in the large outdoor version. More important for comparative analysis are the obvious species differences. Juvenile rabbits were by far the best learners. Laboratory mice, on the other hand, appeared to be rather moderate or poor learners. Acquisition of the task was obvious, but they seemed to be unable to attain a stable performance at a high level of 7–8 consecutive correct responses. Characteristically, they committed, during and particularly towards the end of the training, a higher number of double entries. An analysis of mouse foraging behaviour in semi-naturalistic settings has revealed that the predominant foraging strategy of house mice is a win-stay strategy (Dell'Omo *et al.* 2000). Hence, species-specifics constraints in solving strategies must be taken into account. Nonetheless, radial maze learning of mice in the laboratory appears to predict, at least in selected cases, the performance in a naturalistic and more complex set-up (see below).

Similar to the water maze, impaired radial maze performance has been attributed to a variety of brain systems. In mammals, the classical win-shift task was found to depend critically on the integrity of the rat hippocampus and its afferent or efferent projections (Olton *et al.* 1978b, Harley 1979, Jarrard *et al.* 1984). Likewise, hippocampal lesions disrupted performance in the standard radial maze task in mice (Rossi-Arnaud *et al.* 1991, Ammassari-Teule and De Marsanich 1996). While initial studies reported no working memory deficits after lesions of extrahippocampal systems including the amygdala, frontal cortex, posterolateral neocortex and caudate nucleus (Becker *et al.* 1980), methodologically more refined studies revealed radial maze impairments correlated with damage in the mediodorsal thalamus (Persinger *et al.* 1994), and impairments of the reference memory component after lesions of the cholinergic nucleus basalis (Murray and Fibiger 1985, Dornan *et al.* 1997), septum (Janis *et al.* 1998) and striatum (Packard and White 1990). Damage to the dorsal striatum impaired selectively acquisition of the spatial win-stay task, while damage of the lateral amygdala impaired acquisition of an intramaze cue task but spared learning of the spatial tasks versions (McDonald and White 1993). Lesions of the higher-order association cortex in rats (prefrontal, perirhinal and parietal association cortex) were also affecting radial maze learning and performance to various degrees, e.g., Di Mattia and Kesner (1988), Seamans *et al.* (1995), Joel *et al.* (1997) and Liu and Bilkey (1998). Taken together, the neurobiological data indicate that performance in a radial maze task is sensitive to many kinds of malfunction within a cortico-subcortical network including the hippocampal formation and adjacent neocortical association areas, that is, in the neuronal machinery underlying cognitive processing in higher vertebrates (Lipp and Wolfer 1998). Finally, it should be noted that developmental exposure to PCBs entailed a facilitation of radial maze learning in rats, suggesting a (possibly) strain-dependent mixture of hypo- or hyperthyroidism during postnatal development (Schantz *et al.* 1997, Seo *et al.* 1999).

In conclusion, the radial maze offers several advantages for use in behavioural ecotoxicology, chiefly by the ease of learning, the adaptability to many species and its sensitivity to brain malfunctions. Its main disadvantage is that it is less suitable for screening purposes and that species peculiarities in solving strategies must be taken into account.

6.3.1.4 Other maze-based and open field tests

The following section shall give a short survey of some tests that have been used or are becoming popular in the field of phenotypic analysis of mutant mice. For an in-depth analysis of underlying mechanisms and original novel approaches potentially useful for behavioural ecotoxicology, the reader is referred to the work of Schenk *et al.* (1995).

6.3.1.4.1 Object recognition tests in the open field

A simple variant of open field behaviour analysis consists in placing objects for exploration in a geometrical arrangement. After habituation to the set-up, objects can either be removed or replaced by new ones, which usually elicits increased locomotion and approaches to the novel object. It is also possible to change the locations of the objects alone, which may elicit activity due to an altered spatial configuration. By measuring the increase and directionality of locomotor activities, spatial and nonspatial working memory (the ability to hold temporary context-specific information) can thus be assessed in a single trial (Ennaceur and Meliani 1992). The test works well with rats, and has been successfully modified for use with mice (Janus *et al.* 1995, Fiore *et al.* 1996, Thinus-Blanc *et al.* 1996). Since this adaptation requires much more habituation and removal of fear-inducing factors, chances are good that the test can be adapted for the study of wild rodents as well. Because of its brevity and suitability for computerised analysis, this type of spatially based object recognition is likely to become widely used in many fields of neuroscience.

Selective abolishment of spatial working memory but not of object recognition memory is often interpreted as a sign of hippocampal malfunction. However, studies of hippocampal and amygdalar lesion studies in rats are not free from contradiction (Aggleton *et al.* 1989, Mumby *et al.* 1992). Likewise, it should be kept in mind that hippocampally lesioned animals can show exaggerated activity after introduction of novel objects (Nyakas *et al.* 1983). Because the test assesses simultaneously baseline activities and cognitive functions, it is in any case useful for complementing a cognitive test battery applicable to a wider range of species.

6.3.1.4.2 Spatial alternation tests

A classical set of tests used to assess spatial memory and learning in rodents is based on spatial alternation in T-mazes. During the first trial of a daily session, both arms are baited. Subsequently, food is placed only in the arm opposite to the one visited during the previous trial. By prolonging intertrial times, the task challenges spatial working memory increasingly, a procedure known as delayed spatial alternation (Cory-Slechta 1989). It is thus suitable to assess cognitive deficits related to working memory, and it has been shown to be sensitive to manipulations or lesions of various brain systems, e.g., the noradrenergic projection (Wenk *et al.* 1987), the prefrontal cortex (Winocur 1991), the septo-hippocampal system (Wan *et al.* 1995), and the hippocampus proper (Dalland 1976, Winocur 1991), to name a few.

The task can be applied, with modifications, to a variety of species and has been successfully used to document deficits following PCB exposure in both monkeys (Schantz *et al.* 1991) and rats (Schantz *et al.* 1997). It can be used in mice (Beracochea and Jaffard 1985, Crusio *et al.* 1990, Lalonde *et al.* 1986) (but see below) and can be modified to include more naturalistic situations, making it suitable for nonlaboratory species (Whishaw *et al.* 1995).

It has two drawbacks, however. The interpretation is based heavily on the assumption of a win-shift strategy, the predominant strategy in the rat. Extensions to other species with preferential win-stay strategies such as mice have thus first to assess a potential species-specific bias in choice strategies. The second problem is the length of testing (often exceeding 10 days) which makes it impractical for screening a large number of animals.

6.3.2 TESTS ASSESSING ANXIETY AND FEAR

This category of tests, based on the analysis of fear responses chiefly in rodents, is useful for discovering behavioural alterations due to conflicting behavioural tendencies inherent in most of the tests. They can also be used to discover pathological changes in the rather well-defined brain systems mediating fear and flight responses, such as the caudal hypothalamus, the central grey and the amygdala (LeDoux 1994). In practice, the latter is often attempted but not infrequently marred with problems of interpretation.

6.3.2.1 Contextual and cued fear conditioning

This test is based on the propensity of small rodents to remain immobile (freezing) when confronted with a threatening stimulus. Usually, the animal is placed in a box and its precession activity recorded. Then it receives an unconditioned stimulus (UCS) in the form of successive electric shocks. The UCS is usually combined with a tone as a conditioning stimulus (CS). The mouse or rat then develops a freezing response. After a defined period, the animal is placed in the box again, yet without exposure to the tone. Most subjects recognise the threatening environment and freeze, indicating a memory for the contextual situation. The duration of freezing is measured and taken as an index of contextual fear-related memory. The same animal (or a subgroup of treated animals) may also be placed in a novel environment lacking the fear-inducing properties such a grid floors and other cues, and is confronted with the CS. The duration of freezing in response to the tone is then considered as an index of CS-related memory. The test can be used to assess short- versus long-term memory by testing the subjects 30 min and 24 h after the shock exposure

(Brambilla *et al.* 1997), and to measure extinction. The sensitivity of the test in discovering brain malfunctions can be increased by presenting the CS with a variable delay after shock exposure, a procedure known as trace conditioning.

The conditioned fear procedure is thought to dissociate malfunctions of the hippocampus and the amygdala. In fact, hippocampal lesions usually impair the contextual freezing response yet spare the responses to auditory cues, although this distinction is less clear in mice in which genetic confounding differences have been found (Chen *et al.* 1996, Logue *et al.* 1997). On the other hand, lesions of the amygdala have been found to spare contextual freezing (Phillips and Ledoux 1992), probably because the amygdala receives direct projections from midbrain auditory nuclei. Although not done yet, it would seem applicable to test many nonlaboratory rodents.

On the other hand, it shares the problem of all tests based on assessing the duration of fear-induced freezing or other responses requiring behavioural inhibition: it is sensitive to nonmemnonic confounding factors regulating the activity level. For example, it has been shown that F1 hybrid mice from inbred strains (which generally show excellent learning and memory in complex situations) quickly developed a freezing response, showing this behaviour after 24 h, but started to move much earlier than other mice thought to have deficient long-term memory because of inactivation of an intracellular signalling protein (Gass *et al.* 1998). Likewise, genuinely passive animals tend to show 'better' fear-related memory, while behavioural disinhibition or pathological hyperactivity tend to 'impair' such memory. In contrast to the other more simple paradigms below, the combined use of contextual and cued conditioning offers a (limited) tool to narrow down interpretation.

6.3.2.2 Passive avoidance

This is a classic task probably sharing many features of contextual and cued fear conditioning. Its design is more simple, the potential for theoretical explanations correspondingly reduced. Basically, the animal is shocked after leaving a safe platform (step-down avoidance) or compartment (step-through avoidance). Following a specified interval, the subject is returned to the safe place and the latency to move to the shock place is taken as an indicator of fear-related memory. As a short one-trial task, it is probably better suited to assess species-specific response patterns and can be used for mass screening. The necessary application of electric shocks may entail confounding mechanisms related to endocrine responses and coping with emotional stimuli. This makes interpretation less easy, yet can increase the sensitivity of the test to detect subtle pathological changes in the brain. The classic hippocampal electrolytic lesion syndrome includes impaired passive avoidance (Blanchard and Fial 1968), but this might be due to damaged fibres of passage (Munoz and Grossman 1981). In any event, damage of any system mediating behavioural inhibition is likely to produce deficits in this task.

6.3.2.3 Two-way active avoidance

This is another classic task developed chiefly for psychopharmacological investigations. The most popular version is known as shuttlebox learning. The apparatus consists of two chambers with electrified grid floors for delivery of shock (UCS), which are separated by a connecting hole. The CS is a light or a tone that is activated in the compartment occupied by the animal, as registered by photocells or switches activated by a toggling floor plate. The animal must avoid the shock by crossing to the other side. The correct solution of the task thus involves continuous moving from one side to the other, hence the term shuttlebox. Between trials of CS-UCS presentation there is an intertrial interval (ITI) of random duration. Locomotor activity (crossing to the other side) during the ITI can be permitted or punished by shock. Because shuttleboxes are easily automated, they became a standard tool for pharmacological mass screening in both rats and mice. However, the seemingly simple task takes a fairly long period for learning: in mice up to 5 days with 50–100 trials. This is because it involves several behavioural processes arranged in a response hierarchy (Bolles 1970). First the animals must learn to escape to the other compartment. Afterwards, they face a dilemma as the task requires moving back to a place of previous punishment. This often entails freezing responses that must be overcome. In shuttleboxes that permit crossings during the intertrial interval, animals can show pseudo-learning by adopting a high level of locomotor activity. Often animals develop behavioural strategies minimising exposure to shock but which are counted as failures. For example, they might move, in anticipation of a shock, in front of the connecting hole, wait for shock delivery and escape with one jump to the other side, where they re-initiate this behavioural sequence.

This complexity of processes makes the shuttlebox a sensitive tool to discover pathological changes in many brain systems, but does not permit much interpretation. Improved shuttlebox learning is part of the classic hippocampal lesion syndrome in rodents (Antelman and Brown 1972, Carlson 1970, Olton and Isaacson 1968) but even for this structure, more specific lesions have been reported to impair or spare shuttlebox learning (Munoz and Grossman 1981). A review of the many other brain systems having been reported to mediate shuttlebox learning is not attempted here; it suffices to note that most of the results have been shown to depend additionally on sex, genetic differences, stimulus parameters and training schedules.

6.3.2.4 Tests measuring spontaneous anxiety versus exploration

This class of tests does avoid the confounding effects of punishments and focuses on the spontaneous behaviour of the animals. They tend to be more specific for evaluation of treatment effects on particular brain systems, in particular for the GABA ergic system (Crestani *et al.* 1999). They are also more compatible with the natural behaviour of many small rodents, permit computerised analysis and are of short duration. They are not easily transferred to nonrodent species, however, and interpretation must cope with the fact that behaviour in these tasks usually reflects a balance between fear-related mechanisms and mechanisms governing exploration and locomotor activity. Two tests amenable for videotracking analysis are described.

6.3.2.4.1 Light-dark box

The simple set-up consists of a box with two compartments: one side with a covered box, another with an illuminated surface. The animal is placed in the dark box, and the latency to enter the illuminated compartment is measured. By using videotracking and keyboard input by an observer, the behaviour of the

Table 6.5 Light-dark box variables extracted by analysis of videotracking

Zone variables	Description
ZCL	latency to enter closed zone for the first time (s)
ZCN	number of visits to closed zone (x/min)
ZCV	average duration of visits to closed zone (s)
ZCT	time spent in closed zone (%)
ZON	number of visits to open zone (x/min)
ZOV	average duration of visits to open zone (s)
ZOT	time spent in open zone (%)
ZIL	latency to enter intermediate zone for the first time (s)
ZIN	number of visits to intermediate zone (x/min)
ZIV	average duration of visits to intermediate zone (s)
ZIT	time spent in intermediate zone (%)
Grooming in open zone	
GTL	latency to first grooming (s)
GTN	grooming rate (x/min in open zone)
GTV	average duration of grooming episodes (s)
GTI	% open zone time spent grooming
Rearing in open zone	
RTL	overall latency to first rearing (s)
RTN	overall rearing rate (x/min in open zone)
Motor activity in open zone	
ACT	% time moving
RST	number of rests (x/min)
RTM	time spent resting (%)
RDU	average duration of rests (s)
PTH	path traveled per min spent in open zone (m/min)
SPD	average speed while moving (m/s)

x = counts, m = metre, s = second
Computed by WINTRACK (D.P. Wolfer) using XY-coordinates and sampling frequency
Grooming and rearing entered by keyboard codes

animal in the illuminated side of the box can be analysed further (Table 6.5) for activity levels, conflicting tendencies being expressed as grooming, etc. This test is useful for a quick assessment of basal anxiety and activity levels.

6.3.2.4.2 Elevated null-maze

Elevated mazes such as the elevated cross-maze, null-maze or t-maze (Viana *et al.* 1994) assess the exploratory behaviour of rodents on elevated runways which provide both secure parts and open parts initially avoided by the animals because of their cliff aversion. The elevated cross- (or plus) maze is the more traditional one, consisting of two open and two visually shielded arms. Its behavioural components have been analysed in detail both for rats (File 1993) and mice (Fernandez-Espejo 1997). The null-maze or O-maze (Shepherd *et al.* 1994) is formed by a circular runway with shielded and open parts (Figure 6.2c). This latter variant is better suited for computerised videotracking analysis. Table 6.6 shows the variables that can be extracted from automated observation, while Table 6.7 shows the results of a factor analysis in 104 mice carrying differentially-sized deletions of the prion protein (unpublished data). Three independent factors can be extracted: factor 1 related to basal activity levels including path length, activity and speed, a second factor assessing the exploratory components (chiefly shown by animals lurking into the open zone and exploring the cliffs by lateral head dips), and a factor related to fear as characterised by long periods inside the covers and high levels of defecation. Elevated plus mazes have been successfully used to characterise subtle behavioural effects due to other targeted mutations (Lacroix *et al.* 1998). It must be noted, however, that this task is sensitive to both genetic differences (Trullas and Skolnick 1993) and laboratory environment (Crabbe *et al.* 1999).

6.3.3 TESTS BASED ON SOCIAL AND INGESTIVE BEHAVIOUR

This category of tests includes some tests based on natural behaviour that became popular in the field of phenotypic analysis of genetically modified mice. They are not easily automated and require much manual and observational work. On the other hand, they are potentially useful for behavioural ecotoxicology because they measure behavioural parameters of particular importance for biological fitness. A thorough review about social learning and its implications for neurobiology is provided by Choleris and Kavaliers (1999).

6.3.3.1 Intruder-aggression paradigm

Males from many rodent species react with hostility when faced with an intruder of the same sex. Among laboratory species, this is particularly evident for mice (Alleva 1993, Alleva *et al.* 1995).

Table 6.6 Null-maze variables extracted by digital analysis of videotracking

Latencies	Description
ZIL	latency to enter intermediate zone for the first time (s)
ZOL	latency to enter open zone for the first time (s)
ZCL	latency to enter opposite closed zone for the first time (s)
DDL	latency to very first head dipping (s)
DOL	latency to first open zone head dipping (s)
BBL	latency to first defecation (s)
Trial variables	
BBN	fecal bolus number (x)
PTH	path travelled (m)
SPD	speed during activity (m/s)
ACT	% time active
RST	number of rests (x)
RTM	time spent resting (s)
RDU	average rest duration (s)
DIN	number of head dipping in intermediate zone (x)
DON	number of head dipping in open zone (x)
ZCN	number of visits to closed zone (x)
ZCT	time spent visiting closed zone (s)
ZIT	time spent in intermediate zone (s)
ZCV	average duration of visit to closed zone (s)
ECN	number of excursions outside closed zone (x)
ECT	time spent on excursions outside closed zone (s)
ECV	average duration of excursions outside closed zone (s)
ICN	number of visits to intermediate zone (x)
ICT	time spent visiting intermediate zone (s)
ICV	average duration of visit to intermediate zone (s)
OCN	number of visits to open zone (x)
OCT	time spent visiting open zone (s)
OCV	average duration of visit to open zone (s)
ZCX	number of closed zone crossings (x)
ZIX	number of intermediate zone crossings (x)
ZOX	number of open zone crossings (x)

x = counts, m = metre, s = second
Computed by WINTRACK (D.P. Wolfer) using XY-coordinates and sampling frequency

The more simple version of the task uses a so-called standard opponent, usually a mouse from a relatively docile strain, which is introduced either into the home cage of the resident, or in an observation arena to which the resident had some time to adapt (Denenberg *et al.* 1973). An important variable is the latency to attack during a single encounter. More thorough analyses require repeated videotaped encounters with manual coding of behavioural elements related to offence and defence (Alleva 1993).

The test outcome is influenced by many factors as evidenced by substantial genetic variation (Crusio 1996). Recognised factors are changes in systems

Table 6.7 Factor analysis of elevated null-maze behaviour (104 mice)

Variables (average of 2 sess.)	Factor 1	Factor 2	Factor 3
Path travelled	**0.96**	0.20	−0.03
Activity time	**0.94**	0.20	0.12
Speed during activity	**0.87**	0.25	−0.13
Head dippings intermediate zone	**0.60**	**0.59**	0.13
Time in intermediate zone	0.08	**0.97**	0.02
Total excursion time	0.35	**0.92**	0.02
Number of boli	−0.33	0.06	**0.57**
Time resting	0.28	0.01	**0.83**
% of variance explained	52	17	13
Tentative interpretation	'Activity'	'Exploration'	'Anxiety'

Bold print: factor scores above 0.40 (traditionally considered as relevant)
Varimax rotation with orthogonal solution

controlling behavioural inhibition, notably the GABAergic system (Saudou *et al.* 1994), the serotoninergic system (Chen *et al.* 1994) and the hippocampus (Guillot *et al.* 1994, Sluyter *et al.* 1994). Other factors involved include levels of testosterone (De Ruiter *et al.* 1993) and corticosterone (Stork *et al.* 1997), olfactory cues (Kemble *et al.* 1995), specific pheromones (Novikov 1993), etc. However, changes in aggressive behaviour can occur without alterations in cognitive behaviour (Chen *et al.* 1994), which makes this test useful for discriminating general from specific effects of environmental toxicants (Husain *et al.* 1996).

6.3.3.2 Social transmission of food preference

This test has gained popularity because of the demonstration that it can be selectively abolished by hippocampal lesions, at least in rats (Bunsey and Eichenbaum 1995). Moreover, it is considered as natural learning not marred by operant schedules. It has been used to characterise, with mixed success, effects of deletion of the calcium response binding element protein (CREB) on long-term memory (Kogan *et al.* 1997, Gass *et al.* 1998).

The test is manually conducted. A demonstrator animal is allowed to eat food laced with characteristic scents such as cinnamon or chocolate, and which have previously been shown to be equally attractive for the animals. An observer animal is then kept with the demonstrator and allowed to explore it for some time. A defined period later, usually 24 h, the hungry demonstrator receives two cups with scented food, and the quantity of consumed food is measured. Usually, the food with the scent learnt from the demonstrator is preferred. Equal or reversed preferences are interpreted as missing memory.

As this test is being actually adopted by many groups, future results will show its usefulness and acceptance. In any case, confounding individual variations in food preference must be considered seriously (Galef 1993).

6.3.3.3 Taste aversion

Taste avoidance learning is of particular ecological relevance as it can permit to avoid exposure to toxicants. It can be acquired individually through experience with toxicants, or socially through observation of a demonstrator (Zahorik and Houpt 1981).

The standard method in the laboratory is to measure first saccharin preference in a two-bottle choice test. Animals are then injected intraperitoneally with a solution of lithium chloride. This causes visible illness in rats but few overt signs in mice (Welzl *et al.* 2001). After a defined interval, animals are tested for saccharine preference again. In most cases, the animals develop a conditioned taste aversion (CTA). This one-trial learning allows the convenient assessment of long-term memory and its extinction. By omitting the injection procedure, the test can also be adapted to wild species (Rabin and Hunt 1992). It has proven to be sensitive to targeted mutations in mice (Welzl *et al.* 2001). For an overview of neuroanatomical substrates see Chambers (1990).

6.3.4 TESTS DETECTING ALTERED BEHAVIOURAL PATTERNING

A category of tests potentially capable of detecting most subtle alterations of behavioural organisation analyses changes in behavioural patterning (Cohn and MacPhail 1996). One class of them is assessed by operant conditioning for patterned response schedules. Such tests are suitable for automated measuring, but are mostly tailored to rats and pigeons. Cross-species comparisons can be done with encephalised species such as monkeys and cats, yet their application to wildlife species and even mice is not so easy. Another approach is based on ethological techniques focusing on the statistical analysis of behavioural sequences. This type of analysis requires detailed manual coding of behavioural elements. For both types of analysis, the relation to brain systems is not well understood and has often not been attempted, particularly in ethology. Thus, this section will present only the methodological principles.

6.3.4.1 Schedule-controlled behaviour

The basic set-up consists mostly of test chambers where a lever press (or pecking to a disc for pigeons) delivers a reward. The basic procedure is quickly learned by many animal species. Subsequently, the delivery of reward can be made dependent on the number of lever presses, the so-called fixed ratio (FR) schedules. Likewise, fixed interval (FI) schedules impose a certain delay before reward is delivered (Tilson 1993). The lever press rates are amenable to sophisticated statistical analysis involving time sequences, autocorrelation and Markov chains, and it has been shown that these procedures are able to discover a variety of drug effects (Weiss *et al.* 1989). Interpretation and meaning of these effects remain difficult, however. The technique has also been applied to detect PCB-dependent changes in FI responses of rats treated postnatally with subtoxic doses. These animals showed a slight hyperreactivity reminiscent of animal models of attentional-hyperactivity deficit disorders (Holene *et al.* 1998). However, another study testing for attentional deficits in perinatally treated PCB rats, using operant procedures, found no evidence for attentional deficits but a possible interaction with GABA-mediated pathways (Bushnell and Rice 1999). It would seem that one of the more predictable effects of PCB exposure is an impairment of behavioural inhibition, that becomes particularly evident in tasks that require a response to be withheld such as in FI schedules. In agreement with this notion, distinct impairment of differential reinforcement of low-rate schedule (DRL) was observed in monkeys treated perinatally with PCB, but there were no differences in a spatial discrimination reversal task. DRL is a schedule in which the animals have to wait a relatively long period (often about 30 s) in order to obtain a reward, and each premature lever press postpones reward. Performance in DRL schedules is sensitive to lesions of the hippocampus (Rawlins 1985) and is generally low in genetically defined lines of animals with poor behavioural inhibition (Driscoll and Baettig 1982).

In conclusion, analysis of schedule-controlled behaviour is a useful method for an in-depth analysis of recognised impairments but has limitations as a screening tool, chiefly because of the relatively long time period required to establish reliable baselines in controls. The possible exception are schedules assessing behavioural inhibition. However, it is likely that such impairments will be detected by the tests described above.

6.3.4.2 Ethological sequence analysis

One of the methodological hallmarks of ethology is the sophisticated analysis of behavioural sequences (Cohn and MacPhail 1996). Videotaped sequences of naturalistic behaviour are coded in terms of behavioural units defined by an ethogram. An ethogram classifies the species-specific behaviour into units (e.g., grooming, nibbling, boxing) that do not overlap in time. Ordinarily, behavioural units do not occur in random order but follow certain sequences, e.g., for an open field, walking, rearing, grooming. Thus, the primary computation involves the length and frequency of sequences (Ellenbroek *et al.* 1992). These data can then be subjected to various procedures developed for the analysis of time data (see also the following chapter). One of the models uses Markov chain procedures which compute the probability of occurrence of a certain behavioural unit as a function of the preceding behavioural unit. The result are so-called transition rates between behavioural states which predict the likelihood of behavioural changes. These transition rates may be subjected to ANOVA and factor analysis in order to compare group or treatment effects (Puopolo *et al.* 1999). Other more descriptive models include graphs of transition probabilities which depict the organisation of behaviour in the form of flow diagrams (Baerends 1976).

The main problem of this approach is the interpretation of data in terms of treatment effects, genetic factors, etc. Unfortunately, the most predictable outcome in the case of subtle brain impairments is that the time structure of behaviour becomes either more stochastic, or more stereotyped, in the worst case oscillating chaotically between the extremes. Such chaotic states are easily grasped by intuition, but pose formidable problems if they are analysed in terms of statistically comparable measures obtained from information theory. Even cases of isolated differences, such as the altered transitional probability between two behavioural units, are difficult to interpret even for those who develop such analyses. The reader may have noted that many cognitive deficits in the tests reported before (e.g., for the water maze) are based on problems of selecting and maintaining appropriate solving strategies rather than on memory impairments. It would be highly desirable to have analytical tools assessing transition rates between strategies for comparative analysis. However, for these cases the situation is even worse, because it additionally requires to define unequivocally the start and the end of a particular strategy. Moreover, strategies are highly context-dependent.

Fortunately, there are also some examples of sequence analysis that have been shown to permit the screening of toxicant-dependent changes in behaviour, or have ecological implications. One example is the microanalysis of licking rates in the home cage of rats which showed altered diurnal patterns following ingestion of triethyltin in the water, the daily consumption being unchanged (Evans 1994). Another example of possible relevance for assessing biological fitness is a cost-benefit analysis of the behavioural sequence organisation in mice (Guillot and Meyer 2000). This approach is based on the costs of executing species-specific behavioural units during a circadian rhythm (e.g., locomotion, grooming, nest building), taking into account the time distribution of meals and the total food intake by the mouse. Such behavioural sequences can then be compared against randomly generated artificial sequences for the same time period. These comparisons can indicate whether the actual behavioural sequences show an optimised energy gain, assuming that random sequences will result in suboptimal cost-benefit relations. By recording individual patterns of behaviour in naturalistic environments contaminated with toxicants this analytical procedure may help quantify inefficient behavioural organisation with respect to cost-benefit optimisation. For example, sequence analysis has been used to detect altered parental behaviour in gulls naturally exposed to lead contamination (see description of the Burger and collaborators' studies discussed in Chapter 9). The technical requirements for this approach are available (see below).

6.4 NEW APPROACHES

Behavioural ecotoxicology must assess behavioural changes both in the laboratory and under realistic field conditions (Little 1990). This requires developments in three fields: (1) laboratory assays must be validated under at least naturalistic conditions; (2) tests have to include ecologically relevant species, possibly in their habitats (Scherer 1992); (3) behavioural assessment must be complemented by analysis of brain and physiological markers of exposure to environmental toxicants.

6.4.1 CORRELATING LABORATORY AND FIELD STUDIES

The following example is taken from our laboratory in which we try to determine the functional consequences of targeted mutations in a naturalistic environment. For this purpose, we use two approaches. Mice are released into large outdoor pens, and short- and long-term survival of various targeted mutations are assessed by catching and genotyping the entire mouse population at regular intervals (Vyssotski *et al.* 2000,2002). At the same time, all mice can be tagged with microchips permitting recognition of individuals by means of antennae reading the microchip code (Dell'Omo *et al.* 1998). In order to analyse the learning behaviour of mice living outdoors, we have also developed computer-controlled devices delivering food according to temporal and spatial schedules (Dell'Omo *et al.* 2000).

One example of such studies is summarised in Figure 6.5. The problem was to analyse the nature of the behavioural phenotype observed after targeted forebrain-specific deletion of the gene for the neurotrophin receptor trkB. In the laboratory assays, the behavioural analysis had revealed a perplexing combination of moderately impaired radial maze learning (Figure 6.5a), severely impaired water maze learning (Figure 6.5b), improved two-way avoidance learning, but no or minor effects in contextual fear conditioning and various tests assessing anxiety and locomotion (Minichiello *et al.* 1999). Our interpretation was that the animals were chiefly suffering from an inability to switch strategies rather than from spatial or other memory deficits. A total of 40 mutant mice and controls were then released into an outdoor pen in western Russia, which had been equipped with computer-controlled feeders programmed to deliver one-eighth of the daily food portion only during the first visit of an individual, and to chase illegitimate visitors committing re-entry errors. All experiments and programming in this study was carried out by A. Vyssotski (Vyssotski *et al.* 2002). As he feared that the mutants might starve under a such a rigid food delivery regimen, he allowed free feeding inside the shelters every third day. Both mutants and controls learned the patrolling task quite well, steadily increasing the number of visited feeders. However, once the task was learned, the mutants were unable to switch their patrolling strategy and continued to do so even when free food was available in the shelters. Thus, this study confirmed the tentative diagnosis of reduced behavioural flexibility.

This study also demonstrates that economical mass screening can be done under semi-naturalistic conditions, simply by studying the behaviour of an entire

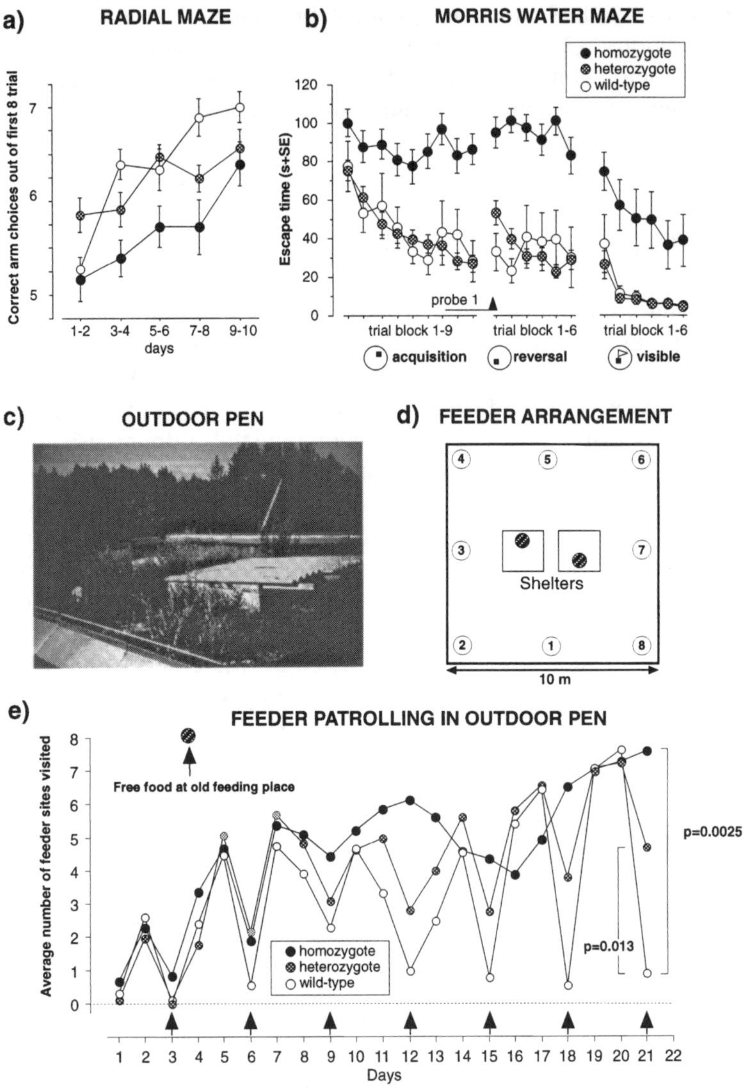

Figure 6.5 Comparative learning performance of mice lacking the neurotrophin receptor trkB in the laboratory and in outdoor settings. a) Slightly impaired learning performance of mutants in a laboratory radial maze (for details see Minichiello *et al*. 1999). b) Severely impaired water maze learning of trkB mutants. Note that many mutant mice were unable to learn the task, chiefly due to excessive thigmotaxis (Minichiello *et al*. 1999); c) outdoor pen in western Russia used to study natural learning of transponder-tagged mice; d) Arrangement of feeders in outdoor pen. After release, mice received food first inside the pens. Afterwards, food was delivered according to a radial maze training schedule, but every third night food was delivered in the former location; e) Development of maze patrolling in outdoor pen. Mutants were able to learn the patrolling task, but were unable to switch their patrolling strategy once they acquired the task (Vyssotski *et al*. 2002).

population over many days. The next generation of test boxes will operate by means of radio control. This will enable us to study the learning behaviour of wild living voles and wood mice by placing battery-operated set-ups in their habitats. The potential extension to ecotoxicological studies is obvious. In principle, the same set-up could also be used to assess cost-benefit optimisation in naturalistic settings (see above).

6.4.2 SELECTING ECOLOGICALLY-RELEVANT TEST SPECIES

Rats and mice may present convenient test models for ecotoxicological studies. Yet, their extreme adaptability to different habitats, including commensalism with humans, is not ideal to assess ecology and behaviour in a systematic way. Hence future investigations with rodents should focus on wild small mammals occupying defined ecological niches. In the context of comparative studies of maze learning abilities in small wild mammals, we have noted that the bank vole (*Clethrionomys glareolus*) might represent an excellent candidate model for behavioural ecotoxicology. It is found all over Europe, its ecological parameters are well investigated, and the bank vole performs excellently in classical laboratory tasks, even when testing wild-caught animals. In the water maze, the performance of bank voles matches and even surpasses that of rats, and is distinctly superior to mouse strains. Figure 6.6 shows one example (out of many) of perfect water maze learning with no performance variation. They are also good radial maze learners, comparable to rats. In addition, captive bank voles can be handled easily.

In comparison, other pan-European species appear less suitable for at least the more complex tasks. Wild-caught wood mice (*Apodemus sylvaticus*) can perform excellently in the water maze, but performance is mostly confounded

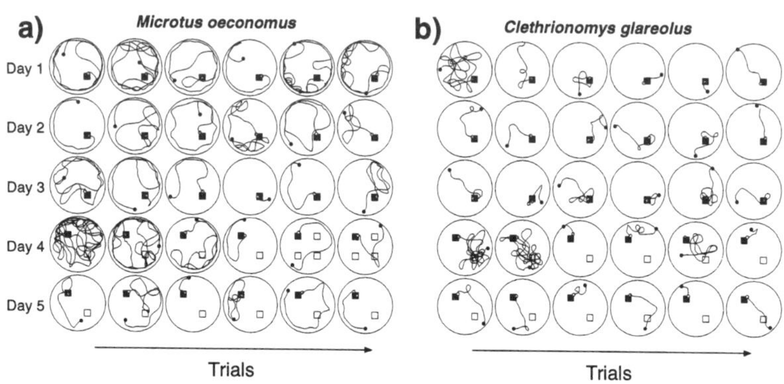

Figure 6.6 Swimming navigation learning in two European species typically investigated in ecological studies (Pleskacheva *et al.* 2000). Note the almost perfect learning of a wild-caught bank vole *Clethrionomys glareolus*, and the much less efficient learning of the root vole *Microtus oeconomus*.

by their strong emotional reactions. The other vole species tested thus far (*Microtus oeconomus*, the root vole) is easier to handle and suited for a variety of laboratory tests. These examples show, together with other studies (Schenk 1987, Kavaliers and Galea 1994, Galea *et al.* 1996, Pleskacheva *et al.* 2000), that tests assessing cognitive behaviour can be applied to ecologically relevant rodent species.

6.4.3 NEUROBIOLOGICAL CORRELATES

A concomitant assessment of structural and neurophysiological changes is indispensable for meaningful interpretation of behaviour. In view of the accumulated body of knowledge, the most promising candidate system for correlative assessment of environmentally-induced cognitive alterations in mammals is the hippocampus. Apart from its undisputed role in memory and learning, it is also of interest for its propensity to accumulate environmental neurotoxicants (Sato *et al.* 1984). Moreover, the hippocampus is equipped with receptors for virtually any circulating hormone, a feature of interest for the study of endocrine disrupters.

6.4.3.1 Neurophysiology

The actual standard for analysing neurophysiological alterations is the measurement of long-term potentiation (LTP) in hippocampal slices or in the intact brain. LTP denotes a long-lasting change in synaptic transmission after a short tetanic electrical stimulation, and is widely perceived as an indicator of synaptic plasticity underlying learning and memory. In rats, developmental exposure to PCB mixtures entailed a persistent decrease of LTP in the adult animals (Gilbert and Crofton 1999). Likewise, impaired hippocampal LTP was observed after lead consumption during different postnatal periods in rats (Gilbert *et al.* 1999). Altered synaptic transmission has also been observed after acute exposure of normal hippocampal slices from rats to PCB and dioxin (Hong *et al.* 1998). Clearly, these approaches are still restricted to laboratory animals. However, technology being developed in our and other laboratories will offer microtelemetrical devices for monitoring LTP and single unit activity in freely moving animals inside and outside the laboratory.

6.4.3.2 Neuroanatomy

The most promising candidate system for monitoring subtle developmental intoxication is the hippocampal mossy fibre system. This is formed by the axons of the granule cells in the dentate gyrus, and differentiates relatively late during the postnatal period of rodents. The distribution of axons ending in the CA3 region of the hippocampus shows a neatly laminar arrangement that can be visualised by means of Timm's stain (Figure 6.7). Staining and morphometry is

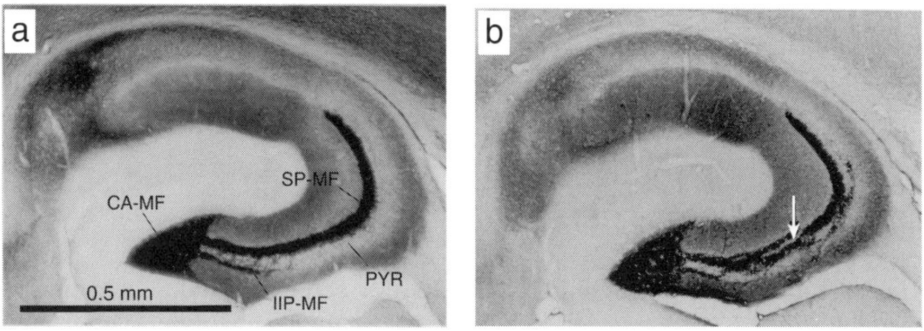

Figure 6.7 The hippocampal mossy fibre system in normal mice (a) and in 101/HY mice (Lil'p *et al.* 1995) characterised by developmental instability (b). Arrow in (b) points to developmental lamination defects in the suprapyramidal mossy fibre layer caused by ectopic pyramidal cells. The size of the infrapyramidal mossy fibre projection is known to correlate with hippocampus-dependent behaviours (Lipp *et al.* 1988). Abbreviations: CA4-MF, mossy fibre field in CA4; IIP-MF, intra/infrapyramidal mossy fibre projection; PYR, pyramidal cell layer; SP-MF, suprapyramidal mossy fibre projection.

easy, even in postmortem brains. This allows comparative studies of trapped dead animals.

Two properties make the mossy fibre system interesting for ecotoxicological studies. The distribution in the CA3 region shows a critical period during postnatal development during which it reacts sensibly to genetic factors, hormonal levels and toxicants (Sato *et al.* 1984, Lipp *et al.* 1988). Afterwards, the distribution remains relatively stable, conserving the traces of even subtle developmental insults. For example, postnatal hyperthyroidism increases the extent of the mossy fibre projection synapsing on the basal dendrites of hippocampal pyramidal cells (the intra/infrapyramidal mossy fibre projection, IIP-MF) actually correlated with a moderate improvement in radial maze learning of adult mice (Schwegler *et al.* 1991). Interestingly, it has been shown recently that pre- and postnatal exposure of rats to PCB reduces the extent of the IIP-MF projection (Pruitt *et al.* 1999). Because many hippocampus-dependent learning behaviours co-vary subtly with the IIP-MF projection, these findings align well with the PCB effects on LTP and impaired performance in spatial learning tasks. However, future studies may harmonise these findings with the observations reporting actually improved radial maze learning after exposure to PCB (Schantz *et al.* 1997).

6.5 CONCLUSIONS AND OUTLOOK

6.5.1 WHICH TEST?

In order to perform economical screening, the test battery presented for screening for cognitive changes in mice can be used selectively. In our experience, a combination of water maze testing, a null-maze test and a test assessing aversively motivated learning are in most cases sufficient to diagnose even subtle impairments due to targeted gene deletions or exposure to toxicants. It should be emphasised again that testing for cognitive deficits must always be accompanied by sensorimotor testing using established test batteries.

6.5.2 HOW TO INTERPRET

The presented tests are definitely designed for screening and should not be over-interpreted with respect to underlying brain mechanisms. Discovered behavioural anomalies then require specific testing according to hypotheses derived from the knowledge of the potentially affected brain systems. One point, often neglected in toxicology, deserves special attention. Any interference with normal brain functions, be this targeted gene deletions, lesions, chronic or acute exposure to environmental toxicants, will entail compensatory masking by a wide range of homeostatic mechanisms, such as up- or downregulation of receptors, or structural rearrangement of brain circuitry when the interference happens during ontogeny. The behavioural phenotype never reflects directly the altered mechanisms, but the sum of compensatory actions that are often unpredictable. Thus, multiple (pleiotropic) effects on behaviour following any manipulation are the rule rather than the exception. Finally, it must be kept in mind that even behavioural changes thought to reflect cognition, learning and memory are nothing more than altered probabilities of acquired or innate motor patterns. As a large part of the mammalian forebrain is involved in translating multiple processing in any part of the brain into modifications of ongoing motor activities (Lipp and Wolfer 1998), disturbed brain functions, whether system-specific or widespread, tend to appear as disorganised behavioural strategies rather than clearly attributable deficits. In species with a more developed prefrontal cortex governing behavioural organisation, the deficits are likely to be attributed, incorrectly, to damage in this region. In mice or rats with a comparatively much larger hippocampal formation, the same symptoms are then, equally questionably, attributed to damage in these structures organising the behavioural activities of rodents.

6.5.3 WHAT NEXT

Conceptually, behavioural ecotoxicology is torn between two seemingly incompatible extremes: precisely controllable laboratory assays which are often not realistic, and more realistic yet uncontrollable field studies. However, the message from the field of targeted gene deletions is clear: even the utmost precise control of the causative agent, elimination of a gene, does not translate into controllable behaviour but results in approximately the same highly variable and unpredictable behavioural effects as the far less controllable toxicological studies. This simply means that one of the main justifications of laboratory studies, control of the noxious agent in order to obtain reproducible and predictable effects, is questionable. A laboratory environment is still useful to produce comparable test populations, but its main *raison d'être* has vanished to a great extent. In consequence, behavioural ecotoxicology can and must shift out of the laboratory. The technology for assessing complex behaviour in the natural habitat of many species is already available, and will doubtlessly improve due to technological advances in microtelemetry. To put it bluntly: the excuses to stay in the laboratory are dwindling and behavioural scientists in ecotoxicology have to face the challenges of real-world situations.

ACKNOWLEDGEMENTS

I wish to thank Giacomo Dell'Omo for continuous help and critical reading, and David P. Wolfer for his instrumental role in data analysis. His program WINTRACK is freeware and available by electronic request from http://www.dpwolfer.ch/wintrack.

This publication was supported by the Swiss National Science Foundation and the NCCR "Neural plasticity and repair".

REFERENCES

Aadland J, Beatty WW and Maki RH (1985) Spatial memory of children and adults assessed in the radial maze. *Developmental Psychobiology*, **18**, 163-172.

Aggleton JP, Blindt HS and Rawlins JNP (1989) Effects of amygdaloid and amygdaloid-hippocampal lesions on object recognition and spatial working memory in rats. *Behavioral Neuroscience*, **103**, 962-974.

Alleva E (1993) Assessment of aggressive behavior in rodents. *Journal of Neuroscience Methods*, **14**, 111-137.

Alleva E, Petruzzi S and Ricceri L (1995) Evaluating the social behavior of rodents: laboratory, semi-naturalistic and naturalistic approaches. In *Behavioural Brain Research in Naturalistic and Semi-Naturalistic Settings: Possibilities and Perspectives*, Alleva E, Fasolo A, Lipp H-P, Nadel L and Ricceri L (eds), Kluwer, Dordrecht, The Netherlands, pp. 359-374.

Ammassari-Teule M and De Marsanich B (1996) Spatial and visual discrimination learning in CD1 mice: partial analogy between the effect of lesions to the hippocampus and the amygdala. *Physiology and Behavior*, **60**, 265-271.

Antelman SM and Brown TS (1972) Hippocampal lesions and shuttlebox avoidance behavior: a fear hypothesis. *Physiology and Behavior*, **9**, 15-20.

Arolfo MP and Brioni JD (1991) Diazepam impairs place learning in the Morris water maze. *Behavioral and Neural Biology*, **55**, 131-136.

Baerends GP (1976) On drive, conflict and instinct, and the functional organization of behavior. In *Progress in Brain Research. Vol. 45: Perspectives in Brain Research*, Corner MA and Swaab DF (eds), Elsevier, Amsterdam, pp. 427-447.

Balda RP and Kamil AC (1988) The spatial memory of Clark's nutcrackers (*Nucifraga columbiana*) in an analog of the radial-arm maze. *Animal Learning and Behavior*, **16**, 116-122.

Becker JT, Walker JA and Olton DS (1980) Neuroanatomical bases of spatial memory. *Brain Research*, **200**, 307-320.

Beracochea D and Jaffard R (1985) Memory deficits subsequent to chronic consumption of alcohol in mice: an analysis based on spontaneous alternation behavior. *Behavioural Brain Research*, **15**, 15-25.

Bignami G (1996) Economical test methods for developmental neurobehavioral toxicity. *Environmental Health Perspectives*, **104** (Suppl. 2), 285-298.

Blanchard RJ and Fial RA (1968) Effects of limbic lesions on passive avoidance and reactivity to shock. *Journal of Comparative and Physiological Psychology*, **66**, 606-612.

Block F and Schwarz M (1997) Correlation between hippocampal neuronal damage and spatial learning deficit due to global ischemia. *Pharmacology, Biochemistry and Behavior*, **56**, 755-761.

Bond AB, Cook RG and Lamb MR (1981) Spatial memory and the performance of rats and pigeons in the radial-arm maze. *Animal Learning and Behavior*, **9**, 575-580.

Bolles RC (1970) Species-specific defense reactions and avoidance learning. *Psychological Reviews*, **77**, 32-48.

Brambilla R, Gnesutta N, Minichiello N, White G, Roylance A, Herron C, Ramsey M, Wolfer DP, Cestari V, Rossi-Arnaud C, Grant SGN, Chapman PF, Lipp H-P, Sturani E and Klein R (1997) A role for the RAS signaling pathway in synaptic transmission and long-term memory. *Nature*, **390**, 281-286.

Bunsey M and Eichenbaum H (1995) Selective damage to the hippocampal region blocks long-term retention of a natural and nonspatial stimulus-stimulus association. *Hippocampus*, **5**, 546-556.

Bushnell PJ and Rice CD (1999) Behavioral assessment of learning and attention in rats exposed perinatally to 3, 3′, 4, 4′, 5-pentachlorobiphenyl (PCB 126). *Neurotoxicology and Teratology*, **21**, 381-392.

Büeler H, Fischer M, Lang Y, Blüthmann H, Lipp H-P, DeArmond SJ, Prusiner SB, Aguet M and Weissmann C (1992) The neuronal cell surface protein PrP is not essential for normal development and behavior of the mouse. *Nature*, **356**, 577-582.

Cain DP, Saucier D, Hall J, Hargreaves EL and Boon F (1996) Detailed behavioral analysis of water maze acquisition under APV or CNQX: contribution of sensorimotor disturbances to drug-induced acquisition deficits. *Behavioral Neuroscience*, **110**, 86-102.

Campbell IL and Gold LH (1996) Transgenic modeling of neuropsychiatric disorders. *Molecular Psychiatry*, **1**, 105-120.

Carlson NR (1970) Two-way avoidance behavior of mice with limbic lesions. *Journal of Comparative and Physiological Psychology*, **70**, 73-78.

Chambers KC (1990) A neural model for conditioned taste aversions. *Annual Review of Neuroscience*, **13**, 373-385.

Chen C, Kim JJ, Thompson RF and Tonegawa S (1996) Hippocampal lesions impair contextual fear conditioning in two strains of mice. *Behavioral Neuroscience*, **110**, 1177-1180.

Chen C, Rainnie DG, Greene RW and Tonegawa S (1994) Abnormal fear response and aggressive behavior in mutant mice deficient for alpha-calcium-calmodulin kinase II. *Science*, **266**, 291-294.

Chen C and Tonegawa S (1997) Molecular genetic analysis of synaptic plasticity, activity-dependent neural development, learning, and memory in the mammalian brain. *Annual Review of Neuroscience*, **20**, 157-184.

Chiarotti F, Alleva E, Bignami G (1987) Problems of test choice and data analysis in behavioral teratology: the case of prenatal benzodiazepines. *Neurotoxicology and Teratology*, **9**, 179-186.

Choleris E and Kavaliers M (1999) Social learning in animals: sex differences and neurobiological analysis. *Pharmacology, Biochemistry and Behavior*, **64**, 767-776.

Cohn J and MacPhail RC (1996) Ethological and experimental approaches to behavior analysis: implications for ecotoxicology. *Environmental Health Perspectives*, **104**, 299-305.
Conrad CD and Roy EJ (1993) Selective loss of hippocampal granule cells following adrenalectomy: implications for spatial memory. *Journal of Neuroscience*, **13**, 2582-2590.
Cory-Slechta DA (1989) Behavioral measures of neurotoxicity. *Neurotoxicology*, **10**, 271-296.
Crabbe JC, Wahlsten D and Dudek BC (1999) Genetics of mouse behavior: interactions with laboratory environment. *Science*, **284**, 1670-1672.
Crawley JN and Paylor R (1997) A proposed test battery and constellations of specific behavioral paradigms to investigate the behavioral phenotypes of transgenic and knockout mice. *Hormones and Behavior*, **31**, 197-211.
Crestani F, Lorez M, Baer K, Essrich C, Benke D, Laurent JP, Belzung C, Fritschy J-M, Lüscher B and Möhler H (1999) Decreased $GABA_A$-receptor clustering results in enhanced anxiety and a bias for threat cues. *Nature Neuroscience*, **2**, 833-839.
Crusio WE (1996) The neurobehavioral genetics of aggression. *Behavior Genetics*, **26**, 459-461.
Crusio WE, Bertholet J-Y and Schwegler H (1990) No correlations between spatial and non-spatial reference memory in a T-maze task and hippocampal mossy fibre distribution in the mouse. *Behavioural Brain Research*, **41**, 251-259.
Dalland T (1976) Response perseveration of rats with dorsal hippocampal lesions. *Behavioral Biology*, **17**, 473-484.
Day LB and Schallert T (1996) Anticholinergic effects on acquisition of place learning in the Morris water task — spatial mapping deficit or inability to inhibit nonplace strategies. *Behavioral Neuroscience*, **110**, 998-1005.
Dell'Omo G, Bryenton R and Shore RF (1997) Effects of exposure to an organophosphate pesticide on behavior and acetylcholinesterase activity in the common shrew *Sorex araneus*. *Environmental Toxicology and Chemistry*, **16**, 272-276.
Dell'Omo G, Pleskacheva M, Wolfer DP, Lipp H-P and Shore RF (2002) Comparative effects of exposure to an organophosphate pesticide on locomotor activity of laboratory mice and five species of wild rodents. *Bulletin of Environmental Contamination and Toxicology*, in press.
Dell'Omo G, Ricceri L, Wolfer DP, Poletaeva II and Lipp H-P (2000) Temporal and spatial adaptation to food restriction in mice under naturalistic conditions. *Behavioural Brain Research*, **115**, 1-8.
Dell'Omo G and Shore RF (1996) Behavioral and physiological effects of acute sublethal exposure to dimethoate on wood mice, *Apodemus sylvaticus* [I - Laboratory Studies]. *Archives of Environmental Contamination and Toxicology*, **31**, 91-97.
Dell'Omo G, Shore RF and Lipp H-P (1998) An automated system based on microchips for monitoring individual activity of wild small mammals. *Journal of Experimental Zoology*, **280**, 97-99.
Dell'Omo G, Wolfer DP, Alleva E and Lipp H-P (1995) Developmental exposure to ozone induces subtle changes in swimming navigation of adult mice. *Toxicology Letters*, **81**, 91-99.
Denenberg VH, Gaulin-Kremer E, Gandelman R and Zarrow MX (1973) The development of standard stimulus animals for mouse (*Mus musculus*) aggression testing by means of olfactory bulbectomy. *Animal Behaviour*, **21**, 590-598.
De Ruiter AJ, Feitsma LE, Keijser JN, Koolhaas JM, Van Oortmerssen GA and Bohus B (1993) Differential perinatal testosterone secretory capacity of wild house mice testes is related to aggressiveness in adulthood. *Hormones and Behavior*, **27**, 231-239.
Di Mattia BV and Kesner RP (1988) Role of the posterior parietal association cortex in the processing of spatial event information. *Behavioral Neuroscience*, **102**, 397-403.

Dornan WA, McCampbell A, Tinkler GP, Hickman LJ, Bannon AW, Decker MW and Gunther KL (1997) Comparison of site specific injections into the basal forebrain on water maze and radial arm maze performance in the male rat after immunolesioning with 192 IgG saporin. *Behavioural Brain Research*, **86**, 181-189.

Driscoll P and Baettig K (1982) Behavioral, emotional and neurochemical profiles of rats selected for extreme differences in active, two-way avoidance performance. In *Genetics of the Brain*, Lieblich I (ed.), Elsevier, Amsterdam, pp. 96-123.

Ellenbroek BA, Hoeks B and Cools AR (1992) Information statistical analysis and the frequential method of data collecting: description and illustration of a new model in the study of animal behaviour. *Behaviour*, **121**, 35-60.

Ennaceur A and Meliani K (1992) A new one-trial test for neurobiological studies of memory in rats. III. Spatial vs. non-spatial working memory. *Behavioural Brain Research*, **51**, 83-92.

Etienne AS, Sitbon S, Dahn HC and Maurer R (1994) Golden hamsters on the eight-arm maze in light and darkness: the role of dead reckoning. *Quarterly Journal of Experimental Psychology*, **47**, 401-425.

Evans HL (1994) Neurotoxicity expressed in naturally occurring behavior. In *Neurobehavioral Toxicity: Analysis and Interpretation*, Weiss B and O'Donoghue J (eds), Raven Press, New York, pp. 111-135.

Fernandez-Espejo E (1997) Structure of the mouse behaviour on the elevated plus-maze test of anxiety. *Behavioural Brain Research*, **86**, 105-112.

File SE (1993) The interplay of learning and anxiety in the elevated plus-maze. *Behavioural Brain Research*, **58**, 199-202.

Fiore M, Dell'Omo G, Alleva E and Lipp H-P (1995) A comparison of behavioural effects of prenatally administered oxazepam in mice exposed to open-fields in the laboratory and the real world. *Psychopharmacology*, **122**, 72-77.

Fiore M, Probert L, Kollias G, Akassoglou K, Alleva E and Aloe L (1996) Neurobehavioral alterations in developing transgenic mice expressing TNF-alpha in the brain. *Brain Behavior and Immunology*, **10**, 126-138.

Fischer W, Chen KS, Gage FH and Björklund A (1992) Progressive decline in spatial learning and integrity of forebrain cholinergic neurons in rats during aging. *Neurobiology of Aging*, **13**, 9-23.

Foreman N and Ermakova E (1998) The radial maze: twenty years on. In Handbook of Spatial Research Paradigms and Methodologies. Vol. 2: Clinical and Comparative Issues, Foreman N and Gillett R (eds), Psychology Press, Hove, UK, pp. 87-143.

Frick KM, Baxter MG, Markowska AL, Olton DS and Price DL (1995) Age-related spatial reference and working memory deficits assessed in the water maze. *Neurobiology of Aging*, **16**, 149-160.

Frynta D (1994) Exploratory behaviour in 12 Palearctic mice species (Rodentia: Muridae): a comparative study using 'free exploration' tests. *Acta Societas Zoologica Bohemica*, **57**, 173-182

Galea LAM, Kavaliers M and Ossenkopp KP (1996) Sexually dimorphic spatial learning in meadow voles *Microtus pennsylvanicus* and deer mice *Peromyscus maniculatus*. *Journal of Experimental Biology*, **199**, 195-200.

Galef BG Jr (1993) Individual differences in responses of Norway rats to social induction of food preferences. *Behavioural Processes*, **30**, 309-316.

Gass P, Wolfer DP, Balschun D, Rudolph D, Frey U, Lipp H-P and Schütz G (1998) Deficits in memory tasks of mice with CREB mutations depend on gene dosage. *Learning and Memory*, **5**, 274-288.

Gilbert ME and Crofton KM (1999) Developmental exposure to a commercial PCB mixture (Aroclor 1254) produces a persistent impairment in long-term potentiation in the rat dentate gyrus *in vivo*. *Brain Research*, **850**, 87-95.

Gilbert ME, Mack CM and Lasley SM (1999) The influence of developmental period of lead exposure on long-term potentiation in the adult rat dentate gyrus *in vivo*. *Neurotoxicology*, **20**, 57-69.

Golani I, Benjamini Y and Eilam D (1993) Stopping behavior: constraints on exploration in rats (*Rattus norvegicus*). *Behavioral Brain Research*, **53**, 21-33.

Granon S and Poucet B (1995) Medial prefrontal lesions in the rat and spatial navigation: evidence for impaired planning. *Behavioral Neuroscience*, **109**, 474-484.

Guillot A and Meyer J-A (2000) A cost-benefit analysis of behavioural sequence organization in laboratory mice. *Ethology Ecology and Evolution*, **9**, 119-132.

Guillot P-V, Roubertoux PL and Crusio WE (1994) Hippocampal mossy fiber distributions and intermale aggression in seven inbred mouse strains. *Brain Research*, **660**, 167-169.

Hany J, Lilienthal H, Roth-Harer A, Ostendorp G, Heinzow B and Winneke G (1999) Behavioral effects following single and combined maternal exposure to PCB 77 (3, 4, 3′, 4′-tetrachlorobiphenyl) and PCB 47 (2, 4, 2′, 4′-tetrachlorobiphenyl) in rats. *Neurotoxicology and Teratology*, **21**, 147-156.

Harley CW (1979) Arm choices in a sunburst maze: effects of hippocampectomy in the rat. *Physiology and Behavior*, **23**, 283-290.

Hilton SC and Krebs JR (1991) Spatial memory of four species of *Parus*: performance in an open-field analogue of a radial maze. *Quarterly Journal of Experimental Psychology*, **42B**, 345-368.

Holene E, Nafstad I, Skaare JU and Sagvolden T (1998) Behavioural hyperactivity in rats following postnatal exposure to sub-toxic doses of polychlorinated biphenyl congeners 153 and 126. *Behavioural Brain Research*, **94**, 213-224.

Hong SJ, Grover CA, Safe SH, Tiffany-Castiglioni E and Frye GD (1998) Halogenated aromatic hydrocarbons suppress CA1 field excitatory postsynaptic potentials in rat hippocampal slices. *Toxicology and Applied Pharmacology*, **148**, 7-13.

Hostetter G and Thomas GJ (1967) Evaluation of enhanced thigmotaxis as a condition of impaired maze learning by rats with hippocampal lesions. *Journal of Comparative and Physiological Psychology*, **63**, 105-110.

Hsiao KK, Chapman P, Nilsen S, Eckman C, Harigaya Y, Younkin S, Yang FS and Cole G (1996) Correlative memory deficits, α-β elevation, and amyloid plaques in transgenic mice. *Science*, **274**, 99-102.

Husain R, Adhami VM and Seth PK (1996) Behavioral, neurochemical and neuromorphological effects of deltamethrin in adult rats. *Journal of Toxicology and Environmental Health*, **48**, 515-526.

Janis LS, Glasier MM, Fulop Z and Stein DG (1998) Intraseptal injections of 192 IgG saporin produce deficits for strategy selection in spatial-memory tasks. *Behavioural Brain Research*, **90**, 23-34.

Janus C, Janus M and Roder J (1995) Spatial exploration in transgenic mice expressing human β-S100. *Neurobiology of Learning and Memory*, **64**, 58-67.

Jarrard LE (1978) Selective hippocampal lesions and spatial discrimination in the rat. *Society for Neuroscience Abstracts*, **3**, **222**.

Jarrard LE (1983) Selective hippocampal lesions and behavior: effects of kainic acid lesions on performance of place and cue tasks. *Behavioral Neuroscience*, **97**, 873-889.

Jarrard LE, Okaichi H, Steward O and Goldschmidt RB (1984) On the role of hippocampal connections in the performance of place and cue tasks: comparisons with damage to hippocampus. *Behavioral Neuroscience*, **98**, 946-954.

Joel D, Tarrasch R, Feldon J and Weiner I (1997) Effects of electrolytic lesions of the medial prefrontal cortex or its subfields on 4-arm baited, 8-arm radial maze, two-way active avoidance and conditioned fear tasks in the rat. *Brain Research*, **765**, 37-50.

Kavaliers M and Galea LAM (1994) Spatial water maze learning using celestial cues by the meadow vole, *Microtus pennsylvanicus*. *Behavioural Brain Research*, **61**, 97-100.
Kemble ED, Garbe CM and Gordon C (1995) Effects of novel odors on intermale attack behavior in mice. *Aggressive Behavior*, **21**, 293-299.
Kimble D and Whishaw IQ (1994) Spatial behavior in the Brazilian short-tailed opossum (*Monodelphis domestica*): comparison with the Norway rat (*Rattus norvegicus*) in the Morris water maze and radial arm maze. *Journal of Comparative Psychology*, **108**, 148-155.
Kogan JH, Frankland PW, Blendy JA, Coblentz J, Marowitz Z, Schütz G and Silva AJ (1997) Spaced training induces normal long-term memory in CREB mutant mice. *Current Biology*, **7**, 1-11.
Kolb B and Whishaw IQ (1991) Mechanisms underlying behavioral sparing after neonatal retrosplenial cingulate lesions in rats: Spatial navigation, cortical architecture, and electroencephalographic activity. *Brain Dysfunctions*, **4**, 75-92.
Kraemer PJ, Brown RW, Baldwin SA and Scheff SW (1996) Validation of a single-day Morris water maze procedure used to assess cognitive deficits associated with brain damage. *Brain Research Bulletin*, **39**, 17-22.
Kruk MR (1997) Measuring behavior in the twenty-first century. *Trends in Neurosciences*, **20**, 187-189.
Lacroix L, Broersen LM, Weiner I and Feldon J (1998) The effects of excitotoxic lesion of the medial prefrontal cortex on latent inhibition, prepulse inhibition, food hoarding, elevated plus maze, active avoidance and locomotor activity in the rat. *Neuroscience*, **84**, 431-442.
Lalonde R, Botez MI and Boivin D (1986) Spontaneous alternation and habituation in a T-maze in nervous mutant mice. *Behavioral Neuroscience*, **100**, 350-352.
Ledoux JE (1994) Emotion, memory and the brain. *Scientific American*, **270**, 50-59.
Lil'p IG, Bizikoeva FZ, Poletaeva II and Ivanov VI (1995) Behavior of adult mice of different strains after exposure to thiophosphamide during organogenesis. *Russian Journal of Genetics*, **31**, 85-92.
Lipp H-P, Pleskacheva MG, Gossweiler H, Ricceri L, Smirnova AA, Garin NN, Perepiolkina OP, Voronkov DN, Kupstov PA and Dell'Omo G (2001) A large outdoor radial maze for comparative studies in birds and mammals. *Neuroscience and Biobehavioral Reviews*, **25**, 83-99.
Lipp H-P, Schwegler H, Heimrich B and Driscoll P (1988) Infrapyramidal mossy fibers and two-way avoidance learning: developmental modification of hippocampal circuitry and adult behavior of rats and mice. *Journal of Neuroscience*, **8**, 1905-1921.
Lipp H-P and Wolfer DP (1998) Genetically modified mice and cognition. *Current Opinion in Neurobiology*, **8**, 272-280.
Little EE (1990) Behavioral toxicology — stimulating challenges for a growing discipline. *Environmental Toxicology and Chemistry*, **9**, 1-2.
Liu P and Bilkey DK (1998) Lesions of perirhinal cortex produce spatial memory deficits in the radial maze. *Hippocampus*, **8**, 114-121.
Logue SF, Paylor R and Wehner JM (1997) Hippocampal lesions cause learning deficits in inbred mice in the Morris water maze and conditioned-fear task. *Behavioral Neuroscience*, **111**, 104-113.
McDonald RJ and White NM (1993) A triple dissociation of memory systems: hippocampus, amygdala, and dorsal striatum. *Behavioral Neuroscience*, **107**, 3-22.
McDonald RJ and White NM (1994) Parallel information processing in the water maze: evidence for independent memory systems involving dorsal striatum and hippocampus. *Behavioral and Neural Biology*, **61**, 260-270.
Means LW, Alexander SR and O'Neal MF (1992) Those cheating rats: male and female rats use odor trails in a water-escape 'working memory' task. *Behavioral and Neural Biology*, **58**, 144-151.

Minichiello L, Korte M, Wolfer DP, Kühn R, Unsicker K, Cestari V, Rossi-Arnaud C, Lipp H-P, Bonhoeffer T and Klein R (1999) Essential role for TrkB receptors in hippocampus-mediated learning. *Neuron*, **24**, 401-414.
Moore FR and Osadchuk TE (1982) Spatial memory in a passerine migrant. In *Avian Navigation*, Papi F and Wallraff HG (eds), Springer-Verlag, Berlin, pp. 319-328.
Morris RGM (1984) Development of a water-maze procedure for studying spatial learning in the rat. *Journal of Neuroscience Methods*, **11**, 47-60.
Morris RGM, Schenk F, Tweedie F and Jarrard LE (1990) Ibotenate lesions of hippocampus and/or subiculum: dissociating components of allocentric spatial learning. *European Journal of Neuroscience*, **2**, 1016-1028.
Moser VC (1991) Applications of a neurobehavioral screening battery. *Journal of the American College of Toxicology*, **10**, 661-669.
Mumby DG, Wood ER and Pinel JPJ (1992) Object-recognition memory is only mildly impaired in rats with lesions of the hippocampus and amygdala. *Psychobiology*, **20**, 18-27.
Munoz C and Grossman SP (1981) Spatial discrimination, reversal and active or passive avoidance learning in rats with KA-induced neuronal depletions in dorsal hippocampus. *Brain Research Bulletin*, **6**, 399-406.
Murray CL and Fibiger HC (1985) Learning and memory deficits after lesions of the nucleus basalis magnocellularis: reversal by physostigmine. *Neuroscience*, **14**, 1025-1032.
Müller U, Cristina N, Li Z-W, Wolfer DP, Lipp H-P, Rülicke T, Brandner S, Aguzzi A and Weissmann C (1994) Behavioral and anatomical deficits in mice homozygous for a modified β-amyloid precursor protein (βAPP) gene. *Cell*, **79**, 755-765.
Nitta A, Itoh A, Hasegawa T and Nabeshima T (1994) β-amyloid protein-induced Alzheimer's disease animal model. *Neuroscience Letters*, **170**, 63-66.
Novikov SN (1993) The genetics of pheromonally mediated intermale aggression in mice: current status and prospects of the model. *Behavior Genetics*, **23**, 505-508.
Nyakas C, De Kloet ER, Veldhuis HD and Bohus B (1983) Hippocampal corticosterone receptors and novelty-induced behavioral activity: effect of kainic acid lesion in the hippocampus. *Brain Research*, **288**, 219-228.
Olton DS and Isaacson RL (1968) Hippocampal lesions and active avoidance. *Physiology and Behavior*, **3**, 719-724.
Olton DS, Handelman GE and Walker JA (1978a) Spatial memory and food-searching strategies. In *Foraging Behavior. Ecological, Ethological, and Psychological Approaches*, Kamil AC and Sargent TD (eds), Garland STPM Press, New York, pp. 333-354.
Olton DS, Walker JA and Gage FH (1978b) Hippocampal connections and spatial discrimination. *Brain Research*, **139**, 295-308.
Overman WH, Pate BJ, Moore K and Peuster A (1996) Ontogeny of place learning in children as measured in the radial arm maze, Morris search task, and open field task. *Behavioral Neuroscience*, **110**, 1205-1228.
Packard MG and White NM (1990) Lesions of the caudate nucleus selectively impair 'reference memory' acquisition in the radial maze. *Behavioral and Neural Biology*, **53**, 39-50.
Peele DB and Vincent A (1989) Strategies for assessing learning and memory, 1978-1987: a comparison of behavioral toxicology, psychopharmacology, and neurobiology. *Neuroscience and Biobehavioral Reviews*, **13**, 33-38.
Peeler DF (1990) Two measures of activity in genetically defined mice as a function of strain, time of the day, and previous experience. *Psychobiology*, **18**, 327-338.
Peeler DF and Nowakowski RS (1987) Genetic factors and the measurement of exploratory activity. *Behavioral and Neural Biology*, **48**, 90-103.

Persinger MA, Bureau YR and Peredery O (1994) Dissociation between conditioned taste aversion and radial maze learning following seizure-induced multifocal brain damage: quantitative tests of serial vs. parallel circuit models of memory. *Physiology and Behavior* **56**, 225-235.

Phillips RG and Ledoux JE (1992) Differential contribution of amygdala and hippocampus to cued and contextual fear conditioning. *Behavioral Neuroscience*, **106**, 274-285.

Pleskacheva MG, Wolfer DP, Kupriyanova IF, Nikolenko DL, Scheffrahn H, Dell'Omo G and Lipp H-P (2000) Hippocampal mossy fibers and swimming navigation learning in two vole species occupying different habitats. *Hippocampus*, **10**, 17-30.

Provost TL, Juarez de Ku LM, Zender C and Meserve LA (1999) Dose- and age-dependent alterations in choline acetyltransferase (ChAT) activity, learning and memory, and thyroid hormones in 15- and 30-day old rats exposed to 1.25 or 12.5 PPM polychlorinated biphenyl (PCB) beginning at conception. *Progress in Neuro-Psychopharmacology and Biological Psychiatry*, **23**, 915-928.

Pruitt DL, Meserve L and Bingman VP (1999) Reduced growth of intra- and infra-pyramidal mossy fibers is produced by continuous exposure to polychlorinated biphenyl. *Toxicology*, **138**, 11-17.

Puopolo M, Santucci D, Chiarotti F and Alleva E (1999) Behavioural effects of endocrine disrupting chemicals on laboratory rodents: statistical methodologies and an application concerning developmental PCB exposure. *Chemosphere*, **39**, 1259-1271.

Rabin BM and Hunt WA (1992) Relationship between vomiting and taste aversion learning in the ferret: studies with ionizing radiation, lithium chloride, and amphetamine. *Behavioral and Neural Biology*, **58**, 83-93.

Rawlins JNP (1985) Associations across time: the hippocampus as a temporary memory store. *Behavioral and Brain Sciences*, **8**, 479-486.

Riekkinen P, Jr, Jäkälä P, Sirviö J and Riekkinen P (1991) The effects of increased serotonergic and decreased cholinergic activities on spatial navigation performance in rats. *Pharmacology, Biochemistry and Behavior*, **39**, 25-29.

Roberts WA (1984) Some issues in animal spatial memory. In *Animal Cognition*, Roitblat HL, Bever TG and Terrace HS (eds), Erlbaum, Hillsdale, NJ, pp. 425-443.

Roberts WA and Van Veldhuizen N (1985) Spatial memory in pigeons on the radial maze. *Journal of Experimental Psychology: Animal Behavior Processes*, **11**, 241-260.

Rogers DC and Hunter AJ (1997) Photothrombotic lesions of the rat cortex impair acquisition of the water maze. *Pharmacology, Biochemistry and Behavior*, **56**, 747-754.

Rossi-Arnaud C, Fagioli S and Ammassari-Teule M (1991) Spatial learning in two inbred strains of mice: genotype-dependent effect of amygdaloid and hippocampal lesions. *Behavioural Brain Research*, **45**, 9-16.

Sato SM, Frazier JM and Goldberg AM (1984) Perturbation of a hippocampal zinc-binding pool after postnatal lead exposure in rats. *Experimental Neurology*, **85**, 620-630.

Saudou F, Amara DA, Dierich A, LeMeur M, Ramboz S, Segu L, Buhot M-C and Hen R (1994) Enhanced aggressive behavior in mice lacking 5-HT_{1B} receptor. *Science*, **265**, 1875-1878.

Schantz SL, Levin ED and Bowman RE (1991) Long-term neurobehavioral effects of perinatal PCB exposure in monkeys. *Environmental Toxicology and Chemistry*, **10**, 747-756.

Schantz SL, Seo B-W, Mosthagian J and Amin S (1997) Developmental exposure to polychlorinated biphenyls or dioxin: do changes in thyroid function mediate effects on spatial learning? *American Zoologist*, **37**, 399-408.

Schenk F (1987) Comparison of spatial learning in woodmice (*Apodemus sylvaticus*) and hooded rats (*Rattus norvegicus*). *Journal of Comparative Psychology*, **101**, 150-158.

Schenk F, Grobéty M-C, Lavenex P and Lipp H-P (1995) Dissociation between basic components of spatial memory in rats. In *Behavioural Brain Research in Naturalistic and Semi-Naturalistic Settings: Possibilities and Perspectives*, Alleva E, Fasolo A, Lipp H-P, Nadel L and Ricceri L (eds), Kluwer, Dordrecht, The Netherlands, pp. 277-300.

Scherer E (1992) Behavorial responses as indicators of environmental alterations—approaches, results, developments. *Journal of Applied Ichthyology*, **8**, 122-131.

Schwegler H, Crusio WE, Lipp H-P, Brust I and Mueller GG (1991) Early postnatal hyperthyroidism alters hippocampal circuitry and improves radial maze learning in adult mice. *Journal of Neuroscience*, **11**, 2102-2106.

Seamans JK, Floresco SB and Phillips AG (1995) Functional differences between the prelimbic and anterior cingulate regions of the rat prefrontal cortex. *Behavioral Neuroscience*, **109**, 1063-1073.

Seo BW, Sparks AJ, Medora K, Amin S and Schantz SL (1999) Learning and memory in rats gestationally and lactationally exposed to 2,3,7,8-tetrachlorodibenzo-p-dioxin (TCDD). *Neurotoxicology and Teratology*, **21**, 231-239.

Shepherd JK, Grewal SS, Fletcher A, Bill DJ and Dourish CT (1994) Behavioural and pharmacological characterisation of the elevated 'zero-maze' as an animal model of anxiety. *Psychopharmacology*, **116**, 56-64.

Silva AJ, Fedorov N, Kogan J, Frankland P, Coblentz J, Lundsten R, Friedman E, Smith A, Cho Y and Giese KP (1996) Genetic analysis of function and dysfunction in the central nervous system. *Cold Spring Harbor Symposia on Quantitative Biology*, **61**, 239-246.

Sluyter F, Jamot L, Van Oortmerssen GA and Crusio WE (1994) Hippocampal mossy fiber distributions in mice selected for aggression. *Brain Research*, **646**, 145-148.

Spencer RL, O'Steen WK and McEwen BS (1995) Water maze performance of aged Sprague-Dawley rats in relation to retinal morphologic measures. *Behavioural Brain Research*, **68**, 139-150.

Spetch ML and Edwards CA (1986) Spatial memory in pigeons (*Columba livia*) in an open-field feeding environment. *Journal of Comparative Psychology*, **100**, 266-278.

Spetch ML and Honig WK (1988) Characteristics of pigeons' spatial working memory in an open-field task. *Animal Learning and Behavior*, **16**, 123-131.

Stork O, Welzl H, Cremer H and Schachner M (1997) Increased intermale aggression and neuroendocrine response in mice deficient for the neural cell adhesion molecule (NCAM). *European Journal of Neuroscience*, **9**, 1117-1125.

Suzuki SW, Augerinos G and Black BA (1980) Stimulus control of spatial behavior on the eight-arm maze in rats. *Learning and Motivation*, **11**, 1-18.

Thinus-Blanc C, Save E, Rossi-Arnaud C, Tozzi A and Ammassari-Teule M (1996) The differences shown by C57BL/6 and DBA/2 inbred mice in detecting spatial novelty are subserved by a different hippocampal and parietal cortex interplay. *Behavioural Brain Research*, **80**, 33-40.

Tilson HA (1993) Neurobehavioral methods used in neurotoxicological research. *Toxicology Letters*, **68**, 231-240.

Trullas R and Skolnick P (1993) Differences in fear motivated behaviors among inbred mouse strains. *Psychopharmacology*, **111**, 323-331.

Viana MB, Tomaz C and Graeff FG (1994) The elevated T-maze: a new animal model of anxiety and memory. *Pharmacology, Biochemistry and Behavior*, **49**, 549-554.

Vyssotski AL, Dell'Omo G, Vyssotski DL, Wolfer DP, Schachner M, Poletaeva II and Lipp H-P (2000) Natural learning: food place reversal in mice overexpressing the cell adhesion molecule L1. In *Symposium: Behavioural Phenotyping of Mouse Mutants*, Schröder H, Steckler T, van der Staay F-J and Walther M (eds), University of Cologne, Cologne, p. 95.

Vyssotski AL, Dell'Omo G, Poletaeva II, Vyssotski DL, Minichiello L, Klein R, Wolfer DP and Lipp H-P (2002) Long-term monitoring of hippocampus-dependent behavior in naturalistic settings: mutant mice lacking the neurotrophin receptor TrkB in the forebrain show spatial learning but impaired behavioral flexibility. *Hippocampus*, **12**, 27-38.

Wan R-Q, Givens BS and Olton DS (1995) Opioid modulation of working memory: intraseptal, but not intraamygdaloid, infusions of β-endorphin impair performance in spatial alternation. *Behavioral and Neural Biology*, **63**, 74-86.

Weiss B, Ziriax JM and Newland MC (1989) Serial properties of behavior and their chemical modification. *Animal Learning and Behavior*, **17**, 83-93.

Welzl H, D'Adamo P and Lipp H-P (2001) Conditioned taste aversion as a learning and memory paradigm. *Behavioural Brain Research*, **125**, 205-213.

Wenk GL, Hughey D, Boundy V, Kim A, Walker R and Olton DS (1987) Neurotransmitters and memory: role of cholinergic, serotonergic, and noradrenergic systems. *Behavioral Neuroscience*, **101**, 325-332.

Whishaw IQ, Coles BLK and Bellerive CHM (1995) Food carrying: a new method for naturalistic studies of spontaneous and forced alternation. *Journal of Neuroscience Methods*, **61**, 139-143.

Wilkie DM and Slobin P (1983) Gerbils in space: performance on the 17-arm radial maze. *Journal of the Experimental Analysis of Behavior*, **40**, 301-312.

Wilkie DM, Spetch ML and Chew L (1981) The ring dove's short-term memory capacity for spatial information. *Animal Behaviour*, **29**, 639-641.

Winocur G (1991) Functional dissociation of the hippocampus and prefrontal cortex in learning and memory. *Psychobiology*, **19**, 11-20.

Wolfer DP and Lipp H-P (1992) A computer program for detailed off-line analysis of Morris water maze behavior. *Journal of Neuroscience Methods*, **41**, 65-74.

Wolfer DP, Madani R, Valenti P and Lipp H-P (2001) Extended analysis of path data from mutant mice using the public domain software Wintrack. *Physiology and Behavior*, **73**, 745-753.

Wolfer DP, Mohajeri HM, Lipp H-P and Schachner M (1998) Increased behavioral flexibility and reduced retention in the Morris water maze in transgenic mice ectopically expressing the neural cell adhesion molecule L1 in astrocytes. *European Journal of Neuroscience*, **10**, 708-717.

Wolfer DP, Stagliar-Bozizevic M, Müller U and Lipp H-P (1997) Assessing the effects of the 129Sv genetic background on swimming navigation learning in transgenic mutants: a study using mice with a modified β-amyloid precursor gene. *Brain Research*, **771**, 1-13.

Zahorik DM and Houpt KA (1981) Species differences in feeding strategies, food hazards, and the ability to learn food aversions. In *Foraging Behavior. Ecological, Ethological, and Psychological Approaches*, Kamil AC and Sargent TD (eds), Garland Press, New York, pp. 289-310.

7

Problems of Data Analysis in Behavioural Ecotoxicology

FLAVIA CHIAROTTI[1] AND TIM SPARKS[2]
[1]Section of Behavioural Pathophysiology, Lab. FOS, Laboratorio di Fisiopatologia di Organo e Sistema, Istituto Superiore di Sanità, Rome, Italy
[2]Centre for Ecology and Hydrology, Monks Wood, Abbots Ripton, Huntingdon, UK

7.1 INTRODUCTION

On the surface it would appear strange for us to be writing a chapter on the analysis of data from studies of behavioural ecotoxicology. Are the data fundamentally different from those from other disciplines? Cannot we just apply standard techniques? We must recognise, however, that behavioural ecotoxicology is a peculiar subject and incorporates elements from a number of disciplines including ecology, toxicology and behavioural pharmacology. Studies may contain an indication of whether an activity has occurred or not (binary data), how many times such an activity has occurred (count data), a subjective assessment of some feature (ordinal or categorical data), together with continuous measurements such as those relating to activity times or contamination levels. We would like to emphasise that this is not a chapter about statistical methods *per se* and that the problems of analysis that we discuss in this chapter are not unique to behavioural ecotoxicology. However, the nature of the data collected can create problems in analysis by traditional statistical methods.

In behavioural ecotoxicology, particularly with field studies, it is not always possible fully to anticipate the volume of recording. Thus the experimental design and the statistical approach may need to be modified according to the type and availability of collected data. As a result, data may be incomplete, nonhomogeneous and require transformation before processing. Where possible, and this is the case with all studies, the choice of the most appropriate design (e.g., randomised block vs. completely randomised) and statistical method (e.g., parametric vs. nonparametric, least squares or generalised linear model, univariate vs. multivariate) should be made in advance. In some circumstances, however, it can be constrained by the number and the interactions of the many factors whose effects are difficult to separate and by the complexity of the design.

We will focus on the problems typically encountered when analysing behavioural data and, in particular, on problems related to:

1. nonindependent observations
2. nonnormal distribution and heterogeneity of variance
3. multiple correlated response variables
4. unbalanced groups and missing values.

7.1.1 NONINDEPENDENT OBSERVATIONS

In *statistical inference* (i.e., the statistics which allow inference about the population based on a sample from that population; Box 7.1) most of the tests are based on the assumption that the statistical units under analysis are independent. This implies that it should not be possible to predict an observation on the basis of other observations collected in the same experiment. In general, these tests are based on statistics that follow a *theoretical distribution* (Box 7.2) under the *null hypothesis*, and are aimed at rejecting the null hypothesis in favour of an *alternative hypothesis* when the latter is true (Box 7.1). When the observations are not independent (i.e., related to each other), the test statistic fails to follow the theoretical distribution (Kenny and Judd, 1986). For this reason, when the correlation between observations is not properly taken into account, the actual significance level may exceed that calculated, producing an underestimation of the *Type I error probability* (Box 7.1).

Box 7.1 Some elements of statistical inference

Statistical inference is the process by which, on the basis of statistical measures computed on sample data, conclusions, or inferences, can be drawn about the population from which the sample is obtained. As an example we could consider the effects derived from an exposure to the toxicants A and B (one of which may represent a control situation). We may want to compare the type and severity of the observations obtained postexposure to determine whether there are differences between the two toxicants. The normal route is to follow a six-step process:

1) define the null hypothesis: H_0;
2) define an alternative hypothesis: H_1 or H_A;
3) decide on an appropriate critical significance level, e.g. 5%;
4) calculate an appropriate test statistic;
5) assess the significance of the test statistic;
6) compare the value of 5) with the value in 3) and draw conclusions.

An example of this process is

1) Null hypothesis H_0: $\mu_A = \mu_B$
2) Alternative hypothesis H_1: $\mu_A \neq \mu_B$, where μ represents the mean of the observations of the population denoted by the subscript, H_0 is the hypothesis we would like to reject and H_1 is the hypothesis under study. An alternative test statistic using the $\neq$ symbol implies a two-tailed test with no prior knowledge about the direction of any effect. One-tailed tests, using $<$ or $>$ symbols, can be applied where an effect in one direction can be expected.
3) Critical significance level: 5%. This value determines how often we are prepared to accept the false rejection of the null hypothesis. The value of 5% is traditionally used in many circumstances. In some variable ecological situations a relaxation to 10% may be appropriate and where incorrect decisions bear a higher cost, such as in drug testing, a stricter level such as 1% may be chosen.
4) An example of the test statistic would be the two-sample *t*-test. The test statistic is computed from the observations collected on the statistical units. Usually, it is based on some measure of location and of variability of the observed data (e.g., mean and standard error of the mean). The test statistic frequently, but not necessarily, involves an estimator of some population parameter. It follows a theoretical distribution which is valid when H_0 is true and the test assumptions are respected.
5) Under the theoretical distribution, the probability of a test statistic equal to or more extreme than that calculated can be obtained, either from statistical tables or internally within a computer package. This probability is called the significance level and is usually denoted by p.
6) If the significance level is smaller than the critical predetermined level in step 3, then the null hypothesis is rejected in favour of the alternative hypothesis, otherwise there is insufficient evidence to reject the null hypothesis.

In performing these tests, two kinds of error can be made, whose probabilities are denoted by α (Type I error probability) and β (Type II error probability; $1 - \beta$ = power of the test). This situation can be summarised as follows:

Truth	Test result	
	H_0 not rejected	H_0 rejected
H_0 true	$1 - \alpha$	α
H_0 false	β	$1 - \beta$

The significance level p is the estimate of the Type I error probability, that is the probability of the error made by rejecting the null hypothesis when it is true.

Nonindependence between observations is typical in the following situations:

1. When more than one observation is collected on the same statistical unit (nonindependence due to sequence). This, for example, occurs when subsequent observations are taken over time from the same unit (e.g., time-course, repeated measures).
2. When observations are collected on individuals from the same group or area (social or geographical groups) (nonindependence due to groups or space). In the first case the grouping is based on belonging to the same nest, litter, familial unit or cage (pairs or triples as in the case of agonistic or sexual challenges). In the second case the grouping is based on the geographical origin or the place from which the subjects originate. In both cases, because of common history, genetics, or more frequent interaction between subjects, it is expected that intragroup variability in the behavioural response will be lower than the intergroup variability.

When observations are related to each other, nonindependence can follow the *compound symmetry pattern*.[1] This means that the correlation between pairs of observations remains constant whatever pair is considered (*sphericity assumption*; Box 7.3). In the case of repeated measures, the link between observations can follow the *moving average model*[2] or the *autoregressive model*.[3] In both cases, dependency is due either to the exogenous causes of the variable under study, or to the influence that each observation can exert on those that follow. This kind of nonindependence is usually measured by the *autocorrelation coefficient*.[4]

When data are collected on groups of correlated statistical units, nonindependence can be modelled using the *nearest neighbour model*,[5] which is similar to the moving average model. The degree of nonindependence is usually measured either by the *intraclass correlation coefficient*,[6] or by an analogous coefficient.

Under the compound symmetry pattern the main effect or interactions of the repeated measures (or the within-subject factors) can be assessed by the appropriate analysis of variance (ANOVA) models for repeated measures or for randomised block designs. On the contrary, when the nonindependence follows patterns other than the compound symmetry, different statistical methods must be adopted. The first one is the multivariate analysis of variance (MANOVA), which can be applied considering the repeated measures as different variables collected on the same unit. The MANOVA is an exact test of the repeated measures factor and its interactions with the between-subject factors, but in some cases it can be much too conservative. Another approach for testing correlated observations when the sphericity assumption is violated is based on the use of the univariate ANOVA, with corrections applied to the *degrees of freedom*[7]. One of the most frequently used of such corrections, denoted by ε (Box 1954; Greenhouse and Geisser 1959; Huynh and Feldt 1976) depends on the degree to which the *covariance matrix*[8] departs from the sphericity assumption. ε can vary from 1 to $1/(r-1)$, where r represents the number of levels of the repeated measures factor. In particular, ε assumes the value $1/(r-1)$ in case of maximum violation of the sphericity assumption, and 1 when the assumption is respected. Under the null hypothesis, the F statistic has approximately an F distribution with degrees of freedom corrected by ε, but the approximation may be poor for small n. Such an approach allows analysis of repeated measures as a split plot design in time with correction for the degrees of freedom.

When the response variable is ordinal or categorical, *parametric tests*[9] may not always be suitable. On the other hand, *nonparametric (or distribution-free) tests*[10] for complex experimental designs are not available: therefore, the experimental design must be simplified, in order to match the requirements of the available nonparametric methods.

As an example, in a study on the effects of different formulations of granular pesticides on birds (Stafford *et al.* 1996) the data collected on treated captive sparrows included plasma and brain ChE values, mortality occurrence and behavioural symptoms. Pesticides were administered using different granular carrier types, sizes and colours. As plasma ChE activity was measured on the same birds on three different days, data could not be considered independent. Therefore, in order to cope with the possible violation of the sphericity assumption, the authors used MANOVA.

In another study on the effects of foliar insecticides on survival of northern bobwhite quails (Palmer *et al.* 1998) the incubation behaviour of birds was monitored in the field before and after exposure to insecticide. In addition, chick survival was estimated 'by flushing the brood and counting the chicks once they were 28 days old'. In this case, the statistical units represented by single birds were considered independent. In the second set of experiments (laboratory trials), either adopted or human-imprinted quail broods were exposed to different foliar insecticides and brain ChE was determined for each chick. In this case, data from chicks belonging to the same brood (rearing conditions) could not be considered independent and were averaged on this grouping element. The brood was also used as a grouping factor for the subsequent analyses.

In a study analysing the grit colour selection by house sparrows and northern bobwhites (Gionfriddo and Best 1996) birds were housed in aviaries in separate treatment groups of 32–35 individuals. The authors argued that 'birds housed together may not have selected grit colours independently' and thus, that data from the same aviary were considered nonindependent, and groups but not individual birds were the statistical unit. By consequence, the individual data, expressed as grit colour proportions, were averaged over the group to which the individuals belonged, and the mean values were used in the statistical analyses. This situation is also commonly encountered in fish studies with several fish housed together in a single tank. Treating each fish separately as a statistical unit inevitably leads to pseudoreplication.

Finally, in a laboratory study on the behavioural effects of prenatally-administered AZT to mice, CD-1 females were randomly assigned to one of three treatments including controls (Calamandrei *et al.* 1999). AZT was administered daily from day 10 of pregnancy to delivery. At birth, litters were culled to four males and four females, and pups were given to foster dams. Many response variables were collected: quantitative variables (weight, food intake, water intake, activity counts in a Varimex test, frequencies and duration of many behaviours in a gender preference test and during agonistic encounters), and ordinal variables (latencies and frequencies of errors in a passive avoidance test, scores in the Fox battery test). In case of violation of the sphericity assumption quantitative variables were analysed using a parametric ANOVA with a specific correction applied to the degrees of freedom.

7.1.2 NONNORMAL DISTRIBUTION OR HETEROGENEITY OF VARIANCE

A *variable*[11] can be qualitative or quantitative. The variable is defined qualitative when it is possible to assess if its values are equal or different, but is not possible to order them from the smaller to the greater. On the contrary, a variable is called quantitative when the values can be ordered and vary in degree along a single dimension. A quantitative variable can be either discrete or continuous: the former always takes exact values, while the latter can take any value in a given range.

When a variable assumes the values included in a specified set with a specified relative frequency or probability, it is called a random variable of variate. A variate is defined by the set of permissible values and the corresponding probability (or density) function, which gives the probability associated to each permissible value. There are many different *theoretical distributions* (Box 7.2), some of them more frequent in biology and toxicology.

Basic assumptions on the distribution of the response variable are required for most of the statistical tests and, in particular, for the parametric tests (Box 7.3). Under such circumstances the grouping factor (e.g. exposure to an environmental toxicant) can affect the mean value of the behavioural response, but not the shape or the variance of its distribution. Unfortunately, this is the case of many different experimental situations. For instance, counts of the stops in movements of beetles dosed by copper (Appendix 2 in Sparks 2000) show a nonnormal distribution (see Figure 7.1a).

Box 7.2 Most common theoretical distributions

The Binomial distribution can be used to describe the probability of dichotomous events in n trials, when the probabilities of the two complementary and exhaustive realisations of the event in any one trial are known, and denoted with p and $q = 1 - p$, respectively.

The Poisson distribution is appropriate to describe the probability distribution of counts, such as the number of events in a specified period of time. This is a very frequent case in behavioural research, where tests for learning or habituation often result in counting the number either of errors or of correct responses given by the animals in the test session. The Poisson is a discrete distribution, which is defined for integer values from 0 to ∞. The Poisson distribution is characterised by one parameter denoted by λ, representing both the mean and the variance of the distribution.

The most famous distribution, even if not the more frequent, is the normal (or Gaussian) distribution, which was first discovered by De Moivre in 1753 as the limiting form of the binomial distribution. It can be used for variates assuming all the real values within a specified interval. It is a continuous, symmetrical, bell-shaped distribution, characterised by one location (mean) and one scale (variance) parameter. Mean, median and mode coincide, and the relative steepness of ascent in the neighbourhood of the mode, measured by kurtosis, takes a specific value, considered as a standard. Distributions for which kurtosis is less than, equal to or greater than this standard value are said to be platykurtic, mesokurtic and leptokurtic.

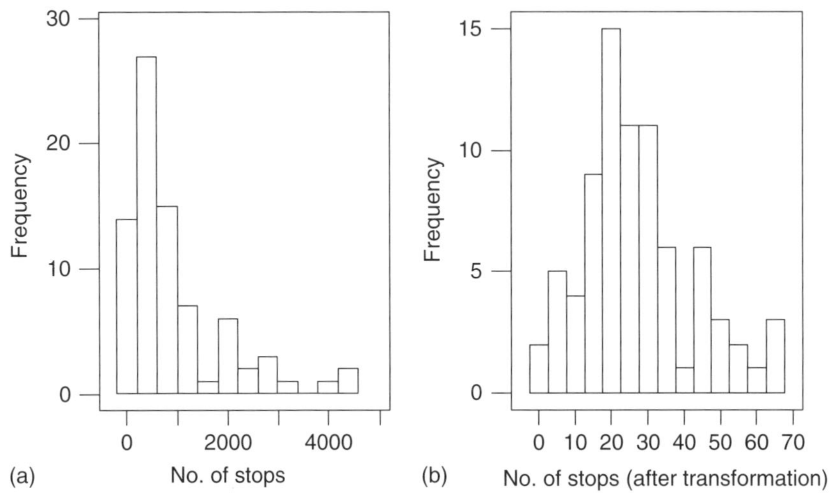

Figure 7.1

Box 7.3 Basic assumptions of statistical tests

Continuity. Almost all of the statistical tests on means (parametric methods) or on medians (nonparametric methods) assume the response variable to be continuously distributed in the population of interest. However, measurements are usually made on a discrete scale, and this conflict between the theoretical distribution and the empirical one must be taken into account. Specifically, when performing nonparametric tests, the probability of tied values in the assumption of continuous response variables is zero. Unfortunately, because of measurement constraints, it may happen to have tied values and, therefore, tied ranks. Appropriate corrections must be adopted to counterbalance the consequent loss of power of the statistical test.

Normality. In many statistical tests, such as ANOVA, *t*-tests and least-squares regression, there is a requirement that the random component of the model (i.e., the residuals), and therefore the response variable, must follow the normal distribution in each subgroup of statistical units and for each measure collected on the units.

Homogeneity of variance or homoscedasticity. When we consider two or more groups of independent statistical units, the variance of the response variable providing an estimator of the error variance, that is the variability of the random component of the model, must be the same for all the groups. When it is true, the variances estimated in the different groups can be pooled to obtain an overall estimator of the error variance, based on more degrees of freedom.

Sphericity. In the case of repeated measures collected on the same statistical units, the *F* statistic for the assessment of the significance of the repeated measures main effect in the parametric ANOVA no longer follows the theoretical *F* distribution, but is positively distorted. Such a distortion does not affect the ANOVA results when the variance-covariance matrix follows the sphericity pattern.

Under small deviations from the normality and homogeneity assumptions the majority of tests give reliable results and are reasonably efficient. This characteristic is called robustness of the statistical test. By contrast, larger deviations can affect substantially the estimators and the statistical results. In particular, this may occur when:

1. the variability is related to the mean (greater means are accompanied by greater variances);
2. the distribution is strongly skewed as in the case of *L*-shaped (with one very long tail) distributions;
3. one or more outliers are present.

In all these situations, parametric tests can be performed after having transformed the data in order to normalise their distribution and homogenise their variances. The *transformation* must be chosen on the basis of the actual distribution of the variable in the population. Unfortunately, this distribution is rarely known, and frequently, subsequent transformations are applied until test significance is achieved. This practice is absolutely and definitely wrong (Box 7.4). Moreover, in some instances quantitative variables may require to be treated as ordinal. For example, this occurs in biochemical assays where the measure of the response variable is limited by the sensitivity of the instrument or the analytical procedure: these limits are defined cut-off values. In behavioural tests the limits can be due to methodological or ethical constraints, as when the experimental subjects should be punished and the punishment must be limited in intensity and duration (e.g. in the passive avoidance and several tests for measuring analgesy). Therefore, if an animal does not respond to the noxious stimulus within a fixed time, the test is stopped and the variable (usually the latency in responding) assumes the cut-off value. Obviously, the introduction of cut-off values, as well as the categorisation of a quantitative variable, produces the loss of part of the information contained in the original variable.

Box 7.4 Transformations

Heterogeneity of variance is often due to the existence of a proportionality between means and standard deviations (SD) in the sampled population; that is the higher the mean, the wider the SD around it.

In order to homogenise the variance between the groups under comparison SDs should be made independent from means. Various transformations can be adopted, according to the distribution of the original response variable. Some examples are given in the following list.

1. The standard deviation is proportional to the square root of mean. This may happen when the response variable is a count, and follows a Poisson distribution. In this case, the appropriate transformation to normalise data distribution and to stabilise variances is $z = \sqrt{x + c}$, where z is the transformed variable, x is the original response variable and c is a constant (usually, c is chosen equal to 1).
2. Standard deviation is proportional to the mean. In this case, an useful transformation is $z = \log(x + c)$, where c must be necessarily added in the case where the original data present at least one zero value (otherwise the transformation can not be applied).
3. Standard deviation is proportional to the squared mean. Original data can be transformed taking the reciprocal, that is $z = 1/x$.
4. When the original data are proportions (that is, x's represent counts out of n trials), they follow a binomial distribution and the appropriate transformation is the inverse sine or angular transformation. Therefore, $z = \sin^{-1} \sqrt{x}/n$.
   x/n should vary between 0 and 1 (if percentage values are used, these should be divided by 100).

In this situation nonparametric tests based on rank transformation are required. This approach is a solution when the response variable is ordinal or can assume cut-off values, but is also particularly suitable when no hypotheses on the distribution in the population can be made.

In a review analysing the effects of pollutants on population dynamics of birds of prey (Newton 1988), the distribution of contaminant concentrations was used to estimate the level of pollutant and associated mortality necessary to cause population decline. This distribution is seldom normal, being usually skewed to the left, with a very long right tail. Therefore, parametric methods based on the normality assumption cannot be used: the most frequently adopted procedure is to transform data taking their $\log_{10}$ value. This transformation normalises the asymmetrical distributions reducing the effect of outliers, thus making the data amenable to a wider range of statistical procedures.

In a study of the response of small mammals to clearcutting, burning and glyphosate application in the Oregon coast range (Cole *et al.* 1998) different species of small mammals were trapped on replicate red Alder sites for three consecutive years, one before and two after the activities of the management of the forest. Data regarding capture rates were classified according to site, treatment, area and trapping season. Data were not normally distributed, but the experimental design was so complex that nonparametric methods could not be used. Therefore, data had to be appropriately transformed, using the square root transformation suitable for Poisson distributed data (Box 7.2). Only species with at least 20 initial captures were considered, the limit being chosen so to have species with enough data. When transformation and other techniques were not successful to normalise and homogenise data, the species was not included in the statistical analysis. As a further example, the beetle stop data is shown Figure 2.1b following square root transformation. Greater symmetry is apparent.

A different approach was followed by Calamandrei *et al.* (1999, see above) in the analysis of latencies in a passive avoidance test. Latencies are usually nonnormal and, in addition, they are cut off due to ethical issues. Therefore, these can be analysed only by nonparametric tests. Unfortunately, nonparametric methods for determining the main effects and interactions of between- and within-subject factors are not available. Thus, complex experimental designs must be simplified (Chiarotti *et al.* 1987). In this case, the latencies in repeated trials during the test and retest phases were synthesised using their medians. Subsequently, medians were analysed applying nonparametric methods. In particular, (1) Kruskal-Wallis ANOVA was applied to the sum of data from the two littermates, to test the main effect of prenatal treatment (three levels: saline, AZT 0.2 mg/ml and AZT 0.4 mg/ml), (2) Wilcoxon test was performed on pairs of data from littermates, to test the main effect of condition (two levels: conditioning and nonreinforcement), and finally (3) Kruskal-Wallis ANOVA was applied to the difference between data from the two littermates, to test the interaction prenatal treatment × condition. Similar problems were met when analysing scores in the Fox battery test. Scores usually range from 0 to 3, representing the different developmental stages (0 = absent behaviour, 1 = uncertain, 2 = incomplete, 3 = fully developed). The number of different scores (4) was small in comparison with the number of repeated measures (usually at least 15, that is the first 2 weeks from birth). Moreover, particularly in the first period, some 'regressions' from one score to the preceding one were possible. For these reasons, each pup's developmental curve was transformed, using the first day of adult-like response that took into account the whole developmental profile of the pup. Nonparametric ANOVA was subsequently applied to the transformed data.

Two important alternatives to transforming to achieve normality are now available. The first is the family of *General Linear Models*[12] that allow error structures for distributions other than normal (e.g. Poisson, binomial). The second alternative, which is likely to gain favour as computer power increases, are randomisation tests based on computer intensive resampling (Sparks and Rothery 1996).

7.1.3 MULTIPLE CORRELATED RESPONSE VARIABLES

The different behavioural variables recorded on a single individual, particularly when mutually exclusive, are generally linked to each other. The use of multiple univariate analyses is not suitable for shedding light on the relationships between the various variables. A suitable approach to understand these intercorrelations is based on the use of multivariate methods.

Principal Component Analysis (PCA)[13] is one of such multivariate methods. The PCA allows a description of set of continuous, intercorrelated behavioural variables by constructing new variables (factors) as linear combinations of the original ones. The new factors are orthogonal (i.e., uncorrelated) and can be interpreted on the basis of the original variables which influence them. This influence is measured by the factor loadings, which represent the correlation between the original variables and the PCA factors. Each statistical unit receives a score on each factor, representing its coordinates in the multidimensional space; these scores can be analysed by the appropriate model of analysis of variance, to assess the effect of 'treatments' on the PCA factors. As an example, six variables in the beetle data set have been subjected to PCA (Figure 7.2). Whilst there is a large amount of overlap between the dose groups, the pattern does not appear to be random and there is more separation of doses on the vertical axis. This axis represents a contrast between velocity (top) and turning rate (below). Analysis of variance of the second axis scores confirms a highly significant difference between dose groups.

When behavioural data consist of sequences of events and the time points at which they occur (transitions between behavioural items), the analysis of sequences can give information about the temporal structure of behaviour. This

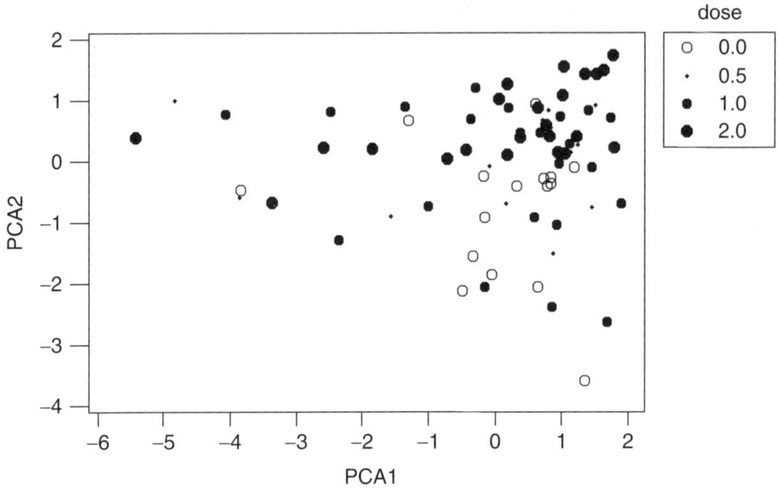

Figure 7.2

can be described by the transition rates between behavioural items and the termination rates of each item. The former represents the chance per time unit that an item is followed by the other one, while the latter represent the chance per time unit that an item is interrupted. Different methods can be used to estimate transition and termination rates, between them the *Continuous Time Markov Chain (CTMC) model* and the *Cox proportional hazards model.*[14]

This approach was followed by Puopolo *et al.* (1999) in a study on behavioural effects of endocrine disrupting chemicals on laboratory rodents. Mice of the CD-1 strain (12 litters of five male pups each) were randomly assigned within litter to one of the five treatments: vehicle solution, low and high doses of PCB-54 and PCB-77. At a juvenile stage each mouse underwent a 20-min social encounter with another animal of the same sex and treatment. Behaviour was videotaped, scored and subsequently analysed for statistical evaluation. The following behaviours were considered: Inactivity, Digging, Self-grooming, Exploring, Social resting, Allogrooming, Social inactivity, Sniffing, Following, Mutual circling, Jumping, Crawling over, Pushing past, Pushing under and Fighting. Nonparametric analysis of variance for randomised block designs performed separately on each behaviour did not reveal any significant difference between control and treated groups. In contrast, the analysis performed on the more frequently displayed behavioural sequences (synthesised by relative transition and termination rates) revealed some interesting significant difference between groups. In particular, with respect to PCB-54, high-dose mice showed a significantly faster transition from Exploring to Self-grooming when compared with control littermates. Moreover, with respect to PCB-77, significantly slower transition rates from Digging and Self-grooming to Exploring were shown by low-dose mice when compared with control littermates, and a significantly slower transition rate from Exploring to Digging characterised high-dose mice in comparison with control littermates. Finally, in PCB-77 animals lower termination rates of Digging and of Exploring were found in low-dose and high-dose mice, respectively, when compared with control mice. Some other differences just missing statistical significance ($0.05 < P < 0.10$) were detected when termination and transition rates were compared between treated and control mice.

7.1.4 MISSING DATA AND UNBALANCING

Behavioural data collected both during laboratory experiments and field work can be either incomplete or unbalanced for the number of the statistical units under comparison. We will discuss separately these two situations.

7.1.4.1 One or more statistical units have missing values

This problem is frequent in the case of repeated measures. In fact, when more observations are collected on the same statistical unit, one or more of them could be missing. In some cases this may happen when one of the experimental animals gets sick or dies, before the end of the experiment. In other cases, such as when assessing the neurobehavioural development, some behavioural items must be collected before the sexual maturation. Thus, it may occur that animals at different periods of their sexual development could contribute with a different number of observations. If the missing data are few (such as, for instance, one per experimental group, and at different time points for the different experimental groups), they can be replaced by values estimated on the basis of the observations collected at different time/condition points on the same unit (Figure 7.3a) and at the same time/condition point on the other units belonging to the same experimental group (Figure 7.3b). In this case the degrees of freedom must be consequently reduced. When a consistent portion of data is missing it could be inappropriate to extrapolate the missing values. If this was the case, it would be better to exclude from the analysis those units more affected by missing values, previously controlling if missing values depend on treatments and/or on experimental conditions (Figure 7.3c).

7.1.4.2 Groups under comparisons are unbalanced

Sometimes the groups under comparison can have a different number of statistical units. This unbalancing can be due to the experimental design or to a constraint emerged when performing the experiment.

The first case is typical when assessing the effect of a condition, such as the presence of a genetical mutation, which cannot be controlled but whose frequency in the sampled population is only approximately known. In some cases it may be useful to collect more data on a new, less experimented, or

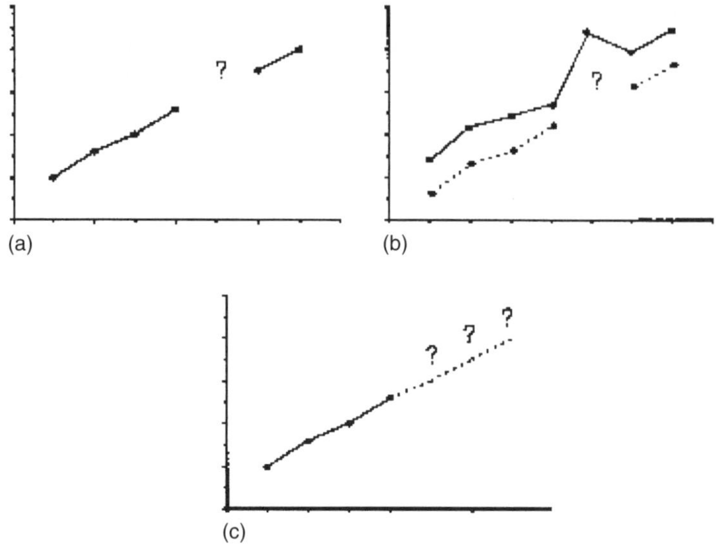

Figure 7.3

more variable treatment/condition than on the standard, control, or less variable treatment/condition.

With regard to unwanted unbalancing, this can result from unexpected, nonsymmetrical reduction (due to mortality, lack of compliance, missing data which can not be replaced by estimated values) in the size of the experimental groups. This situation must be treated appropriately if the reduction is not random, but depends on the treatments, in particular when the experimental design includes more than one factor. The most appropriate approach in dealing with the nonrandom unbalancing in factorial designs is performing the analysis of variance according to the General Linear Model procedure. This method allows an assessment of the incremental effect of one factor over and above the other factors included in the model; thus joint effects are excluded from the assessment of main effects of the factors. Alternative methods allow weighting, e.g. by litter size, to ensure that means based on larger litters are given more weight than those from small litters.

7.2 SUMMARY

Analysis of behavioural ecotoxicological data brings about its own problems. These problems are, however, not unique to this discipline and have been encountered in other areas of the biological sciences. The analysis of count data and binary data are common features of behavioural data, but despite their inherent nonnormality, methods exist to proceed with analysis. A consequence of the variety of data in behavioural studies is that behavioural ecotoxicologists may need to be aware of a wider range of statistical techniques than many of their contemporaries, including specific approaches to analysis of data from low-dose contaminants (Haseman *et al.* 2001). However, the use of nonparametric tests, general linear models with nonnormal error structures and computer intensive resampling techniques ensures that data of a wide variety of forms can be assessed to identify the impact of environmental pollutants on animal behaviour.

ACKNOWLEDGEMENTS

We wish to thank Giacomo Dell'Omo for continuous help and critical reading.

NOTES

[1]Compound symmetry pattern. A number of observations on the same statistical unit follow the compound symmetry pattern if they are equally related to each other with the same correlation coefficient ρ. The compound symmetry is a sufficient, but not necessary, condition for the validity of the F ratio under the general null hypothesis of no treatment effects. Unfortunately, this correlation pattern is frequently violated, causing the ratio of mean squares to be distributed as F with a reduced number of degrees of freedom.

[2]Moving average model (of the i-th order). This model hypothesises that each observation is linked to the i preceding and to the i following observations, while independent from the others.

[3]Autoregressive model (of the i-th order). This model hypothesises that each observation is linked to the preceding i observations according to the following expression: $x_t = a_1 x_{t-1} + a_2 x_{t-2} + \cdots + a_i x_{t-i}$.

[4]Autocorrelation coefficient. Coefficient measuring the internal correlation between members of series of observations ordered in time or space.

[5]Nearest neighbour model. This model hypothesises that the correlation exists among the observations collected on units living near each other, that is, each observation is linked only to the ones which are adjacent in space.

[6]Intraclass correlation coefficient. A measure of correlation within the members of certain natural groups or 'families'; this coefficient is closely allied to a variance ratio in variance analysis.

[7]Degrees of freedom. Number of values of a set of observations which independently contribute to the estimate of a parameter or a statistic.

[8]Covariance matrix. Square matrix whose elements c_{ij} $(i, j = 1, \ldots, n; i \neq j)$ are the covariances between x_i and x_j, while diagonal terms are the variances.

[9]Parametric tests. Tests that are based on hypotheses about the distribution of the response variable (and therefore, about the value of a specific parameter) in the sampled population.

[10]Nonparametric tests. Tests that make no hypothesis about the value of a specific parameter in the sampled population. Distribution-free tests. Tests that make no assumptions about the precise distribution of the variables in the sampled population. The two terms 'nonparametric' and 'distribution-free' are usually treated as synonyms, even if a distinction can be made between the two classes of tests.

[11]When used in a statistical context, the words 'variable' and 'parameter' have a different meaning than in common usage. Specifically, variable means a quantity which varies, taking any one value included in a specified set of values; on the contrary, parameter usually refers to an index synthesising some characteristics of a given population, such as mean and standard deviation, or of a stochastic model, such as intercept and slope in the linear regression.

[12]General Linear Model. Extension of the multiple regression model to a class of maximum likelihood procedures. The dependent (response) variable is supposed to be distributed around a mean, and some function (called the link function) of this mean is supposed to be linearly related to the independent (explanatory) variables. The response variable does not need to be normally distributed, but its distribution must be specified, and the different models including the whole set or subgroups of explanatory variables are compared using the deviance, based on the log likelihood, and asymptotically proportional to a chi-squared variable.

[13]PCA belongs to a large group of multivariate methods (known as Factor analysis) that can be used for extracting factors which are supposed to underlie the observed variables. It is usually chosen for exploratory analyses.

[14]CTMC model. Stochastic process in which several states (e.g. behaviours) are successively visited, and transition rates are used to describe completely the 'velocity' of switching between states, and the tendency to get in or to get out each state. Cox proportional hazard model. Model for estimating the effects of various covariates in survival analysis. The death rates (in our case, the termination and the transition rates) can be expressed as product of the hazard function (i.e., an unknown function of time, equal for all individuals) and log-linear functions of the covariates representing either inherent differences between individuals or different treatment groups.

REFERENCES

Box GEP (1954) Some theorems on quadratic forms applied in the study of analysis of variance problems. II. Effects of inequality of variance and of correlation between errors in the two-way classification. *Annals of Mathematical Statistics*, **25**, 484–498.

Calamandrei G, Venerosi A, Branchi I, Chiarotti F, Verdina A, Bucci F and Alleva E (1999) Effects of prenatal AZT on mouse neurobehavioral development and passive avoidance learning. *Neurotoxicology and Teratology*, **21**, 29–40.

Chiarotti F, Alleva E and Bignami G (1987) Problems of test choice and data analysis in behavioural teratology: the case of prenatal benzodiazepines. *Neurobehavioral Toxicology and Teratology*, **9**, 179-186.
Cole EC, McComb WC, Newton M, Leeming JP and Chambers CL (1998) Response of small mammals to clearcutting, burning, and glyphosate application in the Oregon coast range. *Journal of Wildlife Management*, **62**, 1207-1216.
Gionfriddo JP and Best LB (1996) Grit colour selection by house sparrows and northern bobwhites. *Journal of Wildlife Management*, **60**, 836-842.
Greenhouse SW and Geisser S (1959) On methods in the analysis of profile data. *Psychometrika*, **24**, 95-112.
Haseman JK, Bailer AJ, Kodell RL, Morris R and Portier K (2001) Statistical issues in the analysis of low-dose endocrine disruptor data. *Toxicology Science*, **61**, 201-210.
Huynh H and Feldt LS (1976) Estimation of the Box correction for degrees of freedom from sample data in randomized block and split-plot designs. *Journal of Educational Statistics*, **1**, 69-82.
Kenny DA and Judd CM (1986) Consequences of violating the independence assumption in analysis of variance. *Psychological Bulletin*, **99**, 422-431.
Newton I (1988) Determination of critical pollutant levels in wild populations, with examples from organochlorine insecticides in birds of prey. *Environmental Pollution*, **55**, 29-40.
Palmer WE, Puckett KM, Anderson JR Jr and Bromley PT (1998) Effects of foliar insecticides on survival of northern bobwhite quail chicks. *Journal of Wildlife Management*, **62**, 1565-1573.
Puopolo M, Santucci D, Chiarotti F and Alleva E (1999) Behavioural effects of endocrine disrupting chemicals on laboratory rodents: statistical methodologies and an application concerning developmental PCB exposure. *Chemosphere*, **39**, 1259-1271.
Sparks TH (ed.) (2000) *Statistics in Ecotoxicology*, John Wiley, Chichester.
Sparks TH and Rothery P (1996) Resampling methods for ecotoxicological data. *Ecotoxicology*, **5**, 197-207.
Stafford TR, Best LB and Fischer DL (1996) Effects of different formulations of granular pesticides on birds. *Environmental Toxicology and Chemistry*, **15**, 1606-1611.

Case Studies

8

Altered Behaviour of Invertebrates Living in Polluted Environments

WINDY A. BOYD[1], SANDRA K. BREWER[2] AND PHILLIP L. WILLIAMS[1]

[1]Department of Environmental Health Science, The University of Georgia, Athens, GA, USA

[2]U.S. Fish and Wildlife Service, New Jersey Field Office, Pleasantville, NJ, USA

Illustrations by Emma J. Rosi-Marshall

A fundamental goal of ecotoxicology is to determine how the responses of individuals to pollutants relate to effects at the population level (Truhaut 1975). Organisms may protect themselves against toxicants by changing their behaviour or their behaviour may be directly affected by toxicants. Because behaviour is the cumulative manifestation of genetic, biochemical, physiological and environmental cues, behavioural data may provide a link between individual response and population change, especially for those behaviours that are ecologically manifested as changes in structure and function of the community (Little *et al.* 1985). Toxicant-induced behavioural changes are of interest both for their intrinsic value and because behavioural dysfunctions may be applied in bioassays to detect or monitor the presence of contaminants in the environment.

This chapter is divided into two sections: the first is a general overview of behavioural endpoints that have the potential for ecological manifestations at population and community levels; and the second is a detailed review of selected case studies on invertebrates organisms from aquatic, sediment and terrestrial environments.

It is reasonable to assume any behavioural response may be affected by toxicants, hence the arena for investigations into behavioural ecotoxicology is almost limitless. In reality, however, relatively few responses have been studied in sufficient detail to determine if the response criteria occur rapidly at low toxicant environmental concentrations, in a concentration-response fashion, and are reproducible and reliable. We have selected several behavioural responses to detail based on their degree of use, their ecological relevance to higher-orders of organisation (i.e., populations and communities), and the ease with which the behaviours can be used in monitoring studies.

8.1 PRINCIPAL ENDPOINTS USED IN BEHAVIOURAL CASE STUDIES ON INVERTEBRATES

The following section describes some of the most studied behavioural endpoints described in the literature. A concise overview is presented in Table 8.1.

Table 8.1 Principal behavioural endpoints employed in ecotoxicological studies

Behavioural endpoint	Test media	Types of organisms	Toxicants tested
Burrowing response	Sediment	Crustaceans	PCBs, metals, hydrocarbons
	Terrestrial	Annelids	Textile dyes,
	Sediment	Annelids	PCBs, metals, and pesticides
Phototactic behaviour	Aquatic	Amphipods and daphnids	Copper and lindane
Migratory activity	Terrestrial	Earthworms	Benomyl
Swimming activity	Aquatic	Daphnids, barnacle nauplii, protozoa and rotifers	Variety of organics, salts and metals
Locomotion	Terrestrial	Beetles, nematodes, springtails and woodlice	Variety of metals and organics
Walking activity	Terrestrial	Linyphiid spider	Deltamethrin
Feeding and ingestion rate	Aquatic	Daphnids and rotifers	Variety of metals and pesticides
Valve closure	Aquatic	Molluscs	Variety of metals and pesticides
Case building and net spinning	Aquatic	Caddisflies	Cadmium and copper
Web building	Terrestrial	Spiders	Pesticides
Mate guarding behaviour	Aquatic	Amphipods	Variety of metals and pesticides
Ovipositing behaviour	Aquatic	Midges	Cadmium
Predator–prey interactions	Aquatic	Rotifers	PCP
Infectivity and nictation	Terrestrial	Entomopathogenic nematodes	Pesticides
Learning	Terrestrial	Honey bees	Variety of pesticides

8.1.1 BURROWING BEHAVIOUR

Burrowing is a complex behavioural pattern consisting of a chain of instinctive reflexes (Pynnönen 1996). Because only a minor failure in reflexive response may lead to complete behavioural dysfunction, burrowing may be an extremely sensitive indicator of sediment or soil toxicity. The mechanism of toxicity, however, can be difficult to determine and may include epithelial damage to sensory structures, neurotoxic effects on the nervous system that often lead to a loss of general coordination, avoidance responses to contaminated sediment, or a combination of factors.

Organisms may use burrowing behaviour as an energetic trade-off between other more costly avoidance behaviours and physiological adjustments to unfavourable conditions (Dill 1987). Regulatory behaviours involve conflicts with other behaviours and are likely to be the most sensitive indicators of stress (Pennuto and de Noyelles 1993). For example, drifting behaviour is energetically inexpensive but may pose added predation risks from visually feeding predators, resulting in a conflict if predation pressure is high and environmental conditions provoke an avoidance reaction. In such situations, behaviours that ensure satisfactory avoidance of the contaminant but do not increase predation risks are to be favoured. Burrowing behaviour may be an avoidance behaviour that is favoured when an organism must avoid a topical or water-borne toxicant.

8.1.2 PHOTOTAXIS

The phototactic or migration response of zooplankton is well established and the mechanisms controlling migratory patterns are complex. Typically, decreases in light intensity will cause zooplankton to ascend in the water column and increases in light intensity result in zooplankton descent. The toxicant concentration needed to disrupt this behaviour is often much lower than concentrations that cause lethality (Simonet *et al.* 1978, Saunders *et al.* 1985).

Photo-response of zooplankton is controlled by exogenous factors for both the direction and speed of swimming. Swimming speed has been linked to the rate of change in light intensity (McNaught and Hasler 1964, Ringelberg 1964). The direction of swimming may be controlled by the shape of the Angular Light Distribution (ALD) (Drake and McNaught 1994) and perhaps the degree of polarisation and the angle of the light relative to the zooplankton (Whitney 1941, Ringelberg 1964).

Migratory behaviour may increase either the individual's or the population's chances for survival through increasing resource exploitation, enhancing predator avoidance, optimising metabolism, and possibly providing niche separation (Brooks 1964, Geritsen 1982, Wong *et al.* 1986). Migratory behaviour can be quantified both in the laboratory (Ringelberg 1964, Bohrer 1980) and the field (McNaught and Hasler 1964). Furthermore, ecotoxicity tests have been proposed for both freshwater (Simonet *et al.* 1978, Flickinger *et al.* 1982) and marine organisms (Forward and Costlow 1978, Saunders *et al.* 1985).

Migratory patterns of zooplankton, especially daphnid species, can be separated into motion along the vertical and horizontal planes. Testing for interactions of toxicants on vertical migration patterns requires careful interpretation of the results. Diel changes in direction and velocity necessitate very rigorous testing procedures to control variations. Vertical migration is limited to up, down or no migration. A downward migration might be interpreted as active movement, passive sinking or general lethargy. Likewise, no migration might result from immobilisation of the animal, or a failure to recognise the cues for migration. Expensive equipment is needed to test vertical migration accurately. In contrast, horizontal migration is not believed to be subject to circadian fluctuations (Siebeck 1964,1969), allows for the full 360° range of movement, and proper design can distinguish between immobilisation of the animal or a failure to recognise the cues for migration (Goodrich and Lech 1990). Furthermore, horizontal migration is involved in the same aspects of zooplankton survival (i.e., avoiding predators, locating food and niche separation) and shares the same physiological basis as vertical migration (Huges 1980, Ringelberg 1980, Wong *et al.* 1986). However, horizontal migrations may have different cues and releasing stimuli (Ringelberg 1969).

Phototaxic responses appear to be good behavioural metrics for laboratory and field ecotoxicity studies. Most toxicological work with aquatic invertebrates produces a surplus of organisms that can be used to provide what many toxicological screening strategies lack: a sensitive, sublethal behavioural assay. With the appropriate study design, such behavioural assays can be conducted with minimal time and cost.

8.1.3 FEEDING AND FILTRATION

An organism must be able to obtain enough food for growth, reproduction and metabolic maintenance. Every organism has a threshold concentration of food required for survival; below this limit, the organism will die, above this limit, the organism has additional energy for growth and reproduction. Filtration of food by zooplankton and molluscs, the ingestion of filtered products and the assimilation of ingested particles, has been researched extensively from an ecological perspective.

The filtration and ingestion of food by filter-feeding invertebrates requires movement of appendages and thus coordination of the nervous system. Any neurotoxicant that causes loss of coordination or immobilisation may be revealed by a decrease in filtration rate. Additionally, some chemicals have been shown to induce particle adhesion onto the setae of daphnids, making the movement of filter-feeding appendages more difficult (Day and Kaushik

1987, Stratton and Corke 1981). Several researchers have advocated the use of filtering and ingestion behaviour in the study of ecotoxicology (Flickinger *et al.* 1982, Day and Kaushik 1987). Rates of particle filtration by zooplankton have been observed to decrease after exposure to various chemicals (Reeve *et al.* 1977, Berman and Heinle 1980, Day and Kaushik 1987). Cooley (1977) determined that the filtering activity of *Daphnia retrocurva* found near a canal discharging pulp mill effluent into Nipigon Bay was lower than rates of filtration of the same species found near the middle of Nipigon Bay. Kraak *et al.* (1994) reported that *Dreissena polymorpha* exposed for 48 h to filtered water from polluted sections of the rivers Rhine, Meuse and Amstel had infiltration rates inhibited to 12–58% of reference values. Furthermore, inhibition was higher during low water levels in the rivers Rhine and Meuse than during high water levels, suggesting that contaminants were diluted during high water levels.

Experiments described within the following case studies show that filtration and ingestion rates measured in the field and laboratory are rapid and sensitive indicators of toxic stress and may help to explain other observed changes in the survival, growth and reproduction of nontarget organisms. Like any testing procedure, there are limitations and assumptions made. The assumptions involved in determining filtering rates have been discussed in numerous publications and we refer the reader to Peters (1984) for an excellent summary. Considering the ecological importance of food uptake and the ease and rapidity by which this phenomenon can be evaluated in the laboratory and field, standardised toxicity tests with invertebrates that use feeding behaviour as a test endpoint are already a valuable new approach in the first tier of hazard assessment schemes (Juchelka and Snell 1994).

8.1.4 REPRODUCTIVE BEHAVIOUR

Successful reproduction is of fundamental significance to the preservation of natural populations. Because reproduction is a vulnerable stage in the life cycle, egg and larvae viability are important endpoints in traditional toxicological testing. For animals that lay eggs, it is vital that the female select an egg-laying habitat that is of acceptable quality for egg survival and for normal development of embryos and larvae. Mosquitoes are able to determine water quality for oviposition purposes and larvicidal oils were shown in laboratory and field to be repellent to ovipositing *Culex* mosquitoes (Beehler and Mulla 1996). However, it is important to examine ecologically relevant sublethal responses such as changes in reproductive behaviour that result from toxicant related stress.

One example of such a response is the precopulatory or mate-guarding behaviour exhibited by Amphipoda (Kruschwitz 1972, Parker 1974, Conlan 1991, Aiko 1996). This response is easily quantifiable as male amphipods periodically attach to the dorsal side of females with their gnathopods prior to

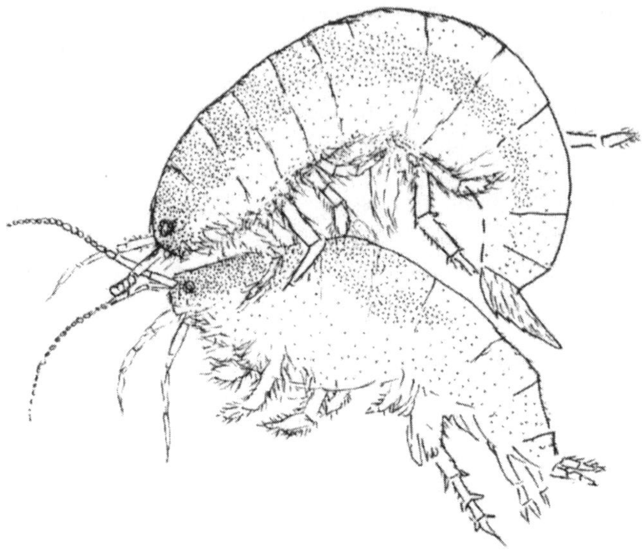

Figure 8.1 Mate-guarding behaviour in Amphipoda.

copulation (Figure 8.1). The pair is able to swim and feed and the attachment may last from several hours to as long as a week depending on the length of time required for the female to moult (Wilder 1940, Strong 1973, Blockwell *et al.* 1998). Mating occurs immediately after the female's moult. Consequently, mate-guarding behaviour is of tremendous ecological importance because it is a requirement to the mating success of the species. Additionally, because the precopulatory guarding is energetically expensive and confers the benefit of ensuring paternity, precopulatory guarding provides a measure of male fitness (Poulton and Pascoe 1990).

Precopulatory mate guarding has been shown to be disrupted (i.e., separation of precopula pairs by body extension or flexion) after exposure to environmental pollutants (Wildish 1972, Linden 1976, Davis 1978) and several methods have been developed to examine this phenomenon. In general, disruption occurs in three stages: (1) male and female swim vigorously; (2) male extends gnathopods to hold the female away from his body and then retracts gnathopods to bring the female back into her former position close to his ventral surface; and (3) male and female simultaneously 'kick' by flexing their bodies until the female is released.

Bioassays based on disruption of precopula coupling, but with different response criteria, have also been used. The two most common methods are induced or direct separation. Induced separation involves exposing precopulatory pairs for a predetermined amount of time and then removing the pairs into a solution of anaesthetic to induce separation and recording the time to separation. The time from the introduction of the anaesthetic to complete separation can be measured and recorded as the IST_{50} (median Induced Separation Time). This test offers the advantage of requiring a predetermined exposure period, which may be an important consideration if the bioassay is to be employed for *in situ* or routine laboratory analyses. In contrast, the direct method involves exposing precopulatory pairs to the toxicant and recording the time to separation. Direct separation methods have the added advantage of allowing the recovery ability to be measured and allowing an opportunity to observe survival, reproductive behaviour, reproductive success and potential generational effects from parental exposure (Davis 1978, McCahon *et al.* 1991, Blockwell *et al.* 1998).

8.1.5 LOCOMOTION

An animal's ability to move is often impaired at sublethal concentrations. The consequences of this impairment may in turn affect any or all of the behaviours described above. Locomotion is a requirement for such activities as migration, food search, reproduction and predator avoidance. Any changes in locomotion may affect the fitness of the organism and, in turn, lead to effects at the population, community, and ecosystem levels (Baatrup and Bayley 1993).

Due to the size of invertebrates and the complexity of their behaviour, initial studies on locomotion concentrated simply on whether an animal was moving or not, yielding only qualitative observations. New computer-automated video tracking systems have allowed for detailed quantitative analyses of movement using such parameters as number of reversals, average rate of movement and number of animals moving (Dusenbery 1992). Computer tracking of invertebrates may serve as a powerful biomonitoring tool, alleviating the need for more lengthy, tedious and expensive toxicity endpoints such as mortality and reproduction. An exhaustive treatment of this argument is given in Chapter 5.

8.2 CASE STUDIES

8.2.1 SEDIMENT ORGANISMS

A concise overview of the sediment organisms and chemicals reported in the following section is provided in Table 8.2. Burrowing behaviour is the main parameter used as an endpoint for assessing the effects of the environmental toxicants on these organisms.

8.2.1.1 Earthworms

Currently, little is known about freshwater oligochaete behavioural responses to contaminated sediment. Burrowing may be an excellent response because

Table 8.2 Sediment organisms

Organism	Species	Toxicants Tested	Reference
Amphipod	*Heterophoxus videns, Monoculodus seabriculosus*	PCBs, metals, hydrocarbons	Lenihan *et al.* (1995)
Crustacean	*Eudorella splendida*	PCBs, metals, hydrocarbons	Lenihan *et al.* (1995)
Heart urchin	*Abatus shackletoni*	PCBs, metals, hydrocarbons	Lenihan *et al.* (1995)
Isopod	*Saduria entomon*	Copper and cadmium	Pynnönen (1996)
Mayfly	*Drunella coloradensis*	Acid	Pennuto and de Noyelles (1993)
Oligochaete	*Stylodrilus heringianus Limnodrilus hoffmeisteri*	Variety, endrin	Keilty *et al.* (1988), White and Keilty (1988)
Polychaete	*Nereis virens Nereis versicolor*	Endosulfan, chlordane, dieldrin, DDT, endrin, parathion, methyl parathion, malathion	McLeese *et al.* (1982) Møhlenbrg and Kiørboe (1983)
Tanaid	*Nototanais dimorphous*	PCBs, metals, hydrocarbons	Lenihan *et al.* (1995)

the endpoint is readily identifiable and quantifiable. Furthermore, observations on burrowing behaviour allow for the collection of nondestructive time series data, making it a promising endpoint for assessing pollutant impacts.

The burrowing behaviour of the freshwater sediment-dwelling earthworm *Stylodrilus heringianus* was studied in response to contaminated deposits containing a variety of toxicants along the Detroit River (White and Keilty 1988). Like its terrestrial counterparts, *Stylodrilus* feeds on deposits within sediments helping to maintain balance in nutrient cycling. In the study, ten worms were added to 25 ml of sediment covered with water. Thereafter, burrowing behaviour was observed at regular intervals up to 96 hours. In control sediments, worms were seen quickly to burrow into the sediments and remain buried throughout the entire experiment. In the most contaminated sediments, on the other hand, only four worms initially burrowed and, by the end of the experiment, eight worms were seen on the sediment surface. In all cases, worms were observed to burrow in a concentration-dependent fashion.

Sediment contaminated by endosulfan, chlordane, dieldrin, DDT, or endrin caused the marine polychaete, *Nereis virens* to emerge (McLeese *et al.* 1982). Similar results occurred for the related species, *N. versicolor* exposed to parathion, methyl parathion and malathion (Møhlenberg and Kiørboe 1983). In this case, exposure resulted in almost complete cessation of burrowing. Keilty *et al.* (1988) exposed the freshwater oligochaetes *Stylodrilus heringianus* and *Limnodrilus hoffmeisteri* to sediments containing between 0 and 1848 μg/g dry wt. endrin. 96 h EC_{50} burrowing responses for both species (mean 17.2 μg/g for *S. heringianus* and 59 μg/g for *L. hoffmeisteri*) were roughly 46 and 150 times lower than the respective 96 h LC_{50}. Both species initially burrowed into the contaminated sediment and then returned to the surface in numbers somewhat proportional to the sediment concentration and the length of exposure. However, *S. heringianus* appeared to be more sensitive to endrin based on the time series analyses: in fact, *S. heringianus* emerged and remained on the surface after only 8 h exposure whereas *L. hoffmeisteri* returned to the surface after 32–48 h. The authors admit that the underlying mechanism for this was not tested but they suspected that the abilities of each species to regulate respiration may be a factor because *L. hoffmeisteri* is an oxy-regulator and *S. heringianus* a partial regulator. Regardless, future use of oligochaete behavioural responses to sublethal sediment contamination is promising.

8.2.1.2 Mayfly

To examine the potential trade-off in behavioural responses, Pennuto and de Noyelles (1993) studied the avoidance responses of the mayfly (*Drunella coloradensis*) to pH stress. Scan sampling of behaviours was performed at 5-min intervals during the first 2 h following acid additions. The behaviours of sitting, ventilating, burrowing (defined for a nymph not visible during the scan), drifting, crawling, and crawling while ventilating were recorded. Acid additions influenced the frequency of behaviours but did not prevent or change any behaviour (i.e., all nymphs exhibited all behaviours under all treatments but at different frequencies). Sitting frequency declined with decreased pH; control nymphs (pH 7.92) were sitting during 75% of the observations, more than both the treatment groups (pH 7.01 and 6.02) combined. Sitting was the most frequent behaviour for the 7.01 pH group. Burrowing behaviour increased at intermediate acidity levels and then declined back to control frequencies at higher acid treatment levels; burrowing behaviour was recorded for 20% of the observations at intermediate pH but less than 10% of the controls and low pH groups. Not surprisingly, drifting behaviour occurred more frequently in the low pH group, where a six-fold increase occurred compared with either the control or intermediate pH group. Drifting was the third most frequent behaviour observed in low pH nymphs (behind crawling and ventilation) but the least frequent in controls. Drift behaviour was not significant at intermediate pH levels. The authors concluded that burrowing behaviour may negate the consequences of increased crawling and drifting due to acidic stress. At intermediate pH, burrowing may be an effective avoidance mechanism from pH-stress, that is risk free in terms of predation if predators are less likely to burrow than are prey. The value of burrowing as an evasion tactic depends on the environmental conditions of the substrate, and the combination of biotic and abiotic factors such as morphology (e.g. animals with fusiform bodies may burrow better than dorso-ventrally flattened organisms), drifting costs, and predation risks.

Although the above work showed the utility of burrowing behaviour in avoiding water-borne toxicity, most studies on burrowing behaviour on aquatic organisms have focused on: (1) sediment or soil toxicity with bivalve molluscs and burrowing crustaceans (Phelps *et al.* 1983) or (2) exposure to toxicants prior to initiation of burrowing activity (Stirling 1975, Eldon and Kristoffersson 1978, McGreer 1979, Eldon *et al.* 1980). Many contaminants including heavy metals, oil, pesticides and phenols have been shown to reduce or completely inhibit burrowing activity in invertebrates (Percy 1976, Møhlenberg and Kiørboe 1983, Olla *et al.* 1983, 1984).

8.2.1.3 Crustaceans

Many burrowing bioassays are carried out in the laboratory and, although laboratory bioassays are a useful tool for evaluating the impacts of sediment and soil toxicology, the relationships between physical, chemical and biological variables under laboratory conditions and in the field have been poorly explored. Lenihan *et al.* (1995) examined the applicability of standard laboratory bioassays as tools for evaluating and predicting the impacts of contaminants on the structure and function of natural communities. They used a model system composed of the infaunal species, communities and sedimentary habitats located along a steep and well-defined pollution gradient of petroleum hydrocarbons at McMurdo Station, Antarctica. The main test organism, the amphipod *Heterophoxus videns*, feeds on other crustaceans and larvae in soft sediments. Burrowing preferences of *H. videns* were monitored, along with the heart urchin *Abatus shackletoni*, the tanaid crustacean *Nototanais dimorphous*, another amphipod *Monoculodes scabriculosus*, and the crustacean *Eudorella splendida*. In a 24-day bioassay, the number of *H. videns* not on or above the sediment surface was recorded every day as buried. *M. scabriculosus* and *A. shackletoni* were assayed identically except that the test period ended after the seventh day. In all cases, the organisms showed some avoidance of contaminated sediments by burrowing less often than animals in control sediments.

Another test in the same study was designed to compare the preference of the organisms for contaminated versus uncontaminated sediments. Test organisms were placed on the contaminated side of a divided container with two sediment types, control and contaminated. The number of burrowed organisms were then observed in each compartment. Although *M. scabriculosus* showed no preference, both *H. videns* and *E. splendida* chose the uncontaminated sediments. In addition, the burrowing behaviour of the tanaid, *N. dimorphus*, was observed. In uncontaminated sediments, all tanaids were buried, while 35% of the animals were not buried in contaminated sediments and were observed either to swim or climb the walls of the aquarium in an attempt to avoid the contaminated sediments.

Pynnönen (1996) demonstrated the possible effects of sediment-bound heavy metals on the burrowing activity of the Baltic Sea scavenging isopod *Saduria entomon* L. Sediments contaminated by copper (10 μg/g dry sediment) or cadmium (35 μg/g) significantly decreased burrowing activity after only 10 min, while iron (200 μg/g) decreased the burrowing activity of the isopod after 60 min exposure. The same metal concentrations were avoided when *S. entomon* was offered a choice between contaminated and uncontaminated sediment. A pre-exposure period of 6 days to 100 μg/L Cu affected the capability of *S. entomon* to detect and avoid 1/10th of the concentration in sediment. Pre-exposure also decreased burrowing activity with significantly more than 30% of the pre-exposed group failing to burrow. *S. entomon* was particularly sensitive to Cu. Even without a pre-exposure period, 17% of the animals exposed to the Cu-contaminated sediment did not burrow, compared with 2-5% of test groups exposed to the other metals. Emergence after initial burrowing was common only for isopods tested with Cu.

Perhaps one of the most interesting studies on horizontal migration involves the semiterrestrial amphipod, *Orchestia bottae* that inhabit the supralittoral zone of the Black Sea. These crustaceans attempt to reach shore when swept out to sea through orientating themselves according to the pattern of sky polarisation (i.e., they exhibit true solar-compass orientation; Papi 1960, Pardi 1960). If, however, they are kept in water for 60-80 min, their initial 'landward' movement is inverted to a 'seaward' migration (Gnyubkin 1995). In a novel experiment, Gnyubkin (1995) developed a bioassay to test the time of inversion for *O. bottae* exposed to Cu ions. The amphipods were collected in the field and immediately placed (500-800 animals/dish) into round glass dishes containing either pure sea water, or 0, 0.1, 0.2, 0.5, 1, or 10 mg/L Cu. Each experiment lasted for 1.5 hours. Throughout the experiment, amphipod reactions were recorded by sketching the outlines of animal aggregations at intervals of 5-15 min. These observations were converted to inversion of orientation or no inversion and the ratios were compared using χ^2. Copper caused an inversion within 1-20 min of exposure; the EC_{50} was between 0.1-0.2 mg/l.

8.2.2 AQUATIC ORGANISMS

A concise overview of the aquatic organisms and chemicals reported in the following section is provided in Table 8.3.

8.2.2.1 Zooplankton

Ciliated protozoa such as the freshwater *Tetrahymena thermophila* are a numerous and represent an ecologically significant portion of an aquatic ecosystem. Because of their simple culturing procedures, laboratory maintenance, and short generation time, protozoans make excellent test organisms. As prey, they serve as a necessary link in aquatic food webs and are commonly used as biomonitors of water quality.

Table 8.3 Aquatic organisms

Organism	Species	Toxicants Tested	Endpoint	Reference
Amphipod	*Orchestia bottae*	Copper	Phototactic behaviour	Gnyubkin (1995)
	Gammarus oceanicus	Aroclor 1254	Mate guarding behaviour	Wildish (1972)
	Gammarus pulex	Cadmium, atrazine, copper, lindane and 3,4-dichloroaniline, ammonia, dissolved oxygen, acid, aluminum, lime	Mate guarding behaviour	Poulton and Pascoe (1990), Pascoe *et al*. (1994), McCahon *et al*. (1991), McCahon and Poulton (1991)
	Anisogammarus pugettensis	Bleached kraft mill effluent	Mate guarding behaviour	Davis (1978)
	Hyallela azteca	Lindane	Mate guarding behaviour	Blockwell *et al*. (1998), (1999)
Barnacle nauplii	*Balanus* sp.	Cadmium, copper	Swimming activity, phototactic behaviour	Lam *et al*. (2000), Lang *et al*. (1980)
Ciliate	*Paramecium aurelia*	Copper, cadmium, mercury, PCP, chlorpyrifos, phenol, naphthol	Feeding and ingestion rate	Juchelka and Snell (1995)
Cladoceran	*Ceriodaphnia dubia*	Copper, cadmium, mercury, PCP, chlorpyrifos, phenol, naphthol	Feeding and ingestion rate	Juchelka and Snell (1995)
Daphnid	*Daphnia pulex*	Carbaryl and kairomone	Swimming activity	Dodson *et al*. (1995)
	Daphnia pulicaria	Sodium bromide	Phototactic behaviour	Drake and McNaught (1994)
	Daphnia magna	Copper, lindane, endosulfan and diazinon	Phototactic behaviour Feeding and ingestion rate	Flickinger *et al*. (1982), Goodrich and Lech (1990)
	Daphnia catawba	SDS	Filtering rate	Jones *et al*. (1991)

Midge	*Chironomus riparius*	Cadmium	Ovipositing behaviour	Williams *et al. (1987)*
Mollusc	*Mytilus edulis*	Cadmium, copper	Valve closure	Sunila (1981), Davenport and Manley (1978), Kramer *et al.* (1989)
	Scrobicularia plana	Zinc	Valve closure	Akberali *et al.* (1981)
	Corbicula fluminea	Chlorine, cadmium and zinc	Valve closure	Doherty *et al.* (1987)
	Sphaerium corneum	Trichlorophenol and hypoxia	Valve closure	Heinonen *et al.* (1997)
	Modiolus sp., *Chlamys opercularis, Crassostrea gigas, Anadara senilis,*	Copper	Valve closure	Manley and Davenport (1979)
	Dreissena polymorpha	Trichloroethylene and PCP, copper and chlorine	Valve closure	Sloof *et al.* (1983), Kramer *et al.* (1989)
Mollusc juvenile	*Villosa iris, Adodonta grandis*	Copper	Valve closure	Jacobson *et al.* (1993)
Protozoa	*Tetrahymena thermophila*	Sodium chloride and guaiacol	Swimming activity	Gilron *et al.* (1999)
Rotifer	*Brachionus calyciflorus*	Copper, PCP, lindane	Swimming activity	Charoy and Janssen (1999)
	Brachionus sp.	Copper, cadmium, mercury, PCP, chlorpyrifos, phenol,	Predator–prey interactions	Preston *et al.* (1999), Juchelka and Snell (1995)
	Asplancha girodi	PCP	Predator-prey interactions	Gómez *et al.* (1997)

One study looked at the chemotactic response of *T. thermophila* to two reference toxicants, sodium chloride and guaiacol (Gilron *et al.* 1999). The behavioural response was the ability of the protozoa to sense and either avoid or move toward the chemicals of interest. A *T*-maze apparatus was used as the testing arena. In each test, one arm contained uncontaminated water while the other arm was filled with the reference solution. Protozoa were then added into the middle stopcock barrel and allowed to migrate to either arm. After 20 min, the stopcocks were closed and cells in each arm were enumerated. Both reference solutions caused consistent responses in that protozoa moved toward sodium chloride and away from guaiacol. Because the test is so easy to control, inexpensive, and gives an indication of sublethal effects, the authors suggested further validation of the method as a possible standard test using pulp mill effluents.

8.2.2.2 Rotifers

Rotifers play an important role in determination of water quality by feeding on algae, recycling nutrients with a high turnover and assimilation rate, and acting as prey for insects and fish (Snell and Janssen 1995). Besides their ecological significance, rotifers lend themselves well to toxicological studies for several reasons including the formation of dormant embryos, called cysts, which allow for synchronised hatching of neonates after prolonged storage. A short life and reproductive cycle, asexual reproduction, and small size also offer advantages in ecotoxicological testing. In fact, American Society for Testing and Materials' (ASTM) standards exist for the freshwater rotifer, *Brachionus calyciflorus* (Figure 8.2), and the marine rotifer, *Brachionus plicatilis* (ASTM 1998a).

In one study, the effects of copper, sodium pentachlorophenol (PCP), and lindane on the swimming behaviour of the freshwater rotifer, *B. calyciflorus* (Charoy and Janssen 1999) were investigated. In rotifers, swimming is the result of the coordinated movement of cilia. This coordinated movement can be changed if either the cilia's movements or the coordination of these movements are affected by toxicants. As with all of the organisms described in these case studies, a change in locomotion may lead to deleterious changes in individual

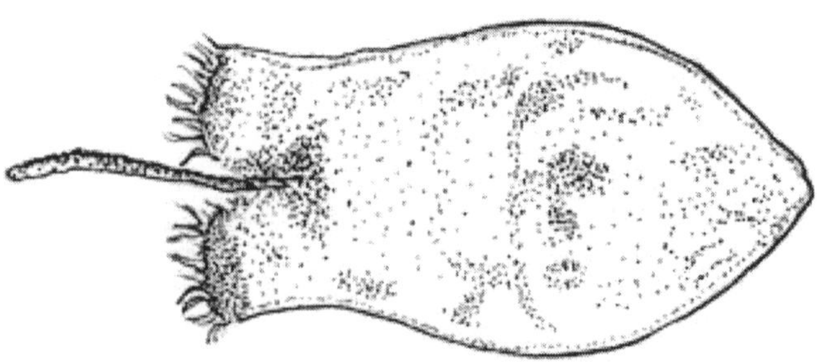

Figure 8.2 *Brachionus calyciflorus* ingesting rod-shaped bacterium.

and population growth, reproduction, escape from predators and, in turn, survival of the population.

In the study, rotifers were exposed to the aforementioned chemicals for 5 min, 2 h, or 6 h. After the exposure period, an individual's movement was recorded with a video camera for 10 minutes. After tracking, the videotape was analysed for two swimming parameters: speed ($mm{\cdot}s^{-1}$) and sinuosity ($rad{\cdot}mm^{-1/2}$). Although times when animals were not moving were not included in the speed and sinuosity parameters, a third test parameter called period(s) of swimming was also calculated. In short, the swimming activity of rotifers was affected, in a concentration-dependent and time-dependent fashion, after at least a 2-h exposure.

Predator–prey interactions can also give estimates of sublethal toxicant effects that might lead to adverse effects at the population and community levels. The swimming behaviour of *B. calyciflorus* was monitored as previously described and related to the predatory behaviour of *Asplancha girodi*, a predatory rotifer (Preston *et al.* 1999). In unpolluted environments, rotifers are capable of avoiding predation by decreasing their swimming speed in the presence of predators. When *B. calyciflorus* was treated with PCP, however, swimming speed was increased, both in the presence and absence of *A. girodi*. Without exposure to PCP, *B. calyciflorus* decreased their swimming speed in the presence of *A. girodi*, as expected. In effect, sublethal PCP exposures led to a 50% increase in predation.

In a related study, the predatory behaviour of *A. girodi* was used as an endpoint of toxicity (Gómez *et al.* 1997). Treatments consisted of PCP exposure of prey and predator, only prey, or only predator. As above, there was a higher chance of encounter if only the prey was exposed. However, when the predator was exposed, encounters decreased in a concentration-dependent fashion. Decreased encounters were also correlated with decreased numbers of captures, ingestions, and times to first ingestion.

8.2.2.3 Amphipods

Wildish (1972) studied interruption of mate guarding behaviour in *Gammarus oceanicus* exposed to 0.021 mg/l of the PCB, Aroclor 1254. This response was more sensitive than acute toxicity tests in which 50% of the animals died at 2 mg/l in 35 days. Linden (1976) separated pairs of *Gammarus oceanicus* and observed that they failed to recouple in oil concentrations of 10 μl/l and higher. Davis (1978) observed direct disruption of precopula by bleached kraft mill effluent of the marine amphipod, *Anisogammarus pugettensis*, in a 117 h bioassay. When dosing ceased and the dissolved oxygen concentration was allowed to return to normal levels, 65% of pairs recoupled within 11 h.

Poulton and Pascoe (1990) created a standardised disruption bioassay (the *Gammarus pulex* Precopula Separation — GaPPS — test) to evaluate the sublethal effects of toxicants. The authors describe both forms (direct and induced separation) but for purposes of brevity, we will discuss only the Induced Separation Test (IST). Only 10% of pairs separated during the 24-h exposure to Cd prior to induced separation. There was a concentration–response relationship with increasing Cd concentration resulting in faster induced separation times. This approach would appear promising in the arena of risk assessment because it allows researchers to relate quickly and at minimal cost traditional, deterministic endpoints (e.g., LC_{50}) to sensitive and nonlethal endpoints (e.g., behaviour) that could result in probabilistic endpoints useful for risk management scenarios. In this case the 96-h LC_{50} value of 44 μg/l Cd corresponded to an IST_{50} of 315 s. The success of such an approach will depend, however, on developing a consistent, highly correlated relationship between lethal toxicity and separation times across a range of toxicant types.

The GaPPS bioassay has been further tested in the laboratory with four toxicants (3,4-dichloroaniline, atrazine, Cu and lindane) and in the field at sites subject to industrial, agricultural, sewage and domestic effluents (Pascoe *et al.* 1994). For each chemical tested, there was a clear reduction in separation time with increasing concentration; there was also a threshold concentration for each chemical, below which treated groups were not significantly different from controls. As for the field validation part of the study, amphipods in the upstream reference section separated in 992 s while those exposed to treated effluent had an IST_{50} of 630 seconds. This pattern was repeated downstream at other effluent sites, which suggests that changes in water quality that appear to be directly related to a variety of effluents may be effectively 'tracked' with this method.

Most of the work on amphipod mate guarding as a bioassay has focused on *Gammarus* spp., especially *pulex*. More recently, however, work has started on species native to North and South America, such as *Hyallela azteca*. Two sublethal toxicity bioassays, modified from the work of Poulton and Pascoe (1990), have been developed for *H. azteca* (Blockwell *et al.* 1998). Blockwell *et al.* (1998) investigated the disruption of precopulatory pairs of *H. azteca* as a direct result of exposure to lindane, indirectly following a timed exposure to lindane, and subsequent separation in an aqueous invertebrate anaesthetic medium. At the end of the 24-h exposure, the percentage of pairs still coupled ranged from 26% in controls to 0% for the two highest exposures. After a 24-h recovery period, however, 20% of the pairs recoupled in the controls and the two lowest treatments.

For the induced separation experiment, the number of pairs following the 4-h exposure to control or lindane was significantly reduced. There was a significant reduction in the IST_{50} with increasing lindane concentration. The LOECs for both the direct and indirect methods was roughly comparable at 17.3 μg/l for induced and 24.4 μg/l for the direct separation. This is approximately twice the sensitivity of a similar bioassay (IST_{50}) using *G. pulex* and lindane (Pascoe *et al.* 1994).

The study by Blockwell *et al.* (1998) looked at acute toxicity of lindane on *H. azteca*. Follow-up work looked at the chronic toxicity of lindane in an attempt to link the effects of toxicants at the individual and population levels by examining the growth, reproductive behaviour and reproductive success of individuals (Blockwell *et al.* 1999). This experiment examined the effects of lindane on *H. azteca* exposed for 35 days. Reproductive behaviour was examined at 16 days and at the conclusion of the study. Lindane exposure significantly reduced mate-guarding behaviour after 16 days of exposure; this effect continued at 35 days of exposure, although there were roughly twice as many mate-guarding pairs at 35 days then at 19 days for both the control and low treatment group. The authors concluded that the reduction in population growth during the study was due to the interactive toxic effects on reproductive behavioural patterns, growth of recruited individuals, and delayed sexual development.

These studies demonstrate that this bioassay appears to be adaptable for use on test species occurring in many habitats of interest and provides rapid information at a low cost and with a minimum of effort. Furthermore, reproductive behavioural studies on species with relatively short generation times, such as many invertebrates, can reveal ecotoxicologically useful knowledge of toxicity processes capable of linking behavioural effects at the individual level to population levels.

8.2.2.4 Daphnids

The freshwater zooplankton *Daphnia pulex* is an integral part of the aquatic ecosystem in that they serve as primary herbivores thus maintaining water quality (Figure 8.3). *Daphnia* were used to test the effects of the pesticide Carbaryl and the naturally occurring kairomone (Dodson *et al.* 1995). Kairomones, released by midge larvae, are a chemical signal from predators that prompt *Daphnia* neonates to develop in a way that makes them more likely to escape predation.

Using a video camera and a mirror, *Daphnia* swimming behaviour was recorded in three dimensions. After taping, videos were digitised and analysed to generate nine different parameters of swimming behaviour. Of those nine, velocity, turning angle, upward hopping angle, downward sinking angle, variance in vertical position and hopping rate were determined to be, in order, the most useful. Three separate experiments were described to test how the chemicals might affect swimming behaviour and escape from predation by fish. In short, while kairomones had no significant effect, carbaryl not only altered swimming behaviour but also resulted in a spinning behaviour that made *Daphnia* more susceptible to being preyed upon by fish.

Drake and McNaught (1994) investigated the effects of the mild sedative sodium bromide on the response of *Daphnia pulicaria* to ALD, the potential steering cue for vertical migration. The authors modified the design of Ringelberg (1964) by mounting the observation chamber (a sealed petri dish with a

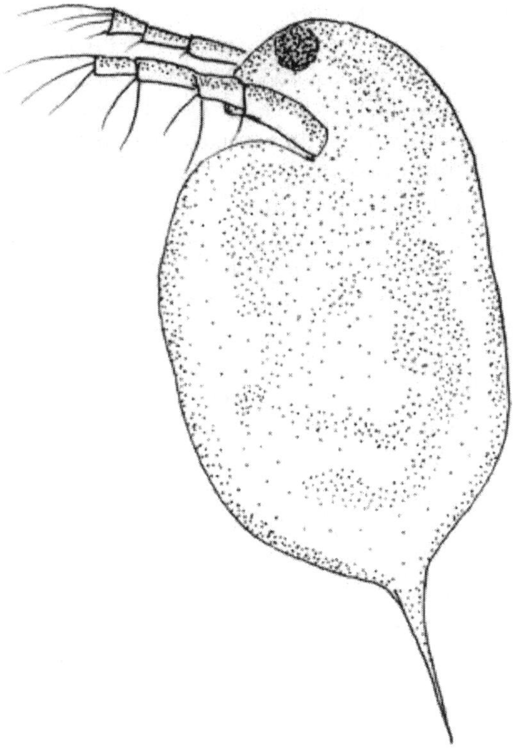

Figure 8.3 *Daphnia pulex*.

small hole located on the edge for filling, draining and loading animals into the chamber) vertically instead of horizontally. The observation chamber was filled with sodium bromide; daphnids were introduced and allowed to acclimate for 10 min. The direction of swimming for each animal was recorded at 5, 10 and 15 min on a standard scoring sheet. This simple system produced an EC_{50} value (based on a reduction of 50% of animals within the 180° window) of 4.9×10^{-6} M, making the bioassay 225 and 56,900 times more sensitive than assays based on reproduction or lethality, respectively (Canton *et al.* 1983).

A much more complex testing system was used by Flickinger *et al.* (1982) to study phototactic behaviour of *Daphnia magna* exposed to Cu for either 88 (first chronic test) or 41 days (second chronic test). The testing chamber was a vertical glass tube, graduated in cm, fitted at the top with a microscope illuminator. Two glass fittings near the top allowed for the introduction of daphnids and the outflow of water, respectively. Behavioural testing was done in a darkened room. Daphnids were allowed 25 min to acclimate to the lighting conditions in the room and then were inserted in the tube at a uniform depth. Thereafter at intervals, depth in the tube was recorded to the nearest cm. After 90 s, a pump was turned on to return the daphnid to the top of the tube for removal and to circulate water before the next daphnid was tested. In six experiments, daphnids exposed to Cu tended to descend less than control daphnids. Copper exposure significantly reduced phototactic response, a result that could not be explained by differences in body size except for 40 μg/l where Cu exposure reduced biomass, as measured by carapace length and mean spine ratio, by 44% and 45% respectively. Changes in phototaxis were apparent within 4 to 6 days of exposures.

Goodrich and Lech (1990) studied changes in orientation of *Daphnia magna* exposed to lindane under laboratory conditions. The modified testing chamber was a circular plexiglass arena divided into eight equal partitions with plexiglass dividers. A polyvinyl chloride (PVC) pipe was inserted into the centre of the chamber to seal the partitions and prevent the early release or escape of the daphnids. A microscope illuminator equipped with a tungsten bulb was positioned to produce a light beam 60° off the horizontal plane, directed to the centre of the chamber. This produced a characteristic and repeatable light. Daphnids exposed to lindane for 24 h were introduced into the testing chamber through the PVC pipe and allowed to acclimate for 10 min in total darkness, after which time the light source was started and the pipe removed. Zooplankton were allowed to swim freely for 1 h and then the pipe was replaced, trapping the zooplankton in either the centre of the tank or in a partition. This procedure resulted in three endpoints: (1) mean angle (average direction for the migration, (2) r-value (an index of the magnitude of random, undirected migration) and (3) percentage of nonresponse (percentage of animals in the centre of the chamber). This approach can also test whether immobilisation or cueing failure occurs when variable periods of illuminance are used. Lindane was found to create a positive concentration-response with a sharp increase in random, undirected migration with increasing concentration. Apparently, daphnids recognised the light cue as evidenced by similar percentages of nonresponse for the control and treated groups. Lindane appeared to affect the physiology that directs migratory direction, resulting in 'misdirection'. Daphnids were able to migrate as seen by high r-values but were not able to migrate in a directed fashion due to a general breakdown in orientative ability as evidenced by the lack of significant differences in grand mean angles for control and exposure groups.

An additional system developed by Dojmi Di Delupis and Rotondo (1988) was employed with the Goodrich and Lech (1990) system to examine if lack of migration is due to immobilisation (kinetic) or failure to respond to the cue (releasing stimuli). In this system, a black PVC chamber (6.3 × 2 × 2.5 cm) divided into two equal cells by a transparent gate was mounted to a stereomicroscope to facilitate visual inspection. One cell had a transparent window at the end through which an optic fibre was attached. The optic fibre connected to an adjustable halogen lamp assures a straight-line diffusion without creating a thermal gradient. The chamber was filled with 20 cc of media appropriate for the invertebrate under study. Organisms were placed in one cell and allowed to migrate for a predetermined time; then, the gate was lowered and the numbers of cells counted.

Ingestion rate is another measure of sublethal toxicant effects. Juchelka and Snell (1995) quantified the ingestion rate of *Ceriodaphnia dubia*, *Paramecium aurelia* and *Brachionus plicatilis* after a 1-h exposure to Cu, Cd, Hg, PCP, chlorpyrifos, phenol and naphthol. A CCD camera interfaced with a microscope captured the magnified images of the zooplankton after ingesting fluorescent beads. The sensitivity of the test varied with the species used. In short, *C. dubia* NOECs for ingestion were less sensitive than published 7-day reproductive tests and 48-h LC_{50} values. For the ciliate, *P. aurelia*, the ingestion NOECs were comparable to published 24-h LC_{50} data. The marine rotifer, *B. plicatilis*, showed less sensitivity with the ingestion test than published 48-h reproductive data but more sensitivity than 24-h LC_{50} values.

Feeding and ingestion rates for both species decreased as Cu concentrations increased. These results are comparable to other studies on *B. calyciflorus* (Capuzzo 1979) and other zooplankton species (Day and Kaushik 1987) exposed to a variety of toxicants.

These studies highlight the main techniques used in assessing filtration starting with the simplest method, visual examination of cell counts. Advances in radiolabelling techniques and electronic particle counters (EPCs) have facilitated the ease of measuring filtration and ingestion in ecotoxicity studies. Additional improvements in video imaging have increased the ease and applicability of visual examination of the organism.

Kaushik *et al.* (1985) used a Haney grazing chamber to study the effects of permethrin on zooplankton and found that individual and community rates of filtration were significantly reduced by concentrations of 0.05 μg/l and greater. Filtration rates of surviving zooplankton in treated containers recovered within one week of treatment. The measurement of changes in filtration rates in the field may be a sensitive indicator of sublethal toxicity and may provide a more rapid monitoring tool for the detection of effects than monitoring changes in population density (Day 1989).

McNaught (1992) compared functional bioassays based on reproduction, respiration and ingestion of various zooplankton exposed to nutrients, heavy metals, pesticides and PCBs, and developed a new ingestion bioassay. In this bioassay, test animals were acclimated to natural food levels of control water for 4 h; meanwhile test concentrations were prepared with control water. An Electronic Particle Counter (EPC) was used to determine particle density for the control and experimental food containers. Zooplankton were transferred from acclimation containers to experimental food containers and allowed to filter for 15 min. Food densities in each container were then determined with the EPC and filtering rate was calculated using the equation of Gauld (1951). McNaught next compared each bioassay type using the following criteria: (1) variability of controls, (2) range of response, and (3) contaminants that altered the response. The ingestion bioassay showed the greatest range of response, detecting differences in pesticides, PCB and heavy metals. In contrast, the reproduction bioassay showed the lowest variability but only detected heavy metals and nutrients. The respiration bioassay also had low variability but could only detect PCB and, even then, only in one case.

Because of advancements in video imaging, it is now feasible actually to measure the Thoracic Appendage Rate (TAR) in daphnids while they feed (Jones *et al.* 1991). To record TAR, it is necessary to tether daphnids to a wire. The mounted daphnid is then placed in the testing chamber and viewed with cold, indirect halogen lighting at a magnification sufficient to determine movement of the thoracic appendage. TAR has been measured by filming the appendages with a video camera coupled to a dissecting microscope. The recorded sequences were replayed slowly enough to observe individual movements of the thoracic appendages within the carapace. Counting for each interval ceased when the *Daphnia* rejected particles or when 50 beats were recorded, whichever happened first. The number of postabdominal rejections in a 3-min interval was also determined for each replicate animal.

Jones *et al.* (1991) used the TAR method to test sublethal and lethal levels of the surfactant, sodium dodecyl sulphate (SDS), on *Daphnia catawba*. In addition, they tested the effect of tethering on filtration ability. Both exposure time and concentration of SDS were associated with significant decreases in TAR of daphnids. After 6 h there were significant declines in TAR with increasing SDS concentration and after 30 h differences were even greater. Not surprisingly, the filtering rates of tethered animals were significantly depressed. The TAR rate for tethered animals was also probably lower but this cannot be tested in free-swimming zooplankton. These results indicate that while the magnitude of the response will be affected by the procedure of tethering, the general relationship between TAR and toxicant level is not.

8.2.2.5 Barnacle nauplii

Another study with zooplankton observed the swimming activity and photobehaviour of nauplii of the acorn barnacle *Balanus improvisus* after exposure to sublethal concentrations of Cu (Lang *et al.* 1980). A similar set-up was used as that described for *Daphnia* except that activity was only recorded from above in two dimensions to derive behavioural responses to a horizontal light stimulus: mean linear velocity (photokinesis) and direction of travel (phototaxis). After 24-h sublethal exposures to Cu, increased concentrations resulted in decreased swimming speeds and increased photoreversals. Normally, nauplii were observed to have a strong positive phototactic response.

A similar study using stage II nauplii of the barnacle *B. amphitrite* investigated the swimming behaviour changes and moulting successes after exposures to Cd (Lam *et al.* 2000). After a 72-h exposure period the percentage of larvae successfully moulted to stage III was determined using a microscope. An increase in sublethal Cd concentrations resulted in a decreased moulting success. Swimming behaviour was assessed using computer tracking coupled to a stereo-microscope after 24 and 48 h. At both exposure times mean curvilinear velocity and mean straight-line velocity decreased during a 1-min test.

8.2.2.6 Chironomids

To determine if *Chironomus riparius* ovipositing adults can detect and avoid Cd, Williams *et al.* (1987) tested several hundred adults. Newly emerged adults were introduced into a chamber containing randomly arranged dishes with cadmium solution. The arrangement of dishes was changed daily to avoid positioning effects. The number of egg ropes oviposited in each Cd solution was recorded every 24 h for 10 days. After 10 days, the cumulative number of egg ropes was under 50 for the two highest concentrations, which are, in fact, lethal to first instar larvae, while the two lowest concentrations and the control were over 100 ropes. The mechanism by which insects such as chironomids discriminate between the available solutions is not clear but Sutcliffe and Carrick (1973) proposed that insect antennae contain chemoreceptors that may be involved in determining water quality.

8.2.2.7 Caddisflies

Décamps *et al.* (1973) were the first to report on structural irregularities in the nets of caddisflies that were exposed to DDT and dieldrin under laboratory conditions. Subsequent studies have shown that symmetry size and regularity of the prey capture net are adversely affected by fenethcarb (Besch *et al.* 1977) and Cu (Besch *et al.* 1979). Pascoe *et al.* (1991) studied net-spinning activity in 146 fifth instar *Hydropsyche angustipennis* and found that net spinning was both velocity and temperature dependent with most nets constructed at 40 cm/s at 20 °C. When Cu (0, 13, 64, 130, and 640 μg/l) was introduced into the test channels net-spinning was reduced in a nearly linear fashion with increasing Cu concentrations. Additionally, Cu concentrations >64 μg/l resulted in nets with a larger mesh size and smaller number of meshes per unit measured.

Besch *et al.* (1979) reported that their laboratory findings were confirmed by field observations but it was not until the work of Petersen and Petersen (1983) that data on the frequencies of net anomalies collected from natural populations were described. These researchers examined 'crossovers', areas in the net '. . .where the rectilinear structure of the net opening is distorted by either strands being fused or added over the orifice'. They found that net-spinning caddisflies (*Hydropsyche*) exhibited anomalies in the structure of their capture nets when collected directly downstream from heavy metal and toxic chemical effluents. These anomalies increased the closer the caddisflies were collected from a point source. Crossover frequency per net was positively correlated with the number of nets showing this anomaly but crossover frequency was negatively correlated to total macroinvertebrate community structure as measured by Shannon diversity.

Most case-building caddisflies increase case size by addition of material as the caddisflies mature; however, certain species such as *Agapetus fuscipes* leave old cases prior to the construction of new ones. This leaves the larvae more vulnerable to predation (Otto and Svensson 1980, Johansson and Johansson 1992) and harmful impacts from toxicant exposure (McCahon *et al.* 1989) during the case-building phase. Furthermore, adverse population effects may occur if case-building speed or efficiency is reduced by toxicants. To test this hypothesis, McCahon *et al.* (1989) tested first, third, and fourth instar *A. fuscipes* exposed to Cd concentrations of 0, 46, 97, and 186 mg Cd/l for 24 h. Caddisflies were removed from their cases during the exposure period and then provided with disrupted case material or fine sand grains and their attempts to rebuild their cases were observed. The ability of caddisflies to construct cases dramatically decreased as the concentration of Cd increased. Moreover, the structure and strength of the new cases also deteriorated, with caddisflies exposed to the highest concentrations creating cases that were actually a loose collection of material.

8.2.2.8 Molluscs

Bivalve molluscs are widely used as indicator organisms in natural settings for monitoring water quality (Sloof *et al.* 1983, Kramer *et al.* 1989, Camusso *et al.* 1994), as well as test organisms in the laboratory (Doherty *et al.* 1987, Mäkelä *et al.* 1991). Several bivalve species are excellent candidates for biomonitoring because they are sedentary, abundant and available throughout the year. They are of a manageable size and hardy enough to be handled and cultured in laboratory settings. Lastly, bivalves offer many possibilities with respect to continuous and automatic detection, reliable and fast response, handling and data interpretation (Koeman *et al.* 1978, Cairns 1979).

Most mussel species keep their shells open continuously for respiration and feeding. Thus valve closure is considered a general stress response in both marine (Manley 1983, Akberali and Trueman 1985) and freshwater bivalves (Kramer *et al.* 1989). Closure isolates soft tissues from the external environment and leads to a gradual decline of oxygen in the mantle cavity water. Although bivalves can lower metabolic rate during anaerobiosis to save energy reserves (Holopainen and Penttinen 1993), valve closure also results in cessation of feeding, which combined with the accumulation of excretory products limits the time the valves can remain closed (Akberali and Trueman 1985).

Valve closure has been studied in *Mytilus edulis* exposed to Cd (Sunila 1981) and Cu (Davenport and Manley 1978) and the marine bivalve *Scrobicularia plana* exposed to Zn (Akberali *et al.* 1981). Subsequently, Doherty *et al.* (1987) showed that Asiatic clams (*Corbicula fluminea*) exposed to chlorine for 14 to 20 days kept their valves sealed for extended periods, which weakened the clams to the point where the action of antifouling molluscicides were enhanced.

The effects of hypoxia and 2,4,5-trichlorophenol (TCP) on the behaviour of the freshwater clam *Sphaerium corneum* (L.) and the accumulation and elimination rates of TCP were studied in the laboratory by Heinonen *et al.* (1997). Valve closure was observed at 30-min intervals during two 12-h monitoring periods. After an initial 'hyperactivity' period, clams under hypoxia held their valves closed for longer than clams maintained under normoxia. At the end of the 36-h exposure period, the number of individuals with closed valves increased under normoxic conditions as well. TCP was rapidly accumulated in clam tissue in normoxia. A significant correlation existed between body burden and length of valve closure during exposure, strongly suggesting that, at least in short-term exposures, bivalves may reduce bioaccumulation by closure of their valves.

The valve movement patterns of immobilised Asiatic clams were monitored during exposure to constant concentrations of Cd or Zn for 24 h after a 24-h acclimation period (Doherty *et al.* 1987). Response duration was concentration-dependent with bivalve closures increasing as the metal concentration increased. Consistent behaviour occurred for both mean times to first closure after exposure and mean time per valve opening over a 24-h exposure. There was no association between duration of valve-closure periods and Cd concentrations; but valve-closure periods were directly correlated with Zn exposure. The authors concluded that at low levels of Cd or Zn contamination, the valve closure response was not likely to interfere with any bioaccumulation processes.

In order for an endpoint to become readily accepted by both the scientific and regulatory community, the endpoint must be easy to discern and record. Automated methods to record behaviour show promise to advance the field of behavioural toxicology. Marceau (1909) first used sooted glass and a mechanical system to record the valve movement of mussels; other comparable systems have been described (Salanki and Balla 1964). Other researchers have developed strain gauges (Djangmah *et al.* 1979, Higgins 1980). Manley and Davenport (1979) used a strain gauge to measure the valve closure activity in the marine bivalves, *Modiolus modiolus*, *M. demissus*, *Chlamys opercularis*, *Crassostrea gigas* and *Anadara senilis* exposed to Cu concentrations. All species, except *C. opercularis*, exhibited an initial valve adduction, followed by a 'water testing' response, defined as a gradual decrease in gape associated with occasional intermittent partings of the valves. Eventually, total closure of the valves occurred. Interestingly, this general behavioural pattern was not observed for *C. opercularis*, the only mobile species tested. For this species, the only discernible response to increasing Cu concentrations was an increase in valve movement. It may well be that *C. opercularis* was attempting to avoid increased Cu concentrations by swimming but, because the animals were fixed in position, was unable to do so.

Increased interest and improved technology has improved the sensitivity and ease of recording bivalve closure activity. Schuring and Geense (1972) were the first to use electromagnetic induction to record valve responses in *Mytilus edulis* and several methods have been developed (Jenner *et al.* 1989).

Kramer *et al.* (1989) extensively tested the valve movement response in mussels with *Mytilus edulis* and *Dreissena polymorpha* under: (1) continuous single toxicant exposure (*Dreissena*), (2) single toxicant exposure with additions of separate spikes in the lab and in the field (*Mytilus*) and (3) an early warning scenario (*Dreissena*). The time of valve closure for *Dreissena* increased with increasing chlorine concentration until virtually no activity was recorded. For *Mytilus*, addition of 37.5 μg/l Cu resulted in a closure within minutes, while an addition of 20 μg/l Cu resulted in a delayed closure response that was less visible. The recording of mussel activity did detect a sudden closure to an unspecified discharge; a discharge that was also detected by other biomonitoring systems.

Systems based on electromagnetic induction allow the valve movement to be recorded. However, to use such a system over a long-term field study (e.g. early detection system), a massive amount of data is needed accurately to record valve movements, and secondly, an analogue signal must be amplified over the distances needed for field studies. Underwater circuitries such as amplifiers always present installation and maintenance difficulties. Englund *et al.* (1994) developed a simple digital system directly to address the shortcomings of previous systems.

These methods work well for adult mussels that are large enough to be handled and sturdy enough to survive the disturbance required to attach monitoring wires. However, in many organisms, early life stages are particularly sensitive to environmental changes. Toxicity testing with juvenile mussels is difficult due to their small size (<1 mm) and limited availability. Standard methods for conducting static acute behavioural toxicity tests with early larval stages of marine bivalves have been developed (ASTM 1989) but comparable methods for freshwater bivalves do not currently exist. A healthy juvenile is typically active with open valves and an extruded foot. Many species have a juvenile stage characterised by a transparent shell that clearly shows many anatomical features. Consequently, the absence or reduction of movement by the foot or cilia can be used as an endpoint but this requires close inspection of each individual and is time intensive.

Jacobson *et al.* (1993) determined the sensitivities of newly metamorphosed juveniles of *Villosa iris* and *Adodonta grandis* to Cu using two methods of postexposure response: direct visual examination and the use of vital staining with neutral red. Individuals were ranked into one of three categories: (1) brightly stained (unaffected by Cu), (2) lightly or partially stained (affected), or (3) unstained (dead). Both species responded to Cu exposures exhibiting valve closure at low concentrations. EC_{50} values of 29 μg/l were calculated for *V. iris* based on vital staining. *A. grandis* juveniles have amber valves that precluded reliable estimates of effective concentrations based on vital staining.

Therefore, the authors concluded that vital staining is an effective means of determining valve closure only for those organisms that have transparent shells.

Donkin and coworkers (1989) have used Clearance Rate (CR) to measure the effects of hydrophobic chemicals and pesticides on the filtering ability of mussels. In this method, each mussel was held in a standard volume of filtered water or filtered water containing the toxicant. A standard aliquot of algae was introduced. A Coulter particle counter at regular intervals (e.g., 20 min) was used to count water samples for a total time period of 80 min. These counts were used to calculate the clearance rate; values obtained from toxicant-exposed mussels were expressed as a percentage of the mean control value and concentration-response curves were created. This method has been shown to be precise and has been successfully used to determine quantitative structure activity relationships (QSARs) for hydrophobic organic chemicals, alkanes and benzene-substituted alkanes (Donkin *et al.* 1989, 1991).

Resultant QSARs showed that the EC_{50} for filter feeding was negatively correlated with log K_{ow} and positively correlated with aqueous solubility, indicating that hydrophobicity has a major influence on toxicity (Donkin *et al.* 1989). Tested compounds with aqueous solubilities $>70\ \mu g/dm$ were deleterious to feeding activity when bioaccumulated to similar concentrations, whereas compounds of lower solubility were less toxic. The exceptions were the PAHs, pyrene and fluoranthene that were less toxic than predicted from their solubilities. The authors confirmed the hypothesis that toxicity cut-off is due to solubility-related phenomena, the effect perhaps being enhanced for aromatic hydrocarbons dosed near to their limits of solubility, by sequestration of crystals within the mussel tissues. This strongly suggests that many organic contaminants detected by chemical analysis of mussels have no direct effect on filter feeding, whereas the less frequently determined volatile compounds are toxic. This approach to developing QSARs has not been as successful for organochlorine pesticides, AChE-inhibiting compounds and pyrethroids (Donkin *et al.* 1997).

Using discontinuous visual observations, Rodgers *et al.* (1980) described an inverse relationship between Cu concentrations and visually assessed filtering activity by Asiatic clams. Belanger *et al.* (1986a,b) observed significant reductions in siphoning by adult and juvenile Asiatic clams exposed to chrysotile asbestos concentrations ranging from 10^2 to 10^8 fibres/l.

8.2.3 TERRESTRIAL ORGANISMS

A concise overview of the terrestrial organisms and chemicals reported in the following section is provided in Table 8.4.

8.2.3.1 Nematodes

Nematodes make up a major proportion of the soil biota helping to maintain the soil ecosystem balance through decay of organic matter and nutrient cycling

Table 8.4 Soil organisms

Organism	Species	Toxicants tested	Endpoint	Reference
Beetle	seven species	Deltamethrin	Locomotion	Wiles and Jepson (1994)
	Poecilus cupreus	Various pesticides	Locomotion	Römbke and Heimbach (1996)
Caddisfly	*Hydropsyche angustipemnis*	Copper, heavy metal and toxic chemical effluents	Net-spinning activity	Pascoe *et al*. (1991), Petersen and Petersen (1983)
	Agapetus fuscipes	Cadmium	Case-building activity	McCahon *et al*. (1989)
Earthworm	*Polypheretima elongata*	Textile dyes	Burrowing preference and avoidance	Ramaswami and Subburam (1992)
	Eisenia foetida	Benomyl, dieldrin, dimethyl phthalate, fluorene	Burrowing preference and avoidance, migratory activity, escape reflexes	Stephenson *et al*. (1998), Mather and Christensen (1998), Drewes (1997)
Honey bee	*Apis mellifera*	Dicofol, endosulfan, decis, baytroid, sevin	Proboscis conditioning	Stone *et al*. (1997), Abramson *et al*. (1999)
Linyphiid spider	*Oedothorax apicatus*	Deltamethrin	Walking activity	Jagers op Akkerhuis (1994)
Nematode	*Caenorhabditis elegans*	Anaesthics, tumor promoters, copper, beryllium, mercury, lead, cadmium, nickel, malathion, vapona, various organic compounds, ethanol	Locomotion	Morgan and Cascorbi (1985), Miwa *et al*. (1982), Williams and Dusenbery (1987), (1990), Terrill and Dusenbery (1996), Avery and Horvitz (1990), Dhawan *et al*. (1999, 2000), Anderson *et al*. (2001)
	Steinernema sp.	Oxamyl and fenamiphos	Locomotion, nictation, and infectivity	Patel and Wright (1996)
Spider	*Araneus diadematus, Larinioides sclopetarius*	Various pesticides	Web building activity	Samu and Vollrath (1992), Lengwiler and Benz (1994)
Springtail	*Folsomia* sp.	Dimethoate	Locomotion	Sørensen *et al*. (1995), Fábián and Petersen (1994)
Woodlouse	*Oniscus asellus*	PVC and polyethylene	Locomotion	Sørensen *et al*. (1997)
	Porcellio scaber	Dimethoate	Locomotion	Bayley and Baatrup (1996)

and are an important phylum for testing in terrestrial ecotoxicology studies. Discussed below are representative species of nematodes that have been used in behavioural toxicity testing.

Caenorhabditis elegans is a nonparasitic nematode that lives in the interstitial water of soils, feeding on bacteria and other microbes. Since the 1970s, the advantages of using *C. elegans* for genetic studies have been known (Brenner 1974). Genetically identical worms can be ordered from the *Caenorhabditis* Genetics Center in Minneapolis, MN at no cost to the researcher. The characterisation of both the neural circuitry and cell lineage is more complete in *C. elegans* than in any other animal (Williams and Dusenbery 1987) and the entire genome (>99%) has been sequenced (*C. elegans* Sequencing Consortium 1998). The nervous system of *C. elegans* consists of only 302 neurons in the hermaphrodite and 381 in the male and the nervous system is highly conserved with higher animals including mammals (Bargmann 1998). Because so much knowledge exists on the physiological and genetic functions, the ability to quantify the toxicity can be supplemented by the possibility of determining the mechanisms associated with the adverse responses (Cressman and Williams 1997). *C. elegans* is becoming an important toxicological test organism and model system for other animals for the aforementioned reasons.

The behaviour of *C. elegans*, continuous feeding while moving in a wave-like motion (Dusenbery 1980), can now be monitored by a video camera interfaced to a computer tracking system that records the rate of locomotion and change of direction of hundreds of subjects simultaneously (Dhawan *et al.* 1999, 2000, Boyd *et al.* 2000, Anderson *et al.* 2001). After exposure in the soil, aquatic or agar media, the worms are rinsed to eliminate debris and any juveniles then transferred with a small volume of media to a thin layer of 1% agar that is poured directly onto a glass plate. Once they are placed on the agar surface, the worms are allowed to disperse evenly in a chamber over water in order to prevent desiccation. Then, the glass plate is placed upside down, directly above a stream of hydrated air and directly below the camera A modified version of the NIH Image Tracker software (Dusenbery 1997) allow the user to manipulate parameters such as the threshold level (maximising the contrast between the background and the worms), the colour spectrum and number of cycles tracked. The data are automatically transferred to an Excel spreadsheet. Information is provided on the number of subjects moving, their individual rate of locomotion and the frequency of change in direction.

Studies have been performed with worms exposed to metals, ethanol and pesticides on agar plates and in liquid media (Williams and Dusenbery 1987, 1990, Boyd *et al.* 2000, Dhawan *et al.* 1999, 2000, Anderson *et al.* 2001). Comparisons of the results of mortality test to behavioural changes suggested that computer tracking provide a reliable quantification of toxicity effects. The behavioural mid-point (expressed as the toxicant concentration where the average movement of exposed worms is 50% of control movement) is generally from 25 to 50 times lower than the LC_{50} value, providing a very sensitive indicator of toxicity.

Using *C. elegans* as an indicator of soil contamination is a promising test for reducing expenses and time. The organism is easy to work with and, given the extensive information base, provides an ideal behavioural model.

Entomopathogenic nematodes are used commercially to parasitise plant pest species. Some researchers have suggested the use of certain pesticides to increase the infectivity and activity of these nematodes. To examine this possibility, infective juveniles (IJs) of *Steinernema carpocapsae* and *S. feltiae* were exposed to two nematicides, oxamyl and fenamiphos (Patel and Wright 1996). A stereomicroscope was used to classify the movement of IJs as normal sinusoidal undulation, coiled and twisted, convulsive or violent twisting, uncoordinated movement or inactive. *S. feltiae* movement was initially stimulated by oxamyl but this effect diminished with time. Oxamyl had no significant effect on the movement of *S. carpocapsae.* Both nematodes' nonsinusoidal movement was significantly and permanently increased by fenamiphos. Similar effects were observed on nictating behaviour, the lifting of the anterior end from the substrate and waving from side to side. Infectivity of wax moths by IJs after 72-h exposures to the pesticides was determined by dissection of moth cadavers. Treatment with both pesticides resulted in decreased infectivity in both species. For these reasons the authors cautioned against using these pesticides or similar pesticides to enhance nictation and infectivity of entomopathogenic nematodes.

8.2.3.2 Earthworms

Earthworms are among the largest and most important of the terrestrial invertebrates. Burrowing through the soil, they serve to mix organic and inorganic portions of the soil. Furthermore, their burrows help to increase soil drainage, which may decrease soil erosion and the occurrence of plant diseases. Three main types of earthworms are recognised including the litter dwellers, the vertical burrowing surface feeders and the horizontal burrowing mineral soil species (Laskowski *et al.* 1998). *Eisenia fetida*, a litter-dwelling species, is a commonly used test organism in many ecotoxicological studies and an ASTM standard has been developed (ASTM 1998b).

In a short communication, Ramaswami and Subburam (1992) reported using earthworms to estimate the toxicity of two textile dyes, Navy blue M3R and Direct brown 2G, common waste products of the textile dye industry. In short, *Polypheretima elongata* failed to burrow into soils contaminated with even low concentrations of the dyes. In essence, these tests led the researchers to conclude that, in the field, the worms would avoid habitats contaminated with the textile dyes. This could prove to be detrimental as the dye effluent in India is sprayed on open lands or used as an irrigant in agricultural lands.

A similar design is the avoidance-response test used to assess the toxicity of contaminated soils. Because earthworms possess chemoreceptors, they are extremely sensitive to chemicals they contact in the soil environment. If they encounter highly contaminated soils, they will simply avoid them by moving to a more favourable environment. A sublethal test ranging from 7–72 h was designed to give a rapid, yet reliable, alternative to acute and chronic tests that range from 7–14 days and 28–55 days, respectively (Stephenson *et al.* 1998). The testing apparatus was a plexiglass container that was divided into six compartments, each of which held 300 g of different dilutions of contaminated or control soils. Holes in the base of the apparatus allowed for worm movement between different compartments. A central compartment without soil was loaded with *E. foetida* or *Lumbricus terrestris*. In both cases, the worms refused to enter only the most contaminated soils. Within 24 h of the start of the test, *E. fetida* had evacuated the two highest levels of contamination while *L. terrestris* took 72 h to do the same. In effect, this sublethal test was found to be more sensitive as well as much quicker than the acute toxicity test since worms avoided the soils that were contaminated at a level below that of the NOAEL calculated from mortality data.

A third study focused on the tendency of earthworms to migrate throughout the soil in response to pesticide applications (Mather and Christensen 1998). This migration away from contaminated soils can seriously affect population density and thus the overall quality of the soil. Specially designed troughs were used in the field to trap worms that moved in from one main direction. Traps were lowered into the troughs so that the top of each trap was level with the soil surface. Benomyl, a carbamate fungicide, was used as the soil contaminant. Of the seven species sampled, a 1.4- to 2.5-fold increase in trappings was observed at the lowest treatment levels of 0.5 and 1.0 kg a.i. ha^{-1}. However, at 2.0 kg a.i. ha^{-1}, there was a 1.3-fold decrease in captures. These differences were seen to be significant as early as 2 days for the two higher concentrations and 4 days for the lowest concentration. In addition, the species composition of trapped worms was changed by the two higher treatments.

Another test system investigated the electrical activity of neural units to give an estimation of the startle reflex, a mean of rapid escape from predators (Drewes and Vining 1984). *E. foetida* was used to test the effects of a neurotoxic insecticide, dieldrin, and two aromatic hydrocarbons, dimethyl phthalate (DMP) and fluorene. Giant nerve fibre conduction velocity was monitored in both the head and tail segments both before and after 48-h exposures. Because dieldrin is neurotoxic, it was shown to cause tonic spasms and hyperresponsiveness to touch. At low doses, dieldrin caused decreased firing of the medial giant nerve fibre (MGF) and, at higher doses, decreased the conduction velocity in both the medial and lateral nerve fibres. Interestingly, fluorene had an opposite effect with low doses decreasing conduction velocity in both nerve fibres and high doses only affecting the MGF. DMP showed only marginal effects even at the highest doses. Decreases in nerve fibre conduction could lead to animals being more susceptible to predation and avoidance of contaminants in the soil. The authors concluded that, although this tool is useful in predicting neurophysiological effects, it was not sensitive to other behavioural effects.

8.2.3.3 Spiders

Spiders represent a diverse class of organisms. Their predatory behaviour involves complex activities such as construction of various web designs to trap and immobilise insects. Because spiders are known to prey on pests, they must be considered beneficial organisms. Historically, research efforts determining the toxicity of pesticides and other toxicants to beneficials have concentrated on mortality as an endpoint. However, a toxicant does not have to be lethal to cause significant changes in behaviour that may affect reproduction or predation and thus population dynamics.

Samu and Vollrath (1992) used the common cross spider *Araneus diadematus* to examine the effects of two fungicides (triadimenol and prochloraz) and two insecticides (cypermethrin and paraffinoil) on orb web building. The pesticides were administered orally and topically. Time to construction of the web was recorded and the web was photographed. Digitised images were used to measure web characteristics such as capture area, total length of radii, total length of spiral and mean mesh size. Only topically applied cypermethrin, a pyrethroid insecticide, delayed the spiders from building webs and affected web characteristics. In addition, spiders did not recycle cypermethrin treated webs due to taste aversion.

Using a similar test, Lengwiler and Benz (1994) investigated sublethal effects of primicarb, deltamethrin, diazinon and dicofol on the grey cross spider *Larinioides sclopetarius*. Web structure was analysed before and after topical application of the pesticides. Digitised images were used to characterise web parameters including web area, median angle, regularity of angles, width of web over length of web, and distances over spiral turns. No pesticide caused significant mortality at the tested concentrations. Deltamethrin and diazinon treatments resulted in significant delays in web building and decreases in web area. No other significant effects on web building were observed. Spiders also recovered from pesticide application quickly.

In an attempt to quantify the effects of deltamethrin, a pyrethroid insecticide, on a nontarget linyphiid spider, a study was carried out on plots of land sprayed with the insecticide (Jagers op Akkerhuis 1994). Environmental conditions such as air temperature, soil and air humidity, and quantity of rain were also monitored because these conditions are known to affect the behaviour, in this case the walking activity, of animals. Linyphiid spiders live most of their time hidden from the direct spray of insecticides, either in cracks in the soil or under the plant canopy. However, as they search for mates or prey, or a place to nest their eggs, walking puts them at greater risk to be exposed to spray or residues. Not only do the activities above depend upon environmental conditions so too does the speed at which the spiders walk. In the case that the spiders come in contact with residues of an insecticide, their walking activity decreases leading to possible death by predation or through water loss.

Round pitfall traps were placed at regular intervals from the outside of each treated or control plot. Spiders were emptied from the traps daily and, after the first 12 h after spraying, every 1.5 h. Insecticide residues were measured at ground level as well as the environmental conditions previously mentioned. Two parameters were developed to estimate the walking activity of the spiders: an instantaneous effect which measured the decline in trap catches and a recovery time which was a measurement relative to the observed instantaneous effect. After deltamethrin spraying, they found a 50–75% decrease in initial effects and recovery half-lives that ranged from 1–3 days.

8.2.3.4 Springtails

Collembolans, or springtails, are among the most abundant of the terrestrial soil invertebrates. Springtails feed on fungi and other microbes and, in highly acidic soils, serve as primary decomposers (Laskowski *et al.* 1998). In addition, springtails serve as prey for many larger soil invertebrates such as spiders, beetles and centipedes. Because of their ecological significance, collembolans are used in a wide range of ecotoxicological tests. In fact, the springtail *Folsomia candida* is used as an international standard test organism for the testing of soil quality (Wiles and Krogh 1998). Similar to rotifers, they also offer the benefits of a rapid life cycle, ease of laboratory cultivation and a high parthenogenic reproduction rate.

Because springtails are prone to desiccation, their locomotory activity allows them to escape adverse environmental conditions. However, a means to measure their activity was difficult before the aid of computer tracking devices. In one study, a computer-automated video-tracking device was used to record and analyse the two-dimensional movement of *F. candida* (Sorensen *et al.* 1995). Dimethoate, an organophosphate pesticide, was used as the contaminant. During the 36-h exposure period, springtails were observed to be increasingly hyperactive and displayed some difficulty in coordinated movement. For the first 10 hours of the exposure, the only significant change was an increased turning rate. However, during the last 10 hours of the exposure, a significant decrease in average velocity and path length as well as increases in number of stops and turning rate were observed.

Other researchers also investigated the effects of dimethoate on another collembolan, *F. fimetaria* (Fábián and Petersen 1994). Unlike the computer tracking studies, visual observations were made once every hour for 24 hours. The activity state of the springtails was noted as at rest, walking, struggling or motionless. The last two descriptors indicated abnormal activity associated with contaminant effects. Another experiment allowed springtails to choose between a contaminated or uncontaminated environment, similar to the earthworm study mentioned previously. Area-choice was recorded every hour for the first 10 hours and once a day for the remaining 7 days. A final experiment observed the dispersion and spatial distribution in a microcosm treated with five concentrations of dimethoate. Collembola location was noted after 1, 3 or 7 days. Abnormal activity (struggling and motionless behaviour) was noted to increase with increasing doses. In the area-choice experiments, springtails either avoided or escaped from the contaminated side of the canisters. About half of the test organisms were located in the control soil within a few hours and, after a few days, nearly all were. However, no significant effect was seen on the spatial distribution or dispersion habits of the springtails in the microcosm test, indicating that complete avoidance of contaminants is unlikely in a natural soil.

8.2.3.5 Aphids

Harrewijn and Kayser (1997) used EPC to study the effect of a novel aphidicide, pymetrozine, applied in a several ways on different phases of stylet penetration and feeding activity on individual aphids. Pymetrozine was either topically applied (0, 50 and 100 ng/mg body weight), directly injected into the haemolymph (7 and 35 ng), added to a chemical diet (50, 100 and 300 μg/l), sprayed onto plants (80–200 mg/l), or plants were allowed to soak for 24 h in treated solution for root uptake (40 and 100 mg/l). Topical applications of 150 ng pymetrozine/mg fresh weight inhibited stylet insertion into the plant. When injected, less than 30 ng/mg was sufficient to produce the same effect. When pymetrozine was systemically applied via plant spraying or root uptake, aphids fed normally. After feeding, they withdrew their stylets and appeared normal. However, at low doses aphids were able to resume feeding whereas aphids exposed to higher doses appeared to have irreversibly disrupted feeding and were unable to re-insert the stylet for feeding. Aphid mobility was not affected up to an estimated haemolymph concentration of 1 mm pymetrozine. Aphids which eventually stopped feeding on treated plants showed EPGs with distorted salivation and ingestion patterns, suggesting that pymetrozine selectively interferes with the nervous regulation of feeding behaviour which results in death due to starvation after a few days.

8.2.3.6 Beetles

In general, beetles feed on the larvae of other insects as well as on adult aphids and collembolans (Laskowski *et al.* 1998). In addition, they serve as prey to larger animals such as birds and mice. While there are a considerable number of beetles that live on or near the soil surface, the following description will focus on a group of cereal aphid predators including ground beetles, rove beetles and a ladybird beetle.

A video camera was used to record the movement of the beetles across a sandy loam soil after treatment with the insecticide deltamethrin (Wiles and Jepson 1994). An index was generated that was intended to reveal the susceptibility of the beetles to pesticide residues. This index consisted of two parameters: an exposure function which gave an estimate of how fast the animal walked, how wide its walking track was, and the proportion of the animal that came in contact with the soil surface; and a susceptibility function that was an LD_{50} after topical treatment with the insecticide. In this manner, a high index would indicate that an insect was vulnerable to contact with lethal doses of the toxicant while a low index would be indicative of a decline in contact and thus uptake and susceptibility. The large ground beetles, *Pterostichus melanarius* and *Nebria brevicollis*, medium-sized ground beetle, *Agonum dorsale*, small ground beetles, *Demetrias atricapillus* and *Bembidion lampros*, small rove beetle, *Tachyporus hypnorum*, and ladybird beetle *Coccinella septempunctata* were used as the test species in the study. First, the topical LD_{50} for each species was determined for the susceptibility function. Next, the mean walking speed for each species was determined by recording the movement of five individuals at a time with a video camera. After recording, the walking paths were traced onto acetate that covered the video monitor. The walking paths were then measured with an ipsometer. Finally, contact area and track width was measured by placing animals on kymograph paper covered with soot. As the animals moved across the paper, a clean trail was left behind on the sooted paper and image analysis was used to estimate the area swept clean. Although the index did not predict all of the beetles' susceptibility in the correct order, it was able to predict the extreme susceptibilities. The authors suggest other factors that must be taken into account before the index is able to predict mortality reliably.

The International Organization for Biological and Integrated Control of Noxious Animals and Plants (IOBC) has published an official guideline detailing the Carabid Beetle Laboratory Test using *Poecilus cupreus* to assess effects of pesticides before registration (Heimbach 1992). This test was used to compare the effects of 126 compounds on beetle behaviour. Test chemicals were misted onto a sand surface including beetles and their food (Römbke and Heimbach 1996). No significant mortality was observed with any substance tested. Abnormal behaviour began as impaired movement of the mandible followed by uncoordinated walking and finally 'lay(ing) on their backs with shivering legs'. Feeding rate of the beetles was also recorded during testing. However, the majority of tests showed no significant difference in behaviour or feeding.

8.2.3.7 Bees

Learning, an aspect of behaviour, can also be affected by toxicants. An excellent review of learning in invertebrates is recommended to the reader desiring a more comprehensive coverage (Abramson 1994). As with spiders, toxic effects of insecticides on beneficial insects such as honey bees are of significant importance. Although most toxicity tests have been based solely on mortality data, effects on behaviour in bees could lead to lower honey yields as well as decreased pollination.

The toxicity of dicofol, a chlorinated hydrocarbon acaricide thought to be relatively nontoxic, was estimated using the Africanized honey bee, *Apis mellifera* (Stone *et al.* 1997). Classical conditioning of proboscis extension was used to measure sublethal toxic effects on learning. Again, dicofol was mixed with a sucrose solution and offered as a food source. Three seconds of a cinnamon odour served as a conditioned stimulus (CS) scent. After the CS, touching the antennae stimulated proboscis extension. Honey was then delivered as the unconditioned stimulus (US). Bees were considered to learn if they extended their proboscis after delivery of the CS but before the US. Although dicofol-exposed bees did learn to some degree, a decreased response to CS as well as a shorter persistence of learning was observed when compared with the controls.

In a similar test, Abramson *et al.* (1999) tested the effects of endosulfan, decis, baytroid and sevin, four pesticides commonly used on cotton crops in Brazil. The study showed that not only were *A. mellifera* not repelled by the pesticides, they consumed the pesticides when mixed with a sucrose solution. Proboscis conditioning was again used to test the ability of the bees to learn. At the recommended concentrations, all pesticides except decis were observed to disrupt learning of *A. mellifera*. Besides inhibition of learning, the insecticides also led to tremors and vomiting in the bees.

8.2.3.8 Woodlice

Woodlice act as primary decomposers of dead plant material and other organic material. Before fully digesting, they forward the organic matter back into the soil, making it available to other soil fauna (Laskowski *et al.* 1998). Woodlice and other isopods search for cool humid places for shelter during the heat of the day. If their ability to move to these shelters is impeded or slowed, the possibility of desiccation and thus lethality is increased.

Sørensen *et al.* (1997) also performed a study on woodlice by use of a similar system to the one used for springtails. In this case, the effects of the resulting pollutants formed by a plastics fire were observed in the woodlouse *Oniscus asellus*. Suspected and known contaminants included polyvinyl chloride (PVC) and polyethylene (PE). The activity of six animals was recorded for 4 h and digitised to obtain the path length, velocity, total number of moves/walked metre, turning rate and turn bias. Woodlice collected directly from the plastic layer walked about half as far as those collected from a reference site. A decreased average velocity and time spent in activity were also observed. However, in woodlice collected from the edge of the plastic layer, there were no significant differences in behavioural parameters. Juveniles hatched from the collected woodlice did not exhibit any significant changes in behaviour. Similar studies on dimethoate with the woodlouse *Porcellio scaber* are discussed in Chapter 5 (Bayley 1995, Bayley and Baatrup 1996).

REFERENCES

Abramson CI (1994) *A Primer of Invertebrate Learning — The Behavioral Perspective*, American Psychological Association, Washington, DC.

Abramson CI, Aquino IS, Ramalho FS and Price JM (1999) The effect of insecticides on learning in the Africanized honey bee (*Apis mellifera* L.). *Archives of Environmental Contamination and Toxicology*, **37**, 529-535.

Aiko M (1996) Precopulatory mate guarding behavior of caprellid amphipods observed in Amakusa, Kysushu, Japan. *Publications from the Amakusa Marine Biological Laboratory*, **12**, 71-78.

Akberali HB and Trueman ER (1985) Effects of environmental stress on marine bivalve molluscs. *Advances in Marine Biology*, **22**, 102-183.

Akberali HB, Wong TM and Trueman ER (1981) Behavioral and siphonal tissue responses of *Scrobicularia plana* (Bivalvia) to zinc. *Marine Environmental Research*, **5**, 251-264.

American Society for Testing and Materials (1989) Standard practice for conducting static acute toxicity tests with larvae of four species of bivalve molluscs, E724-89. In *Annual Book of ASTM Standards* vol 11.04. Philadelphia, pp. 272-296.

American Society for Testing and Materials (1998a) Standard guide for acute toxicity test with the rotifer *Brachionus*. In *Annual Book of ASTM Standards* Volume 11.05, Philadelphia, pp. 837-843.

American Society for Testing and Materials (1998b) Standard guide for conducting laboratory soil toxicity or bioaccumulation tests with the lumbricid earthworm *Eisenia fetida*. In *Annual Book of ASTM Standards* Volume 11.05, Philadelphia, pp. 1056-1074.

Anderson GL, Boyd WA and Williams PL (2001) Assessment of sub lethal endpoints for toxicity testing with the nematode *Caenorhabditis elegans*. *Environmental Toxicology and Chemistry*, **20**, 833-838.

Avery L and Horvitz R (1990) Effects of starvation and neuroactive drugs on feeding in *Caenorhabditis elegans*. *Journal of Experimental Zoology*, **253**, 263-270.

Baatrup E and Bayley M (1993) Quantitative analysis of spider locomotion employing computer-automated video tracking. *Physiology and Behavior*, **54**, 83-90.

Bargmann CI (1998) Neurobiology of the *Caenorhabditis elegans* genome (Review). *Science*, **282**, 2028-2032.

Bayley M (1995) Prolonged effects of the insecticide dimethoate on locomotor behaviour in the woodlouse, *Porcellio scaber* Latr. (Isopoda). *Ecotoxicology*, **4**, 79-90.

Bayley M and Baatrup E (1996) Pesticide uptake and locomotor behaviour in the woodlouse: an experimental study employing video tracking and ^{14}C-labelling. *Ecotoxicology*, **5**, 35-45.

Beehler JW and Mulla MS (1996). Larvicidal oils modify the oviposition behavior of *Culex* mosquitoes. *Journal of Vector Ecology*, **21**, 60-65.

Belanger SE, Cherry DS and Cairns J Jr (1986a) Uptake of chrysotile asbestos fibers alters growth and reproduction of Asiatic clams. *Canadian Journal of Fisheries and Aquatic Science* **43**, 43-52.

Belanger SE, Cherry DS and Cairns J Jr (1986b) Seasonal, behavioral, and growth changes of juvenile *Corbicula fluminea* exposed to chrysotile asbestos. *Water Research* **20**, 1243-1250.

Berman MS and Heinle DR (1980) Modification of the feeding-behavior of marine copepods by sublethal concentrations of water-accommodated fuel-oil. *Marine Biology*, **56**, 59-64.

Besch WK, Schreiber J and Herbst D (1977) Der *Hydropsyche* Toxizitätstest, erprobt an Fenethcarb. *Schweizerische Zeitschrift für Hydrologie — Swiss Journal of Hydrology*, **39**, 69-85.

Besch WK, Schreiber J and Magnin E (1979) Influence du sulfate de cuivre sure la structure du filet des larves d'*Hydropsyche* (Insecta, Trichoptera). *Annales de Limnologie*, **15**, 123-138.

Blockwell SJ, Maund SJ and Pascoe D (1998) The acute toxicity of lindane to *Hyallela azteca* and the development of a sublethal bioassay based on precopulatory guarding behavior. *Archives of Environmental Contamination and Toxicology*, **35**, 432-440.

Blockwell SJ, Maund SJ and Pascoe D (1999) Effects of the organochlorine insecticide lindane (γ-$C_6H_6Cl_6$) on the population responses of the freshwater amphipod *Hyallela azteca*. *Environmental Technology and Chemistry*, **18**, 1264-1269.

Bohrer R (1980) Experimental studies on diel vertical migration: In *Evolution and Ecology of Zooplankton Communities*, Kerfoot WC (ed.), University Press of New England, Hannover, NH, pp. 111-121.

Boyd WA, Anderson GL, Dusenbery DB and Williams PL (2000) Computer Tracking Method for Assessing Behavioral Changes in the Nematode *Caenorhabditis elegans*. In *Environmental Toxicology and Risk Assessment (9th Volume) ASTM 1381*, Price F, Brix K and Lane K (eds), American Society for Testing and Materials, West Conshohocken, PA, pp. 225-238.

Brenner S (1974) The genetics of *Caenorhabditis elegans*. *Genetics*, **77**, 71-94.

Brooks JL (1964) The relationship between the vertical distribution and seasonal variation of limnetic species of *Daphnia*. *Verhandlungen. Internationale Vereinigung fur theoretische und angewandte Limnologie/Proceedings, International Association of Theoretical and Applied Limnology/Travaux*, **15**, 684-690.

C. elegans Sequencing Consortium (1998) Genome sequence of the nematode *Caenorhabditis elegans*: a platform for investigating biology. *Science*, **282**, 2012-2018.

Cairns J (1979) Biological monitoring — concept and scope. In *Environmental Biomonitoring, Assessment, Prediction and Management*, Cairns J, Patil GP and Waters WE (eds), Int. Co-op Publ. House, Burtonsville, MA, pp. 3-20.

Canton JH, Wester PW and Mathijssen-Spickman EAM (1983) Study on the toxicity of sodium bromide to different freshwater organisms. *Food and Chemical Toxicology*, **21**, 369-378.

Camusso M, Balestrini R, Muriano F and Mariani M (1994) Use of freshwater mussel *Dreissena polymorpha* to assess trace metal pollution in lower river Po (Italy). *Chemosphere*, **29**, 729-745.

Capuzzo JM (1979) Effects of halogen toxicants on survival, feeding and egg-production of the rotifer *Brachionus plicatilis*. *Estuarine and Coastal Marine Science*, **8**, 307-316.

Charoy C and Janssen CR (1999) The swimming behaviour of *Brachionus calyciflorus* (Rotifer) under toxic stress. *Chemosphere*, **38**, 3247-3260.

Conlan KE (1991) Precopulatory mating behaviour and sexual dimorphism in the amphipod Crustacea. *Hydrobiologia*, **223**, 255-282.

Cooley JM (1977) Filtering rate performance of *Daphnia retrocurva* in pulp-mill effluent. *Journal of the Fisheries Research Board of Canada*, **34**, 863-868.

Cressman III CP and Williams PL (1997) Reference toxicants for toxicity testing using *Caenorhabditis elegans* in aquatic media. In *Environmental Toxicology and Risk Assessment: Modeling and Risk Assessment (Sixth Volume), ASTM STP 1317*, Dwyer James F, Doane Thomas R and Hinman Mark L (eds), American Society for Testing and Materials, West Conshohocken, PA, pp. 518-532.

Davenport J and Manley A (1978) The detection of heightened sea-water copper concentrations by the mussel *Mytilus edulis*. *Journal of the Marine Biology Association, UK*, **58**, 843-850.

Davis JC (1978) Disruption of precopulatory behaviour in the amphipod *Anisogammarus pugettensis* upon exposure to bleached pulpmill effluent. *Water Research*, **2**, 273–275.

Day KE (1989) Acute, chronic and sublethal effects of synthetic pyrethroids on freshwater zooplankton. *Environmental Toxicology and Chemistry*, **8**, 411–416.

Day K and Kaushik NK (1987) The adsorption of fenvalerate to laboratory glassware and the alga *Chlamydomonas reinhardii*, and its effect on uptake of the pesticide by *Daphnia galeata mendotae*. *Aquatic Toxicology*, **10**, 131–142.

Décamps H, Besch WK and Vobis H (1973) Influence des produits toxiques sur la construction de filet des larves d'*Hydropsyche* (Insecta, Trichoptera). *Comptes Rendues de l'Académie des Sciences, Paris*, Série D, **276**, 375–378.

Dhawan R, Dusenbery D and Williams P (1999) Comparison of lethality, reproduction and behavior as toxicological endpoints in the nematode *Caenorhabditis elegans*. *Journal of Toxicology and Environmental Health (Part A)*, **58**, 451–462.

Dhawan R, Dusenbery D and Williams P (2000) Comparison of metal-induced lethality and behavioral responses in the nematode *Caenorhabditis elegans*. *Environmental Toxicology and Chemistry*, **19**, 3061–3067.

Dill LM (1987) Animal decision making and its ecological consequences: the future of aquatic ecology and behavior. *Canadian Journal of Zoology*, **65**, 803–811.

Djangmah JS, Shumway SE and Davenport J (1979) Effects of fluctuating salinity on the behaviour of the west African blood clam *Anadara senilus* and on the osmotic and ionic concentrations of the haemolymph. *Marine Biology*, **50**, 209–213.

Dodson SI, Hanzato T and Gorski PR (1995) Behavioral responses of *Daphnia pulex* exposed to carbaryl and *Chaoborus* kairomone. *Environmental Toxicology and Chemistry*, **14**(1), 43–50.

Doherty FG, Cherry DS and Cairns J Jr (1987) Valve closure responses of the Asiatic clam *Corbicula fluminea* exposed to cadmium and zinc. *Hydrobiologia*, **153**, 159–167.

Dojmi Di Delupis G and Rotondo V (1988) Phototaxis in aquatic invertebrates: possible use in ecotoxicity tests. *Ecotoxicology and Environmental Safety*, **16**, 189–193.

Donkin P, Widdows J, Evans SV, Worral CM and Carr M (1989) Quantitative structure activity relationships for the effect of hydrophobic organic chemicals on rate of feeding by mussels (*Mytilus edulis*). *Aquatic Toxicology*, **14**, 277–294.

Donkin P, Widdows J, Evans SV and Brinsley MD (1991) QSARs for the sublethal responses of marine mussels (*Mytilus edulis*). *Science of the Total Environment*, **109**, 461–476.

Donkin P, Widdows J, Evans SV, Staff FJ and Yan T (1997) Effect of neurotoxic pesticides on the feeding rate of marine mussels (*Mytilus edulis*). *Pesticide Science*, **49**, 196–209.

Drake DC and McNaught DC (1994) Behavioral assays employing *Daphnia* for detection of sublethal effects: responses to ALD. *Verhandlungen. Internationale Vereinigung fur theoretische und angewandte Limnologie/Proceedings, International Association of Theoretical and Applied Limnology/Travaux*, **25**, 1999–2003.

Drewes CD (1997) Sublethal effects of environmental toxicants on oligochaete escape reflexes. *American Zoologist*, **37**, 346–353.

Drewes CD and Vining EP (1984) Non-invasive electrophysiological monitoring: a sensitive method for detecting sublethal neurotoxicity in earthworms. *Environmental Toxicology and Chemistry*, **3**, 599–607.

Dusenbery D (1980) A behavior of free-living nematodes. In *Nematodes as Model Biological Systems*, Academic Press, New York.

Dusenbery D (1992) Simultaneous tracking of multiple subjects. In *Image Analysis in Biology*, Hader D (ed.), CRC Press, Boca Raton, FL, pp. 329–341.

Dusenbery D (1997) *NIH Image*, Version 1.59, Computer Tracking Program, as modified by D. Dusenbery.

Eldon J and Kristoffersson R (1978) Factors affecting the burrowing activity of *Macoma balthica* (L.) *Annales Zoologici Fennici*, **15**, 127-131.

Eldon J, Pekkarinen M and Kristoffersson R (1980) Effects of low concentrations of heavy metals on the bivalve *Macoma balthica* (L.). *Annales Zoologici Fennici*, **17**, 233-242.

Englund VPM, Heino MP and Melas G (1994) Field method for monitoring valve movements of bivalved molluscs. *Water Research*, **28**, 2219-2221.

Fábián M and Petersen H (1994) Short-term effects of the insecticide dimethoate on activity and spatial distribution of a soil inhabiting collembolan *Folsomia fimetaria* Linné (Collembola: Isotomidae). *Pedobiologia*, **38**, 289-302.

Flickinger AL, Bruins RJF, Winner RW and Skillings JH (1982) Filtration and phototactic behavior as indices of chronic copper stress in *Daphnia magna* straus. *Archives of Environmental Contamination and Toxicology*, **11**, 457-463.

Forward RB Jr and Costlow JD Jr (1978) Sublethal effects on insect growth regulators upon crab larval behavior. *Water Air and Soil Pollution*, **9**, 227-238.

Gauld DT (1951) The razing rate of planktonic copepods. *Journal of the Marine Biological Association of the United Kingdom*, **29**, 695-706.

Geritsen J (1982) Behavioural response of *Daphnia* to rate of temperature change: possible enhancement of vertical migration. *Limnology and Oceanography*, **27**, 254-261.

Gilron G, Gransden SG, Lynn DH, Broadfoot J and Scroggins R (1999) A behavioral toxicity test using the ciliated protozoan *Tetrahymena thermophila*. 1. Method description. *Environmental Toxicology and Chemistry*, **18**, 1813-1816.

Gnyubkin VF (1995) Inversion of orientation in supralittoral amphipods as a response to environmental conditions. *Russian Journal of Marine Biology*, **21**, 135-137.

Gómez A, Cecchine G and Snell TW (1997) Effect of pentachlorophenol on predator-prey interaction of two rotifers. *Aquatic Toxicology*, **37**, 271-282.

Goodrich MS and Lech JJ (1990) A behavioral screening assay for *Daphnia magna*: a method to assess the effects of xenobiotics on spatial orientation. *Environmental Toxicology and Chemistry*, **9**, 21-30.

Harrewijn P and Kayser H (1997) Pymetrozine, a fast-acting and selective inhibitor of aphid feeding. *In situ* studies with electronic monitoring of feeding behaviour. *Pesticide Science*, **49**(2), 130-140.

Heimbach U (1992) Standard laboratory method to test effects of pesticides on *Poecilus cupreus* (Coleoptera, Carabidae). *IOBC/WPRS Bulletin, Bulletin OILB/SROP*, **XV**, 103-109.

Heinonen J, Kukkonen J, Pentitnen O-P and Holopainen IJ (1997) Effects of hypoxia on valve-closure time and bioaccumulation of 2,4,5-trichlorophenol by the freshwater clam *Spaerium corneum* (L.). *Ecotoxicology and Environmental Safety*, **36**, 49-56.

Higgins PJ (1980) Effects of food availability on the valve movements and feeding behaviour of juvenile *Crassostreea virginica* (Gmelin). I. Valve movements and periodic activity. *Journal of Experimental Marine Biology and Ecology*, **45**, 229-244.

Holopainen IJ and Penttinen O-P (1993) Normoxic and anoxic heat output of the freshwater bivalves *Pisidium* and *Sphaerium*. *Oecologia*, **93**, 215-223.

Huges RN (1980) Strategies for survival of aquatic organisms. In *Fundamentals of Aquatic Ecosystems*, Barnes RK and Mann KH (eds), Blackwell Scientific Publications, Oxford, pp. 162-184.

Jacobson PJ, Farris JL, Cherry DS and Neves RJ (1993) Juvenile freshwater mussel (Bivalvia: Unionidae) responses to acute toxicity testing with copper. *Environmental Toxicology and Chemistry*, **12**, 879-883.

Jagers op Akkerhuis GAJM (1994) Effects of walking activity and physical factors on the short term toxicity of deltamethrin spraying in adult epigeal money spiders (Linyphiidae). In *Ecotoxicology of Soil Organisms*, Donker MH, Eijsackers H and Heimbach F (eds), CRC Press, Boca Raton., FL, pp. 323-338.

Jenner HA, Noppert F and Sikking T (1989) A new system for the detection of valve movement response of bivalves. KEMA scientific and technical report 1989-7-2. ISSN-016708590, KEMA, Arnhem, the Netherlands, **7**, 91-98.

Johannson A and Johansson F (1992) Effects of two different caddisfly case structures on predation by a dragonfly larva. *Aquatic Insects*, **14**, 73-84.

Jones M, Folt C and Guarda S (1991) Characterizing individual, population and community effects of sublethal levels of aquatic toxicants — An experimental case-study using daphnia. *Freshwater Biology*, **26**, 35-44.

Juchelka CM and Snell TW (1994) Rapid toxicity assessment using rotifer ingestion rate. *Archives of Environmental Contamination and Toxicology*, **26**, 549-554.

Juchelka CM and Snell TW (1995) Rapid toxicity assessment using ingestion rate of cladocerans and ciliates. *Archives of Environmental Contamination and Toxicology*, **28**, 508-512.

Kaushik NK, Stephenson GL, Solomon KR and Day KE (1985) Impact of permethrin on zooplankton communities in limnocorrals. *Canadian Journal of Fisheries and Aquatic Sciences*, **42**, 77-85.

Keilty TJ, White DS and Landrum PF (1988) Short-term lethality and sediment avoidance assays with endrin-contaminated sediment and two oligochaetes from Lake Michigan. *Archives of Environmental Contamination and Toxicology*, **17**, 95-101.

Koeman JH, Poels CLM and Sloof W (1978) Continuous biomonitoring systems for detection of toxic levels of water pollutants. In *Aquatic Pollutants: Transformation and Biological Effects*, Hutzinger O, van Lelyveld LH and Zoeteman BCJ (eds), Pergamon, London, pp. 339-347.

Kraak MHS, Toussaint M, Lavy K and Davids C (1994) Short-term effects of metals on the filtration-rate of the zebra mussel *Dreissena polymorpha*. *Environmental Pollution*, **84**, 139-143.

Kramer KJM, Jenner HA and de Zwart D (1989) The valve movement response of mussels: a tool in biological monitoring. *Hydrobiologia*, **188/189**, 433-443.

Kruschwitz LG (1972) An analysis of the reproductive behavior patterns of the amphipod *Hyallela azteca*. *American Zoologist*, **12**, 658-659.

Lam PKS, Wo KT and Wu RSS (2000) Effects of cadmium on the development and swimming behavior of barnacle larvae *Balanus amphitrite* Darwin. *Environmental Toxicology*, **15**, 8-13.

Lang WH, Forwar RB Jr, Miller DC and Marcy M (1980) Acute toxicity and sublethal behavioral effects of copper on barnacle nauplii (*Balanus improvisus*). *Marine Biology*, **58**, 139-145.

Laskowski R, Kramarz P and Jepson P (1998) Selection of species for soil ecotoxicity testing. In *Handbook of Soil Invertebrate Toxicity Tests*, Løkke H and Van Gestel CAM (eds), John Wiley and Sons Ltd, Chichester, pp. 21-32.

Lengwiler U and Benz G (1994) Effects of selected pesticides on web building behaviour of *Larinioides sclopetarius* (Clerck) (Araneae, Araneidae). *Journal of Applied Entomology*, **117**, 99-108.

Lenihan HS, Kiest KA, Conlan KE, Slattery PN, Konar BH and Oliver JS (1995) Patterns of survival and behavior in Antarctic benthic invertebrates exposed to contaminated sediments: field and laboratory bioassay experiments. *Journal of Experimental Marine Biology and Ecology*, **192**, 233-235.

Linden O (1976) Effect of oil on the reproduction of the amphipod *Gammarus oceanicus*. *Ambio*, **5**, 36-37.

Little EE, Flerov BA and Ruzhinskaya NN (1985) Behavioral approaches in aquatic toxicity investigations: a review. In *Toxic Substances in the Aquatic Environment: an International Aspect*, Merle PM Jr, Gray RH and Kendall RL (eds), American Fisheries Society, Water Quality Section, Bethesda, MD, pp. 72-98.

Mäkelä TP, Petäinen T, Kukkonen J and Oikari AOJ (1991) Accumulation and depuration of chlorinated phenolics in the freshwater mussel (*Anodonta anatina* L). *Ecotoxicology and Environmental Safety*, **22**, 153-163.

Manley AR (1983) The effects of copper on the behaviour, respiration, filtration and ventilation activity of *Mytilus edulis*. *Journal of the Marine Biology Association, UK*, **63**, 205-222.

Manley AR and Davenport J (1979) Behavioural responses of some marine bivalves to heightened seawater copper concentrations. *Bulletin of Environmental Contamination and Toxicology*, **22**, 739-744.

Marceau F (1909) Recherche sur la morphologie, et l'histologie, et la physiologie comparées des muscles adducteurs des mollusques acephales. *Archives de Zoologie Experimentale et Générale* (Series 5), **2**, 295-469.

Mather JG and Christensen OM (1998) Earthworm surface migration in the field: influence of pesticides using benomyl as test chemical. In *Advances in Earthworm Ecotoxicology*, Sheppard SC, Bembridge JD, Holmstrup M and Posthuma L (eds), SETAC Press, Pensacola, FL, pp. 327-340.

McCahon CP, Whiles AJ and Pascoe D (1989) The toxicity of cadmium to different larval instars of the trichopteran larvae *Agapetus fuscipes* Curtis and the importance of life cycle information to the design of toxicity tests. *Hydrobiologia*, **185**, 153-162.

McCahon CP and Poulton MJ (1991) Lethal and sub-lethal effects of acid, aluminium and lime on *Gammarus pulex* during repeated simulated episodes in a Welsh stream. *Freshwater Biology*, **25**, 169-178.

McCahon CP, Poulton MJ, Thomas PC, Xu Q, Pascoe D and Turner C (1991) Lethal and sublethal toxicity of field simulated farm waste episodes to several fresh-water invertebrate species. *Waste Research*, **25**, 661-671.

McGreer ER (1979) Sublethal effects of heavy metal contaminated sediments on the bivalve *Macoma balthica* (L.). *Marine Pollution Bulletin*, **14**, 57-60.

McLeese DW, Burridge LE and Van Dinter J (1982) Toxicities of five organochlorine compounds in water and sediment to *Nereis virens*. *Bulletin of Environmental Contamination and Toxicology*, **28**, 216-220.

McNaught DC (1992) Zooplankters as indicators of ecosystem health: past finding and future directions. *Journal of Ecosystem Health*, **1**, 271-281.

McNaught DC and Hasler AD (1964) Rate of movement of populations of *Daphnia* in relation to changes in light intensity. *Journal of Fisheries Research Board of Canada*, **21**, 291-315.

Miwa J, Tabluse Y, Furwsana M and Yamasaki H (1982) Tumour promoters specifically and reversibly disturb development and behavior of *Caenorhabditis elegans*. *Journal of Cancer Research and Clinical Oncology*, **104**, 81-87.

Møhlenberg F and Kiørboe T (1983) Burrowing and avoidance behaviour in marine organisms exposed to pesticide-contaminated sediment. *Marine Pollution Bulletin*, **14**, 57-60.

Morgan PG and Cascorbi HF (1985) Effect of anesthetics and a convulsant on normal and mutant *Caenorhabditis elegans*. *Anesthesiology*, **62**, 738-744.

Olla BL, Bejda AJ and Pearson WH (1983) Effects of oiled sediment on the burrowing behaviour of the hard clam, *Mercaneria mercaneria*. *Marine Environmental Research*, **9**, 183-193.

Olla BL, Bejda AJ, Studholme AJ and Pearson WH (1984) Sublethal effects of oiled sediment on the sand worm *Nereis* (*Neanthes*) *virens*: Induced changes in burrowing and emergence. *Marine Environmental Research*, **13**, 121-139.

Otto C and Svensson BS (1980) The significance of case material selection for the survival of caddis larvae. *Journal of Animal Ecology*, **49**, 855-865.

Papi F (1960) Orientation by night: the moon. *Cold Spring Harbor Symposia on Quantitative Biology*, **25**, 475-480.

Pardi L (1960) Innate components in the solar orientation of littoral amphipods. *Cold Spring Harbor Symposia on Quantitative Biology*, **25**, 395-401.

Parker GA (1974) Courtship persistence and female guarding as male time investment strategies. *Behaviour*, **48**, 147-184.

Pascoe D, Gower DE, McCahon CP, Poulton MJ, Whiles AJ and Wulfhorst J (1991) Behavioural responses to pollutants — application in freshwater bioassays. In *Bioindicators and Environmental Management*, Jeffrey DW and Madden B (eds), Academic Press, London, pp. 245-254.

Pascoe DT, Kedwards TJ, Maund SJ, Muthi E and Taylor EJ (1994) Laboratory and field evaluation of a behavioural bioassay — the *Gammarus pulex* (L.) precopula separation (GaPPS) test. *Water Research*, **28**, 369-372.

Patel MN and Wright DJ (1996) The influence of neuroactive pesticides on the behaviour of entomopathogenic nematodes. *Journal of Helminthology*, **70**, 53-61.

Pennuto CM and de Noyelles F Jr (1993) Behavioral responses of *Drunella coloradensis* (Emphemeroptera) nymphs to short-term pH reductions. *Canadian Journal of Fisheries and Aquatic Sciences*, **50**, 2693-2697.

Percy JA (1976) Responses of arctic marine crustaceans to crude oil and oil-tainted food. *Environmental Pollution* **10**, 155-162.

Peters RH 1984. Methods for the study of feeding, grazing and assimilation by zooplankton. In *A Manual on Methods for the Assessment of Secondary Productivity in Fresh Waters*, Downing JA and Rigler FH (eds), Blackwell Scientific Publications, Oxford, pp. 336-412.

Petersen LB-M and Petersen RC Jr (1983) Anomalies in hydropsychid capture nets from polluted streams. *Freshwater Biology*, **13**, 185-191.

Phelps HL, Hardy JT, Pearson WH and Apts CW (1983) clam burrowing behaviour: inhibition by copper-enriched sediment. *Marine Pollution Bulletin*, **14**, 452-455.

Poulton M and Pascoe D (1990) Disruption of precopula in *Gammarus pulex* (L.)-development of a behavioral bioassay for evaluating pollutant and parasite induced stress. *Chemosphere*, **20**, 403-415.

Preston B, Cecchine G and Snell TW (1999) Effects of pentachlorophenol on predator avoidance behavior of the rotifer *Brachionus calyciflorus*. *Aquatic Toxicology*, **44**, 201-212.

Pynnönen K (1996) Heavy metal-induced changes in the feeding and burrowing behaviour of a Baltic isopod, *Saduria* (*Mesidotea*) *entomon* L. *Marine Environmental Research*, **41**, 145-156.

Ramaswami V and Subburam V (1992) Effect of selected textile dyes on the survival, morphology, and burrowing behavior of the earthworm *Polypheretima elongata*. *Bulletin of Environmental Toxicology and Chemistry*, **48**, 249-252.

Reeve MR, Walter MA, Darcy K and Ikeda T (1977) Evaluation of potential indicators of sublethal toxic stress on marine 300 plankton (feeding, fecundity, respiration, and excretion) — Controlled ecosystem pollution experiment. *Bulletin of Marine Science*, **27**, 105-113.

Ringelberg J (1964) The positively phototactic reaction of *Daphnia magna* Straus: a contribution to the understanding of diurnal vertical migration. *Netherland Journal of Sea Research*, **213**, 319-406.

Ringelberg J (1969) Spatial orientation of planktonic crustaceans 2. The swimming in the vertical plane. *Verhandlungen. Internationale Vereinigung fur theoretische und angewandte Limnologie/Proceedings. International Association of Theoretical and Applied Limnology/Travaux*, **17**, 841-847.

Ringelberg J (1980) Introductory remarks: causal and teleological aspects of diurnal vertical migration. In *Evolution and Ecology of Zooplankton Communities*, Kerfoot WC (ed.), University Press of New England, Hanover, NH, pp. 65-68.

Rodgers JH Jr, Cherry DS, Graney RL, Dickson KL and Cairns J Jr (1980) Comparison of heavy metal interactions in acute and artificial stream bioassay stream techniques for the Asiatic clam (*Corbicula fluminea*). In *Aquatic Toxicology*, Eaton JG, Parrish PR and Hendricks AC (eds), STP 707 American Society for Testing and Materials, Philadelphia, pp. 266-280.
Römbke J and Heimbach U (1996) Experiences derived from the Carabid Beetle Laboratory Test. *Pesticide Science*, **46**, 157-162.
Salanki J and Balla L (1964) Ink-lever equipment for continuous recording of activity in mussels. *Annales Instituti Biologici (Tihany)*, **31**, 117-121.
Samu F and Vollrath F (1992) Spider orb web as bioassay for pesticide side effects. *Entomologia Experimentalis et Applicata*, **62**, 117-124.
Saunders JP, Trieff NM, Kalmaz EE and Uchida T (1985) Effect of mercuric ion on attraction to light of *Artemia* sp. *nauplii*. *Ecotoxicology and Environmental Safety*, **9**, 112-120.
Schuring BJ and Geense MJ (1972) Een electronische schakeling voor het registreren van opeingshoek van de mossel *Mytilus edulis* L. TNO-Rapport CL 72.47, 8 pp.
Siebeck O (1964) Research on the behavior of planktonic crustaceans in the littoral. *Verhandlungen. Internationale Vereinigung fur theoretische und angewandte Limnologie/Proceedings, International Association of Theoretical and Applied Limnology/Travaux*, **15**, 746-751.
Siebeck O (1969) Spatial orientation of planktonic crustaceans 1. The swimming behavior in a horizontal plane. *Verhandlungen. Internationale Vereinigung fur theoretische und angewandte Limnologie/Proceedings, International Association of Theoretical and Applied Limnology/Travaux*, **17**, 831-840.
Simonet DE, Knausenberger WI, Townsend LH Jr and Turner EC Jr (1978) A biomonitoring procedure utilizing negative phototaxis of first instar *Aëdes aegypti* larvae. *Archives of Environmental Contamination and Toxicology*, **7**, 339-347.
Sloof W, de Zwart D and Marquenie JM (1983) Detection limits of a biological monitoring system for chemical water pollution based on mussel activity. *Bulletin of Environmental Contamination and Toxicology*, **30**, 400-405.
Snell TW and Janssen CR (1995) Rotifers in ecotoxicology: a review. *Hydrobiologia*, **313/314**, 231-247.
Sørensen FF, Bayley M and Baatrup E (1995) The effects of sublethal dimethoate exposure on the locomotor behavior of the collembolan *Folsomia candida* (Isotomidae). *Environmental Toxicology and Chemistry*, **14**, 1587-1590.
Sørensen FF, Weeks JM and Baatrup E (1997) Altered locomotory behavior in woodlice (*Oniscus asellus* (L.)) collected at a polluted site. *Environmental Toxicology and Chemistry*, **16**, 685-690.
Stephenson GL, Kaushik A, Kaushik NK, Solomon KR, Steele T and Scroggins RP (1998) Use of an avoidance-response test to assess the toxicity of contaminated soils to earthworms. In *Advances in Earthworm Ecotoxicology*, Sheppard SC, Bembridge JD, Holmstrup M and Posthuma L (eds), SETAC Press, Pensacola, FL, pp. 67-81.
Stirling EA (1975) Some effects of pollutants on the behaviour of the bivalve *Tellina tenuis*. *Marine Pollution Bulletin*, **6**, 122-124.
Stone JC, Abramson CI and Price JM (1997) Task-dependent effects of dicofol (kelthane) on learning in the honey bee (*Apis mellifera*). *Bulletin of Environmental Contamination and Toxicology*, **58**, 177-183.
Stratton GW and Corke CT (1981) Interaction of permethrin with *Daphnia magna* in the presence and absence of particulate material. *Environmental Pollution Series A*, **24**, 135-144.
Strong DR (1973) Amphipod amplexus, the significance of ecotypic variation. *Ecology*, **54**, 1383-1388.

Sunila I (1981) Reproduction of *Mytilus edulis* L. (Bivalvia) in a brackish water area, the Gulf of Finland. *Annales Zoologici Fennici*, **18**, 121-128.

Sutcliffe DW and Carrick TR (1973) Studies on mountain streams in the English Lake District. I. pH, calcium and the distribution of invertebrates in the River Duddon. *Freshwater Biology*, **3**, 437-462.

Terrill WF and Dusenbery DB (1996) Threshold chemosensitivity and hypothetical chemoreceptor function of the nematode *Caenorhabditis elegans*. *Journal of Chemical Ecology*, **22**, 1463-1475.

Truhaut R (1975) Ecotoxicology — a new branch of toxicology: a general survey of its aims, methods, and prospects. In *Ecological Toxicology Research. Effects of Heavy Metal and Organohalogen Compounds*, McIntyre AD and Mills CF (eds), Plenum Press, NY, pp. 3-23.

White DS and Keilty TJ (1988) Burrowing avoidance assays of contaminated Detroit River sediments, using the freshwater oligochaete *Stylodrilus heringianus* (Lumbriculidae). *Archives of Environmental Contamination and Toxicology*, **17**, 673-681.

Whitney LV (1941) The angular distribution of characteristic diffuse light in natural waters. *Journal of Marine Research*, **4**, 122-131.

Wilder J (1940) The effects of population density upon growth, reproduction and survival of *Hyallela azteca*. *Physiological Zoology*, **13**, 439-461.

Wildish DJ (1972) Polychlorinated biphenyls (PCB) in sea water and their effect on reproduction of *Gammarus oceanicus*. *Bulletin of Environmental Contamination and Toxicology*, **7**, 182-187.

Wiles JA and Jepson PC (1994) An index of the intrinsic susceptibility of nontarget invertebrates to residual deposits of pesticides. In *Ecotoxicology of Soil Organisms*, Donker MH, Eijsackers H and Heimbach F (eds), CRC Press, Boca Raton, FL, pp. 287-301.

Wiles JA and Krogh PH (1998) Tests with the collembolans *Isotoma virdis*, *Folsomia candida* and *Folsomia fimetaria*. In *Handbook of Soil Invertebrate Toxicity Tests*, Løkke H and Van Gestel CAM (eds), John Wiley and Sons Ltd, Chichester, pp. 131-156.

Williams KA, Green DW, Pascoe D and Gower D (1987) The effects of cadmium on oviposition and egg viability in *Chironomus riparius* (Diptera-Chironomidae). *Bulletin of Environmental Contamination and Toxicology*, **38**, 86-90.

Williams PL and Dusenbery DB (1987) A screening test for neurotoxins using *Caenorhabditis elegans*. In *Model Systems in Neurotoxicology: Alternative Approaches to Animal Testing*, Alan R. Liss Inc., New York, pp. 163-170.

Williams PL and Dusenbery DB (1990) A promising indicator of neurobehavioral toxicity using the nematode *Caenorhabditis elegans*. *Toxicology and Industrial Health*, **6**, 425-440.

Wong CK, Ramcharan CW and Sprules WG (1986) Behavioral responses of a herbivorous calanoid copepod to the presence of other zooplankton. *Canadian Journal of Zoology*, **64**, 161-169.

9

Effects of Environmental Pollutants on Avian Behaviour

JOANNA BURGER[1,2], KURUNTHACHALAM KANNAN[3], JOHN P. GIESY[3], CHRISTIAN GRUE[4] AND MICHAEL GOCHFELD[2,5]

[1]Division of Life Sciences, Rutgers University, Piscataway, NJ, USA
[2]Environmental and Occupational Health Sciences Institute, Consortium for Risk Evaluation with Stakeholder Participation, and Graduate Program in Toxicology, Rutgers University, Piscataway, NJ, USA
[3]National Food Safety and Toxicology Center, Department of Zoology, Institute for Environmental Toxicology, Michigan State University, East Lansing, MI, USA
[4]School of Fisheries, University of Washington, Seattle, WA, USA
[5]Environmental and Community Medicine, UMDNJ-Robert Wood Johnson Medical School, Piscataway, NJ, USA

While the vast majority of ecotoxicology studies with animals have utilised death as an endpoint, there is growing recognition that understanding sublethal effects is critical for assessing environmental effects, and as an early warning of impending environmental problems. Defining and understanding the role of contaminants in the disruption of behaviour and reproductive success of animals is essential for maintaining healthy ecosystems (Hose and Guillette 1995). Understanding the effects of chemicals and radionuclides on behaviour of birds provides useful information for predicting population, community and ecosystem effects.

Our overall objective in this chapter is to provide a context for examining the effects of chemicals on avian behaviour. Neurobehavioural effects involving sensation, perception, cognition, integration, coordination and motor function can lead to deficits in feeding, predator avoidance and reproductive success, resulting in shifts in population dynamics and population declines. Behavioural effects may result directly from the action of neurotoxic chemicals on one or more targets of the nervous system, or secondarily from chemical effects on the endocrine system that in turn modify behaviour.

While rather sophisticated behavioural screening tests have been developed for assessing the effects of pharmaceutical and medical compounds, particularly

Behavioural Ecotoxicology. Edited by Giacomo Dell'Omo.

on mammals (MacPhail and Tilson 1995), the use of behaviour to assess sublethal and potentially widespread population effects in the wild is not well developed, particularly for birds. For example, a recent book on developmental neurotoxicology examined behavioural assessment only for rodents and primates (Slikker and Chang 1998). Similarly, recent reviews of the neurobehavioural and cognitive effects of mercury and lead (Gilbert and Grant-Webster 1995, Rice 1996) contained no information on birds, although some other reports have included information on the behavioural effects of chemicals on other vertebrates (Fiedler *et al.* 1998).

Birds are useful for ecotoxicology studies because they are large, conspicuous, diurnal and of interest to the general public. The latter point cannot be overestimated since most ecotoxicological studies require public support and interest, particularly for long-term monitoring. Furthermore, people are much more concerned when there is a massive die-off of birds, than of many other organisms, such as invertebrates. Because they are so conspicuous, a die-off is obvious and receives immediate media attention. Moreover, most birds are primarily diurnal, and communicate visually and acoustically, which means that investigators can conduct both laboratory and field studies during daylight, when behavioural disruptions are more easily observed, and can observe birds as part of normal activities. Scientists and the general public have an 'idea' about how birds behave in nature and are likely to notice unusual behaviour that may be indicative of the effects of chemicals. Disruptions in incubation and brooding behaviour, as well as lowered reproductive success, can be observed as endpoints, particularly in large birds such as hawks and eagles, and in colonial birds such as cormorants, gulls, terns and pelicans.

Birds are also useful for ecotoxicology because they are relatively long-lived, allowing time for smaller neurotoxicological effects to accumulate, leading to more obvious behavioural effects on foraging, migration and reproduction. Some chemicals bioaccumulate such that concentrations increase with age. Thus birds can be used for chemical biomonitoring, and because some species are long-lived, marked individuals can be monitored for many years. Birds can also be used for multigenerational studies of neurotoxic effects, particularly in laboratory or semi-wild settings. General features of animal models for neurotoxicity are discussed by Walsh (1997).

Birds exhibit a range of dietary habits, allowing assessment of multiple pathways of exposure, at multiple trophic levels. Bird species that are relatively low on the food chain can be used to assess contaminants in seeds and other plants. Such birds are usually quite common and can be collected without undue harm to their populations. Indeed, granivorous species, such as blackbirds and starlings, are often nuisance birds requiring some population control in agricultural areas. Birds that are intermediate on the food chain can be used as indicators of exposure at multiple trophic levels. They may also be sufficiently common to allow appropriate observation and specimen collection.

Further, their diet may more closely match a human diet, which increases their usefulness as bioindicators of human exposure and effect.

Carnivorous or piscivorous birds at the tops of food chains are useful because they bioaccumulate some chemicals, allowing for assessment of potential exposure and effects. Thus, they are particularly useful as sentinel species. The rapid declines of populations of fish-eating birds or birds of prey that have accompanied chemical exposure are very noticeable to scientists and the public alike (Gilman *et al.* 1977, Walsh 1990).

Birds are particularly useful for behavioural studies because they use visual and vocal cues for communication, and they have a developmental period where the young are dependent upon their parents for food and protection. While birds have been used extensively in ecotoxicology studies, their use in risk assessment is less common. One key aspect of risk assessment is interspecific extrapolation. There is controversy over how to extrapolate from toxicologic studies of rodents and other mammals to humans, and the problem is amplified for extrapolating from birds to mammals. Moreover, when extrapolation is performed, different scaling factors are required. For mammals, dose extrapolations use weight raised to the 0.6 to 0.7 power, however evidence for birds indicates that the appropriate scaling factor ranges from 1 to 1.5 (Mineau *et al.* 1996), suggesting that there is considerable work necessary before birds can be modelled in the course of human risk assessment. Appropriate extrapolation can only be applied when there are a sufficient number of studies that examine the effects of toxic chemicals in the field and in the laboratory, with known exposures, and known tissue concentrations associated with specific effects (Burger and Gochfeld 1997a).

9.1 A BRIEF HISTORY OF BIRDS AND ECOTOXICOLOGICAL STUDIES

Birds have been important in the history of ecotoxicological studies, particularly in their role as sentinel species for monitoring of chemical exposure. While death from toxic chemicals had been noted for many years, laboratory studies of the effects of chemicals on wildlife started only after the introduction of DDT (*p,p′* dichlorodiphenyl trichloroethane) in 1943, largely by Stickel and colleagues (Stickel 1946). Despite laboratory studies, however, it was the dramatic mortality of passerine birds such as robins (*Turdus migratorius*), and the decline of birds of prey such as bald eagles (*Haliaeetus leucocephalus*), that led to public concern (Carson 1962). Public outcry also followed the understanding that the use of lead shot led to high mortality of migrant ducks (Bellrose 1959), eventually leading many years later to the cessation of its use for waterfowl hunting.

Laboratory tests indicated that birds fed a sublethal dose of DDT and other chlorinated hydrocarbon insecticides suffered reproductive impairments that could lead to lowered reproductive success (Dewitt 1955). At the same time, declines in populations of fish-eating birds and raptors led to concern about the widespread and often indiscriminate use of persistent and bioaccumulating pesticides. DDT and its degradation products caused eggshell thinning to the point that when the parents sat on their eggs to incubate them they broke. This resulted in complete failures in some years (Ratcliffe 1967, Hickey and Anderson 1968, Hickey 1969, Blus *et al.* 1972). Laboratory experiments indicated that eggshell thinning occurred because of the inhibition of calcium-dependent ATPase in the oviduct, which reduced the transport of calcium to the site of shell formation (Peakall *et al.* 1973). In doves, DDT also induced the production of enzymes which altered endogenous oestrogen levels (Peakall 1970, NRC 1999). A detailed review of the mechanisms of action is given by Feyk and Giesy (1998) and Kendall *et al.* (1998).

The DDT-eggshell thinning example illustrates the importance of field observations and careful laboratory experiments to determine the mechanisms underlying complex effects. It also served as an early warning of the harmful effects of chemicals on wildlife populations, providing people with an example of the potential for adverse effects of human activities on birds. While the effects of DDT and other organochlorines have been rather dramatic, in terms of decreases in reproductive success and declines in populations, there is now concern for more subtle cumulative effects of PCBs and other chemicals as endocrine disrupters (Mineau *et al.* 1984, Environment Canada 1991, Colborn and Clement 1992, NRC 1999). Endocrine disruptive chemicals are being linked to declines in several vertebrate populations in the Great Lakes (Fry 1996), as well as elsewhere (NRC 1999). The mass mortality of seabirds through oiling draws mass media attention (Dunnet 1982), but there are sublethal behavioural effects from oil that lead to decreased feeding and ultimately weight losses (Burger 1997a, Burger and Tsipoura 1998) among birds that survive the acute event.

Behavioural deficits in birds and other wildlife can occur at exposure concentrations one to two orders of magnitude lower than lethal levels. These neurobehavioural deficits can, under some conditions, be manifest as changes in incubation and chick-rearing behaviour, increased vulnerability to predators (Hoffman 1995), thermoregulation, and individual recognition (Burger and Gochfeld 1993a, 1995a, 1997a).

A number of bird populations have declined significantly since primordial times. There are a number of reasons for these declines and in some cases extinctions. These include complex factors such as habitat loss, hunting, changes in nutrition and population genetics, among others. Some of the effects, however, have been caused by exposure to synthetic chemicals (Giesy *et al.* 1994a). Birds, particularly carnivorous and especially piscivorous birds, have been and continue to be exposed to a number of toxic, synthetic compounds (Giesy *et al.* 1994b). In particular, they have been exposed to a number of persistent and bioaccumulative organic compounds, many of which are halogenated. Specifically, it has been the persistent chlorinated hydrocarbons that have accumulated to the greatest concentrations and have been related to the most severe adverse effects (Ludwig *et al.* 1995). These exposures have resulted in a number of adverse effects. Organochlorine pesticides, such as dieldrin, are highly toxic. However, the organophosphorus insecticides are even more acutely toxic, and have caused widespread mortalities of adult birds. Other chemicals have had more subtle effects on the reproductive potential of birds (Kubiak *et al.* 1989), such as deformities and lethality of embryos (Gilbertson 1983, Giesy *et al.* 1994b). These effects have, in turn, caused declines in populations (Peakall 1986). It is these effects that have been referred to as neuro-endocrine disrupting effects (Colborn *et al.* 1996). While studies on the toxic effects of chemical contaminants in birds have been discussed above, very few studies have discussed the neurobehavioural effects of organic compounds in birds. In this chapter we describe the effects of environmental toxicants on behavioural alterations in birds.

9.2 MAJOR ISSUES WITH STUDIES OF BEHAVIOURAL EFFECTS

There are a number of issues surrounding the study of the effects of toxic chemicals on behaviour. These include the duration of studies, behaviour that affects growth and reproductive success versus survival, individual versus population effects, and multigenerational effects. Each of these should be considered when planning studies of the effects of chemicals on avian behaviour.

The duration of a behavioural study depends not only upon the overall objectives, but on the life history of the species studied. Behavioural studies can be short-term, examining acute effects, or they can be longer term, examining chronic or long-lasting effects from an acute exposure. Both types of studies are essential to understand behavioural effects, particularly on neurobehavioural development in young animals. Multigenerational studies are particularly important for a complete understanding of the subtle behavioural effects of toxic chemicals. Heinz (1974, 1975, 1979) found multigenerational effects of mercury in ducks (see below).

Behavioural endpoints should be selected that are biologically relevant, for example, to feeding, finding shelter, recognising mates or avoiding predators. Some of these ultimately depress growth, reduce reproductive success or lower survival. This requires information on the ecology and life history of animals in the wild. Ideally, appropriate selection of behavioural endpoints also involves consideration of the applicability of the tests in both the laboratory and the wild.

Finally, while ecotoxicological testing involves individuals, the overall effect of behavioural deficits on wildlife populations is always paramount. Thus, it is important to report not only the average effect, but the distribution of effects possible from a given dose. For example, Scheuhammer (1988) found that, among zebra finches (*Poephila guttata*) fed diets containing methylmercury chloride, those that died had levels of mercury in the brain similar to levels in those that merely showed signs of impairment and did not die.

9.3 MAJOR CLASSES OF CHEMICALS OF CONCERN

Birds are exposed to a wide range of xenobiotic chemicals that include metals and metalloids, organochlorines, polychlorinated di-aromatic compounds (PCBs, dioxins), DDT, cholinesterase-inhibiting pesticides, petroleum products (complex mixtures of aliphatic and organic compounds) and solvents (short chain chlorinated aliphatics such as tri- and tetrachloroethylene, and formerly carbon tetrachloride) and aromatic solvents such as toluene and xylene. The sources, environmental fate and the mode of action of these chemicals are described in more detail by Barron and colleagues in Chapter 4.

9.4 EFFECTS OF HEAVY METALS ON AVIAN BEHAVIOUR

Heavy metals can elicit a wide range of effects on behaviour and neurobehavioural development in birds. The effects of the metals depend on their state: organic vs. inorganic, and differ among metals. Different metals may cause the same effects even though the mechanisms are not fully understood and may be different. Further, the combined effects of different heavy metals are unclear. Most information on the effects of heavy metals on neurobehavioural development in birds exists for mercury and lead, largely because of their ubiquitousness in the environment and their neurotoxicity (Fimreite 1979, Heinz 1996a, Burger and Gochfeld 1997a,b). Therefore, the following discussion focuses on these metals. One cautionary note, both lead and mercury poisoning can lead to severe, potentially fatal effects, many of which are behavioural in nature. To understand behavioural effects it is useful to test birds with sufficiently low levels that the animals do not become clinically sick, yet show cognitive or other deficits.

9.4.1 LEAD AND NEUROBEHAVIOURAL DEVELOPMENT

Lead poisoning has been known for at least 2500 years (Paine 1995), although it gained prominence because of the morbidity and mortality in waterfowl caused by lead shot (Bellrose 1959). There is a range of toxic effects of lead that may lead to death, including loss of weight (Koh and Harper 1988), greater susceptibility to hunters (Bellrose 1959), decreased gonadal weight (Kendall *et al.* 1982), reductions in eggshell thickness (Haegele *et al.* 1974), smaller clutch size (Finley *et al.* 1976), increased egg-laying interval (Kendall *et al.* 1982), reduced hatching success (Hoffman *et al.* 1985) and a variety of physiological effects (Beyer *et al.* 1988).

Examination of the sublethal behavioural effects of lead have largely been conducted in the laboratory where dose and confounding factors can be controlled. Detailed observations on the behaviour of individuals can be made. The most common behavioural deficit noted for lead is depressed or impaired feeding behaviour. This has been observed for starlings (*Sturnus vulgaris*, Osborn *et al.* 1983), willow ptarmigan (*Lagopus lagopus*, Fimreite 1984), common terns (*Sterna hirundo*, Burger and Gochfeld 1988a), herring gulls (*Larus argentatus*, Burger and Gochfeld 1988b, 1995a) and chickens (Simpson *et al.* 1970). Hyperactivity in starlings exposed to lead was observed by Osborn *et al.* (1983), while Bellrose (1959) noted that lead resulted in shorter migration in ducks. In contrast, Frederick (1976) observed no differences in open-field locomotion behaviour of lead-treated and control mallards, using a test arena and a motion sensor. That study did not examine the frequency or intensity of different behaviours in the open field. Lead was observed to depress serial position performance in rock doves (*Columbia livia*), one of the earliest clear demonstrations of cognitive dysfunction as a result of exposure of birds to lead (Dietz *et al.* 1979).

9.4.2 LEAD AND NEUROBEHAVIOURAL DEVELOPMENT IN LARIDS

Lead is one of the most common heavy metals found in both natural and heavily-contaminated ecosystems (ATSDR 1999). The elimination of lead from gasoline resulted in a reduction of lead in the environment and in humans (Schwartz 1994), but concentrations of lead are still elevated in many inner cities and in the environment generally. Exposure to current concentrations of lead is thought to be causing neurobehavioural, haematologic, nephrotoxic and reproductive effects in humans and other animals (Cory-Slechta *et al.* 1983, Cory-Slechta 1990; Needleman *et al.* 1990, Rice 1996). Embryos and developing young are particularly vulnerable (Eisler 1988).

Because of the behavioural effects of lead on developing young, Burger and Gochfeld embarked on a research programme in the early 1980s to understand the effect of lead on neurobehavioural development, using herring gulls and common terns (Laridae) as models. These species were chosen because they are common and widespread, have been shown to be sensitive to PCBs and other chemicals in the Great Lakes (Mineau *et al.* 1984), and their behaviour and ecology is well understood in the wild (Burger and Gochfeld 1991, Pierotti and Good 1994). While early work with common terns proved useful, latter work concentrated on herring gulls because they are sufficiently abundant to be considered pests in many places, thus making it feasible to collect birds for measurement of contaminant levels in the wild and for assessment of behavioural differences in the laboratory. Moreover, as is clear from the descriptions that follow, both species responded similarly to lead exposure.

The overall paradigm was to achieve environmentally relevant concentrations of lead in blood, to examine the effect of varying time of exposure and dose on neurobehavioural development in the laboratory, and to examine the effects of lead in nature. Tests employed in the laboratory were designed to examine behaviour that is relevant to survival in the wild, and included balance, locomotion, thermoregulation, endurance, begging and feeding, and individual recognition.

9.4.3 OVERALL PROTOCOL

Birds sequester heavy metals in their feathers during their formation (Goede and de Bruin 1984, Furness *et al.* 1986). Feathers have an active blood supply during formation, which atrophies thereafter, leaving the feather as a record of circulating blood concentrations when the feather was formed. The concentrations of lead in feathers can be used to monitor the exposure of birds in the wild, and to determine the upper limit of exposure for laboratory tests (Burger and Gochfeld 1990, Burger *et al.* 1994, Burger 1997b). There is usually a significant correlation between concentrations of lead in feathers and those in internal tissues, including blood, and concentrations of lead in feathers are a good predictor of integrated internal dose (Burger 1993). The larids in our experiments had lead levels in their feathers, at the time of fledging, that were equivalent to the high end of exposure for feathers of wild herring gull fledglings (Burger and Gochfeld 1990, Burger *et al.* 1994, Burger 1997b).

Gulls and terns were injected intraperitoneally 1 to 3 times with lead acetate, in doses of 0, 25, 50 or 100 mg/kg, depending upon the experiment. Lead was not provided in the food because it is difficult to control how much birds eat, which makes it impossible to control or predict internal dose. Depending upon the experiment, chicks were tested daily, every other day, or every third day. Behavioural tests were performed by trained personnel who were blind to the exposure status of birds. Further descriptions of experiments for common terns can be found in Burger and Gochfeld (1985, 1988a), and for herring gulls in Burger (1990, 1998) and Burger and Gochfeld (1993a, b, 1995a, b, 1997a, 2000).

9.4.4 LABORATORY EXPERIMENTS

Locomotion, balance and the ability to right themselves when chicks fall over small cliffs, from low shrubs, down sand dunes or over vegetation are critical for survival in the wild, particularly when pursued by predators. Lead decreases the ability of both common terns and herring gulls to right themselves. Common tern chicks injected at 5 days take 1.2 s (SE = 0.3), compared with only 0.8

s (SE = 0.2) for controls (Burger and Gochfeld 1985). Although common tern chicks can right themselves in less time than herring gulls, due to smaller size and greater agility, the differences between controls and lead-exposed chicks were clear for both species.

Herring gull chicks injected at 2 days took 2.7 s (SE = 0.8) to right themselves, while those injected at 6 days took 1.0 s (SE = 0.0) and those injected at 12 days took 3.0 s (SE = 0.1) (compared with about 2 s for controls of all ages, Burger and Gochfeld 1995a, b). The deficit at 12 days is biologically important because by then herring gulls are moving about their territories and could easily fall (Pierotti and Good 1994, Burger and Gochfeld 1997a).

9.4.4.1 Balance

When tested on an appropriately narrow balance beam for each species, lead-treated chicks of both species remained on the balance beam for less time than did controls, and this varied by the age of injection. When injected with lead at 5 days, common tern chicks were able to remain on the balance beam for only 6 s (SE = 2), compared with 12 s (SE = 3) for controls (Burger and Gochfeld 1985).

For herring gulls, chicks injected with lead at 2 days remained on the beam for 10.1 s (SE = 0.8); those injected at 6 days remained on for 7.5 s (SE = 0.8); those injected at 12 days remained on for 5.5 s (SE = 0.2), compared with about 10 s for controls (Burger and Gochfeld 1995a, b). Thus, the effect of lead was greater when older herring gull chicks were injected, compared with younger chicks. This is surprising since generally, younger animals are more affected by exposure to toxicants than are older ones. Ecologically, the older herring gull chicks were impaired when they were beginning to move about their territories. The results may also reflect the greater agility of older chicks since when they begin to explore, if their balance is impaired by lead, they fall off quicker.

9.4.4.2 Depth perception

Depth perception is particularly important for species that nest on cliffs or in trees where misjudging the edge may result in falling considerable distances to death or starvation because parents no longer feed them outside of the nest. Lead adversely affected depth perception on a 'visual cliff' apparatus in both herring gull and common tern chicks (Figure 9.1). A greater proportion of lead-injected chicks fell off the edge of an elevated platform, compared with controls (Burger and Gochfeld 1985, 1995a). For herring gulls, the fewest chicks fell off the 'cliff' at 12 days postinjection. While most common terns and most herring gulls nest on flat ground, there are cliff-nesting herring gulls, as well as some that nest in trees or low shrubs (Burger and Gochfeld 1991, Pierotti and Good 1994).

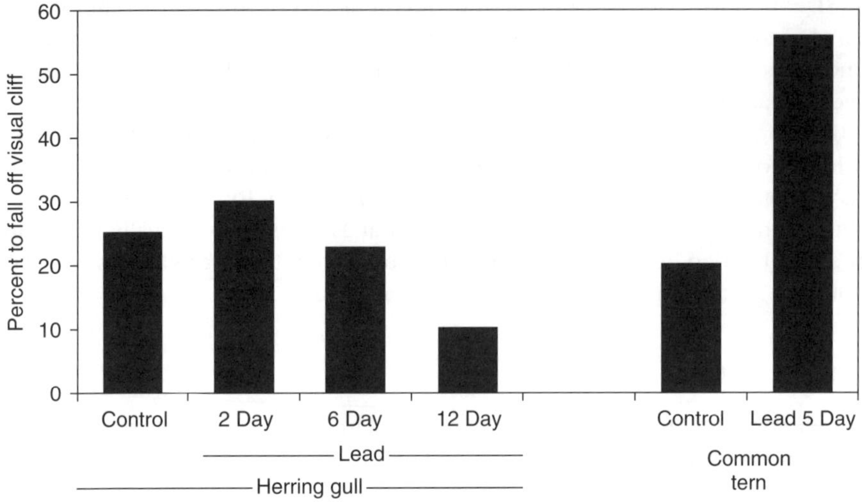

Figure 9.1 Percentage of herring gull chicks falling off a visual cliff apparatus as a function of age of exposure. Control chicks are a combination of controls for each of the experiments (after Burger and Gochfeld 1995b, Burger 1998).

9.4.4.3 Behavioural thermoregulation

Although young birds are often brooded by parents, particularly during inclement weather, predators or other disturbances can force parents from their nests, exposing chicks to thermal stress. These disturbances can sometimes extend for several minutes or hours, which requires chicks to seek the cover of vegetation for protection from inclement weather.

On a hot sunny day, chicks were tested by placing them in an apparatus that provided shade only on one side, and noting the time required for them to find the shade, and to remain in the shade (after Burger and Gochfeld 1985, 1995a, b). Exposure to lead increased the time required to find shade, and decreased the time the chicks remained in the shade, for both herring gull and common tern chicks.

9.4.4.4 Begging and feeding

Many species of birds, including most gulls and terns, normally produce broods of more than one chick, making it essential for each chick to compete successfully with siblings for food brought by parents. A chick that begs less strongly is less likely to obtain food when faced with a more competitive sibling. Begging was tested by scoring the intensity of their calls and other behaviours when soliciting food. Feeding was tested by noting how often they dropped fish or how rapidly they swallowed fish.

Common tern chicks injected with lead at 5 days of age took much longer to swallow fish than control chicks (0.9 vs. 4.3 s, $P < 0.05$, after Burger and

Gochfeld 1988a). Moreover, control common tern chicks could swallow five fish in less than 5 s, while lead-injected chicks required nearly 14 s to do so. In nature, this means that a lead-impaired chick may obtain less food each time a parent brings back a fish, because it would take so long to swallow a fish that it could easily be stolen by a neighbour or hungry siblings (Burger and Gochfeld 1991).

Similarly, herring gull chicks injected with lead at 2 days of age and tested at 45 days of age, required about 16 s to swallow one fish compared with 8 s for controls (Burger 1990). Thus, the effect of lead on food handling remained until fledging.

9.4.4.5 Individual recognition

Individual recognition of parents usually develops at the time that chicks first begin to move around their territory. Once chicks become mobile, it is critical for chicks to recognise their parents and siblings. Unless chicks can recognise their nest site, parents, or siblings they are unable to find their way back to the nest (Burger 1984). When disturbed by predators, adults fly from the nest, giving alarm calls, sending the chicks scurrying for cover. After the disturbance, the chicks must find their way back to the nest, guided by the calls of their parents (Beer 1969).

Individual recognition in herring gulls was investigated by providing them with a choice between their daily caretaker and a noncaretaker of similar appearance at opposite ends of an arena (Burger 1990, Burger and Gochfeld 1995a, 1997a), while sibling recognition was investigated by providing them a choice of one of their siblings or a nonsibling (Burger 1998). Fewer chicks exposed to lead recognised the appropriate stimulus than did control chicks (Burger and Gochfeld 1995b, 1997a, Burger 1998). Lead injected at 2-5 days depressed the percentage of correct responses (39-43% correct responses), while lead injected at 6 or 12 days had little effect on their choices (62-65 % correct responses) compared with controls (56-77% correct responses, depending on age at testing).

Moreover, the time required to make a choice varied significantly depending upon the time of exposure (Burger and Gochfeld 1995b). Normally control chicks take about a second to decide which stimulus to approach. When chicks are exposed to lead at 2 or 4 days of age the latency of response was about 2-3 s, while lead exposure at 6 days resulted in a response latency of 4 s. When chicks are exposed to lead at 12 days there is no difference in the response latency between experimentals and controls, and their response is as accurate.

These observations suggest that there is a decoupling of accuracy from response time. When exposed to lead at 2 to 4 days of age, lead-injected chicks exhibited both a loss of accuracy and an increased latency to respond. When exposed at 6 days they experienced no loss of accuracy but latency increased. When exposed at 12 days of age they showed neither a decrease in accuracy nor an increase in latency.

9.4.4.6 Neurobehavioural effects of lead in the wild

One important aspect of ecotoxicological studies conducted in the laboratory is to extrapolate to the wild, and to be able to predict effects on populations and communities. However, for most behavioural studies this is difficult because pens cannot be large enough to house birds in a truly wild situation, or the birds will not breed in semicaptivity. However, herring gulls afforded the opportunity to study behaviour in the field, as well as the laboratory, because they breed in large colonies near habitation, and can easily be manipulated and then observed from blinds.

The overall protocol was to examine the same neurobehavioural traits in the wild that had been studied in the laboratory, using chicks that were raised in the wild by their parents (Burger and Gochfeld 1994, 1996). The protocols used in the field were the same as those employed in the laboratory. In each nest, at 2 days of age, one chick was injected with the same dose as applied in the laboratory (100 mg/kg), one chick received a saline control injection, and one chick did not receive anything (herring gulls normally lay three egg-clutches). Chicks were banded with colour bands for identification, and the observer was blind to the exposure identity of the chicks. Daily observations were made from the edge of the colony. The gulls were habituated to human presence. The chicks were entirely under the care of their parents until they fledged at about 45 days of age.

There were no differences in behaviour of the two types of control chicks (saline injected or no treatment), however, lead injected chicks were impaired with respect to begging scores, walking scores and the number of times they stumbled when walking during the week following injection (Figure 9.2, Burger and Gochfeld 1994). Further, when pecking at their parents bill to stimulate feeding, lead-injected chicks missed more often than did control

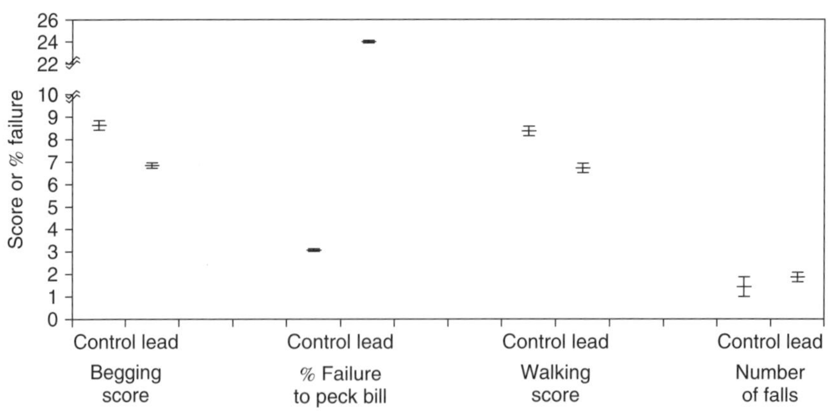

Figure 9.2 Behaviour of control and lead-injected herring gull chicks in the wild. Shown are begging and walking scores (where 10 is the best), percentage failure to peck at their parents bills when begging for food, and number of times chicks fell when running 1 m toward parents (after Burger and Gochfeld 1994, 1997a).

chicks. Although there was some recovery, by 7 weeks of age, some behavioural deficit persisted (Burger and Gochfeld 1994).

Without disturbing the chicks in the wild, it was impossible to test their individual recognition abilities in the same manner as in the laboratory. However, the chicks in the field did wander about their nests and territories, and came back to the nests when their siblings or parents called. In the wild, 10% of the lead-injected chicks were killed by their neighbours when they wandered into neighbouring territories compared with only 1% of controls (Burger and Gochfeld 1997a). This indicates that these chicks did not recognise or respond to the calls or appearances of their parents or siblings.

Although the behavioural deficits remained until fledging, parents managed to close the weight gap between lead-injected and control chicks by providing them with extra food so that by fledging there were no significant differences in weight (Burger and Gochfeld 1996). In the laboratory there are significant weight differences at fledging, with lead-treated chicks weighing less than control chicks (Burger and Gochfeld 1988b). Remarkably, at fledging, lead-injected chicks in the wild had lower concentrations of lead in the blood than did lead-injected chicks raised in the laboratory. This difference may be partly due to weight differences or to differences in diet or exercise (Burger and Gochfeld 1996).

The behaviour of the lead-injected chicks in nature was similar to that observed in the laboratory (Figure 9.3, after Burger and Gochfeld 1994, 1997a, 2000). This indicates that neurobehavioural deficits observed in the laboratory can be used to extrapolate to wild conditions, with care. It does suggest that similar behavioural traits should be tested in the field and the laboratory. These experiments indicate that lead affects the behaviour of herring gulls and common terns, although the effects differ depending upon both the dose and the time course of exposure. Not all behaviours are most severely affected at the same age. While behavioural deficits are generally considered sublethal, some lead-injected chicks in the wild were killed by neighbours, presumably due to delayed individual recognition caused by lead exposure.

One aspect that requires further study is the effect of long-term, chronic exposure of lead on behavioural development. To date we have not completed these studies, although they are critical to our understanding of the effects of lead on populations in the wild. Effects that linger between generations have been shown for mercury (Heinz 1974, 1975, 1979), and similar studies should be conducted with lead. Finally, the results of our experiments indicate that behavioural effects may be more sensitive than physiological effects because the dose used to elicit the responses was lower, the exposure was for a very brief period of time, and we did not find a difference in haematocrit levels.

9.4.4.7 Mercury and neurobehavioural effects

Although mercury occurs in both inorganic and organic (mainly methylmercury) forms in nature, methylmercury is the most toxic to birds (Eisler 1987, Wren

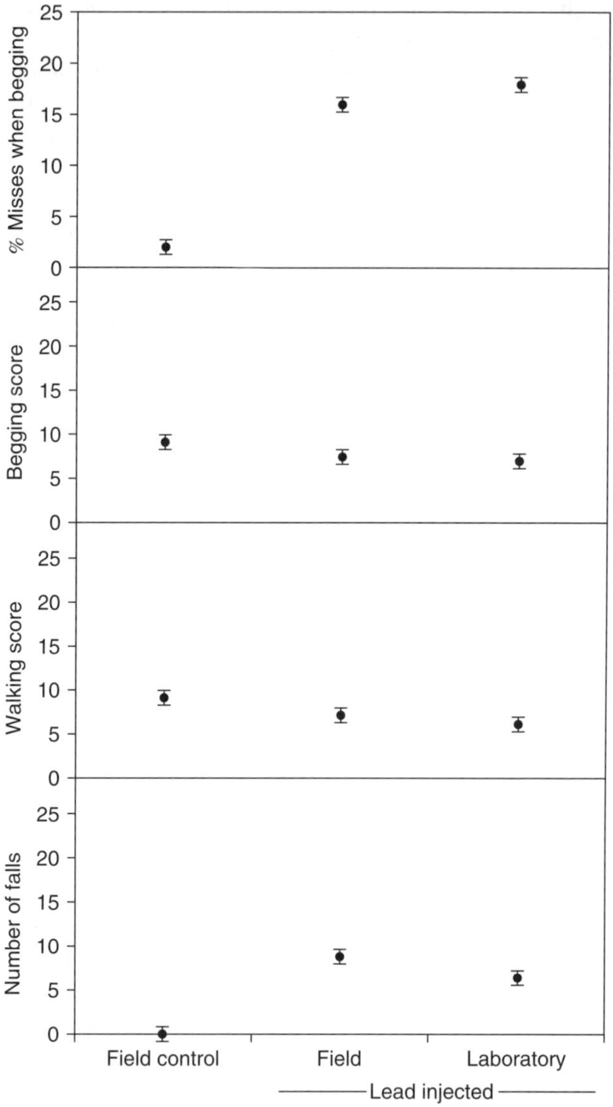

Figure 9.3 Comparison of behaviour of herring gulls raised in the field (lead-injected and controls) with lead-injected herring gulls raised in the laboratory. In all cases the lead-injected chicks differed significantly from the field controls (after Burger and Gochfeld 1994, 1997a, 2000).

et al. 1995, Elliott *et al.* 1996). Methylmercury is able to cross the blood-brain barrier (Backstrom 1969), which results in both lethality and neurobehavioural effects. Methylmercury is distributed globally in the environment, and is listed as one of the six most dangerous chemicals in the world's environment by The

International Program of Chemical Safety (Gilbert and Grant-Webster 1995). The results of laboratory studies indicate that there are behavioural effects of mercury on birds (Burger and Gochfeld 1997b), and as with lead, the fetus, embryo and developing young are most vulnerable (Gilbert and Grant-Webster 1995).

Gary Heinz conducted some of the classic studies of the effects of methylmercury on birds, including behavioural effects. He studied the behaviour of mallard (*Anas platyrhynchos*) ducklings from parents dosed with 0.5 to 3 ppm methylmercury in the feed (Heinz 1974, 1975, 1976a, b). Chicks received their exposure from the eggs (levels in the eggs were between 6 and 9 ppm, Heinz 1974, 1975). Ducklings that had been exposed during embryonic development spent less time at the end of a runway where their parent's calls were being played than did controls, and they were also hyperresponsive to a frightening stimulus (Heinz 1975). Although not conducted as a choice recognition experiment, this result indicates some impairment of recognition behaviour.

The mallards were studied for three generations (Heinz 1976a, 1976b, 1979). Mallards were fed a control diet or one containing 0.5 ppm mercury. A greater percentage of females fed mercury laid their eggs outside of their nests, which Tejning (1967) also reported for chickens fed organic mercury. Treated ducks also produced fewer ducklings, than did controls. All three generations of ducklings from females fed mercury were less responsive to recordings of parental calls and were more responsive to a frightening stimulus. However, they exhibited no differences among treatments during open-field testing (Heinz 1979). The avoidance response also can be suppressed by mercury exposure, as it was for Japanese quail (*Coturnix coturnix*, Kreitzer and Heinz 1974).

Finley and Stendell (1978) conducted similar experiments with black ducks (*Anas rubripes*), although they only followed one group of hens for two seasons. Effects included lowered clutch size, egg production, hatchability and survival of ducklings in those fed a diet containing mercury. Some of these reproductive measures could be caused, in part, by neurobehavioural deficits. For example, hens fed mercury incubated less well, and their chicks exhibited tremors and loss of coordination.

When captive zebra finches (*Poephila guttata*) were fed diets containing 5 ppm methylmercury chloride for 76 days, 25% died, and 40% exhibited neurological effects, such as lethargy, fluffing of feathers, and difficulty balancing on their perches (Scheuhammer 1988). Zebra finches fed 1 or 2.5 mg/kg methylmercury in the diet for the same time period did not exhibit signs of mercury poisoning. Neurological effects of mercury poisoning did not occur until mercury reached a concentration of 15 ppm in the brain and 30–40 ppm in the liver or kidney. However, the levels in birds that died were similar to those merely showing signs of impairment. These results indicate that a critical level of mercury had to be reached in the brain before there were neurological and behavioural effects and that there was only a narrow margin between these signs and death. It was further concluded that the levels necessary to cause neurological effects and death were similar in different species, despite differences in body size (Scheuhammer 1988).

In conclusion, studies of the effects of mercury on birds have used death or declines in reproductive success as endpoints (see reviews in Fimreite 1979, Heinz 1996a). More subtle behavioural effects that are components contributing to lowered reproductive success are studied less often, and there is a need for additional laboratory and field studies.

9.4.4.8 Other metals and neurobehavioural effects

There is little research on the potential behavioural effects of other metals, largely because laboratory studies have concentrated on acute effects, rather than chronic, low-level effects.

9.4.4.9 Chromium and Manganese

The behavioural effects of chromium and manganese were examined by Burger and Gochfeld (1995c). Two-day old herring gulls were injected with manganese (25 mg/kg), chromium (50 mg/kg), or a saline solution. There were significant differences in weight gain, with the controls gaining weight faster than either treated group. Of 14 behavioural endpoints tested, there were significant differences in righting time, balance, thermoregulation and visual cliff; controls performed the best or equally so on all measures. Some of the behaviours that have direct relevance in the field, and influence survival (see section above on lead, Figure 9.4), show that chromium and manganese-impaired birds would likely be less able to compete successfully for food, find shade quickly during thermal stress, and move about their territory without falling.

There has been little other work with either metal, except that Heinz and Haseltine (1981) found no differences between control and chromium-exposed young black ducks in avoidance behaviour to a frightening stimulus.

9.4.4.10 Selenium

There has been considerable interest in the effects of selenium, particularly on survival and reproductive success, largely because of the effects noted at Kesterson Reservoir in California, where high selenium levels resulted from agricultural drainage water (Ohlendorf *et al.* 1988). Relatively high concentrations of selenium in the kidneys and liver were associated with symptoms such as hepatic lesions, liver changes and congenital malformations (Ohlendorf *et al.* 1988, 1990), as well as lowered reproductive success and adult mortality (Ohlendorf *et al.* 1986). Similar effects have occurred elsewhere in the western United States (Skorupa and Ohlendorf 1991) and in Texas (King *et al.* 1994). Similar reproductive effects were also obtained in controlled laboratory conditions (Eisler 1985a, Heinz and Fitzgerald 1993, Heinz 1996b). The wide-ranging

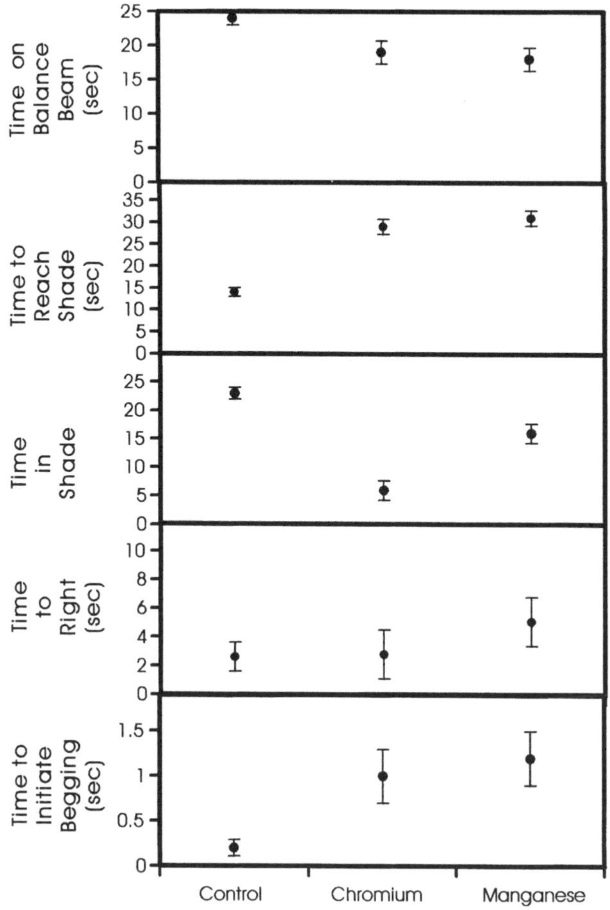

Figure 9.4 Behavioural responses of herring gulls exposed to chromium, manganese, and a saline solution at 2 days of age and tested within the next week (after Burger and Gochfeld 1995c).

effects of selenium on reproductive success suggests that there are subtle behavioural effects from selenium, but this needs to be carefully examined in laboratory studies.

9.4.4.11 Cadmium, Aluminum, Tin

There is relatively little research on the effects of other metals on the behaviour of birds. Most research with cadmium in birds has dealt with teratogenic, pathological and physiological effects, as well as acute toxicity (Cain *et al.* 1983, Eisler 1985b, Wren *et al.* 1995). Young birds seem to be more susceptible to cadmium poisoning (Wren *et al.* 1995), suggesting that they may also be more susceptible to neurobehavioural deficits during this time, but this requires experimentation.

Interest in aluminium in birds largely stems from the observation of reduced reproductive success on acidified lakes (Rattner *et al.* 1987). Birds foraging in acidified lakes suffer from increased effects of aluminium due to high demands of calcium (Nybo 1996). Nybo (1996) examined the effects of six different diets where aluminium was combined with calcium and phosphorus, but was interested in bone kinetics rather than in behavioural differences.

Controlled laboratory experiments on the effect of tin on birds has indicated a range of effects, from weight loss and feeding suppression, to collapse, convulsions and death (Eisler 1989). Few studies have examined more subtle behavioural effects, although Idemudia and McMillan (1986) reported that four intramuscular injections of triethyltin (2 weeks apart, at 1.75 mg/kg body weight) resulted in suppression of pecking behaviour, although by 27 h the birds had recovered. In mammals, the hippocampus is among the targets of organotin as well as organolead, resulting in hyperactivity of limbic system responses (Brown *et al.* 1979); this needs to be examined in the avian brain.

9.5 ORGANOPHOSPHATES AND CARBAMATES AND NEUROBEHAVIOURAL EFFECTS IN BIRDS

Use of organophosphate (OP) insecticides for agricultural purposes increased as the use of more persistent and ecologically harmful chlorinated hydrocarbons has been restricted since the early 1970s. Many OP insecticides, however, are hazardous to wildlife for short periods following application and have caused mortalities of some species of birds. Although OPs and carbamates are relatively nonpersistent, they can be acutely toxic. Several cases of inadvertent poisoning resulting in deaths of nontarget birds have occurred (Stone 1979, White *et al.* 1979). As many as 2.9 million song birds were killed during the fenitrothion-phosphamidon spray programme for spruce budworm (Pearce *et al.* 1976). In addition to passerine birds, poisoning of raptors such as bald eagles, red-tailed hawks (*Buteo jamaicensis*) and great horned owl (*Bubo virginianus*) by OP compounds have been reported (Henny *et al.* 1987).

Exposure to OP and carbamates can be primary, through direct contact with the pesticide spray, or secondary, through ingestion of contaminated, dead or struggling insects or birds. Although OP compounds and carbamates do not biomagnify in the food chain, it is possible for birds to be exposed to toxic concentrations through contaminated prey items. Concentrations of fenitrothion in insects, spiders, bud worm larvae, and moss capsules 3 h after aerial application of 210 g active ingredient (a.i.) per hectare ranged from 0.01 to 17 mg/kg, wet wt (Forsyth and Martin 1993). Grasshoppers collected following aerial application of carbofuran at a rate of 134 g a.i./ha contained 6.1 to 38 mg/kg, wet wt (Martin *et al.* 1996). Similarly, concentrations of chlorpyrifos and dimethoate in grasshoppers were in the range of several parts-per-million immediately following their application in grasshopper control (Martin *et al.* 1996). Application of carbaryl at 1.44 kg a.i per ha resulted in accumulation in killdeer (*Charadrius vociferus*) gizzard at concentrations ranging from 0.4 to 1.86 mg/kg, wet wt (Fair *et al.* 1995). These results suggest that contaminated prey items can be a major source of exposure of birds to OPs and carbamates.

Both OPs and carbamates are known to cross the blood-brain barrier and thus they can affect brain neurochemistry. Several studies have established a relationship between exposure to OP and carbamate pesticides and behavioural changes in birds. Behavioural changes in birds due to OP exposure have been observed in laboratory studies. Physiological distress, reduced singing activity, memory effects, anorexia and susceptibility to predation have been reported as some of the physiological and behavioural responses of common grackles (*Quiscalus quiscala*) exposed to sublethal concentrations of anti-acetylcholinesterase agents (Grue *et al.* 1991). Dietary exposure of ring-necked pheasants, northern bobwhites and Japanese quail to sublethal concentrations of OP insecticides caused reductions in egg production, fertility, hatchability and survival (Shellenberger *et al.* 1966, Rattner *et al.* 1982, Martin *et al.* 1996). Reductions in hormone concentrations, which possibly resulted from altered hypothalamo-hypophysical function or steroid metabolism, also were observed in bobwhite quail fed diets treated with ethyl parathion (Rattner *et al.* 1982). In a semi-field investigation of spraying methyl parathion in home ranges of California quail (*Callipepla californica*), no measurable effects for nest attentiveness, use of treated areas, nest abandonment or recruitment were found. However, California quail exposed to low levels of methyl parathion exhibited poor reproductive performance (Kilbride *et al.* 1992).

9.5.1 CHOLINESTERASE INHIBITION

The toxicity of OP and carbamate pesticides is primarily due to the inhibition of acetylcholinesterase (AChE), an enzyme necessary for hydrolysing the neurotransmitter acetylcholine in the cholinergic synapses of the nervous system (see also Chapter 3 and Ecobichon 1997 for a review). Excessive accumulation of acetylcholine results in blockage of neurotransmission and death likely occurs when the brain's respiratory centre or the myoneural junctions of respiratory muscles are blocked. Measurement of brain and blood acetylcholinesterase (AchE) inhibition has been used as a biomarker of exposure to a wide ranges of OP insecticides and carbamates including aminocarb, mexacarbate, dimethoate, acephate, carbaryl and trichlorfen (Holmes and Boag 1990). In birds plasma butyrylcholinesterase (BChE) and other plasma and liver microsomal carboxyl esterase (CbEs) have been also suggested as alternative biomarkers of OP and carbamate exposure (Fossi *et al.* 1994). Studies have shown that a fall in brain AChE activity to 50% of normal can bring about significant changes in bird behaviour (Mineau and Peakall 1987, Hart 1993, Fryday *et al.* 1996). A semifield exposure study involving feeding of chlorpyrifos, dimethoate and carbofuran contaminated grasshoppers fed to pheasants inhibited AChE activity (Martin *et al.* 1996). Brain cholinesterase levels were inhibited in bobwhite exposed to methyl parathion and chlorpyrifos (Bussiere *et al.* 1989, Cairns *et al.* 1991).

9.5.1.1 Hiding and decreased activity

Reduced activity and lethargy are common symptoms of birds exposed to OPs (Grue 1982). European starlings exposed to chlorfenvinphos in captivity often conceal themselves (Fryday *et al.* 1996). It has been suggested that reduced behavioural activity of birds exposed to OP compounds could bias measures of exposure in field studies if the affected individuals were underrepresented in samples collected by shooting or netting (Millikin and Smith 1990). When poisoned birds seek cover, it is more difficult to find causalities in field trials and the probability that dead birds will be found and recorded in wildlife casuality monitoring schemes is reduced. This sampling bias could result in an underestimation of the actual effects of OP compounds in birds.

In contrast to that observed in OP-exposed birds, mourning doves and bobwhite quails receiving a lethal dose of aldicarb, a carbamate insecticide, are limited in their cover-seeking ability due to the rapid onset of physical impairment (ataxia) associated with this compound (Hawkes *et al.* 1996). In fact, birds that exhibited overt symptoms of carbamate exposure remained at the release site until obvious manifestations of carbamate poisoning subsided or mortality occurred. Birds that become intoxicated while feeding on a carbamate pesticide treated agricultural field may remain motionless until toxic effects decrease or mortality occurs. Their ability to find cover if threatened by a predator is severally impaired.

White-throated sparrows (*Zonotrichia albicollis*) exposed to fenitrothion spent less time hopping, singing, perching, bill wiping, preening and pecking activities than controls (Forsyth and Martin 1993). Operantly conditioned key-pecking in ring doves (*Streptopelia risoria*) decreased following intermuscular injection of mevinphos (Lewis *et al.* 1973). A substantial reduction in singing activity of zebra finches was recorded in fenitrothion-treated spruce fields and this was suggested to be due to adult mortality (Pearce *et al.* 1976).

9.5.1.2 Susceptibility to predation

Northern bobwhite quail dosed with methyl parathion and exhibiting cholinesterase depression were more susceptible to predation by cats than were birds without ChE depression (Galindo *et al.* 1985). House sparrows (*Passer domesticus*) dosed with fenthion were more susceptible to predation by American kestrels (*Falco sparverius*, Hawkes *et al.* 1996).

9.5.1.3 Food avoidance

OP compounds can influence feeding behaviour by inhibiting cholinesterase in the central nervous system. This, in turn suppresses appetite and reduces avoidance learning and results in peripheral blockade, which reduces food-seeking or motivational behaviour (Adams 1977). House sparrows avoided foods treated with temephos or bromophos (Hill 1971). Similarly, ring-necked pheasants, in free-choice feeding trials, avoided foods treated with diazinon and shifted food preferences when preferred foods were treated (Bennett and Prince 1981). The effect of methyl parathion on dietary discrimination ability was assessed in 2-week-old northern bobwhite (Bussiere *et al.* 1989). After exposure to an initial oral dose of methyl parathion, feeding behaviour and taste discrimination ability of bobwhite chicks was impaired due to exposure to methyl parathion. Decreases in dietary discrimination ability and food-seeking activity were correlated with a concomitant decrease in brain AChE activity (Bussiere *et al.* 1989). Behavioural and physiological responses of grackles exposed to dietary concentrations of dicrotophos, methyl parathion, fenthion and fenitrothion suggested that the mortality was due to pesticide-induced anorexia (Grue 1982).

Effects of intoxication on feeding activity are suggested to be context dependent because animals are adapted to detecting and avoiding toxicants in the diet. Reduced food intake by birds exposed to OPs has been described as anorexia and attributed to physiological distress (Grue *et al.* 1991), but this response may be more complex. Captive birds avoid food treated with pesticides and feed selectively on untreated food if it is available. In the absence of an alternative or if treated and untreated foods are not readily distinguished, birds offered food containing OPs appear to starve themselves to death, although they continue to eat small quantities and resume normal intake levels rapidly when untreated food is restored (Hart 1993).

9.5.1.4 Orientation

Laboratory and field experiments have demonstrated that birds utilise several environmental cues such as stars, geomagnetic field and polarised light at sunset to assist them in proper orientation during migration (Able 1991). Dietary exposure of migratory white-throated sparrows to acephate at environmentally relevant concentration (256 mg/kg) affected migratory orientation. It has been suggested that exposure to acephate could affect the memory of the migratory route and wintering ground (Vyas *et al.* 1995).

9.6 ORGANOCHLORINES AND NEUROBEHAVIOURAL EFFECTS IN BIRDS

Polyhalogenated aromatic hydrocarbons (PAHs) such as DDT, polychlorinated biphenyls (PCBs), polychlorinated dibenzo-*p*-dioxins (PCDDs) and polychlorinated dibenzofurans (PCDFs) are widespread environmental contaminants.

Due to their persistent and bioaccumulative properties, high concentrations of these compounds can accumulate in predators at the top of a food chain. During the past few decades a number of environmental studies have shown links between the accumulation of these organochlorine compounds in bird tissues and a variety of effects observed in fish-eating water birds (Gilbertson *et al.* 1991, Bosveld and Van den Berg 1994, Giesy *et al.* 1994a,b). While embryotoxic effects and reproductive failure caused by organochlorines have been the subject of extensive investigation (Kubiak *et al.* 1989, Barron *et al.* 1995), relatively less information is available regarding neurobehavioural effects of these compounds. However, a number of these compounds have been demonstrated to cause neurobehavioural effects in humans and wildlife (Giesy and Kannan 1998). Some of these compounds are known to be neurotoxic. Evidence for neurobehavioural effects of organochlorines are available from epidemiological and laboratory rodent exposure studies. Accidental exposure to PCBs via contaminated cooking oil in Japan and Taiwan resulted in developmental and psychomotor delays in children born to exposed women (Tilson and Kodavanti 1997). Laboratory animal exposure studies have shown that PCBs can cause neurobehavioural effects in exposed monkeys and rodents (Tilson *et al.* 1990). In general, experiments with laboratory animals support the conclusion that PCBs are developmental neurotoxicants producing effects on activity, neurological development and cognitive function.

Field observations of reproductive failure in birds have suggested that ingestion of organochlorines has affected their behaviour. For instance, exposure to organochlorines has been associated with cessation of nest building and egg incubation, and destruction and cannibalism of eggs (Ratcliffe 1970).

The behaviour of birds can be influenced by exposure to certain organochlorine pollutants (Peakall and Peakall 1973, Peakall 1985). Few field studies of behavioural effects have been conducted because it is difficult to make observations and to isolate the potential causative agents. However, there are some correlative studies that have been conducted. The results of these studies need to be considered as speculative, since there might be a range of chemical and nonchemical factors that are all correlated with the observed effects. One such study was that conducted by Mora *et al.* (1993). In that investigation, site tenacity of Caspian terns (*Sterna caspia*) in the North American Great Lakes was studied. Breeding colony site tenacity was found to be inversely correlated with concentrations of PCBs in the blood of the terns. The authors speculated that this may have been due to direct neurobehavioural effects, but it could also have been due to depressed reproduction at the locations where concentrations of PCBs were greater. It is known that birds tend to migrate from areas if they are unable to reproduce successfully. In the case of some species of birds on the Great Lakes, concentrations of certain PCB congeners have been correlated with both embryo-lethality and greater rates of congenital deformities (Giesy *et al.* 1994a,b). Thus, the effects on reproduction may have been responsible for the decreased site tenacity at those locations. A field study in Green Bay,

Wisconsin, USA, found that Forster's tern (*Sterna forsteri*) eggs from a PCB contaminated location took longer to hatch than those from a less contaminated location (Kubiak *et al.* 1989). Inadequate parental care was implicated as the cause of poorer hatching success of herring gulls and Forster's terns breeding on the Great Lakes (Fox *et al.* 1978). Nevertheless, establishment of a cause - effect relationship based on field studies of pollutant-related behavioural changes is difficult because of other confounding factors such as weather, changes in food supply and human disturbances. Moreover, the contaminant to which the birds are exposed is rarely a single compound and the composition and concentration of the mixture is likely to vary during the course of the experiment.

9.6.1 PCBs

Several laboratory exposure studies have reported behavioural effects in birds exposed to organochlorine compounds. Behavioural effects caused by exposure to PCBs in birds include delayed growth, decreased parental attentiveness, impaired courtship behaviour and impaired avoidance behaviour (Dahlgren *et al.* 1972, Peakall and Peakall 1973). Exposure of mourning doves (*Zenaida macroura*) to 10 mg/kg Aroclor 1254 in the diet for 42 days significantly inhibited courtship behaviour intensity and delayed nesting (Tori and Peterle 1983). The observed effect of PCBs on courtship behaviour and nesting was thought to be due to the imbalance of gonadotrophic hormones such as oestrogen and testosterone, an effect already reported for the American kestrel (Lincer and Peakall 1970). Dietary exposure to 10 mg/kg of Aroclor 1254 of ring doves resulted in reduced brain neurotransmitters such as dopamine and norepinephrine concentrations by 39% and 45% relative to control values, respectively (Heinz *et al.* 1980). These results suggested that depletion of brain neurotransmitters might result in abnormal behaviour in sensitive avian species. Ring doves exposed to 3,3′,4,4′-tetrachlorobiphenyl (PCB 77) exhibited altered retinol (vitamin A) synthesis (Spear *et al.* 1989). Vitamin A deficiency in wildlife has been earlier associated with reduced growth, and visual and nervous disorders (Brouwer *et al.* 1989). Dietary exposure of 7-days old Japanese quail to 200 mg/kg of Aroclor 1254 for 8 days resulted in suppression of avoidance response to a moving silhouette (Kreitzer and Heinz 1974). When European robins were exposed to Clophen A50, a technical PCB mixture, they exhibited increased nocturnal activity compared with unexposed individuals (Ulfstrand *et al.* 1971). Exposure to a mixture of organochlorines including PCBs, *p,p*′-DDE, mirex and photomirex in the diet of ring doves altered the nature and duration of courtship behaviour in a dose-dependent fashion. The concentrations of oestradiol and testosterone in both males and females were altered following the exposure (McArthur *et al.* 1983).

Construction of nests by birds requires complex behaviours that are controlled by hormone status (Logan and Wingfield 1995). Since exposure to some environmental residues can lead to changes in hormone status and thus, potentially behaviour (McArthur *et al.* 1983). Construction of bird nests is analogous to construction of webs by spiders, which has been used for a long time to study the effects of drugs and environmental contaminants on behaviour. By monitoring the construction of nests by birds, the overall behaviour and coordination of birds can be monitored. If either behaviour or locomotor coordination is impaired, changes in the form and quality of nests might be expected.

Tree swallows have a limited home range and, if appropriate nest boxes are provided, will nest in them. Tree swallows eat insects, and when nesting near water, eat primarily emergent aquatic insects. For this reason, they are a good sentinel species for monitoring for the potential adverse effects of contaminants accumulated into sediments (Ankley *et al.* 1993, Jones *et al.* 1993, Froese *et al.* 1998). For this reason, in some locations, tree swallows have been exposed to and accumulated relatively great concentrations of some residues, including organochlorines. Two studies found relatively great concentrations of PCBs accumulated from sediments in tree swallows accompanied by changes in the nests construction (McCarty and Secord 1999, Secord *et al.* 1999). Neuro-behavioural effects of exposure to PCBs were inferred because the mass of the nest and the general quality of nests of birds exposed to greater concentrations of PCBs were less than those in areas where exposure to PCBs was lower.

Because the concentrations of co-occurring residues, such as DDTs were relatively small in the tree swallows, it is unlikely that the observed neuro-behavioural effects were caused by other residues. However, the age of the bird is a confounding factor in nest construction. Younger birds generally build smaller nests of lesser quality. In the field study reported by McCarthy and Secord (1999) the birds at the location where exposure to PCBs was high, were younger than the birds at the reference location where exposure to PCBs was low. Thus, there was a confounding factor that makes it impossible to conclude that there was a direct behavioural effect of PCBs alone. However, this study indicates the types of studies that can be used to monitor for subtle behavioural effects of exposures to residues in the environment. To be a more useful tool, an atlas of nest characteristics for a range of potential sentinel species would need to be established.

To be a useful indicator of adverse effects on birds, responses such as nest architecture should be correlated with a more tangible measure of population fitness. In the case of tree swallows, nest architecture is related to reproductive performance (Lombardo 1994).

9.6.2 DDT

Several studies have reported that aberrant reproductive behaviour reduced fecundity in birds exposed to *p,p′*-DDE. Dietary exposure of ring doves to *p,p′*-DDE inhibited courtship and nesting behaviours, which resulted in reduced reproductive performance (Tori and Peterle 1983). Circulating concentrations of luteinising hormone in ring doves fed *p,p′*-DDE increased at a lesser rate than that of unexposed individuals. This effect ultimately delayed nesting and egg-laying (Richie and Peterle 1979). Field studies have shown that exposure to great concentrations of DDT (212 mg/kg in brain, 838 mg/kg in liver) in wild bald eagles have caused tremors prior to death (Garcelon and Thomas 1997). A low degree of nest defence among merlins (*Falco columbarius*) having thin eggshells and high *p,p′*-DDE concentrations in eggs compared with those of less affected pairs was observed (Fox and Donald 1980). Based on this a suggested threshold of 8 mg/kg, wet wt, *p,p′*-DDE in eggs was proposed, above which abnormal behaviour could be seen (Fox and Donald 1980).

9.6.3 OTHER ORGANOCHLORINE PESTICIDES

Northern bobwhite quail (*Colinus virginianus*) trained to peck a green lighted key to receive food were monitored for 14 days (Gesell *et al.* 1979). They were then dosed with five different concentrations of 0.05 to 0.30 mg/kg dieldrin per day for 42 days. All the concentrations of dieldrin altered the cage behaviour of bobwhites. Doses greater than 0.1 mg of dieldrin resulted in slower and less accurate responses to operant conditioning. Some birds from 0.2 to 0.3 mg dose-group died before the 28-days test ended. Mean concentrations of dieldrin in the brain of bobwhites that received 0.05, 0.1, 0.2, 0.25 or 0.3 mg dieldrin were 2.58, 4.01, 11.1, 11.1 or 11.8 mg/kg, wet wt, respectively. Cage-spacing and vocalisation of all treated quail ceased within 15 days after treatment with all doses. Neural breakdown of fatally exposed quail resulted in body tremors and an inability to remain upright.

Mallards fed 4 to 30 mg/kg of dieldrin in diet from hatching for up to 75 days resulted in a dose-dependent decrease in brain neurotransmitters such as seratonin, norepinephrine and dopamine (Sharma *et al.* 1971, 1976). Exposure to dieldrin produced measurable changes in aggressive behaviour even at a dose of 4 mg/kg, the lowest concentration tested. Exposure to dieldrin has resulted in decreased avoidance behaviour, altered visual cliff performance in pheasants (*Phasianus colchicus*) and Japanese quail (Dahlgren and Linder 1974, Kreitzer and Heinz 1974). Similarly, exposure of birds to other cyclodiene insecticides such as endrin and chlordane and toxaphene has resulted in decreased avoidance behaviour and increases in errors during discrimination acquisition experiments (Kreitzer and Heinz 1974, Kreitzer 1980)

9.6.4 DIOXINS AND RELATED COMPOUNDS

Piscivorous birds inhabiting areas contaminated with 2,3,7,8-tetrachloro-dibenzo-*p*-dioxin (TCDD) and related chemicals are chronically exposed during embryonic development via the yolk. If TCDD-like chemicals have antioestrogenic effects (Janz and Bellward 1996) in birds, *in ovo* exposure to these compounds during the perinatal period may be responsible for certain behavioural characteristics of the reproductive dysfunction observed in wild fish-eating bird populations. However, *in ovo* exposure of chickens, pigeons and great blue herons (*Ardea herodias*) to TCDD did not affect oestrogen receptor and plasma oestradiol concentrations in female birds at doses that caused minimal embryo mortality (Janz and Bellward 1996). These results suggest that *in ovo* doses of TCDD likely to result in significant effects on oestrogen homeostasis are in the lethal range for the early developing embryo. Similarly no significant differences in parental attentiveness or food provisioning were observed in ospreys (*Pandion haliaetus*) from a TCDD-contaminated and reference location in Wisconsin, USA (Woodford *et al.* 1998).

Despite the documented behavioural effects of organochlorines in birds, the mechanism for the observed effects remains obscure. Compounds like dieldrin and PCBs can alter the concentrations of brain neurotransmitters following exposure, and thus the effects could be direct. Other organochlorines may cause behavioural effects through mechanisms involving endocrine disruption (disruption of steroid hormone action), thyroid hormone metabolism or by affecting homeostasis of vitamin A.

Organochlorines such as TCDD and TCDD-like chemicals can affect endocrine system of humans and wildlife through a variety of mechanisms (Gasciewicz 1997). These compounds, referred to also as xenooestrogens, can alter sex hormone concentrations and metabolism by mimicking the action of endogenous oestrogen or by modulating the pathways of steroid hormone metabolism. Organochlorine insecticides such as *o,p′*-DDT, kepone and methoxychlor can bind to the chicken oestrogen receptor and produce oestrogenic effects *in vivo* (Eroschenko and Palmiter 1980, Kupfer and Bulger 1980). For instance, early administration of sex hormone can have profound and permanent sex reversing effects on adult behaviour in birds (Adkins-Regan 1987, see also Chapter 2 in this book). When male Japanese quail embryos are injected with 1 μg oestradiol during a critical period of incubation, sex-dependent behaviour can be reversed. As adults they fail to mount, crow, or strut; they are completely demasculinised and are behaviourally indistinguishable from females. Female quail treated with an antioestrogen before hatching can be masculinised, and will mount other females as adults (Adkins-Regan 1987). The critical period for the sexual differentiation of behaviour varies with the biology of each bird species. In order for the avian embryo to be exposed to xenooestrogens during the critical period, the oestrogenic substances must be passed from the female to eggs. Some oestrogenic pollutants of concern, such as *o,p′*-DDT and some PCB congeners are readily transferred to lipid-rich yolk. It is quite possible for fish-eating birds to acquire substantial exposure to such residues from their diet, and then to transfer these residues to egg yolk.

Organochlorine-induced oestrogenic effects in avian populations have also been suggested based on field observation of increase in the incidence of female-female pairings in gull populations in regions contaminated by DDT (Fry *et al.* 1987, Fox 1992). The incidence of female-female pairings in a colony is usually estimated by documenting the number of nests which contain an abnormally large number of eggs, or 'supernormal clutches'. A single female gull typically lays one to three eggs. The situations by which four-egg clutches arise are variable and are in many cases unclear, and often are not the result of female-female pairings (Conover *et al.* 1979). However, nests with five or more eggs are almost always the result of multiple-female associations. Documentation of the percentage of nests with five or more eggs provides a reasonable assessment of the incidence of female-female pairing in a colony, although it is probably an underestimation of the actual incidence. This was suggested to be due to a skew in sex ratio of gulls in North America. However, direct evidence for the hypothesis that environmental contaminants may have played a role in the sex ratio skew is lacking, and any mechanism which could cause differential male mortality in gull population needs further investigation.

The effects of organochlorines on thyroid hormone can have profound effects on neurological function, which are similar to the endocrine disruption mechanism for the observed behavioural changes in birds. Thyroid hormones have been shown to act on neurological development by increasing the rate of neuronal proliferation in the cerebellum and then to stimulate differentiation. PCBs and TCDD can alter thyroid function by mimicking active thyroid hormones T3 and T4, which all share similar structural properties. They are all halogenated and have two phenyl rings. PCBs, TCDD and thyroid hormones all have similar protein binding characteristics and bind to cytosolic Ah-receptor and to thyroid hormone binding proteins such as transthyretin and thyroid hormone receptor (Porterfield 1994). Thus, PCBs and TCDD-related chemicals can alter thyroid function, which can ultimately result in neurological damage in developing embryo. However, *in ovo* exposure of TCDD in chicken, pigeon and great blue heron did not affect plasma thyroid hormone levels (Janz and Bellward 1996).

9.7 OTHER ORGANIC COMPOUNDS

9.7.1 OIL

Increased oil transport operations and off-shore drilling have resulted in oil spills in coastal areas (Burger 1997c). Sea birds and shorebirds are often affected by these oil spills because they concentrate and forage along the ocean (Rosie and Barnes 1983). Several studies have examined the effects of oil on reproduction and other physiological parameters in birds (Rosie and Barnes 1982). However, relatively few studies have reported on the behavioural effects of oils in birds. Studies have shown that oiling of seabirds increases metabolic rate and decreases growth (Butler *et al.* 1986, Boersma *et al.* 1988). Similarly, exposure of birds to oil can result in changes in steroid hormones (Fowler *et al.* 1995). Mallard ducklings exposed by aerial spray to crude oil via eggs had decreased avoidance behaviour (Albers and Heinz 1983). A recent study reported that oiled sanderlings (*Calidris alba*) spent more time preening and bathing and less time resting than control birds (Burger and Tsipoura 1998). In addition, oiled sanderlings were also more aggressive than unoiled birds.

9.7.2 TRIPHENYL PHOSPHITE

Triphenyl phosphite (TPP) is a triphenyl ester derivative of phosphorus acid and serves as a chemical intermediate in the manufacture of vinyl and other types of plastics, epoxy resins and certain pharmaceuticals (Tanaka *et al.* 1992). It has been established that TPP is a potent neurotoxicant with exposure resulting in hyperexcitability, bidirectional circling, ataxia, spasticity in cats and rodents (Veronesi *et al.* 1986). Domestic chickens exposed to a single subcutaneous injection of 1000 mg TPP/kg body weight exhibited ataxia and paralysis after 7 days, due to the effect of TPP on central nervous system (Tanaka *et al.* 1992).

9.8 CONCLUSIONS

Birds are useful models for studying subtle neurobehavioural effects, because, as in humans, their primary modes of communication are visual and acoustic. This facilitates study designs including direct observations of behaviour. Moreover, since ethologists have provided extensive information on bird behaviour, and since extensive ethograms have been established for many species, there is a solid foundation on which to build studies of chemical perturbation of behaviour. What remains largely unknown is the selectivity of different chemicals for different targets within the nervous system, as well as differential sensitivity of these targets among species, gender and age classes. Other factors influencing response to neurotoxicants can include endocrine state which in turn varies with season. So far, most of the studies described have been conducted in the laboratory, mainly to assess the toxicity of pesticides currently in use. Field studies have been mainly correlative, linking behavioural changes to concentrations of contaminants in environmental media, from which exposure and dose are sometimes inferred. Only a few studies have been brought from the laboratory into the field, or vice-versa, and these offer great promise for understanding not only dose – response relationships for specific endpoints, but the impact of chemical exposures on behaviour, reproduction and survival under natural conditions.

Chronic exposure to most heavy metals has some effects on the nervous system, but mercury, lead and manganese are among the best known. Although the organochlorines have some direct neurotoxic properties, their main impacts on behaviour appear to be mediated by the endocrine system, and the recent resurgence of interest in endocrine disrupting compounds (EDs), promises to provide new insights into the mechanisms of behavioural impairment (see also Chapter 11). The OP and carbamate toxicity has been well-studied in birds and other organisms, particularly with regard to lethality, and in many cases observations on behaviour are made incidentally at sublethal doses. Studies particularly aimed at understanding the behavioural effects of cholinesterase inhibiting chemicals on behaviour, have demonstrated important consequences which are not surprising, given that acetylcholine is the pre-eminent neurotransmitter effecting not only central, but peripheral and autonomic functions. If only the autonomic system were effected, there would still be behavioural consequences related to parasympathetic dysfunction, particularly. The role of acetylcholine and the impact of cholinesterase inhibition in the central nervous system, still requires extensive study.

ACKNOWLEDGEMENTS

We thank the following organisations for funding over the years to support our research: NIMH (JB), NIEHS ESO 5022, 5059 (JB/MG), NSF (JB), Environmental Protection Agency (JB/ MG), Consortium for Risk Evaluation with Stakeholder Participation (CRESP) through the Department of Energy (AI # DE-FC01-95EW55084 DE-FG 26-00NT 40938, JB/MG) and New Jersey Department of Environmental Protection (JB).

REFERENCES

Able KP (1991) The development of migratory orientation mechanisms. In *Orientation in Birds*, Berthold P (ed.), Birkhauser Verlag, Basel, pp. 166-179.

Adams PM (1977) Effects of anticholinergic and cholinesterase blocking drugs on appetitive behavior under different deprivation conditions. *Life Science*, **21**, 129-136.

Adkins-Regan E (1987) Sexual differentiation in birds. *Trends in Neurosciences*, **10**, 517-522.

Albers PH and Heinz GH (1983) FLIT-MLO and No.2 fuel oil: effects of aerosol applications to mallard eggs on hatchability and behavior of ducklings. *Environmental Research*, **30**, 381-388.

Ankley GT, Niemi GJ, Lodge KB, Harris HJ, Beaver DL, Tillitt DE, Schwartz, TR, Giesy JP, Jones PD and Hagley C (1993) Uptake of planar polychlorinated biphenyls and 2,3,7,8-substituted polychlorinated dibenzofurans and dibenzo-*p*-dioxins by birds nesting in the lower Fox River and Green Bay, Wisconsin. *Archives of Environmental Contamination and Toxicology*, **24**, 332-344.

ATSDR (1999) Toxicological profile for lead. Agency for Toxic Substances and Disease Registry. U.S. Public Health Service, Atlanta.

Backstrom J (1969) Distribution studies of mercuric pesticides in quail and some freshwater fishes. *Acta Pharmacologica et Toxicologica*, **27** (Suppl. 3), 1-103.

Barron MG, Galbraith H and Beltman D (1995) Comparative reproductive and developmental toxicology of PCBs in birds. *Comparative Biochemistry and Physiology*, **112C**, 1-14.

Beer CC (1969) Laughing gull chicks: recognition of their parents' voices. *Science*, **166**, 1030-1032.

Bennett RS and Prince HH (1981) Influences of agricultural pesticides on food preferences and consumption by ring-necked pheasants. *Journal of Wildlife Management*, **45**, 74-82.

Bellrose FC (1959) Effects of ingested lead shot upon waterfowl populations. *Transactions of the North American Wildlife Conference*, **16**, 121-135.

Beyer WN, Spaan JW, Sileo L and Franson JC (1988) Lead poisoning in six captive avian species. *Archives of Environmental Contamination and Toxicology*, **17**, 121-130.

Blus LJ, Gish CD, Belisle AA and Prouty RM (1972) Logarithmic relationship of DDE residues to eggshell thinning. *Nature*, **235**, 376-378.

Boersma PD, Davies EM and Reid W (1988) Weathered crude oil effects on chicks of fork-tailed storm petrels. *Archives of Environmental Contamination and Toxicology*, **17**, 527-531.

Bosveld ATC and Van den Berg M (1994) Effects of polychlorinated biphenyls, dibenzo-*p*-dioxins and dibenzofurans on fish-eating birds. *Environmental Review*, **2**, 147-166.

Brouwer A, Reijnders PJH and Koeman JH (1989) Polychlorinated biphenyl (PCB) contaminated fish induces vitamin A and thyroid hormone deficiency in the common seal (*Phoca vitulina*). *Aquatic Toxicology*, **15**, 99-106.

Brown AW, Aldridge WN, Street BW and Verschoyle RD (1979) The behavioral and neuropathologic sequelae of intoxication by trimethyltin compounds in the rat. *American Journal of Pathology*, **97**, 59-82.

Burger J (1984) Pattern, mechanism, and adaptive significance of territoriality in herring gulls (*Larus argentatus*). *Ornithological Monograph*, **34**, 1-91.

Burger J (1990) Behavioral effects of early postnatal lead exposure in herring gulls (Larus argentatus). *Pharmacology Biochemistry and Behavior*, **36**, 7-13.

Burger J (1993) Metals in avian feathers: bioindicators of environmental pollution. *Reviews in Environmental Toxicology*, **5**, 203-311.

Burger J (1997a) Effects of oiling on feeding behavior of sanderlings and semipalmated plovers in New Jersey. *Condor*, **99**, 290-298.

Burger J (1997b) Heavy metals and selenium in herring gull (*Larus argentatus*) nesting in colonies from eastern Long Island to Virginia. *Environmental Monitoring Assessment*, **48**, 285-296.

Burger J (1997c) *Oil spills*. Rutgers University Press, New Brunswick, NJ.

Burger J (1998) Effects of lead on sibling recognition in young herring gulls. *Toxicology Science*, **43**, 155-160.

Burger J and Gochfeld M (1985) Early postnatal lead exposure: behavioral effects in common tern chicks (*Sterna hirundo*). *Journal of Toxicology and Environmental Health*, **16**, 869-886.

Burger J and Gochfeld M (1988a) Effects of lead on growth and feeding behavior of young common terns (*Sterna hirundo*). *Archives of Environmental Contamination and Toxicology*, **17**, 513-517.

Burger J and Gochfeld M (1988b) Effects of lead on growth in young herring gulls (*Larus argentatus*). *Journal of Toxicology and Environmental Health*, **25**, 227-236.

Burger J and Gochfeld M (1990) Tissue levels of lead in experimentally exposed herring gull (*Larus argentatus*) chicks. *Journal of Toxicology and Environmental Health*, **29**, 219-233.

Burger J and Gochfeld M (1991) *The Common Tern: Breeding Biology and Social Behavior*. Columbia University Press, New York.

Burger J and Gochfeld M (1993a) Lead and behavioral development in young herring gulls: effects of timing of exposure on individual recognition. *Fundamental and Applied Toxicology*, **21**, 187-195.

Burger J and Gochfeld M (1993b) Lead and cadmium accumulation in eggs and fledgling seabirds in the New York bight. *Environmental Toxicology and Chemistry*, **12**, 261-267.

Burger J and Gochfeld M (1994) Behavioral impairments of lead-injected young herring gulls in nature. *Fundamental and Applied Toxicology*, **23**, 553-561.

Burger J and Gochfeld M (1995a) Behavior effects of lead exposure on different days for gull (*Larus argentatus*) chicks. *Pharmacology Biochemistry and Behavior*, **50**, 97-105.

Burger J and Gochfeld M (1995b) Effects of varying temporal exposure to lead on behavioral development in herring gull (*Larus argentatus*) chicks. *Pharmacology Biochemistry and Behavior*, **52**, 601-608.

Burger J and Gochfeld M (1995c) Growth and behavioral effects of early postnatal chromium and manganese exposure on herring gull (*Larus argentatus*) chicks. *Pharmacology Biochemistry and Behavior*, **50**, 607-612.

Burger J and Gochfeld M (1996) Lead and behavioral development: parental compensation for behaviorally impaired chicks. *Pharmacology Biochemistry and Behavior*, **55**, 339-349.

Burger J and Gochfeld M (1997a) Lead and neurobehavioral development in gulls: a model for understanding effects in the laboratory and the field. *NeuroToxicology*, **18**, 279-287.

Burger J and Gochfeld M (1997b) Risk, mercury levels, and birds: relating adverse laboratory effects to field biomonitoring. *Environmental Research*, **75**, 160-172.

Burger J and Gochfeld M (2000) Effects of lead on larids: a review of laboratory and field studies. *Journal of Toxicology and Environmental Health* (Part B), **3**, 59 -78.

Burger J and Gochfeld M. Effects of lead and exercise on performance and endurance in young herring gulls.

Burger J and Tsipoura N (1998) Experimental oiling of sanderlings (*Calidris alba*): behavior and weight changes. *Environmental Toxicology and Chemistry*, **17**, 1154-1158.

Burger J, Shukla T, Benson T and Gochfeld M (1994) Lead levels in exposed herring gulls: differences in the field and laboratory. In *Hazardous waste: impacts on human and ecological health*, Johnson BL, Xintaras C and Andrews JS Jr (eds), Princeton Scientific Publ. Co., Princeton, pp. 115-123.

Bussiere JL, Kendall RJ, Lacher Jr, TE and Bennett RS (1989) Effect of methyl parathion on food discrimination in northern bobwhite (*Colinus virginianus*). *Environmental Toxicology and Chemistry*, **8**, 1125-1131.

Butler RG, Peakall FB, Leighton FA, Borthwick J and Harmon RS (1986) Effects of crude oil exposure on standard metabolic rate of Leach's storm-petrel. *Condor*, **88**, 248-249.

Cain BW, Sileo L, Franson C and Moore J (1983) Effects of dietary cadmium on mallard ducklings. *Environmental Research*, **32**, 286-297.

Cairns MA, Maguire CC and Williams BA (1991) Brain cholinesterase activity of bobwhite acutely exposed to chlorpyrifos. *Environmental Toxicology and Chemistry*, **10**, 657-664.

Carson R (1962) *Silent Spring*. Houghton Mifflin, Boston, 368 pp.

Colborn T and Clement C (eds) (1992) *Chemically Induced Alterations in Sexual and Functional Development: the Wildlife/Human Connection*, Princeton Sci. Publ. Princeton.

Colborn T, Dumanowski D and Peterson MJ (1996) *Our Stolen Future. Are we threatening our fertility, intelligence and survival — A scientific detective story*. Penguin Books, Inc., USA, 306 pp.

Conover MR, Miller DE and Hunt GLJ (1979) Female-female pairs and other unusual reproductive associations in Ring-billed and California Gulls. *Auk*, **96**, 6-9.

Cory-Slechta DA (1990) Exposure duration modifies the effect of low-level lead on fixed interval performance. *NeuroToxicology*, **11**, 427-442.

Cory-Slechta DA, Weiss B and Cox C (1983) Delayed behavioral toxicity of lead with increasing exposure concentration. *Toxicology and Applied Pharmacology*, **71**, 342-352.

Dahlgren RB and Linder RL (1974) Effects of dieldrin in penned pheasants through the third generation. *Journal of Wildlife Management*, **34**, 320-330.

Dahlgren RB, Linder RL and Carlson CW (1972) Polychlorinated biphenyls: their effects on penned pheasants. *Environmental Health Perspectives*, **1**, 89-101.

Dietz DD, McMillan DE and Mushak P (1979) Effects of chronic lead administration on acquisition and performance of serial position sequences by pigeons. *Toxicology and Applied Pharmacology*, **47**, 377-384.

Dewitt JB (1955) Effects of chlorinated hydrocarbon insecticides upon quail and pheasants. *Journal of Agricultural Food and Chemistry*, **3**, 672-683.

Dunnet JC (1982) Oil pollution and seabird populations. *Philosophical Transactions of the Royal Society of London B*, **316**, 513-534.

Ecobichon DJ (1997) Anticholinesterase insecticides. In *Comprehensive Toxicology Vol. 11: Nervous System and Behavioral Toxicology*, Sipes IG, McQueen CA and Gandolfi AJ (eds), Pergamon, New York, pp. 447-456.

Eisler R (1985a) *Selenium hazards to fish, wildlife, and invertebrates: a synoptic review*. US Fish and Wildlife Service: Biological Report 85 (1.5), Washington, DC.

Eisler R (1985b) *Cadmium hazards to fish, wildlife, and invertebrates: a synoptic review*. US Fish and Wildlife Service: Biological Report 85 (1.2), Washington, DC.

Eisler R (1987) *Mercury hazards to fish, wildlife, and invertebrates: a synoptic review*. US Fish and Wildlife Service: Biological Report 85 (1.10), Washington, DC.

Eisler R (1988) *Lead hazards to fish, wildlife, and invertebrates: a synoptic review*. US Fish and Wildlife Service, Biological Report 85 (1.14), Washington, DC.

Eisler R (1989) *Tin hazards to fish, wildlife, and invertebrates: a synoptic review*. US Fish and Wildlife Service, Biological Report 85 (1.15). Washington, DC.

Elliott JE, Norstrom RJ and Smith GEJ (1996) Patterns, trends and toxicological significance of chlorinated hydrocarbon and mercury contaminants in bald eagle eggs from the Pacific coast of Canada, 1990-1994. *Archives of Environmental Contamination and Toxicology*, **31**, 354-367.

Environment Canada (1991) Toxic Chemicals in the Great Lakes and Associated Effects. *Vol. 11 Part 2: Effects of contaminants on wildlife species*. Environment Canada,Toronto, Ontario.

Eroschenko VP and Palmiter RD (1980) Estrogenicity of kepone in birds and mammals. In *Estrogens in the Environment*, McLachlan JA (ed.), Elsevier/North-Holland, New York, pp. 305-324.

Fair JM, Kennedy PL and McEwen LC (1995) Effects of carbaryl grasshopper control on nesting killdeer in North Dakota. *Environmental Toxicology and Chemistry*, **14**, 881-890.

Feyk LL and Giesy JP (1998) Xenobiotic modulation of endocrine function in birds. In *Principles and Processes for Evaluating Endocrine Disruptors in Wildlife*, Kendall RJ, Dickerson RL, Giesy JP and Suk WA (eds), SETAC Press, Pensacola, FL, pp. 121-140.

Fiedler N, Burger J and Gochfeld M (1998) Neurobehavioral toxicity. In *Maxcy-Rosenau-Last Public Health and Preventive Medicine*, Wallace RB (ed.), Appleton & Lange, Stamford, CO, pp. 427-435.

Fimreite N (1979) Accumulation and effects of mercury on birds. In *The Biochemistry of Mercury in the Environment*, Nriagu JO (ed.), Elsevier/North Holland Press, Amsterdam, pp. 601-627.

Fimreite N (1984) Effects of lead shot ingestion in willow grouse. *Archives of Environmental Contamination and Toxicology*, **33**, 121-126.

Finley MT, Dieter MP and Locke LN (1976) Lead in tissues of mallard ducks dosed with two types of lead shot. *Bulletin of Environmental Contamination and Toxicology*, **16**, 261-268.

Finley MT and Stendell RC (1978) Survival and reproductive success of black ducks fed methyl mercury. *Environmental Pollution*, **16**, 51-63.

Forsyth DJ and Martin PA (1993) Effects of fenitrothion on survival, behavior and brain cholinesterase activity of white-throated sparrows (*Zonotrichia albicollis*). *Environmental Toxicology and Chemistry*, **12**, 91-103.

Fossi MC and Leonzio C (eds) (1994) *Nondestructive Biomarkers in Vertebrates*. Lewis, Boca Raton, pp. **345**.

Fowler GS, Wingfield JC and Boersma PD (1995) Hormonal and reproductive effects of low levels of petroleum fowling in magellanic penguins (*Spheniscus magellanicus*). *Auk*, **112**, 382-389.

Fox GA (1992) Epidemiological and pathobiological evidence of contaminant-induced alterations in sexual development in free-living wildlife. In *Chemically-induced Alterations in Sexual and Functional Development: the Wildlife/Human Connection*, Colborn T and Clement C (eds.), Princeton Scientific Publishing Co., Princeton, pp. 147-158.

Fox GA and Donald T (1980) Organochlorine pollutants, nest defense behaviour and reproductive success in merlins. *Condor*, **82**, 81-84.

Fox GA, Gilman AP, Peakall DB and Anderka FW (1978) Behavioral abnormalities of nesting Lake Ontario herring gulls. *Journal of Wildlife Management*, **42**, 477-483.

Frederick RB (1976) Effects of lead nitrate ingestion on open-field behavior of mallard ducklings. *Bulletin of Environmental Contamination and Toxicology*, **16**, 739-742.

Froese KL, Verbrugge DA, Ankley GT and Giesy JP (1998) PCBs in sediments, aquatic insects and tree swallow eggs and nestlings from the Saginaw River watershed. *Environmental Toxicology and Chemistry*, **17**, 484-492.

Fry DM (1996) Vulnerability of avian populations to environmental pollutants. *Comments on Toxicology*, **5**, 401-413.

Fry DM, Toone CK, Speich SM, and Peard RJ (1987) Sex ratio skew and breeding patterns of gulls: demographic and toxicological considerations, *Studies in Avian Biology*, **10**, 26-43.

Fryday SL, Hart ADM and Langton SD (1996) Effects of exposure to an organophosphorus pesticide on the behavior and use of cover by captive starlings. *Environmental Toxicology and Chemistry*, **15**, 1590-1596.

Furness RW, Muirhead SJ and Woodburn M (1986) Using bird feathers to measure mercury in the environment: relationship between mercury content and moult. *Marine Pollution Bulletin*, **17**, 27-30.

Galindo J, Kendall RJ, Driver CJ and Lacher T Jr. (1985) The effect of methyl parathion on the susceptibility of bobwhite quail (*Colinus virginianus*) to domestic cat predation. *Behavioral and Neural Biology*, **43**, 21-36.

Garcelon DK and Thomas NJ (1997) DDE poisoning in an adult bald eagle. *Journal of Wildlife Diseases*, **33**, 299-303.

Gasiewicz TA (1997) Dioxins and the Ah receptor, probes to uncover processes in neuroendocrine development. *NeuroToxicology* **18**, 393-414.

Gesell GG, Robel RJ, Dayton AD and Frieman J (1979) Effects of dieldrin on operant behavior of bobwhites. *Journal of Environmental Science and Health*, **B14**, 153-170.

Giesy JP and Kannan K (1998) Dioxin-like and non-dioxin-like effects of polychlorinated biphenyls: Implications for risk assessment. *Critic. Rev.* **28**, 511-569.

Giesy JP, Ludwig JP and Tillitt DE (1994a) Dioxins, dibenzofurans, PCBs and colonial fish-eating water birds. In *Dioxins and Health*, Schecter A (ed.), Plenum Press, New York, pp. 249-307.

Giesy JP, Ludwig JP and Tillitt DE (1994b) Deformities in birds of the Great Lakes region: assigning causality. *Environmental Science Technology*, **28**, 128A-135A.

Gilbert SG and Grant-Webster KS (1995) Neurobehavioral effects of developmental methylmercury exposure. *Environmental Health Perspectives*, **103**, 135-142.

Gilbertson M, Kubiak TJ, Ludwig JP and Fox G (1991) Great Lakes embryo mortality, edema and deformity syndrome (GLEMEDS) in colonial fish-eating birds: similarity to chick edema disease. *Journal of Toxicology and Environmental Health*, **33**, 455-520.

Gilbertson M (1983) Etiology of chick edema disease in herring gulls in the lower Great Lakes. *Chemosphere*, **12**, 357-370.

Gilman AP, Peakall DB, Hallett DJ, Fox FA and Norstrom RJ (1977) Herring gulls (*Larus argentatus*) as monitors of contamination in the Great Lakes. In *Animals as Monitors of Environmental Pollution*, National Academy Press, Washington, DC. pp. 280-289.

Goede AA and de Bruin M (1984) The use of bird feather parts as a monitor for metal pollution. *Environmental Pollution*, **8**, 281-298.

Grue CE (1982) Response of common grackles to dietary concentrations of four organophosphate pesticides. *Archives of Environmental Contamination and Toxicology*, **11**, 617-626.

Grue CE, Hart ADM and Mineau P (1991) Biological consequences of depressed brain cholinesterase activity in wildlife. In *Cholinesterase-Inhibiting Insecticides*, Vol. 2, *Chemicals in Agriculture*, Mineau P (ed.), Elsevier, New York, pp. 151-209.

Haegele MA, Tucker RK and Hudson RH (1974) Effect of dietary mercury and lead on eggshell thickness in mallards. *Bulletin of Environmental Contamination and Toxicology*, **16**, 261-269.

Hart ADM (1993) Relationships between behavior and the inhibition of acetylcholinesterase in birds exposed to organophosphorus pesticides. *Environmental Toxicology and Chemistry*, **12**, 321-336.

Hawkes AW, Brewer LW, Hobson JF, Hooper MJ and Kendall RJ (1996) Survival and cover-seeking response of northern bobwhites and mourning doves dosed with aldicarb. *Environmental Toxicology and Chemistry*, **15**, 1538-1543.

Heinz G (1974) Effects of low dietary levels of methyl mercury on mallard reproduction. *Bulletin of Environmental Contamination and Toxicology*, **11**, 386-392.

Heinz G (1975) Effects of methylmercury on approach and avoidance behavior of mallard ducklings. *Bulletin of Environmental Contamination and Toxicology*, **13**, 554-564.

Heinz G (1976a) Methylmercury: second-year feeding effects on mallard reproduction and duckling behavior. *Journal of Wildlife Management*, **40**, 82-90.

Heinz G (1976b) Methylmercury: second-generation reproductive and behavioral effects on mallard ducks. *Journal of Wildlife Management*, **40**, 710-715.

Heinz G (1979) Methylmercury: reproductive and behavioral effects of three generations of mallard ducks. *Journal of Wildlife Management*, **43**, 394-401.

Heinz G (1996a) Mercury poisoning in wildlife. In *Noninfectious Diseases of Wildlife*, (2nd edn), Fairbrothers A, Locke LN and Hoff GL (eds), Iowa State University Press, Ames, IA, pp. 118-127.

Heinz G (1996b) Selenium in birds. In *Selenium in Birds*, Beyer WN, Heinz GH and Redmon-Norwood (eds), Lewis Publishers, Boca Raton, FL, pp. 447-458

Heinz G and Fitzgerald MA (1993) Reproduction of mallards following overwintering exposure to selenium. *Environmental Pollution*, **81**, 117-122.

Heinz G and Haseltine SD (1981) Avoidance behavior of young black ducks treated with chromium. *Toxicology Letters*, **8**, 307-310.

Heinz GH, Hill EF and Contrera JF (1980) Dopamine and norepinephrine depletion in ring doves fed DDE, dieldrin and Aroclor 1254. *Toxicology and Applied Pharmacology*, **53**, 75-82.

Henny CJ, Kolbe EJ, Hill EF and Blus LJ (1987) Case histories of bald eagles and other raptors killed by organophosphorus insecticides topically applied to livestock. *Journal of Wildlife Diseases*, **23**, 292-295.

Hickey JJ (1969) *The Peregrine Falcon Populations: Their Biology and Decline*. University Wisconsin Press, Madison.

Hickey JJ and Anderson DA (1968) Chlorinated hydrocarbons and eggshell changes in raptorial and fish-eating birds. *Science*, **162**, 271-273.

Hill EF (1971) Toxicity of selected mosquito larvicides to some common avian species. *Journal of Wildlife Management*, **36**, 635-639.

Holmes SB and Boag PT (1990) Effects of the organophosphorus pesticide fenitrothion on behavior and reproduction in zebra finches. *Environmental Research*, **53**, 62-75.

Hoffman DJ (1995) Wildlife toxicity testing. In *Handbook of Ecotoxicology*, Hoffman DJ, Rattner BA, Burton GL Jr and Cairns J Jr (eds), Lewis Publishers, Boca Raton, FL, pp. 47-69.

Hoffman DJ, Franson JC, Pahee OH, Bunck CM and Murray HC (1985) Biochemical and hematological effects of lead ingestion in nestling American kestrel (*Falco sparverius*). *Comparative Biochemistry and Physiology*, **80**, 431-439.

Hose JE and Guillette LJ (1995) Defining the role of pollutants in the disruption of reproduction in wildlife. *Environmental Health Perspectives*, **103**, 87-91.

Idemudia SA and McMillan DE (1986) Effects of chemicals on delayed matching behavior in pigeons. III. Effects of triethyltin. *NeuroToxicology*, **7**, 75-88.

Janz DM and Bellward GD (1996) In *ovo* 2,3,7,8-tetrachlorodibenzo-*p*-dioxin exposure in three avian species. 2. Effects on estrogen receptor and plasma sex steroid hormones during the perinatal period. *Toxicology and Applied Pharmacology*, **139**, 292-300.

Jones PD, Giesy JP, Newsted JL, Verbrugge DA, Beaver DL, Ankley GT, Tillitt DE, Niemi GJ and Lodge KB (1993) 2,3,7,8-Tetrachlorodibenzo-*p*-dioxin equivalents in tissues of birds at Green Bay, Wisconsin, USA. *Archives of Environmental Contamination and Toxicology*, **24**, 345-354.

Kendall RJ, Scanlon PR and Di Giulio RT (1982) Toxicology of ingested lead shot in ringed turtle doves. *Archives of Environmental Contamination and Toxicology*, **11**, 259-263.

Kendall RJ, Dickerson RL, Giesy JP and Suk WA (eds) (1998) *Principles and Processes for Evaluating Endocrine Disruptors in Wildlife*. SETAC Press, Pensacola, FL, 491 pp.

King KA, Custer TW and Weaver DA (1994) Reproductive success of barn swallows nesting near a selenium-contaminated lake in east Texas, USA. *Environmental Pollution*, **84**, 53-58.

Kilbride KM, Crawford JA and Williams BA (1992) Response of female California quail (*Callipepla califonica*) to methyl parathion treatment of their home ranges during the nesting period. *Environmental Toxicology and Chemistry*, **11**, 1337-1343.

Koh TS and Harper MJ (1988) Lead-poisoning in black swans, *Cygnus atratus*, exposed to spent lead shot at Bool Lagoon Game Reserve, South Australia. *Australian Wildlife Research*, **15**, 395-403.

Kreitzer JF (1980) Effects of toxaphene and endrin at very low dietary concentrations on discrimination acquisition and reversal in bobwhite quail, *Colinus virginianus*. *Environmental Pollution*, **23**, 217-230.

Kreitzer JF and Heinz G (1974) The effects of sublethal dosages of five pesticides and a polychlorinated biphenyl on the avoidance response of *Coturnix* quail chicks. *Environmental Pollution*, **6**, 21-29.

Kubiak TJ, Harris HJ, Smith LM, Schwartz TR, Stalling DL, Trick JA, Sileo L, Docherty DE and Erdman TC (1989) Microcontaminants and reproductive impairment of the Forster's tern on Green Bay, Lake Michigan-1983. *Archives of Environmental Contamination and Toxicology*, **18**, 706-727.

Kupfer D and Bulger WH (1980) Estrogenic properties of DDT and its analogs, In *Estrogens in the environment*, McLachlan JA (ed.), Elsevier/North-Holland, New York, pp. 239-262.

Lari L, Massi A, Fossi MC, Casini S, Leonzio C and Focardi S (1994) Evaluation of toxic effects of the organophosphorus insecticide azinphos-methyl in experimentally and naturally exposed birds. *Archives of Environmental Contamination and Toxicology*, **26**, 234-239.

Lewis MF, Mertens HW and Steen JA (1973) Behavioral changes from chronic exposure to pesticides used in aerial applications: effects of phosdrin on performance of monkeys and pigeons on variable interval reinforcement schedules. *Aerospatial Medicine*, **44**, 290-293.

Lincer JL and Peakall DB (1970) Metabolic effects of polychlorinated biphenyls in the American kestrel. *Nature*, **228**, 783-784.

Logan CA and Wingfield JC (1995) Hormonal correlates of breeding status, nest construction and parental care in multiple-brooded northern mockingbirds, *Mimus polyglottos*. *Hormones and Behavior*, **29**, 12-30.

Lombardo MP (1994) Nest architecture and reproductive performance in tree swallows (*Tachycineta bicolor*). *Auk*, **111**, 814-824.

Ludwig JP, Auman HJ, Weseloh DV, Fox GA, Giesy JP and Ludwig ME (1995) Evaluation of the effects of toxic chemicals in the Great Lakes Cormorants, has causality been established? *Colonial Waterbirds*, **18**, 60-69.

MacPhail RC and Tilson HA (1995) Behavioral screening tests, past, present and future. In *Neurotoxicology: Approaches and Methods*, Academic Press, New York, pp. 231-238.

Martin PA, Johnson DL and Forsyth DJ (1996) Effects of grasshopper-control insecticides on survival and brain acetylcholinesterase of pheasant (*Phasianus colchicus*) chicks. *Environmental Toxicology and Chemistry*, **15**, 518-524.

McArthur MLB, Fox GA, Peakall DB and PhilogPne BJR (1983) Ecological significance of behavioral and hormonal abnormalities in breeding ring doves fed an organochlorine mixture. *Archives of Environmental Contamination and Toxicology*, **12**, 343-353.

McCarty JP and Secord AL (1999) Nest building behavior in PCB-contaminated tree swallows. AUK, **116**, 55-63.

Millikin RL and Smith JNM (1990) Sublethal effects of fenitrothion on forest passerines. *Journal of Applied Ecology*, **27**, 983-1000.

Mineau P and Peakall DB (1987) An evaluation of avian impact assessment techniques following broad scale forest insecticide sprays. *Environmental Toxicology and Chemistry*, **6**, 781-791.

Mineau P, Fox GA, Norstrom RJ, Weseloh DV, Hallet DJ and Ellenton JA (1984) Using the herring gull to monitor levels and effects of organochlorine contamination in the Canadian Great Lakes. *Advanced Environmental Science Technology*, **14**, 425-452.

Mineau P, Collins BT and Baril A (1996) On the use of scaling factors to improve interspecies extrapolation of acute toxicity in birds. *Regulatory Toxicology and Pharmacology*, **24**, 24-29.

Mora MA, Auman HJ, Ludwig JP, Giesy JP, Verbrugge DA and Ludwig ME (1993) PCB's and chlorinated insecticides in plasma of Caspian terns: relationships with age, productivity and colony-site tenacity in the Great Lakes. *Archives of Environmental Toxicology and Chemistry*, **24**, 320-331.

National Research Council (NRC) (1999) *Hormonally-active Agents in the Environment*, National Academy Press, Washington, DC.

Needleman HL, Schell A, Bellinger D, Leviton A and Allred EN (1990) The long-term effects of exposure to lead in childhood. An 11 year follow-up report. *New England Journal of Medicine*, **322**, 83-88.

Nybo S (1996) Effects of dietary aluminum on chicks *Gallus gallus domesticus* with different dietary intake of calcium and phosphorus. *Archives of Environmental Contamination and Toxicology*, **31**, 177-183.

Ohlendorf HM, Hothem RL, Bunck CM, Aldrich TW and Moore JF (1986) Relationships between selenium concentrations and avian reproduction. *Transactions of the North American Wildlife Natural Resources Conference*, **51**, 330-342.

Ohlendorf HM, Hothem RL, Bunck CM and Marois KC (1990) Bioaccumulation of selenium in birds at Kesterson Reservoir, California. *Archives of Environmental Contamination and Toxicology*, **19**, 495-507.

Ohlendorf HM, Kilness AW, Simmons JL, Stroud RD, Hoffman DJ and Moore JF (1988) Selenium toxicosis in wild aquatic birds. *Journal of Toxicology and Environmental Health*, **24**, 67-92.

Osborn D, Enoy WJ and Gull KR (1983) The toxicity of trialkyl lead compounds to birds. *Environmental Pollution*, **31A**, 261-275.

Paine DJ (1995) Lead in the environment. In *Handbook of Ecotoxicology*, Hoffman DJ, Rattner BA, Burton GA Jr and Cairns J Jr (eds), Lewis Publishers, Boca Raton, FL, pp. 356-391.

Peakall DB (1970) p,p′-DDT: effect on calcium metabolism and concentration of oestradiol in the blood. *Science*, **168**, 592-594.

Peakall DB (1985) Behavioral responses of birds to pesticides and other contaminants. *Residue Reviews*, **96**, 45-77.

Peakall DB (1986) Accumulation and effects on birds. In *PCBs in the Environment*, Aaid JS (ed), Vol. II, CRC Press, Boca Raton, FL, pp. 31-47.

Peakall DB, Lincer JL, Risebrough RW, Pritchard JB and Kinter WB (1973) DDE-induced eggshell thinning: structural and physiological effects in three species. *Comparative and General Pharmacology*, **4**, 305-315.

Peakall DB and Peakall ML (1973) Effects of a polychlorinated biphenyl on the reproduction of artificially and naturally incubated dove eggs. *Journal of Applied Ecology*, **10**, 863-868.

Pearce PA, Peakall DB and Erskine AJ (1976) Impact on forest birds of the 1975 spruce budworm spray operation in New Brunswick. *Canadian Wildlife Servive Program Notes*, **62**, 1-7.

Pierotti RJ and Good TP (1994) Herring gull (*Larus argentatus*). *Birds of North America*, **124**, 1-21.

Porterfield SP (1994) Vulnerability of the developing brain to thyroid abnormalities: environmental insults to the thyroid system. *Environmental Health Perspectives*, **102**, 125-130.

Ratcliffe DA (1967) Decreases in eggshell weight in certain birds of prey. *Nature*, **215**, 208-209.

Ratcliffe DA (1970) Changes attributable to pesticides in egg breakage frequency and egg shell thickness in some British birds. *Journal of Applied Ecology*, **7**, 67-115.

Rattner BA, Sileo L and Scanes CG (1982) Oviposition and the plasma concentrations of LH, progesterone and corticosterone in bobwhite quail (*Colinus virginianus*) fed parathion. *Journal of Reproduction and Fertility*, **66**, 147-155.

Rattner BA, Haramis GM, Chu D, Bunck CM and Scanes CG (1987) Growth and physiological conditions of black ducks reared on acidified wetlands. *Canadian Journal of Zoology*, **65**, 2953-2958.

Rice DC (1996) Behavioral effects of lead: commonalities between experimental and epidemiologic data. *Environmental Health Perspectives*, **104**, 337-351.

Richie PJ and Peterle TJ (1979) Effect of DDE on circulating luteinizing hormone levels in ring doves during courtship and nesting. *Bulletin of Environmental Contamination and Toxicology*, **23**, 220-226.

Rosie D and Barnes SN (1983) *The Effects of Oil on Birds*. Tri-state bird rescue & research, Inc., Wilmington, DE, p. 178.

Scheuhammer AM (1988) Chronic dietary toxicity of methylmercury in the zebra finch *Poephila guttata*. *Bulletin of Environmental Contamination and Toxicology*, **40**, 123-130.

Schwartz J (1994) Societal benefits of reducing lead exposure. *Environmental Research*, **66**, 105-124.

Secord AL, McCarty JP, Echols KR, Meadows JC, Gale RW and Tillitt DE (1999) Polychlorinated biphenyls and 2,3,7,8-tetrachlorodibenzo-p-dioxin equivalents in tree swallows from the upper Hudson River, New York State, USA. *Environmental Toxicology and Chemistry*, **18**, 2519-2525.

Sharma RP, Winn DS and Low JB (1976) Toxic, neurochemical and behavioral effects of dieldrin exposure in mallard ducks. *Archives of Environmental Contamination and Toxicology*, **5**, 43-53.

Sharma RP, Winn DS and Shupe JL (1971) Neurologic and behavioral effects of dietary dieldrin in juvenile mallard ducks. *Toxicology and Applied Pharmacology*, **19**, 405-406.

Shellenberger TE, Newell GW, Adams RF and Barbaccia J (1966) Cholinesterase inhibition and toxicologic evaluation of two organophosphate pesticides in Japanese quail. *Toxicology and Applied Pharmacology*, **8**, 22-28.

Simpson CF, Damron BL and Harms RH (1970) Abnormalities of erythrocytes and renal tubules of chicks poisoned with lead. *American Journal of Veterinary Research*, **31**, 515-523.

Skorupa JP and Ohlendorf HM (1991) Contaminants in drainage water and avian risk thresholds. In *The Economics and Management of Water and Drainage in Agriculture*, Dinar A and Zilberman (eds), Kluwer Academic Publishers Norwell, MA, pp. 345-368.

Slikker W Jr and Chang LW (1998) *Handbook of Developmental Neurotoxicology*, Academic Press, New York.

Spear PA, Bourbonnais DH, Peakall DB and Moon TW (1989) Dove reproduction and retinoid (vitamin A) dynamics in adult females and their eggs following exposure to 3,3′,4,4′-tetrachlorobiphenyl. *Canadian Journal of Zoology*, **67**, 908-913.

Stickel LF (1946) Field studies of a *Peromyscus* population in an area treated with DDT. *Journal of Wildlife Management*, **10**, 216.

Stone WB (1979) Poisoning of wild birds by organophosphate and carbamate pesticides. *New York Fish and Game Journal*, **26**, 37-47.

Tanaka D Jr., Bursian SJ and Lehning EJ (1992) Neuropathological effects of triphenyl phosphite on the central nervous system of the hen (*Gallus domesticus*). *Fundamental and Applied Toxicology*, **18**, 72-78.

Tejning S (1967) Biological effects of methylmercury dicyandiamide-treated grain in the domestic fowl Gallus gallus. *Oikos*, **8** (Suppl.), 1-116.

Tilson HA, Jacobson JL and Rogan WJ (1990) Polychlorinated biphenyls and the developing nervous system: cross-species comparisons. *Neurotoxicology and Teratology*, **12**, 239-248.

Tilson HA and Kodavanti PRS (1997) Neurochemical effects of polychlorinated biphenyls: an overview and identification of research needs. *NeuroToxicology*, **18**, 727-744.

Tori GM and Peterle TJ (1983) Effects of PCBs on mourning dove courtship behavior. *Bulletin of Environmental Contamination and Toxicology*, **30**, 44-49.

Ulfstrand S, S'dergren A and Rab'l J (1971) Effect of PCB on nocturnal activity in caged robins, *Erithacus rubecula*. *Nature*, **231**, 467-468.

Veronesi B, Padilla S and Newland D (1986) Biochemical and neuropathological assessment of triphenyl phosphite in rats. *Toxicology and Applied Pharmacology*, **83**, 203-210.

Vyas NB, Kuenzel WJ, Hill EF and Sauer JR (1995) Acephate affects migratory orientation of the white-throated sparrow (*Zonotrichia albicollis*). *Environmental Toxicology and Chemistry*, **14**, 1961-1965.

Walsh PM (1990) The use of seabirds as monitors of heavy metals in the marine environment. In *Heavy Metals in the Marine Environment*, Furness RW and Rainbow PS (eds), CRC Press, Boca Raton, FL, pp. 183-204.

Walsh TJ (1997) *In vivo* systems: animal models of neurological diseases. In *Comprehensive Toxicology Vol. 11: Nervous System and Behavioral Toxicology*, Sipes IG, McQueen CA and Gandolfi AJ (eds), Pergamon, New York, pp. 419-429.

White DH, King KA, Mitchell CA, Hill EF and Lamont TG (1979) Parathion causes secondary poisoning in a laughing gull breeding colony. *Bulletin of Environmental Contamination and Toxicology*, **23**, 281-284.

Woodford JE, Karasov WH, Meyer MW and Chambers L (1998) Impact of 2,3,7,8-TCDD exposure on survival, growth and behavior of ospreys breeding in Wisconsin, USA. *Environmental Toxicology and Chemistry*, **17**, 1323-1331.

Wren CD, Harris S and Harttrup NA (1995) Ecotoxicology of mercury and cadmium. In *Handbook of Ecotoxicology*, Hoffman DJ, Rattner BA, Burton GA Jr and Cairns J Jr (eds), Lewis Publishers, Boca Raton, FL, pp. 392-423.

10

Effects of Endocrine Disrupters on Behaviour and Reproduction

PAOLA PALANZA[1] AND FREDERICK VOM SAAL[2]
[1]*Dipartimento di Biologia Evolutiva e Funzionale, Parma University, Parma, Italy*
[2]*Division of Biological Sciences, University of Missouri-Columbia, Columbia, MO, USA*

10.1 INTRODUCTION

A relatively new area of ecotoxicological research is the study of the capacity of certain man-made contaminants (herbicides, fungicides, pesticides and industrial chemicals) to disrupt the endocrine system of animals and humans. These environmental chemicals are referred to as endocrine disrupters (EDs) (Colborn *et al.* 1993), because they can disrupt endocrine function via many different mechanisms; EDs can: (1) mimic the effects of endogenous hormones, (2) antagonise the effects of endogenous hormones, (3) alter the synthesis and metabolism of endogenous hormones, and (4) alter plasma hormone binding proteins and hormone receptor levels in target tissues (Soto *et al.* 1995, Nagel *et al.* 1997).

During critical periods in development (both foetal and neonatal life in rodents) endocrine signals serve to regulate developmental events, and disruption of these signals can permanently, and thus adversely, alter the course of development and the subsequent functioning of organs and systems. The primary focus of endocrine disrupter research with animals has thus been on developmental effects (Colborn *et al.* 1993), although only a few studies focusing on developmental effects of EDs in humans have been conducted (e.g., Jacobson and Jacobson 1996, Lonky *et al.* 1996, Brouwer *et al.* 1998). Embryos and foetuses appear to be especially vulnerable to exposure to EDs (Colborn *et al.* 1993). Mothers can pass contaminants to their offspring via deposition in eggs (amphibians, reptiles, birds) or transplacentally (mammals), and after birth by breastfeeding newborns. Even though adult animals exposed to a contaminant at a particular dose of an ED may not show any adverse effects, their offspring may have lifelong behavioural, reproductive and immune abnormalities, such as altered aggressive behaviour, reduced sperm production and decreased immune responsiveness (Porter *et al.* 1984, vom Saal *et al.* 1998, Palanza *et al.* 1999b).

A few epidemiological studies in humans, and numerous wildlife studies looking at mammals, birds, reptiles, amphibians and fish, as well as laboratory studies primarily using rodents, have linked exposure of a developing embryo to environmental contaminants with many permanent adverse health effects detected in later adulthood. These effects include: abnormal blood hormone levels; reduced fertility; altered sexual and social behaviour; modified immune system; masculinisation of females; feminisation of males (reduced testes and penis size); undescended testicles; cancers of the female and male reproductive tract; malformed fallopian tubes, uterus and cervix; and altered bone density and structure (see Colborn and Clement 1992, Colborn *et al.* 1998 for a review).

For a long time the primary focus of research regarding the effects of man-made chemicals was on their capacity to act as mutagens or to induce gross abnormalities after administration of a dose which has typically been much higher than would be encountered in food, water or air outside of an industrial workplace. However, with regard to EDs, the major concern is with functional changes in tissues due to exposure to low, environmentally-relevant doses during critical periods in organ development (we refer to 'environmentally-relevant doses' as levels that could be encountered by wildlife and humans in the environment). The issue of dose in toxicological testing for effects of EDs is critical. We will discuss evidence that qualitatively different, and thus totally unexpected, effects of EDs can occur at doses currently labelled as 'safe' for humans based on methods of toxicological testing that only require examination of a few doses which are thousands or even millions of times higher than those encountered in the environment.

There are some environmental chemicals that react with DNA, which can lead to mutations as well as initiate DNA repair mechanisms. However, there is no evidence that within an environmentally relevant dose range, any oestrogenic EDs bind covalently to DNA and result in changes in the genetic message (Ben-Jonathan and Steinmetz 1998). An important feature that distinguishes exposure to endocrine disrupters from chemicals that bind to DNA, resulting in mutations, is that there is no recognition that an abnormal event has occurred due to the action of an ED. Since there is no recognition of an abnormal event, there are no 'repair' systems operating in foetuses to correct the effects of EDs. Thus, the consequence of binding of an oestrogenic ED, such as bisphenol A, to oestrogen receptors in foetuses is an alteration of a normal developmental process; there is thus no cellular recognition that anything abnormal has happened and no 'protective' or 'repair' processes are available to be engaged.

At this time, the best characterised endocrine-disrupting chemicals are those able to bind to oestrogen receptors in cells. Chemicals which are capable of disrupting the endocrine systems of animals fall into other categories as well. For example, there are environmental antioestrogens (e.g. dioxin, some polychlorinated biphenyls (PCBs) (Safe *et al.* 1991, Jansen *et al.* 1993), environmental anti-androgens (e.g. Vinclozolin, p,p′DDE, methoxychlor) (Kelce *et al.* 1995, Gray *et al.* 1999a,b), toxicants that reduce steroid hormone levels (e.g. fenarimol and other fungicides, endosulphan), toxicants that interfere with thyroid hormone levels in blood (Brouwer *et al.* 1998, Porterfield and Hendry 1998), toxicants that act through effects on the CNS (e.g. dithiocarbamate pesticides, methanol), and other toxicants that affect hormonal status (cadmium, benzidine based dyes).

10.2 ENDOCRINE DISRUPTING CHEMICALS WITH OESTROGENIC ACTIVITY

Substances encountered in the environment that act like oestrogenic hormones in organisms are also referred to as environmental oestrogens. Some of these compounds, called phytoestrogens, occur naturally in plants such as clover, soybeans and other legumes, whole grains and many fruits and vegetables (Welshons *et al.* 1990). Others are synthetic chemicals made commercially for a specific purpose or produced as a by-product of manufacturing processes. These man-made oestrogenic chemicals are found in: herbicides and pesticides (insecticides such as o,p′-DDT, a contaminant in dichlodiphenyltrichloroethane — DDT — endosulfan, dieldrin, methoxychlor, phenothrin, kepone, dicofol, toxaphene and chordane); products associated with plastics (bisphenol A and nonylphenol); pharmaceuticals (ethinyl oestradiol, diethylstilboestrol — DES — cimetidine); ordinary household products (breakdowns products of detergents and associated surfactants, including nonylphenol and octylphenol); and industrial chemicals, such as PCBs (Soto *et al.* 1995, Nagel *et al.* 1998, Go *et al.* 1999).

Oestrogen is a critical hormone with regard to functioning of the reproductive system in adult females, but oestrogen is now also recognised to play an important role in normal foetal development and the functioning of male reproductive organs (vom Saal 1989, Hess *et al.* 1997, vom Saal *et al.* 1997). So far, scientific interest in the effects of exposure of man and animals to environmental oestrogens has focused on the potential for effects on male fertility and gross reproductive disturbances, while the wider effects of exposure to these compounds has received less attention. A primary interest is that during foetal life, sex steroids, such as oestradiol, have marked effects on the development of the neuroendocrine system and subsequent behaviour, and we have been examining whether oestrogenic EDs can disrupt the normal course of these developmental events.

Most EDs that bind to oestrogen receptors show a significantly lower biological activity relative to oestradiol using *in vitro* or *in vivo* assays (Welshons *et al.* 1999). As a result, oestrogenic EDs are typically referred to as being 'weak' oestrogens and thus, of little concern with regard to exposure to doses typical of those based on use of products containing the EDs (here we exclude the much higher exposures that can occur in the workplace, such as in factories that manufacture the EDs).

However, as is true for many developmentally important signalling molecules, oestradiol regulates the course of development at much lower concentrations than had previously been appreciated. Alteration of the developmental programme (for example, of prostate development) can occur as a result of changing the free serum concentration of oestradiol by as little as 0.1 pg/ml (0.1 parts per trillion) in mice (vom Saal *et al.* 1997). This disruption of the normal developmental programme can lead to irreversible changes in the functioning of organ systems throughout the rest of the life (vom Saal *et al.* 1998, Welshons *et al.* 1999). This appears to occur without altering the genetic code in cells by mutations and, instead, involves altering the processes involved in turning on and off specific genes, as well as setting the rate of activity of genes that are turned on during the developmental period of cell differentiation (Holliday 1987). The consequence of endogenous hormones acting at very low concentrations is that even though EDs are, in fact, typically weaker than the endogenous hormones, they are still very potent. For example, an oestrogenic ED that is 1000-times weaker than oestradiol would be expected to show oestrogenic activity at a free circulating concentration of 100 pg/ml (100 parts per trillion; Welshons *et al.* 1999). Doses of EDs, such as the oestrogenic ED, bisphenol A used to make plastic, that result in low circulating levels within the range of 100 parts per trillion (Taylor *et al.* 1999) had previously been considered 'safe' (vom Saal and Sheehan 1998), but have now been shown in numerous studies to be able to produce adverse effects (Nagel *et al.* 1997, Steinmetz *et al.* 1997, Howdeshell *et al.* 1999, Markey *et al.* 1999, Ohsako *et al.* 1999, Gupta, 2000a).

10.2.1 OESTROGENIC ENDOCRINE DISRUPTERS: EVOLUTIONARY CONSERVATION OF OESTROGEN RECEPTORS

Selective pressures appear to have operated during vertebrate evolution such that all vertebrates share a set of homologous neuroendocrine control mechanisms mediating reproductive functions and socio-sexual behaviours (Nelson 1995). The underlying mechanisms of the action of hormones such as oestradiol are fundamentally identical across vertebrates (Katzenellenbogen *et al.* 1979, Pakdel *et al.* 1989, vom Saal 1995).

The hormone oestradiol is identical in all vertebrates; all steroid hormones are identified by their precise structure, unlike protein hormones, which can vary in amino acid sequences but have the same name and function in different species. In addition, the region of the classical oestrogen receptor (ER alpha) that binds oestradiol in fish is highly conserved and is thus very similar in birds and in women (Katzenellenbogen *et al.* 1979, Pakdel *et al.* 1989, White *et al.* 1994). Species and tissue distribution, and binding characteristics of the recently discovered ER beta, are now being investigated, as are the possibility of unique mechanisms of response to oestrogens and other steroids in different vertebrates (Paech *et al.* 1997, Kuiper *et al.* 1997). However, the high degree of conservation of the alpha form of the oestrogen receptor over hundreds of millions of years of vertebrate evolution has profound implications with regard to oestrogenic endocrine disrupters (vom Saal 1995). The most parsimonious assumption is that if a chemical can bind to the oestrogen receptor in one vertebrate, it should be expected to bind to oestrogen receptors in any other vertebrate, including humans, until this hypothesis is falsified. This is not intended to suggest that the outcome of binding to the receptor will be the same in different species. In addition, this does not imply that in different tissues within a species, or even within a tissue at different times in life, the effects of binding of a chemical to oestrogen receptors will be the same.

The outcome of binding of any oestrogenic chemical to the oestrogen receptor depends on the conformational change induced in the receptor, the interaction of the receptor with tissue-specific proteins associated with the transcriptional apparatus, and the specific genes associated with oestrogen response elements (EREs) to which the transcriptional regulating complex of ligand, receptor and associated proteins binds, thus regulating the process of transcription (Katzenellenbogen *et al.* 1996). The central issue is that there will be some physiological consequence of the event of binding of a chemical to oestrogen receptors. It is the fact that some change in cell function will occur as a consequence of having oestrogen receptors occupied by an environmental oestrogenic endocrine disrupter that is of such importance when this occurs during critical periods in development. A change in what would have been the natural (undisturbed) course of development will occur, and it is only the specific nature of the change (disruption) that will differ from species to species, from tissue to tissue, and as a function of the time in development that exposure occurs.

10.2.2 THE DES SYNDROME AS A MODEL FOR OESTROGENIC ENDOCRINE DISRUPTION IN ANIMALS AND HUMANS

There is an extensive literature concerning the long-term effects of exposure to the synthetic oestrogens, such as DES (diethylstilboestrol), during foetal/neonatal life in human (Takasugi and Bern 1988) and animal studies (Vannier and Raynaud 1980). Daughters of mothers who took the synthetic oestrogen DES during pregnancy to prevent miscarriage have higher rates of reproductive problems, reproductive cancer (vagina, cervix) and malformed reproductive organs (uterus, cervix; Nollar *et al.* 1990). Sons may also face higher rates of malformed or small penises, undescended testicles and abnormal sperm (Gill *et al.* 1976), although these men have not shown reduced fertility (Raloff 1995). Unfortunately, relatively little information is available for DES sons relative to DES daughters and for neural and psychological effects of exposure (Swan and vom Saal 2000). Psychological studies by systematic semistructured interviews have, however, shown increased rates of bisexual or homosexual orientation in DES-exposed women in comparison to several control groups (Ehrhardt *et al.* 1985, Meyer-Bahlburg *et al.* 1995). DES is not only a potent oestrogen, similar in strength to the natural estrogen oestradiol, but it also has the unique ability to concentrate in target tissues, such as the reproductive tracts of birds, reptiles and other animals, during development and cause abnormalities (Shah and McLachlan 1976). This drug serves as an example of what potent oestrogenic compounds can do and may illustrate the health effects that other environmental oestrogens can produce (Newbold 1995).

10.3 THE HORMONAL CONTROL OF DEVELOPMENT

Steroid hormones interact to regulate the normal development of the reproductive system and other tissues; during critical periods in cell differentiation, hormones are involved in 'imprinting' (permanently turning on or off) specific genes in cells with receptors for the hormones (vom Saal *et al.* 1992). For example, in male foetuses, testosterone acts as a prohormone that is converted into the more potent androgen, 5α-dihydrotestosterone (DHT), in the urogenital sinus; DHT acts in conjunction with oestradiol to regulate the development of the prostate from the embryonic urogenital sinus in mice (vom Saal *et al.* 1997, Gupta *et al.* 2000a). In contrast, in most mammalian species that have been studied, testosterone is not converted into DHT during differentiation of organs that develop from the Wolffian ducts (epididymidis, vas deferens and seminal vesicles), although this conversion does occur after the foetal period of differentiation, and there is no evidence for a synergistic interaction between androgen and oestrogen during the development of these organs (vom Saal *et al.* 1992, 1998). The effects of endocrine disrupters with oestrogenic activity are thus different in organs that differentiate from the urogenital sinus and Wolffian ducts in males (vom Saal *et al.* 1998).

Sex steroids exert potent influences on the nervous system during critical developmental periods and on adulthood by organising and reorganising the neuronal circuitry involved in neuroendocrine and behavioural functions (Matsumoto 1991). The development of sexual dimorphisms in behaviour and cognitive function and of sexual dimorphic brain areas in vertebrates depends on the epigenetic action of gonadal hormones, androgens and oestrogens. For sexual differentiation of higher vertebrates, mammals and birds, sex determining genes control the development of the gonads. Gonadal sex determination then leads to the sex differences in the production and release of gonadal hormones, androgens and oestrogens, according to either a male or a female pattern. The action of androgens and oestrogens specifies subsequently the male- or female-typical development of the brain, which is initially monomorphic between the sexes (Phoenix *et al.* 1959, Whalen 1974, McEwen 1981). Gonadal steroids affect brain development by binding to intracellular steroid receptors, the androgen receptors (AR), oestrogen receptors (ER) and progesterone receptors (PR). After steroid-dependent sexual differentiation of the brain during ontogeny, oestrogens, androgens and progestins are important for the activation of socio-sexual behaviours during later stages of life. Typically, in adulthood hormone-induced alterations of the brain strengthen the degree of sexual dimorphism of brain areas achieved during ontogeny. Sex differences in the brain are thought to explain sex differences in behaviour and brain functions (for review: Crews 1993, Gahr 1994). There are a wide array of behavioural differences, in addition to sexual behaviours, between males and females due to differences in sex hormones during foetal life.

Specific behavioural differences between males and females include differences in infant play, aggression, learning, exploration, activity level, food intake and preference, novelty seeking, as well as many other behaviours. While some of these differences reflect activational effects of oestradiol and testosterone in the blood of adult males and females, many are due to differences in brain functions, which were organised during early life by differential actions of sex steroids. In rodents, as in other mammals, most nonreproductive behaviours have been described as showing sex differences in quantity of performance rather than being present in one sex and absent in the other (Goy and McEwen 1980). An example of a behaviour that is organised by sex hormones during the period of brain development in rodents and primates is aggressiveness towards adults and infants (vom Saal 1983, 1984). Similarly, maze learning and the acquisition of shuttle-box avoidance responses also are 'organized' by sex hormones during development (Beatty 1979). At this time, however, few studies have addressed the role of sex steroids exposure during perinatal life in rodents in influencing behaviours typically studied using the ethological approach, such as responses to changes in the socio-ecological environment and, more generally, behavioural strategies in coping.

The developmental effects of steroids are typically irreversible and are referred to as 'organizational', while effects in adults are typically reversible and are referred to as 'activational' (vom Saal 1983, Arnold and Breedlove 1985). Specifically, oestrogen or aromatizable androgens (that are converted to oestrogen by the enzyme aromatase in cells) play a significant role in regulating neuronal development and neural circuit formation during the perinatal period, and these organisational actions of sex steroids can induce permanent sexual dimorphisms in certain brain regions, in synaptic formation, in dendritic length, in the distribution patterns of serotoninergic fibres and in neuronal connectivity (Matsumoto 1991).

An interesting phenomenon that has revealed the exquisite sensitivity of developing tissues, including the brain, to gonadal steroids is the intrauterine position phenomenon, which occurs in litter-bearing mammals. Naturally occurring variation in the levels of testosterone and oestradiol in female mouse, rat and gerbil foetuses (due to being positioned *in utero* between male or female foetuses) leads to marked differences in a wide range of reproductive and behavioural traits: genital morphology, timing of puberty, length of oestrus cycle, aggressive and territorial behaviour, sexual behaviour and sexual attractiveness; a wide range of reproductive and behavioural traits in males is also influenced by prior intrauterine position (vom Saal 1989, vom Saal *et al.* 1999). These behaviours and related physiological systems have been proposed to have evolved via the action of sexual selection. Animals that share a common phylogeny show similar morphological, physiological and behavioural traits with common adaptive functions that were therefore likely subjected to the same selection pressures.

The timing of exposure (in addition to the dose) of developing tissues to hormones or endocrine disrupters is critical. During the period when the reproductive organs and the central nervous system are undergoing rapid developmental changes (before homeostatic-protective-mechanisms have developed), endocrine disrupting chemicals, at environmentally-relevant concentrations within the range of exposure of human and wildlife populations, can lead to irreversible alterations in development. This can occur during development at exposure levels that might produce little effect in an adult (Bern 1992, vom Saal 1995, vom Saal *et al.* 1995, vom Saal and Sheehan 1998).

10.4 EFFECTS OF ENDOCRINE DISRUPTERS ON WILDLIFE

A substantial literature has reported that EDs may alter development, leading to altered behaviour and reproductive capacity in wildlife. Researchers over the last few years have unearthed a variety of ED-related effects in wildlife: interrupted sexual development; thyroid system disorders; inability to breed; reduced immune response; and abnormal mating and parenting behaviour, to the disappearance of entire populations. Reproductive system abnormalities in wildlife have been related to endocrine disrupters in fish, alligators and turtles, birds and mammals (Colborn *et al.* 1993, 1998).

Much of the evidence of detrimental effects of exposure to EDs emerged from studies on wildlife in the Great Lakes region of North America — animals living in or along the shorelines of the Lakes and dependent upon food resources from the Lakes. In the 1960s and 1970s a number of studies reported reproductive failure in Great Lakes herring gulls and terns (Keith

1966, Ludwig and Tomoff 1966, Gilbertson 1974, Gilbertson and Hale 1974). Mortality and reproductive failure of ranch mink fed a diet of contaminated Great Lakes fish was reported, and attributed to the sensitivity of mink to PCBs; a dose-response relationships between PCB concentration in food and reproductive failure, measured as reduced litter size, increased pup mortality, and reduced pup weight was found (Aulerich *et al.* 1973). Most salmonids native to the region disappeared and many of the stocked fish were either unable to reach sexual maturity or could not produce viable offspring; increased egg and fry mortality was also reported (Leatherland 1992). Great Lakes salmon showed thyroid enlargement and dysfunction (Leatherland 1998). The number of bald eagles producing offspring along the shoreline of the Great Lakes had been decreasing, and in 1970 no shoreline bald eagle pairs were successful (Colborn 1991). Water birds nesting in colonies showed declines in populations and abnormalities in a large number of animals (Gilbertson 1974, Gilman *et al.* 1977, Fox and Weseloh 1987, Peakall and Fox 1987). In every instance, the troubled populations were those dependent upon fish from the Great Lakes.

Alterations of reproductive and socio-sexual behaviours were reported in herring gulls nesting in the Great Lakes; gulls in these colonies had a high incidence of supernormal clutches and female-female pairing, exhibited disinterest and delays in breeding and reproduction (Shugart 1980), abnormal incubation behaviour (Fox *et al.* 1978), loss of expression of territorial defence, poor parenting, and feminisation and demasculinisation (Shugart 1980, Rattner *et al.* 1984, Fry *et al.* 1987). The result was a severely reduced reproductive success. These effects were associated with high levels of organochlorine contaminants, such as DDT and PCBs, and their *in vivo* metabolites (Fry and Toone 1981, Rattner *et al.* 1984). The Foster's tern, which eats fish from Green Bay in Lake Michigan, has experienced severely reduced reproductive success, a reduction attributed to the organochlorine bioaccumulation in the fish (Kubiak *et al.* 1989). The compounds also cause eggshells of some species to thin, increasing mortality.

The American alligator of Lake Apopka, Florida, is one of the most cited examples of a wildlife population affected by environmental pollutants, including endocrine disrupters. Alligators living in Florida's Lake Apopka were exposed to the oestrogenic pollutants dicofol, DDT and its metabolites, DDD and DDE, when a nearby chemical plant had an extensive spill in 1980. Years later, alligators showed a severe population decline in the lake, and researchers found higher-than-normal mortality among eggs and newborn alligators. They also found that adolescent females had severe ovarian abnormalities and had blood oestrogen levels two times higher than normal. The male juvenile alligators had smaller than normal penises, abnormal testes, higher oestrogen levels, and lower testosterone levels in their blood than control males of the same age from another lake in Florida. The characteristics of the Lake Apopka males reveal a profile indicative of feminisation processes during foetal sexual differentiation. This lead to the conclusion that chemicals from the spill not only killed developing eggs outright, but also altered the foetus' endocrine system (hormone levels and sexual development), which severely limited the alligator's ability to reproduce (Guillette 1995, Guillette *et al.* 1994). Heinz *et al.* (1991) found DDT metabolites (as well as other pesticides and the PCB mixture, Aroclor) in Lake Apopka and also in alligator eggs laid by females living in the lake. Recent studies suggest that p,p′-DDE, in conjunction with other compounds, may be responsible for abnormal phallus and other reproductive abnormalities observed in the alligators (Guillette *et al.* 1996). Similarly, the Florida red-belly turtle (*Chrysemys nelsoni*) was found to have suffered deleterious effects, including the development of ovo-testes, from DDT contamination of Lake Apopka (Guillette *et al.* 1994).

Sex reversal effects of endocrine disrupting chemicals on organisms with temperature-dependent sex determination have been reported. Oestrogens are involved in normal female development of many organisms with temperature-dependent sex determination (TSD), including turtles (Crews 1993). In the red-eared slider turtle (*Trachemys scripta elegans*), inhibition of oestrogen synthesis results in males, and application of oestrogens to eggs containing presumptive males results in females; application of oestrogenic chemicals and oestradiol have demonstrated sex-reversal effects similar to effects of endogenous oestrogens (Crews *et al.* 1989). For example, when oestrogenic hydroxylated PCBs are applied to eggs incubated at all-male or male-biased temperatures, females result (Bergeron *et al.* 1994). Recently, Willingham and Crews (1998) demonstrated sex reversal effects of environmentally relevant doses of *trans*-Nonachlor, *cis*-Nonachlor, Aroclor 1242, p,p′-DDE, and chlordane, either when assayed alone or applied together, on the red-eared slider turtle. Guillette (1996) has demonstrated that *trans* -Nonachlor, dicofol, and p,p′-DDD, but not p,p′-DDE, are oestrogenic in the American alligator, and caused sex reversal from male to female. It is important to note that the oestrogenic activity of the different compounds may be species- or genus-specific. For example, Clark *et al.* (1998) found that in the tiger salamander (*Ambystoma tigrinum*), p,p′-DDE acts as an oestrogen, but technical-grade DDT (80% p,p′-DDT and 20% o,p′-DDT) acts as an antioestrogen. Podreka *et al.* (1998) did not find any changes in sex ratio, hatchling size or hatchling weight in the green sea turtle *Chelonia mydas* after administration of p,p′-DDE.

Reproductive and developmental abnormalities have also been observed in several populations of fish exposed to effluents from sewage treatment plants and paper mills in England. Male fish living near municipal sewage outlets in England had both male and female sex characteristics and their livers produced vitellogenin, a female egg-yolk protein not normally found in males (Jobling and Sumpter 1993). The fish living close to the sewage outlets had severe abnormalities, while the fish living farther downstream had less severe symptoms. Several different chemicals, especially the alkylphenols, chemicals found in detergents and plastics, are suspected of causing the feminising effects.

A recent study by Allen *et al.* (1999) showed that male fish living in four United Kingdom estuaries had significantly elevated levels vitellogenin in their blood, suggesting an ongoing exposure to an oestrogenic substance. Researchers sampled flounder from the Tyne, Crouch, Thames and Mersey estuaries. Fish from the Tyne and Mersey had four to six orders of magnitude higher concentrations of vitellogenin than did fish living in the Adle River, the control estuary. Seventeen per cent of male fish (five out of 30) from the Mersey had eggs (oocytes) in their testes. These biological effects of pollutants may impair breeding and adversely affect fish populations, especially in species more responsive to oestrogenic chemicals and those who spend sensitive periods in development in the polluted environments.

10.5 LABORATORY AND HUMAN STUDIES

The endocrine system of all vertebrates has many common characteristics, and chemicals that can alter the endocrine system in wildlife thus have the potential to alter the endocrine system in humans. Recent studies are now confirming that the same chemicals implicated in some of the adverse effects observed in wildlife are also related to detrimental effects in humans (Jacobson and Jacobson 1996, Lonky *et al.* 1996, Daly *et al.* 1998, Brouwer *et al.* 1998, Porterfield and Hendry 1998).

10.5.1 REPRODUCTIVE EFFECTS

There are trends in genital abnormalities in men, such as a 50-year decline in semen quality (Carlsen *et al.* 1995, Swan *et al.* 1997, Swan and vom Saal 2000), a 20-year steady increase in genital tract malformations, such as cryptorchidism (undescended testes; John Radcliffe Group 1986), hypospadias (malformed penis and urethra; Paulozzi *et al.* 1997), and testicular and prostate cancer (Sharpe and Skakkebaek 1993, Hass and Sakr 1997). As yet, studies have not been conducted to determine whether there is a relationship between any of these trends and endocrine disrupters. However, a decrease in testicular sperm production, and developmental changes and cancer in the epididymides and prostate, as well as abnormalities such as cryptorchidism, have been experimentally produced in laboratory animals as a result of exposure during development to oestrogenic chemicals (McLachlan *et al.* 1975, Grocock *et al.* 1988, Sharpe *et al.* 1996, Nagel *et al.* 1997, vom Saal *et al.* 1997, 1998). The exposure of developing rodents to high doses of environmental oestrogens advances puberty and alters their subsequent reproductive functions (Gray 1992). Exposure of female rats to high doses of o,p′DDT during development results in precocious puberty as well as acceleration of the loss of fertility, referred to as the delayed anovulatory syndrome (Welch *et al.* 1969, Heinrichs *et al.* 1971). In male rats, exposure to a high dose of o,p′DDT during early life leads to marked impairment of fertility and reduced weight of prostate and seminal vesicles (Gellert *et al.* 1974).

In addition, abnormalities in daily sperm production per gram of testis, and epididymis, prostate and seminal vesicles have been produced by prenatal exposure to doses of the oestrogenic endocrine disrupter, bisphenol A and methoxychlor, within an environmentally relevant range for human exposure (Nagel *et al.* 1997, vom Saal *et al.* 1998, Welshons *et al.* 1999). Howdshell *et al.* (1999) also found that prenatal exposure to a dose of bisphenol A (2.4 μg/kg), comparable to human exposure levels, accelerated postnatal growth rate and advanced puberty in female mice, although individual differences in endogenous oestradiol resulting from natural variations (intrauterine position) influenced the responsiveness of the female to bisphenol A. This finding demonstrates that oestrogenic endocrine disrupters interact with endogenous oestrogens to determine the course of development of oestrogen-responsive tissues.

Recently, Moccarelli and coworkers (2000) reported that high serum concentrations of dioxin in parents from Seveso, Italy (where the population was exposed to a rel dioxin in 1976) were linked to a relative increase in the number of female births. More specifically, the lowered sex ratio in offspring was significantly linked to exposure of fathers, rather than mothers, to dioxin. The median concentration of dioxin in fathers in this study is similar to doses that induce epididymal impairment in rats and is about 20 times the estimated average concentration of dioxin currently found in human beings, as well as in mammal wildlife, in industrialised countries. These findings could have important public health and ecological implications. Indeed, variation in sex ratio can significantly impact the behavioural ecology of a population, as mating patterns adopted by different species or even different populations of the same species are related to sex ratio. For example, female-biased sex ratios are considered to increase female — female competition and to decrease mating opportunities for females (Berglund *et al.* 1993).

10.5.2 NEUROLOGICAL AND BEHAVIOURAL EFFECTS IN HUMAN INFANTS

A connection has been hypothesised between prenatal and postnatal exposure to EDs and disturbances in neurological and behavioural development. Most of the available human data on the neurological effects of EDs are limited to PCBs and dioxins (Schantz 1996, Tilson *et al.* 1990b). Developmental effects such as altered learning and motor skills have also been linked with prenatal exposure to

EDs. Studies conducted on highly exposed populations (i.e., the rice oil incidents in Japan in 1968, and in Taiwan in 1979, 'Yucheng-disease') have shown that *in utero* exposure of human infants to PCBs resulted in pre- and postnatal growth retardation, neuromuscular effects (hypotonia), as well as behavioural changes like hyperactivity, lower cognitive capacities and a greater incidence of behavioural problems (Rogan *et al.* 1988, Chen and Hsu 1994). Since then, a number of studies have been initiated to study the neurobehavioural effects of exposure to the lower levels of PCBs present in the environment. Two long-term human studies have examined the relationship between parental consumption of Great Lakes fish and neurobehavioural sequelae in their offspring. The Jacobsons' studies, showed an association between neurobehavioural deficits in the infants and their mother's consumption of Lake Michigan PCB-contaminated fish, based on umbilical cord blood levels of PCBs (Jacobson *et al.* 1989). Children with highest PCB levels processed information more slowly, had lower quantitative and visual discrimination memory (poorer short-term memory — Jacobson *et al.* 1985), a poorer performance in the McCarthy verbal and memory scales test (i.e., reduced verbal skills — Jacobson *et al.* 1990) and an average 6.2 IQ point deficit when 11 years old (Jacobson and Jacobson 1996). A number of methodological concerns have been raised by the Jacobson study, including issues related to exposure assessment, sample selection and control of potential confounding variables. A second major study on the effects of Great Lakes fish consumption on behavioural functioning is still in its early stages but, newborns from women which consumed high amount of fish were impaired on reflexes, autonomic and habituation clusters of the neonatal *Behavioural Assessment Scale* (BAS) relative to newborns from women which consumed low or no fish (Lonky *et al.* 1996).

Other studies have tried to associate maternal and/or foetal levels of PCBs and other endocrine disrupters with neurobehavioural development of children, by using levels of PCBs and dioxin in cord and maternal plasma as an index of prenatal exposure. The study conducted by Rogan and colleagues did not find any evidence of deficits in memory function nor in cognitive performance in children in relation to their *in utero* exposure to PCBs (Rogan *et al.* 1986, Gladen *et al.* 1988). More recently, however, in the 'Dutch cohort' study, *in utero* exposure to background PCB concentrations was associated with adverse neonatal neurological development and with poorer cognitive functioning in preschool children (Patandin *et al.* 1998, 1999).

In spite of these discrepancies, certain commonalties exist between different human studies, which all report alterations in the *Brazelton Neonatal Behavioural Assessment Scale*, suggesting that exposure to environmental contaminants, including PCBs and endocrine disrupters, may induce subtle, transient alterations in maturation of the human CNS (Stewart *et al.* 2000). However, human studies, while potentially compelling, are limited to a correlational approach. Furthermore, a common problem in these epidemiological studies is that contaminated fish contain a large number of putative developmental neurotoxicants (e.g., methylmercury, p,p′DDE, PCBs, and several pesticides), thus rendering almost impossible to determine which contaminant(s) may be responsible for the observed associations between fish consumption and neurobehavioural effects. Laboratory derived data, including studies in animals exposed to environmentally-relevant doses of different PCBs and other EDs as well as mixtures, can greatly contribute to understand the observed effects in human populations.

10.5.3 NEUROLOGICAL AND BEHAVIOURAL EFFECTS IN EXPERIMENTAL ANIMALS

A number of studies with rodents have shown that exposure to relatively low levels of PCBs, which consist of a mixture of chemicals (there are 209 potential PCB congeners), can result in the formation of *in vivo* metabolites with an affinity for oestrogen receptors. PCBs produce a variety of effects *in vivo*, including changes in oestrogens, androgens and corticosteroids. These endocrine changes are associated with subtle behavioural and neurological deficits, particularly if exposure occurs during development. It has been reported that pre- or perinatal exposure to PCBs results in hyperactivity in mice and rats (Tilson *et al.* 1990a, Daly *et al.* 1989, 1998) and impaired discrimination learning in rats (Holene *et al.* 1995). Exposure to TCDD or PCBs throughout gestation and lactation in monkeys produced changes on several tasks assessing learning and memory (Schantz and Bowman 1989, Schantz *et al.* 1991, 1996). Brouwer *et al.* (1995) present evidence of polyhalogenated aromatic hydrocarbons affecting neurological functions in animals and human infants. Exposure to PCBs *in utero* and postnatally in rats also produced deficits on a number of behavioural tasks, such as locomotor activity, active avoidance tasks and temporal patterning of operant conditioning (Lileinthal *et al.* 1990, Schantz *et al.* 1995, Rice and Hayward 1999).

Daly and her collaborators have conducted a series of studies on rats eating contaminated fish; they have shown that consumption of Great Lakes fish has produced robust behavioural effects in adult animals. Specifically, adult rats fed ground rat chow with 30% of Ontario Lake salmon for 30 days showed an increase in responsivity to negative stimuli relative to rats fed ground rat chow with 30% Pacific Ocean salmon or no salmon. This has been shown using several behavioural measures, including reactivity to electric shock and unpredictable aversive nonreward (Daly *et al.* 1989), and reductions in anticipated food reward (Daly 1991). This change in responsivity is specific to negative events *per se*, given that response to appetitive reinforcers does not differ from control rats (Daly 1991). A more recent study has examined the effects of parental consumption of Lake Ontario salmon on the behaviour of the offspring tested at adulthood. Adult rats which were the biological offspring of, or nursed by, dams consuming a 30% Lake Ontario diet showed a much greater depression effect following the removal of a food reward in the runway task (Daly *et al.* 1998).

Little is known about possible neurological and behavioural effects of other EDs. In male mice exposure to a high dose of o,p′DDT during early life leads to marked neurobehavioural effects, such as a change in locomotor activity associated with a change in muscarinic cholinergic receptors in the cerebral cortex (Eriksson *et al.* 1992). More specifically, this study showed an increase in spontaneous motor behaviour in a novel environment and a decrease in muscarinic cholinergic receptors density in the cerebral cortex of adult male mice receiving a single oral dose of DDT (0.5 mg) at the age of 10 days. DDT-exposed mice showed a delayed habituation to the novel environment, suggesting disruption of this simple, nonassociative learning process (i.e., habituation). Since spontaneous behaviour can be considered to involve inhibition of responses, Eriksson and coworkers (1992) suggested that the observed changes in the cerebral cortex of mice reaching adult age may be the result of a lack of appropriate ability to inhibit responses, thereby allowing a hyperactive condition to develop in adult animals. Early interference DDT with the muscarinic cholinergic neurotransmission in the neonatal mouse during sensitive developmental periods (in this case limited to the developmental phase around day 10) thus develop permanent disturbances in behaviour.

Exposure during early life to doses of methoxychlor below those currently labelled as 'safe' for daily consumption by humans can result in changes in territory marking behaviour in male mice (vom Saal *et al.* 1995). In this study pregnant CF-1 mice were fed with different doses of DES, which served as a positive control for oestrogen action, and two insecticides, o,p′DDT and methoxychlor. The male offspring were examined at adulthood for number of urine marks deposited when placed into a novel environment for 60 minutes. The rate of urine marking increased with low doses of DES (from 20 ng/kg maternal body weight/day; 20 parts per trillion, ppt) and then decreased at the highest dose administered prenatally (200 μg/kg/day), thus resulting in an inverted U function (Figure 10.1). An increased rates of urine marking were also observed in males whose mothers were fed o,p′DDT and methoxychlor. Marking the environment with urine is a common behaviour in the house mouse and has different social functions like, for instance, to advertise the dominant territorial status of a male to other male competitors and to potential mates (e.g., Bronson 1979). A possible explanation of this finding is that prenatal exposure to low levels of oestrogenic chemicals increased a male's territory defence potential. Alternatively, it is possible that habituation to a novel environment (i.e., the

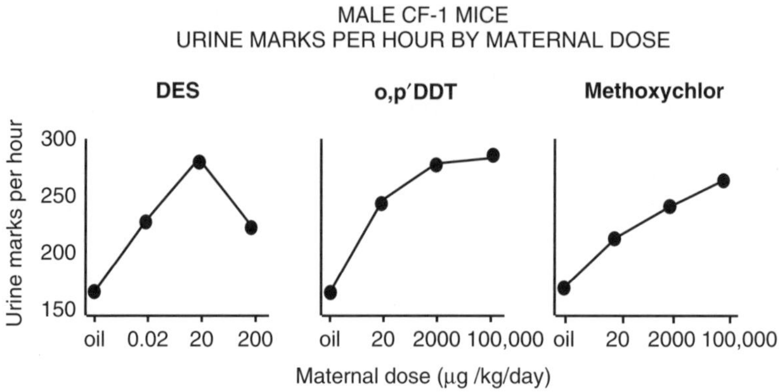

Figure 10.1 Mean number of urine marks deposited on filter paper during a 1-h test in a novel cage by adult male offspring of pregnant mice fed diethylstilboestrol, o,p′DDT or methoxychlor. The doses indicated are per kilogram body weight of the pregnant mice fed a chemical from gestation day 11–17. Urine of mice contained chemicals, fluorescent under ultraviolet light, which allowed individual urine marks to be counted. From: vom Saal *et al*. 1995.

urine-marking test cage) was delayed in exposed males. In follow-up studies, Palanza and coworkers (1999a,b) found several subtle behavioural changes associated with prenatal exposure to low doses of oestrogenic pesticides in mice (see 10.6.1 for a more detailed description of these findings).

In a series of experiments in rats, Farabollini and collaborators (1999) showed that perinatal exposure to low doses of bisphenol A influenced various aspects of nonsocial behaviour, differently in males and females, confirming that exposure to a weak environmental oestrogen in the period of sexual differentiation of the brain can influence adult behaviour. In these experiments, bisphenol A was administered orally to female rats either at a very low concentration (40 μg/kg body weight) from 10 days before mating until the weaning of the pups (i.e., chronic exposure) or at a higher dosage (400 μg/kg body weight) from day 14 of gestation to day 6 after birth (i.e., acute exposure during a critical period for brain organisation). The offspring of the treated mothers were tested as adults in the holeboard and the elevated plus maze, which measures forms of nonsocial behaviour related to anxiety, exploration and locomotion (see Chapter 6 for a detailed description of these testing procedures). The factor analysis indicated that in the treated male offspring both anxiety and the motivation to explore were reduced, while in females, motor activity and motivation to explore were depressed by perinatal exposure to bisphenol A. No substantial differences between the two modalities of bisphenol A administration were found, thus suggesting that the prolonged treatment with the low dosage compensates for the higher dosage given during the shorter steroid-sensitive period (Farabollini *et al.* 1999).

10.6 BEHAVIOUR AS A TARGET OF EDs EXPOSURE

There has been increasing interest, both at the scientific and regulatory level, in the use of animal behavioural methods for evaluating neural effects of environmental chemicals, and neurobehavioural toxicity is now acknowledged as a significant outcome of chemical exposure (see Kulig *et al.* 1996 for a review). Behavioural indices may be particularly sensitive to perturbation of hormonal systems because they represent the endpoint of integrated systems. Even subtle alterations in any of the component systems are likely to be reflected in the disruption of behaviour. Importantly, disturbances in behaviour (revealed by studies focusing on effects on individuals) may be of biological significance in both human and animal populations, due to impaired responsiveness to environmental demands that could result in a reduced social adaptability. However, so far scientific interest in the effects of exposure of animals to environmental EDs has, to a large degree, focused on effects on male fertility and gross reproductive disturbances, while effects in humans have focused on cognitive effects (see above; Jacobson and Jacobson 1996). Unfortunately, the wider behavioural effects of exposure to these compounds has been largely unexplored.

The functional capacity of the central nervous system cannot be determined by histological or even physiological studies independent of behavioural analysis. A central nervous system deficit may become evident only upon a specific kind of behavioural challenge, and consequences of exposure to environmental chemicals can be subtle. This is a critical issue since toxicological testing for developmental effects has focused on teratology (gross visible damage) and not on outcomes that could not be detected on gross physical examination (vom Saal and Sheehan 1998). Examination of both learned and unlearned (reflex and phylogenetically specialised) behaviours may reveal subtle deficits in CNS function, which may or may not be accompanied by demonstrable tissue pathology.

Of particular importance to understanding possible adverse effects of EDs are those behaviours critical for survival and reproduction, such as competitive aggression, exploration, sexual behaviour and parental behaviour. Both the developmental organisation and adult expression of these behaviours are regulated by the neuroendocrine-gonadal axis. Depending on the species, testosterone can influence the development and expression of behaviour in females as well as males, and oestradiol can influence the development and expression of behaviour in males as well as females (vom Saal 1989, vom Saal *et al.* 1992). The male and female neuroendocrine systems have been shaped by evolutionary processes to maximise fitness. In this context, it is important to apply strategies employed in ethological analysis to the study of effects of oestrogenic EDs on behaviour. By this we mean that experiments should be designed based on whether they can help to predict if a behavioural alteration might be adverse from the perspective of the adaptive significance of the behaviour under investigation. This focus on adaptation is quite different from the traditional experimental approach of behavioural toxicology, in which animals are typically used as 'tools' to detect alterations in neural or endocrine mechanisms. This focus on animals as 'detectors' has not involved considering whether the social and environmental situations in which animals are tested are ethologically appropriate and thus relevant in terms of the adaptive function of the behaviour being examined. We have proposed that the context and function of behaviour are of paramount importance when studying the underlying neural substrates, and we have referred to this approach as *ethotoxicology* (Parmigiani *et al.* 1998).

10.6.1 THE HOUSE MOUSE AS A MODEL FOR BEHAVIOURAL EFFECTS OF EDs

We use as a model animal for our studies an outbred stock of CD-1 house mice (*Mus musculus domesticus*). Previous research has shown that, relative to many other laboratory stocks of mice, CD-1 mice are very similar to wild mice in their behaviour and social organisation in seminatural environments (Parmigiani *et al.* 1989). The house mouse is widely distributed throughout the world and thus has been subjected to varied ecological pressures. Although there are reports of feral populations of this species living totally apart from humans, most commonly house mice live associated with human communities as commensals of man, thus potentially being exposed to the same environmental factors, including pollutants (Bronson 1979, Berry 1989).

In the studies reviewed below, we examined the effects of exposure during foetal life to a wide range of doses of three oestrogenic chemicals (DES, o,p′DDT, and methoxychlor) on behavioural development of mice challenged with several behavioural tests at different developmental stages: after birth, at weaning, at puberty, and as adults. It is critical to note here that in assessing the safety of these chemicals, toxicologists have only examined very high doses and used models to predict 'safe' exposure levels that, to our knowledge, have never before been directly tested to determine whether the models were actually predictive of 'safe' doses. We report here on effects of exposure during critical periods in brain and reproductive organ development to low, environmentally relevant doses of o,p′DDT and methoxychlor, within the range of exposure of humans and animals and below the doses currently predicted to be without risk of adverse effects (i.e., 'safe' doses) for each of these chemicals (ATSDR 1993, 1994).

DDT is one of the most commonly detected environmental pollutants in human tissues. It has been known for over 40 years that the *in vivo* metabolite of DDT, p,p′-DDE accumulates in body fat, and reaches particularly high levels in breast milk. The o,p′-isomer of DDT accounts for 11–29% of DDT. o,p′-DDT appears to be the primary oestrogenic component of technical grade DDT. p,p′-DDE is the highly persistent *in vivo* metabolite found in animal (including human) tissues (Kreiss *et al.* 1981), which is supposed to act as an androgen receptor antagonist (Kelce *et al.* 1995), and only shows very weak partial oestrogenic activity at a dose just below that which is acutely toxic (Soto *et al.* 1995). p,p′-DDE is not thought to contribute to the oestrogenic activity that is related to exposure to o,p′-DDT associated with the use of DDT as an insecticide.

Methoxychlor [bis-p-methoxy DDT; 1,1,1-trichloro-2,2-bis(p-methoxyphenyl)ethane] is an analogue of DDT, but methoxychlor and its *in vivo* metabolites are far less persistent in animal tissues. Methoxychlor exposure is due to its use as an insecticide on pets, in home gardens, and on crops and livestock. Methoxychlor only has oestrogenic effects *in vivo* after demethylation in the liver to mono-hydroxy-methoxychlor (30% of administered dose) or bis hydroxy-methoxychlor (23% of administered dose; Kapoor *et al.* 1970), with the most potent oestrogenic metabolite being bis-hydroxy-methoxychlor (Welch *et al.* 1969, Bitman and Cecil 1970, Bulger and Kupfer 1983).

Pregnant females were fed different doses of these chemicals (methoxychlor and o,p′-DDT: 20 - 2000 ng/g body weight/day; DES: 0.02 - 200 ng/g body weight/day) for 7 days during gestation (days 11 - 17), during the time that the foetal reproductive system and neuroendocrine system begin differentiating (vom Saal *et al.* 1992). The offspring were examined for effects on postnatal development. The main findings can be summarised as follows. (1) Animals prenatally exposed to methoxychlor showed changes in the development of reflexes. Exposure to a very low dose of methoxychlor (below the current dose deemed 'safe'; ATSDR 1994) appeared to produce an increased reactivity during early postnatal life, and the time of first appearance and subsequent complete maturation of various reflexes and responses was accelerated in the exposed pups as compared to controls (Palanza *et al.* 2001). (2) Methoxychlor exposed adolescent mice (28 - 35 days old) showed a decreased reaction time in exploring both a novel environment and a novel object (Palanza *et al.* 1999a). (3) The onset of male intrasex aggression appeared to be delayed in males prenatally exposed to low doses of methoxychlor, since exposed males showed low levels of aggressive interactions during early adolescence but not after they reached adulthood (Palanza *et al.* 1999a). (4) The proportion of both males and females attacking a same-sex conspecific was increased in mice prenatally exposed to low doses of DES and, marginally, to o,p′-DDT (Palanza *et al.* 1999b). This effect appeared to be related to a decreased latency to attack (Figure 10.2). However, males prenatally exposed to o,p′DDT displayed a decreased intensity of aggression. This finding suggested that reactivity to a same-sex conspecific was influenced, but that aggressiveness, *per se*, may not have been directly affected. Future research will be required to address this question.

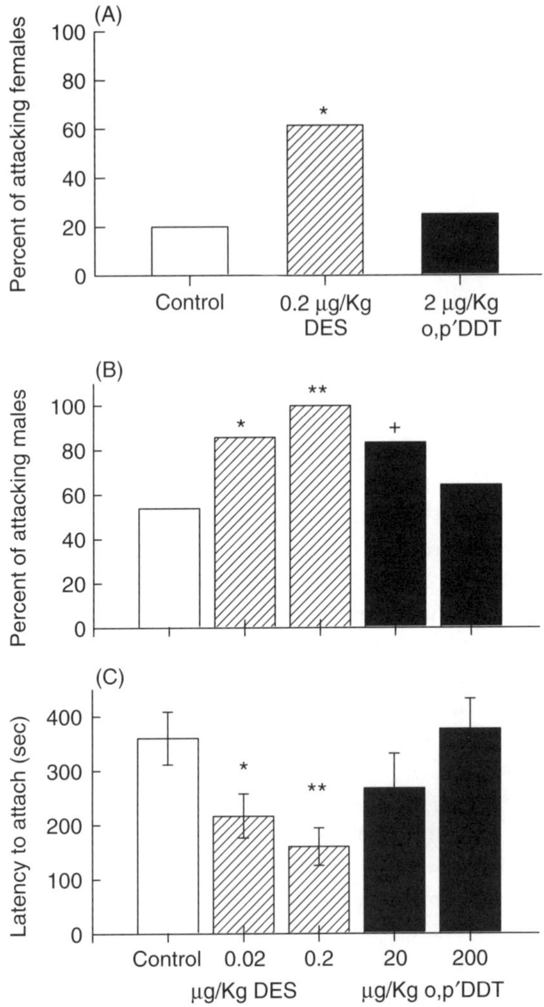

Figure 10.2 Percentage of females (A) attacking a same-sex intruder and percentage of attack (B) and mean (±SEM) latency to attack (C) same-sex intruders by adult males. Females and males were the offspring of mice fed DES, o,p′ DDT, or no chemical (control) from gestation day 11–17. From Palanza *et al.* 1999b. * = $P < 0.05$, ** = $P < 0.01$ vs. controls.

Prenatal exposure to very high (toxic) vs. low (environmentally-relevant) doses of these chemicals can produce different effects on a number of behavioural measures. A consistent finding is that prenatal exposure to low doses of a number of oestrogenic chemicals, such as DES, o,p′-DDT, methoxychlor, can produce increased reactivity at different developmental stages and in different experimental paradigms. In a variety of new situations, such as encountering a novel environment or a novel object (including an unfamiliar conspecific), the prenatally exposed animals were more reactive than controls. A possible explanation is that prenatal exposure to estrogenic chemical produces a nonspecific increase in locomotor activity. On the other hand, a more reactive profile to changes in the environment could be related to differences in emotionality. Steroid hormones are involved during foetal life in setting number and sensitivity of brain receptors for steroids and neurotransmitters (McEwen and Alves 1999). Endocrine disrupters might interfere with the normal development of serotonergic and GABA-ergic receptor systems, thus altering responses to stressful situations. Specific tests to measure anxiety responses and an ethopharmacological approach can help to identify the mechanisms mediating behavioural changes due to developmental exposure to environmental chemicals.

10.7 THE NEED TO DIRECTLY STUDY LOW, ENVIRONMENTALLY RELEVANT DOSES OF ENDOCRINE DISRUPTING CHEMICALS

The range of outcomes attributable to exposure to environmentally-relevant doses of EDs that are within the range of exposure of wildlife and humans is not well known for all but a very few chemicals. In particular, functional changes, such as changes in behavioural responses or changes in organ function, as opposed to gross toxicity or teratology, have not typically been examined in toxicological studies. However, a series of important findings derived from both epidemiological and laboratory studies suggests that: (1) exposure to EDs can induce significant reproductive, neurological and behavioural dysfunctions in wildlife, laboratory animals, and humans, particularly following exposure during critical developmental periods; (2) the neurochemical actions of EDs depends on the developmental status at the time of exposure and on the species-specific sensitivity; and (3) the mechanisms responsible for these changes may involve alterations in basic cellular signalling processes and endocrine functions that influence the synthesis and activity of important CNS neurotransmitters, the organisation of the developing brain, and thus, ultimately, the behavioural effects of EDs.

Explanations of the effects of prenatal exposure to EDs are complicated by the fact that hormones and hormone-mimicking chemicals do not show linear dose - response relationship through a wide dose range. Instead, nonmonotonic functions, such as U-shaped or inverted U-shaped curves, are found (vom Saal *et al.* 1997). The issue of doses is critical, since only a few (typically three) high doses are usually examined in toxicological studies. Thus, the safety of low doses of environmental chemicals, that are within the range of human exposure, is not directly tested, but, instead, is inferred. This process of estimating the risk posed by exposure to environmentally relevant doses of chemicals, known as risk assessment, is based on models that assume a monotonic dose - response relationship. The assumption is that responses either increase or stay the same (asymptote) as dose increases, but responses never first increase and then decrease with dose. In fields of science in which receptor-mediated effects are studied (such as responses to neurotransmitters or hormones), it is well recognised that inhibition of response can occur at high doses of a ligand, in part, due to receptor downregulation. This literature needs to be taken into consideration when studying effects of EDs. The fact that we observed permanent effects on behaviour at doses within the range of human exposure (and thus environmentally relevant) should be of concern for those involved in the regulatory process and for toxicologists, whose method of studying the same chemicals would have predicted no effect at these low doses.

The current process of assessing the safety of environmental chemicals has been challenged based on findings showing that nonmonotonic dose-response curves are commonly found in endocrinology (vom Saal and Sheehan 1998). Specifically, in studies in which pregnant female mice were administered a five-log range of doses of DES, we obtained a nonmonotonic dose-response function for anogenital distance and body weight at birth, reflex development (Palanza *et al.* in preparation), urine marking rates (vom Saal *et al.* 1995), and adult prostate size in male offspring (vom Saal *et al.* 1997, Welshons *et al.* 1999). However, nonmonotonic dose-response curve do not occur for all responses to EDs. The conclusion from our studies, as well as from other findings, is that responses to EDs cannot be assumed to be monotonic across a wide dose range and outcomes in response to low, environmentally-relevant doses of EDs may not be predicted by effects at higher doses.

10.8 CONCLUSIONS AND PERSPECTIVES

It is generally assumed that natural selection operates to create a phenotype that is optimum for a particular environment. If exposure to EDs changes that phenotype, leading to a less-than-optimum set of traits, such as an altered level of reactivity to stimuli or altered aggressiveness, a negative impact on those individuals in the population is likely to occur, and changes in population dynamics will likely follow. While reproduction and sociosexual behaviours of a particular species are adapted to specific environmental conditions (Krebs and Davies 1981), there are many factors that give rise to individual differences in these behaviours (vom Saal 1989). Consequently, there is an evolving range of social and reproductive behaviours that occurs among animals within any population due to variation in genotype, hormone levels, experience, etc. Shifts in the proportion of animals within a population that show specific traits, such as increased aggressiveness or exploration, can influence social structure and dispersion, thus affecting population dynamics (vom Saal 1984, Brain and Parmigiani 1990, Cowell *et al.* 1998).

It is well recognised in toxicology that not all individuals are equally sensitive to environmental chemicals, and there has long been a concern with sensitive subpopulations that show a greater response to chemical exposure than the rest of the population. Thus, when exposed to EDs during development, some individuals will be shifted in phenotype to a greater degree than will others, a finding recently reported by Howdeshell *et al.* (1999). One consequence could be a loss of phenotypes in the population, thus possibly decreasing population variance; alternatively, an increase in variance for some traits could occur. In either case, this could lead to a 'feed forward' process that can change genotype within the population, that is, a change in the course of evolution. The genetically predisposed individuals that show the greatest change in phenotype in response to EDs could have had, in an uncontaminated environment, the optimum phenotype for the environment. However, the phenotype resulting from exposure of these sensitive individuals to EDs may now be the optimum.

The ethotoxicological approach can help to take into account the importance of evolutionary processes for understanding the potential effects both on individuals and on populations, due to exposure to EDs.

ACKNOWLEDGEMENTS

This work was supported by Italian Ministry of University and Scientific Research (MURST-COFIN 2000) and CNR (National Council of Research) grants to PP.

REFERENCES

Allen Y, Scott AP, Matthiessen P, Haworth S, Thain JE and Feist S (1999) Survey of estrogenic activity in United Kingdom estuarine and coastal waters and its effects on gonadal development of the flounder *Platichthys flesus*. *Environmental Toxicology and Chemistry*, **18**, 1791-1800.

Arnold AP and Breedlove SM (1985) Organizational and activational effects of sex steroids on brain and behavior: a reanalysis. *Hormones and Behavior*, **19**, 469-498.

ATSDR (1993) *Toxicological Profile for DDT*, Agency for Toxic Substances and Disease Registry, U.S. Department of Health and Human Services, Atlanta, Georgia.

ATSDR (1994) *Toxicological Profile for Methoxychlor*, Agency for Toxic Substances and Disease Registry, U.S. Department of Health and Human Services, Atlanta, Georgia.

Aulerich RJ, Ringer RK and Iwamoto S (1973) Reproductive failure and mortality in mink fed on Great Lakes fish. *Journal of Reproduction and Fertility*, **19**, 365-376.

Beatty W (1979) Gonadal hormones and sex differences in non reproductive behaviors in rodents: organizational and activational influences. *Hormones and Behavior*, **12**, 112-163.

Ben-Jonathan N and Steinmetz R (1998) Xenestrogens: the emerging story of bisphenol A. *Trends in Endocrinology and Metabolism*, **9**, 124-128.

Bergeron JM, Crews D and McLachlan JA (1994) PCBs as environmental estrogens: turtle sex determination as a biomarker of environmental contamination. *Environmental Health Perspectives*, **102**, 780-781.

Berglund A, Magnhagen C, Bisazza A, Koenig B and Huntingford F (1993) Female-female competition over reproduction. *Behavioral Ecology*, **4**, 184-187.

Bern HA (1992) The fragile fetus. In *Chemically Induced Alterations in Sexual and Functional Development: The Wildlife/Human Connection*. Colborn T and Clement C (eds) Princeton Scientific Publishing, Princeton, pp. 9-16.

Berry RJ (1989) Genes, behaviour and fitness in mice: concepts and confusions. In *House Mouse Aggression*, Brain P, Mainardi D and Parmigiani S (eds), Harwood Academic Publishers, Chur, pp. 23–48.

Bitman J and Cecil H (1970) Estrogenic activity of DDT analogs and polychlorinate biphenyls. *Journal of Agricultural Food & Chemistry*, **8**, 1108–1120.

Brain PF and Parmigiani S (1990) Variation in aggressiveness and social structure in the house mouse populations. *Biological Journal of the Linnean Society*, **41**, 257–269.

Bronson FH (1979) The reproductive ecology of the house mouse. *Quarterly Review of Biology*, **54**, 246–299.

Brouwer A, Ahrlborg UG, van de Bergh M, Birnbaum LS, Boersma ER, Bosveld B, Denison MS, Gray LE, Hagmar L, Holene E, Huisman M, Jacobson SW, Jacobson JL, Koopman-Essemboom C, Koppe JG, Kulig Bm, Morse DC, Muckle G, Peterson RE, Sauer PJJ, Seefal RF, Smits-van Proije AE, Touwen BCL, Weisglas Kuperus N and Winnecke G (1995) Functional aspects of developmental toxicity of polyhalogenated aromatic hydrocarbons in experimental animals and human infants. *European Journal of Pharmacology*, **293**, 1–40.

Brouwer A, Morse DC, Lans MC, Schuur AG, Murk AJ, Klasson-Wehler E, Bergman A and Visser TJ (1998) Interactions of persistent environmental organohalogens with the thyroid hormone system: mechanisms and possible consequences for animal and human health. *Toxicology Industrial Health*, **14**, 59–84.

Bulger WH and Kupfer D (1983) Estrogenic action of DDT analogs. *American Journal of Industrial Medicine*, **4**, 163–173.

Carlsen E, Giwereman A, Keding N and Skakkebaek N (1995) Declining semen quality and increasing incidence of testicular cancer: is there a common cause? *Environmental Health Perspectives*, **103**, 137–139.

Chen YC and Hsu CC (1994) Effects of prenatal exposure to PCBs on the neurological function of children: a neuropsychological and neurophysiological study. *Developmental Medicine and Child Neurology*, **36**, 312–320.

Clark EJ, Norris DO and Jones RE (1998) Interactions of gonadal steroid and pesticides (DDT, DDE) on gonaduct growth in larval tiger salamanders, *Ambystoma tigrinum. General and Comparative Endocrinology*, **109**, 279–289.

Colborn T (1991) Epidemiology of Great lakes bald eagles. *Journal of Toxicology and Environmental Health*, **33**, 395–453.

Colborn T and Clement C (eds) (1992) Chemically-induced alterations in sexual and functional development: the wildlife/human connection. *Advances in Modern Environmental Toxicology*, vol. 21, Princeton Scientific Publishing, Princeton, NJ, 401 pp.

Colborn T, vom Saal FS and Soto AM (1993) Developmental effects of endocrine disrupting chemicals in wildlife and humans. *Environmental Health Perspectives*, **101**, 378–84.

Colborn T, Smolen MJ and Rolland R (1998) Environmental neurotoxic effects: the search for new protocols in functional teratology. *Toxicology and Industrial Health*, **14**, 9–23.

Cowell LG, Crowder LB and Kepler TB (1998) Density-dependent prenatal androgen exposure as an endogenous mechanism for the generation of cycles in small mammal populations. *Journal of Theoretical Biology*, **190**, 93–106.

Crews D (1993) The organizational concept and vertebrates without sex chromosomes. *Brain and Behavioral Evolution*, **42**, 202–214.

Crews D, Wibbels T and Gutzke WHN (1989) Action of sex steroid hormones on temperature-induced sex determination in the snapping turtle (*Chelydra serpentina*). *General and Comparative Endocrinology*, **76**, 159–166.

Daly H (1991) Reward reductions found more adversive by rats fed environmentally contaminated salmon. *Neurotoxicology and Teratology*, **13**, 449–453.

Daly HB, Hertzler DR and Sargent DM (1989) Ingestion of environmentally contaminated Lake Ontario salmon by laboratory rats increased avoidance of unpredictable aversive nonreward and mild electric shock. *Behavioural Neuroscience*, **103**, 1356-1365

Daly HB, Stewart PW, Lunkenheimer L and Sargent D (1998) Maternal consumption of Lake Ontario salmon in rats produces behavioral changes in the offspring. *Toxicology and Industrial Health*, **14**, 25-39.

Ehrhardt A, Meyer-Bahlburg HFL, Rosen LR, Feldman JF, Veridiano NP, Zimmerman I and McEwan BS (1985) Sexual orientation after prenatal exposure to exogenous estrogen. *Archives of Sexual Behavior*, **14**, 57-77.

Eriksson P, Ahlbom J and Fredriksson A (1992) Exposure to DDT during a defined period in neonatal life induces permanent changes in brain muscarinic receptors and behaviour in adult mice. *Brain Research*, **582**, 277-281.

Eroschenko VP and Cooke PS (1990) Morphological and biochemical alterations in reproductive tracts of neonatal female mice treated with the pesticide methoxychlor. *Biology of Reproduction*, **42**, 573-583.

Farabollini F, Porrini S and Dessì-Fulgheri, F (1999) Perinatal exposure to the estrogenic pollutant bisphenol A affects behavior in male and female rats. *Pharmacology Biochemistry and Behavior*, **64**, 687-694.

Fox GA, Gilman AP, Peakall DB and Anderka FW (1978) Behavioral abnormalities of nesting Lake Ontario herring gulls. *Journal of Wildlife Management*, **42**, 477-483.

Fox GA and Weseloh DV (1987) Colonial waterbirds as bio-indicators of environmental contamination in the Great Lakes. *ICBP Technical Publishing*, **6**, 209-216.

Fry DM, Toone CK, Speich SM and Peard RJ (1987) Sex ratio skew and breeding patterns of gulls: demographic and toxicological considerations. *Studies on Avian Biology*, **10**, 26-37.

Fry DM and Toone CK (1981) DDT-induced feminization of gull embryos. *Science*, **213**, 922-924.

Gahr M (1994) Brain structure: causes and consequences of brain sex. In: *The Differences between the Sexes*, Short RV and Balaban E (eds), Cambridge University Press, Cambridge, pp. 273-299.

Gellert R, Heinrichs W and Swerdloff R (1974) Effects of neonatally-administered DDT homology on reproductive functions in male and female rats. *Neuroendocrinology*, **16**, 84-94.

Gilbertson M (1974) Pollutants in breeding herring gull. *Canadian Field-Naturalist*, **88**, 273-280.

Gilbertson M and Hale R (1974) Early embryonic mortality in a herring gull colony in lake Ontario. *Canadian Field-Naturalist*, **88**, 354-356.

Gill WB, Schumacher GFB and Bibbo M (1976) Structural and functional abnormalities in the sex organs of male offspring of mothers treated with diethylstilbestrol (DES). *Journal of Reproductive Medicine*, **16**, 147-153.

Gilman AP, Fox GA, Peakall DB, Teeple SM, Carroll TR and Haymes GT (1977) Reproductive parameters and egg contaminant levels of Great Lakes herring gulls. *Journal of Wildlife Management*, **41**, 458-468

Gladen BC, Rogan WJ, Hardy P, Thullen J, Tingelstad and Tully M (1988) Development after exposure to polychlorinated biphenyls and dichlorodiphenyl dichloroethene transplacentally and through human milk. *Journal of Pediatrics*, **113**, 991-995.

Go V, Garey J, Wolff MS and Pogo BGT (1999) Estrogenic potential of certain pyrethroid compounds in the MCF-7 human breast carcinoma cell line. *Environmental Health Perspective*, **107**, 173-177.

Goy R and McEwen BS (1980) *Sexual Differentiation of the Brain*. MIT Press, Cambridge, MA.

Gray LE (1992) Chemical induced alterations of sexual differentiation: a review of effects in humans and rodents. In *Chemically Induced Alterations in Sexual and Functional Development: The Wildlife/Human Connection*, Colborn T and Clement C (eds), Princeton Scientific Publishing, Princeton, pp. 203-230.

Gray J, Earl L, Ostby J, Cooper RL and Kelce WR (1999a) The estrogenic and antiandrogenic pesticide methoxychlor alters the reproductive tract and behavior without affecting pituitary size or LH and prolactin secretion in male rats. *Toxicology and Industrial Health*, **15**, 37-47.

Gray J, Earl L, Wolf C, Lambright C, Mann P, Price M, Cooper RL and Ostby J (1999b) Administration of potentially antiandrogenic pesticides (procymidone, linuron, iprodione, chlozolinate, p,p′-DDE, and ketoconazole) and toxic substances (dibutyl- and diethylhexyl phthalate, PCB 169, and ethane dimethane sulphonate) during sexual differentiation produces diverse profiles of reproductive malformations in the male rat. *Toxicology and Industrial Health*, **15**, 94-118.

Grocock CA, Charlton HM and Pike MC (1988) Role of the fetal pituitary in cryptorchidism induced by exogenous maternal estrogen during pregnancy in mice. *Journal of Reproduction and Fertility*, **83**, 295-300.

Guillette L (1990) Postnatal exposure to a mixture of low chlorinated PCBs in rats. *Fundamental and Applied toxicology*, **15**, 457-467.

Guillette LH (1995) Endocrine disrupting environmental contaminants and developmental abnormalities in embryos. *Human and Ecological Risk Assessment*, **1**, 25-36.

Guillette LH, Gross TS, Masson GR, Matter JM, Percival HF and Woodward AR (1994) Developmental abnormalities of the gonad and abnormal sex hormone concentrations in juvenile alligators from contaminated and control lakes in Florida. *Environmental Health Perspectives*, **102**, 680-688.

Guillette LH, Pickford DB, Crain DA, Rooney AA and Percival HF (1996) Reduction in penis size and plasma testosterone concentrations in juvenile alligators living in a contaminated environment. *General Comparative Endocrinology*, **101**, 32-42.

Gupta C (2000) Reproductive malformation of the male offspring following maternal exposure to estrogenic chemicals. *Proceedings of the Society of Experimental Biology and Medicine*, **224**, 61-68.

Hass GP and Sakr WA (1997) Epidemiology of prostate cancer. *Cancer Journal Clinics*, **47**, 273-287.

Heinrichs WL, Gellert RJ, Bakke JL and Lawrence NL (1971) DDT administered to neonatal rats induces persistent estrus syndrome. *Science*, **173**, 642-643.

Heinz GH, Percival HF and Jennings ML (1991) Contaminants in American alligator eggs from Lake Apopka, Lake Griffin and Lake Okeechobe, Florida. *Environmental Monitoring Assessment*, **16**, 227-285.

Hess RA, Bunick D, Ki-Ho L, Bahr J, Taylor JA, Korach KS and Lubahn DB (1997) A role for estrogens in the male reproductive system. *Nature*, **390**, 509-512.

Holene E, Nafstadt I, Skaare JU, Bernhoft A, Engen P and Sagvolden T (1995) Behavioral effects of pre- and postnatal exposure to individual polychlorinated biphenyl congeners in rats. *Environmental and Toxicological Chemistry*, **14**, 967-976.

Holliday R (1987) The inheritance of epigenetic defects. *Science*, **238**, 163-170.

Howdeshell KL, Hotchkiss AK, Thayer KA, Vandenbergh JG and vom Saal FS (1999) Exposure to bisphenol A advances puberty. *Nature*, **401**, 763-764.

Jacobson J and Jacobson SW (1996) Intellectual impairment in children exposed to polychlorinated biphenyls *in utero*. *New England Journal of Medicine*, **335**, 783.

Jacobson J, Jacobson SW and Humphrey HEB (1990) Effects of *in utero* exposure to polychlorinated biphenyls and related contaminants on cognitive functioning in young children. *Journal of Pediatrics*, **116**, 38-45

Jacobson JL, Humphrey HEB, Jacobson SW, Schantz SL, Mullin MD and Welch R (1989) Determinants of PCBs, PBBs and DDT levels in the sera of young children. *American Journal of Public Health*, **79**, 1364-1374.

Jacobson SW, Hfein GG, Jacobson JL, Schwartz PL and Dowler JK (1985) The effect of intrauterine PCB exposure on visual recognition memory. *Child Development*, **56**, 853-860.

Jansen HT, Cooke PS, Porcelli J, Liu TC and Hansen LG (1993) Estrogenic and antiestrogenic actions of PCBs in the female rat: *in vitro* and *in vivo* studies. *Reproductive Toxicology*, **7**, 1-11.

Jobling JS and Sumpter JP (1993) Detergent component in sewage effluent are weakly estrogenic to fish: an *in vitro* study using rainbow trout (*Oncorynchus mykiss*) hepatocytes. *Aquatic Toxicology*, **27**, 361-372.

John Radcliffe Group (1986) Cryptorchidism: an apparent substantial increase since 1960. *British Medical Journal*, **293**, 1401-1404.

Kapoor IP, Metcalf RL, Nystrom RF and Sangha GK (1970) Comparative metabolism of methoxychlor, methiochlor, and DDT in mouse, insects, and in model ecosystem. *Journal of Agricultural Food Chemistry*, **18**, 1145-1152.

Katzenellenbogen BS, Katzenellenbogen JA and Mordecai D (1979) Zearalenones: characterization of the estrogenic potencies and receptor interactions of a series of fungal-resorcylic acid lactones. *Endocrinology*, **105**, 33-40.

Katzenellenbogen JA, O'Malley BW and Katzenellenbogen BS (1996) Tripartite steroid hormone receptor pharmacology: interaction with multiple effector sites as a basis for the cell- and promoter-specific action of these hormones. *Molecular Endocrinology*, **10**, 119-131.

Keith JA (1966) Reproduction in a population of herring gulls (*Larus argentatus*) contaminated by DDT. *Journal of Applied Ecology*, **3**, 57-70.

Kelce WR, Stone R, Law SC, Gray LE, Kemppainen JA and Wilson EM (1995) Persistent DDT metabolite p,p′-DDE is a potent androgen receptor antagonist. *Nature*, **375**, 581-585.

Krebs JR and Davies NB (1981) *An Introduction to Behavioral Ecology*. Sinauer Association Inc. Publishers, Sunderland.

Kreiss K, Sack M, Kimbrough RL, Needha LL, Smrek AL and Jones BT (1981) Cross-sectional study of a community with exceptional exposure to DDT. *Journal of the American Medical Association*, **45**, 1926-1930.

Kubiak TJ, Harris HJ, Smith LM, Schwartz TR, Stalling DL, Trick JA, Sileo L, Docherty DE and Erdman DC (1989) Microcontaminants and reproductive impairment of the Foster's tern on Green Bay, Lake Michigan-1983. *Archives of Environmental Contamination and Toxicology*, **18**, 706-727.

Kuiper GJM, Carlsson B, Grandien K, Enmark E, Häggblad J, Nilsson S and Gustafsson JA (1997) Comparison of the ligand binding specificity and transcript tissue distribution of estrogen receptors alpha and beta. *Endocrinology*, **138**, 863-870.

Kulig B, Alleva E, Bignami G, Cohn J, Cory-Slechta D, Landa V, O'Donoghue J and Peakall D (1996) Animal behavioral methods in neurotoxicity assessment: SGOMSEC joint report. *Environmental Health Perspectives*, **104** (Suppl. 2), 193-204.

Leatherland JF (1992) Endocrine and reproductive functions in Great Lakes salmon. In *Chemically Induced Alterations in Sexual and Functional Development: The Wildlife/Human Connection*, Colborn T and Clement C (eds), Princeton Scientific Publishing, Princeton, pp. 129-145.

Leatherland JF (1998) Changes in thyroid hormone economy following consumption of environmentally contaminated Great Lakes fish. *Toxicology and Industrial Health*, **14**, 41-57.

Lilienthal H, Neuf M, Munoz C and Winneke G (1990) Behavioral effects of pre plus postnatal exposure to a mixture of low chlorinated PCBs in rats. *Fundamental and Applied Toxicology*, **15**, 457-467.

Lonky E, Reihman J, Darvill T, Mather J and Daly H (1996) Neonatal behavioral assessment scale performance in humans influenced by maternal consumption of environmentally contaminated Lake Ontario fish. *Journal of Great Lakes Research*, **22**, 198-212.

Ludwig JP and Tomoff CS (1966) Reproductive success and insecticide residues in Lake Michigan herring gulls. *Jack Pine Warbler*, **44**, 77-84.

Matsumoto A (1991) Synaptogenic action of sex steroids in developing and adult neuroendocrine brain. *Psychoneuroendocrinology*, **16**, 25-40.

McEwen BS (1981) Neural gonadal steroid actions. *Science*, **211**, 1301-1303.

McEwen BS and Alves SE (1999) Estrogen actions in the central nervous system. *Endocrinology Review*, **20**, 279-307.

McLachlan J, Newbold R and Bullock B (1975) Reproductive tract lesions in male mice exposed prenatally to diethylstilbestrol. *Science*, **190**, 991-992.

Meyer-Bahlburg HF, Ehrhardt AA, Rosen LR, Gruen RS, Veridiano SP, Vann FH and Neuwalder HF (1995) Prenatal estrogen exposure and the development of homosexual orientation. *Developmental Psychology*, **31**, 12-21.

Mocarelli P, Gerthoux PM, Ferrari E, Patterson DG, Kieszak SM, Brambilla P, Vincoli N, Signorini S, Tramacere P, Carreri V, Samp EJ, Turner WE and Needham LL (2000) Paternal concentrations of dioxin and sex ratio of offspring. *Lancet*, **355**, 1858-1863.

Nagel SC, vom Saal FS, Thayer KA, Dhar MG, Boechler M and Welshons WV (1997) Relative binding affinity-serum modified access (RBA-SMA) assay predicts the relative *in vivo* bioactivity of the xenestrogens bisphenol A and octylphenol. *Environmental Health Perspectives*, **105**, 70-76.

Nagel SC, vom Saal FS and Welshons WV (1998) The effective free fraction of estradiol and xenestrogens in human serum measured by whole cell uptake assays: physiology of delivery modifies estrogenic activity. *Proceedings of the Society of Experimental Biology and Medicine*, **217**, 300-309.

Nelson RJ (1995) *An Introduction to Behavioral Endocrinology*. Sinauer Association Inc. Publishers, Sunderland.

Newbold R (1995) Cellular and molecular effects of developmental exposure to diethylstilbestrol: implications for other environmental estrogens. *Environmental Health Perspectives*, **103**, 83-87.

Nollar KL, O'Brien PC, Colton T, Kaufman R and Melton LJ (1990) Medical and surgical diseases associated with *in utero* exposure diethylstilbesterol. In *Clinical Practice of Gynecology*, Nollar KL (ed.), Elsevier Science Publishing, New York, pp. 1-7.

Paech K, Webb P, Kuiper GJM, Nilsson S, Gustafsson JA, Kushner PJ and Scanlan TS (1997) Differential ligand activation of estrogen receptors by ERα and ERß at AP1 sites. *Science*, **277**, 1508-1510.

Pakdel F, Le Guellec C, Vaillant C, Le Roux MG and Valotaire Y (1989) Identification and estrogen induction of two estrogen receptors (ER) messenger ribonucleic acids in the rainbow trout liver: sequence homology with other ERs. *Molecular Endocrinology*, **3**, 44-51.

Palanza P, Parmigiani S and vom Saal FS (2001) Prenatal exposure to diethylstilbestrol, o,p′-DDT and methoxychlor alters early postnatal development in mice. *Hormones and Behavior*, **40**, 252-265.

Palanza P, Morellini F, Parmigiani S and vom Saal FS (1999a) Prenatal exposure to endocrine disrupting chemicals: effects on behavioral development. *Neuroscience and Biobehavioral Reviews*, **23**, 1011-1127.

Palanza P, Parmigiani S, Liu H and vom Saal FS (1999b) Prenatal exposure to low doses of the estrogenic chemicals diethylstilbestrol and o′,p′ DDT alters aggressive behavior of male and female house mice. *Pharmacology Biochemistry and Behavior*, **64**, 665-672

Parmigiani S, Brain PF and Palanza P (1989) Ethoexperimental analysis of different forms of intraspecific aggression in the house mouse. In *Ethoexperimental Approaches to the Study of Behavior*, Blanchard RJ, Brain PF, Blanchard DC and Parmigiani S (eds), Kluwer Academic Publishers, Dordrecht, pp. 418-431.

Parmigiani S, Palanza P and vom Saal S (1998) Etho-toxicology: an evolutionary approach to the study of environmental endocrine disrupting chemicals. *Toxicology and Industrial Health*, **14**, 333-339.

Patandin S, Koopman-Esseboom C, de Ridder MA, Weisglas-Kuperus N and Sauer PJ (1998) Effects of environmental exposure to polychlorinated biphenyls and dioxins on birth size and growth in Dutch children. *Pediatric Research*, **44**, 538-545.

Patandin S, Lanting Ci, Mulder PG, Boersman ER, Sauer PJ and Weisgls-Kuperus N (1999) Effects of environmental exposure to polychlorinated biphenyls and dioxins on cognitive abilities in Dutch children of 42 months of age. *Journal of Pediatrics*, **134**, 33-41.

Paulozzi LJ, Erickson JD and Jackson RJ (1997) Hypospadias trends in two US surveillance systems. *Pediatrics*, **100**, 831-834.

Peakall DB and Fox GA (1987) Toxicological investigation of pollutant-related effects in Great Lakes gulls. *Environmental Health Perspectives*, **102**, 125-130.

Phoenix CH, Goy RW, Gerall AA and Young WC (1959) Organizing action of prenatally administered testosterone propionate on the tissues mediating mating behavior in the female guinea pig. *Endocrinology*, **65**, 369-382.

Podreka S, Georges A, Maher B and Limpus CJ (1998) The environmental contaminant p,p′-DDE fails to influence the outcome of sexual differentiation in the marine turtle *Chelonia mydas*. *Environmental Health Perspectives*, **106**, 185-188.

Porter WP, Hinsdill RD, Fairbrother A, Olson LJ, Jaeger J, Yuill T, Bisgaard S, Hunter WG and Nolan K (1984) Toxicant-disease-environment interactions associated with suppression of immune system, growth and reproduction. *Science*, **224**, 1014-1017.

Porterfield S and Hendry LB (1998) Impact of PCBs on thyroid hormone directed brain development. *Toxicology and Industrial Health*, **14**, 103-20.

Raloff J (1995) DES sons face no fertility problems. *Science News*, **147**, 324.

Rattner BA, Eroschenko VP, Fox GA, Fry DM and Gorsline J (1984) Avian endocrine responses to environmental pollutants. *Journal of Experimental Zoology*, **232**, 683-689.

Rice D and Hayward S (1999) Effects of exposure to 3,3′,4,4′,5-pentachlorobiphenyl (PCB 126) throughout gestation and lactation on behavior (concurrent random interval-random interval and progressive ratio performance) in rats. *Neurotoxicology and Teratology*, **21**, 679-687.

Rogan WJ, Gladen BC, McKinney JD, Carreras N, Hardy P, Thullen J, Tingelstad and Tully M (1986) Neonatal effects of transplacental exposure to PCBs and DDE. *Journal of Pediatrics*, **109**, 335-341.

Rogan WJ, Gladen BC, Hung KL, Koong Sl, Shia Ly, Taylor JS, Wu YC, Yang D, Ragan NB and Hsu CC (1988) Congenital poisoning by polychlorinated biphenyls and their contaminants in Taiwan. *Science*, **241**, 334-336.

Safe S, Astroff B, Harris M, Zacharewski T, Dickerson R, Romkes M and Biegel L (1991) 2,3,7,8-Tetrachlorodibenzo-p-dioxin (TCDD) and related compounds as antiestrogens: characterization and mechanism of action. *Pharmacology and Toxicology*, **69**, 400-409.

Schantz SL (1996) Developmental neurotoxicity of PCBs in humans: what do we know and where do we go from here? *Neurotoxicology and Teratology*, **18**, 217-227.

Schantz SL and Bowman RE (1989) Learning in monkeys exposed to 2,3,7,8-tetrachlorodibenzo-p-dioxin (TCDD). *Neurotoxicology and Teratology*, **11**, 13-19.

Schantz SL, Levin ED and Bowman RE (1991) Long-term neurobehavioral effects of perinatal polychlorinated biphenyl (PCB) exposure in monkeys. *Environmental Toxicology and Chemistry*, **10**, 747-756.

Schantz SL, Moshtaghian J and Ness DK (1995) Spatial learning deficits in rats exposed to ortho-substituted PCB congeners during gestation and lactation. *Fundamental and Applied Toxicology*, **26**, 117-126.

Schantz SL, Seo BW, Moshtaghian J, Peterson RE and Moore RW (1996) Effects of gestational and lactational exposure to TCDD or coplanar PCBs on spatial learning. *Neurotoxicology and Teratology*, **18**, 305-313.

Shah HC and McLachlan JA (1976) The fate of diethylstilbestrol in the pregnant mouse. *Journal of Pharmacology and Experimental Teratology*, **197**, 687-696.

Sharpe RM and Skakkebaek N (1993) Are estrogens involved in falling sperm count and disorders of the male reproductive tract? *Lancet*, **341**, 1392-1395.

Sharpe RM, Fisher JS, Millar MM, Jobling S and Sumpter JP (1996) Gestational and lactational exposure of rats to xenestrogens results in reduced testicular size and sperm production. *Environmental Health Perspectives*, **103**, 1136-1143.

Shugart GW (1980) Frequency and distribution of polygyny in herring gulls in 1978. *Condor*, **90**, 933-935.

Soto AM, Sonnesnchein C, Chung KL, Fernandez MF, Olea N and Serrano F (1995) The E-SCREEN assay as a tool to identify estrogens: an update on estrogenic environmental pollutants. *Environmental Health Perspectives*, **103**, 113-122.

Steinmetz R, Brown NG, Allen DL, Bigsby RM and Ben-Jonathan N (1997) The environmental estrogen bisphenol A stimulates prolactin release *in vitro* and *in vivo*. *Endocrinology*, **138**, 1780-1786.

Stewart P, Reihman J, Lonky E, Darvill T and Pagano J (2000) Prenatal PCB exposure and neonatal behavioral scale (NBAS) performance. *Neurotoxicology and Teratology*, **22**, 21-29.

Swan SH and vom Saal FS (2000) Alterations in male reproductive development: the role of endocrine disrupting chemicals. In *Endocrine Disruptors in the Environment*. Metzler M (ed), Springer Verlag, Heidelberg.

Swan SH, Elkin EP and Fenster L (1997) Have sperm densities declined: a reanalysis of global trend data. *Environmental Health Perspectives*, **105**, 1228-1232.

Takasugi N and Bern HA (1988) Introduction: abnormal genital tract development in mammals following early exposure to sex hormones. In *Toxicity of Hormones in Perinatal Life*, Mori T and Nagasawa H (eds), CRC Press, Boca Raton, pp. 1-7.

Taylor JA, Judy BM, Rottinghaus BA, Blackwell KJ, Rottinghaus GE, Allworth LC, vom Saal, FS and Welshons WV (1999) Bisphenol A bioaccumulates in the serum of pregnant female mice. Paper presented at the Estrogens in the Environment meeting, Tulane University, New Orleans, October 1999.

Tilson HA, Davis GJ, McLachlan JA and Lucier GW (1990a) The effects of polychlorinated biphenyls given prenatally on the neurobehavioral development of mice. *Environmental Research*, **19**, 466-474.

Tilson HA, Jacobson JL, Rogan WJ (1990b) Polychlorinated biphenyls and the developing nervous system: cross-species comparisons. *Neurotoxicology and Teratology*, **12**, 239-248.

Vannier B and Raynaud J (1980) Long-term effects of prenatal estrogen treatment on genital morphology and reproductive function in the rat. *Journal of Reproduction and Fertility*, **59**, 43-49.

vom Saal FS (1983) Models of early hormonal effects on intrasex aggression in mice. In *Hormones and Aggressive Behavior*, Svare B (ed), Plenum, New York, pp. 197-222.

vom Saal FS (1984) The intrauterine position phenomenon: effects on physiology, aggressive behavior and population dynamics in house mice. In *Biological Perspectives on Aggression*, Flannelly K, Blanchard R and Blanchard D (eds), *Progress in Clinical Biology Research*, Vol. 169, Liss, New York, pp. 135-179.

vom Saal FS (1989) Sexual differentiation in litter bearing mammals: influence of sex of adjacent fetuses *in utero*. *Journal of Animal Sciences*, **67**, 1824-1840.

vom Saal FS (1995) Environmental estrogenic chemicals: their impact on embryonic development. *Human Ecological Risk Assessment*, **1**, 3-15.

vom Saal FS and Sheehan DM (1998) A challenge to risk assessment posed by low-dose effects of endocrine disrupting environmental chemicals. *Forum for Applied Research and Public Policy*, **13**, 11-18.

vom Saal FS, Montano MM and Wang HS (1992) Sexual differentiation in mammals. In *Chemically Induced Alterations in Sexual and Functional Development: the Wildlife/Human Connection*, Colborn T and Clement C (eds), Princeton Scientific Publishing, Princeton, pp. 17-83.

vom Saal FS, Nagel SC, Palanza P, Boechler M, Parmigiani S and Welshons WV (1995) Estrogenic pesticides: binding relative to estradiol in MCF-7 cells and effects of exposure during fetal life on subsequent territorial behavior in male mice. *Toxicology Letters*, **77**, 343-350.

vom Saal FS, Timms BG, Montano MM, Palanza P, Thayer KA, Nagel SC, Dhar D, Ganjam VK, Parmigiani S and Welshons WV (1997) Prostate enlargement in mice due to fetal exposure to low doses of estradiol or diethylstilbestrol and opposite effects at high doses. *Proceedings of the National Academy of Sciences USA*, **94**, 2056-2061.

vom Saal FS, Cooke PS, Palanza P, Thayer KA, Nagel S, Parmigiani S and Welshons WV (1998) A physiologically-based approach to the study of bisphenol-A and other estrogenic chemicals on the size of reproductive organs, daily sperm production, and behavior. *Toxicology and Industrial Health*, **14**, 239-260.

vom Saal FS, Clark MM, Galef BG, Drickamer LC and Vandenbergh JG (1999) The intrauterine position (IUP) phenomenon. In *Encyclopedia of Reproduction*, Vol. 2, Knobil E and Neill J (eds), Academic Press, New York, pp. 893-900.

Welch R, Levin W and Conney A (1969) Estrogenic action of DDT and its analogs. *Toxicology and Applied Pharmacology*, **14**, 358-367.

Welshons WV, Rottinghaus GE, Nonneman DJ Dolan-Timpe M and Ross PF (1990) A sensitive bioassay for detection of dietary estrogens in animal feeds. *Journal of Veterinarian Diagnostic Investigation*, **2**, 268-273.

Welshons WV, Nagel SC, Thayer KA, Judy BM and vom Saal FS (1999) Low dose bioactivity of xenestrogens in animals: fetal exposure to methoxychlor and other xenestrogens increases adult prostate size in mice. *Toxicology and Industrial Health*, **15**, 12-25.

White R, Jobling S, Hoare SA, Sumpter JP and Parker MG (1994) Environmentally persistent alkylphenolic compounds are estrogenic. *Endocrinology*, **135**, 175-182.

Whalen RE (1974) Sexual differentiation: models, methods, and mechanisms. In *Sex Differences in Behavior*, Friedman RC, Richart RH and Wiele RLV (eds), John Wiley & Sons, New York, pp. 467-184.

Willingham E and Crews D (1999) Sex reversal sex effects of environmentally relevant xenobiotic concentrations on the red eared slider turtle, a species with temperature-dependent sex determination. *General and Comparative Endocrinology*, **113**, 429-435.

Perspectives

11

Behavioural Measures of Injuries to Fish and Aquatic Organisms: Regulatory Considerations

EDWARD E. LITTLE
US Geological Survey, Columbia Environmental Research Center, Columbia, MO, USA

11.1 INTRODUCTION

Regulations concerned with the environmental safety of chemical substances in the United States range from the determination of concentration criteria for specific compounds to the assessment of site-specific hazards resulting from the release of hazardous substances into aquatic habitats. Thus, certain regulations are based on calculated not-to-be-exceeded concentrations derived primarily from data on mortality and bioaccumulation whereas others determine pathways of exposure and measure the cumulative impact of that exposure to natural systems. Numerous kinds of biological information can be utilised in the regulatory process ranging from the subcellular to the ecosystem level. Biomarkers of exposure can, for example, indicate previous exposure to certain compounds and be useful to establishing pathways of exposure. Biomarkers of effects can indicate injury to critical functions that will limit survival or reproduction. Among such critical functions is behaviour, a phenomenon that is both a process as well as an objective endpoint.

Behaviour is the cumulative manifestation of physiological processes critical for the performance of essential life functions. Behaviour enables continuous adjustments to external stimuli and to internal cues that maintain the organism under beneficial circumstances to grow and to reproduce. Thus, behaviour is a selective response that ensures adaptation to environmental conditions and, in concert with morphological and physiological adaptations, permits the best opportunity for survival and reproductive success by enabling organisms

Behavioural Ecotoxicology. Edited by Giacomo Dell'Omo. © 2002 John Wiley & Sons Ltd efficiently to exploit resources within the defined parameters of their habitat (Little and Brewer 2001).

Behaviour is a sequence of quantifiable actions that operate through the central and peripheral nervous system (Keenleyside 1979). These behavioural actions reflect the integration of biochemical, genetic and physiological processes that maintain the steady state of the organism (Olla *et al.* 1980, Beitinger 1990). Species-typical behavioural patterns are the result of adaptation to environmental variables and enable an animal to seek environmental situations favourable to survival and to avoid those that are unfavourable. In this manner adaptive behaviours ensure the organism's survival as well as the viability of its population. Behaviour is a structured and predictable sequence of activities of significant adaptive value essential to an organism's existence (Little *et al.* 1993a).

Behavioural investigations are particularly relevant when evaluating the effects of hazardous substances because adaptive behavioural function is crucial in the determination of environmental injury. Behavioural toxicity occurs when a contaminant or other stressful condition induces a behavioural change that exceeds the normal range of variability (Marcucella and Abramson 1978). A number of behavioural studies have been conducted to determine the effects of sublethal contamination and other harmful environmental conditions on fish behaviour. Reviews by Atchison *et al.* (1987), Beitinger (1990), Henry and Atchison (1991), Blaxter and Hallers-Tjabbes (1992) and Little *et al.* (1993a) reveal that virtually any behavioural response can be affected by hazardous substances. These responses vary in complexity from simple reflexes to dynamic social interactions. Much of the published data indicate that behavioural functions are usually quite sensitive to toxicants and often provide an early indication of toxicosis. In addition, avoidance or attraction responses and feeding preferences and aversions can significantly influence the level or duration of exposure to harmful environmental conditions actually experienced by an organism through contact with their environment or diet (Beitinger 1990).

Numerous behaviours may be affected, either simultaneously or consecutively, during exposure to hazardous substances, depending on the type of chemical, concentration and duration of exposure. During a 10-day exposure of juvenile bluegill (*Lepomis macrochirus*) to an organophosphorus (OP) insecticide, fonofos, four different behavioural responses were altered (Fairchild and Little 1999). Aggression was inhibited on the first day of exposure, and abnormal swimming, spasms and feeding inhibition were evident by the second day of exposure at 11 ug/l. The appearance of a particular abnormality varied with concentration. For example, reduced feeding appeared after 2 days of exposure at 11 ug/l; after 3 days at 6 ug/l; and after 10 days at 2 ug/l. As with other toxicological endpoints, the magnitude or severity of behavioural response varies with the duration and concentration of exposure.

11.2 APPLICATIONS OF BEHAVIOURAL DATA IN ENVIRONMENTAL REGULATION

The applicability of behavioural measures in environmental regulation generally increases as the regulatory focus becomes increasingly site-specific. For example, the Toxic Substances Control Act (TSCA) of 1976 regulates the production transport and use and disposal of chemicals. TSCA is primarily concerned with substances that might be released into the environment during their registered use. Primary testing guidelines for aquatic tests include acute toxicity, early lifestage exposures, and bioaccumulation using designated fish species. Behavioural data are not included under TSCA.

The Federal Insecticide, Fungicide, and Rodenticide Act (FIFRA 1947) regulates chemicals that are intended to be applied directly into the environment for the purposes of controlling insects, vegetation, fungi, and rodents. FIFRA specifies a tiered testing approach including (1) limited chemical fate and acute toxicity; (2) multiple acute toxicity tests with freshwater and marine species; (3) early lifestage and chronic toxicity tests; and (4) controlled field tests in experimental or natural systems. These regulations determine the sensitivity of a limited number of species and lifestages relative to expected environmental concentrations. Safety factors are used within each test tier and if risk is inferred then the next tier of testing is required. Behavioural data are not specified for aquatic organisms under FIFRA.

The Federal Water Pollution Control Act of 1948 (Revised as the Clean Water Act, CWA, of 1972) is the primary regulatory authority for compliance monitoring of effluents from point source discharges within the National Pollution Discharge and Elimination System (NPDES). Chemical and biological testing are required. Protective concentrations, or water quality criteria, are based on data from at least eight aquatic species including fish, invertebrates and plants that are used to interpolate a concentration that is protective for 95% of the aquatic species. The CWA also mandates the development of community-based biocriteria using fish and invertebrates to monitor stream quality. Infrequently a description of behavioural data has been included in the discussion of water quality criteria for certain chemicals. For example, the water quality criteria for malathion describe adverse behavioural reactions to exposure in addition to mortality and biochemical changes that result from exposure. The most frequent use of behavioural measures associated with CWA has been in effluent monitoring by industry to provide an early warning system for failure in effluent treatment processes. The most successful of the behavioural measures for such monitoring has been the ventilatory response which continuously monitors respiratory rate, regularity and gill purge or cough responses of fish exposed to the effluent stream (Drummond and Carlson 1977, Diamond *et al.* 1990). Other behavioural measures used in effluent monitoring have included fish locomotor activity and orientation to water flow (rheotaxis). In swimming activity, an alarm function is generated following an increase or decrease in the frequency of movement that exceeds a predetermined alarm threshold (Morgan 1979, Smith and Bailey 1988). Orientation against water flow, the rheotaxic response, has also been used in the behavioural monitoring of effluents (Besch *et al.* 1977, Poels, 1977). In the course of exposure to a toxic substance the fish turn away either to escape the noxious stimuli or as a result of diminished swimming capacity caused by toxicological impairments from the exposure. As with the respiratory monitor, the locomotory and orientation responses are monitored in real time using automated electronic monitoring systems. By monitoring deviations in behavioural response a utility or factory can ensure CWA compliance with their effluent treatment.

Section 404 of the CWA authorises the Army Corps of Engineers to issue permits for the discharge of dredged or fill materials into navigable waters. The Environmental Protection Agency (EPA) is authorised under Section 404 to prohibit use of a site as disposal site if discharges have unacceptable effects on fishery areas. Dredge disposal activities can result in the reintroduction of hazardous substances and can also cause major impacts to the physical habitat. Therefore, the collection of behavioural data pursuant to discharge permitting would be especially effective in understanding the habitat needs of the resident population at the site of discharge as well as the sensitivity of these species to potential contamination. For example, behavioural data might be necessary to demonstrate the avoidance of 'chemical barriers' by migrating fishes.

The Comprehensive Environmental Response, Compensation and Liability Act of 1980 (CERCLA) provides for the recovery of damages for injury to natural resources, including fish, as a result of the release of hazardous substances. 'Damage' refers to the monetary value of lost resources whereas 'injury' refers to a measurable adverse change, loss, or destruction of natural resources. In contrast to TSCA, FIFRA, CWA regulations, environmental investigations of Natural Resource Damage Assessments (NRDA 1986) directed by CERCLA are site specific and focused on the populations or communities unique to the site. Essential aspects of such investigations are (1) in documenting the pathway of contamination from the site of the spill to the resource suspected to be impacted by the release, and (2) in determining the extent and magnitude of resultant injury to the resource, which includes air, soil and water resources, in addition to fish and wildlife. Site remediation is a third important aspect of these regulations.

The actual measure of injury to aquatic resources is covered under the Type B Technical Information Document within Section 301 of the CERCLA. Fish assessments, in addition to other forms of assessment, may be used to measure injury in the NRDA process and must meet four criteria: (1) the biological response measured must be frequently observed as the result of exposure to oil or other hazardous substances, (2) exposure to oil or other hazardous substances is known to cause the biological response in free-ranging organisms, (3) exposure to oil or other hazardous substances is known to cause the biological response in controlled experiments, and (4) the biological response must be practical to perform and must produce scientifically valid results meeting these criteria. Measures of injury under NRDA include (1) mortality in natural environments, *in-situ* cages, and laboratory toxicity tests, (2) reproductive effects, (3) skin and liver neoplasia, (4) brain cholinesterase inhibition, (5) external fin erosion, and (6) behavioural avoidance and clinical indications of toxicity. These injuries carry rebuttable presumption which gives them the status of *de facto* environmental injuries. Other endpoints, including many biochemical and immunological indicators and other behavioural responses, may be included in future revisions of NRDA as measures of injury or as measures providing weight of evidence to the exposure pathway or injury.

The Oil Pollution Act of 1990 (OPA) provides for enhanced capabilities for oil spill response and natural resource damage assessments and provides for research and development. In addition, OPA provides for the evaluation of effectiveness and environmental impacts of bioremediation technologies as well as scientific and monitoring plans for ecologically sensitive areas. Like NRDA, the OPA regulations are site specific for the purpose of exposure pathway determination and injury. In contrast to the measures of injury codified under the NRDA Regulations, OPA follows a weight-of-evidence approach in determining causality between the release of oil and environmental injury. A range of behavioural responses, in addition to physiological, biochemical, and immunological endpoints, may be considered as weight of evidence of environmental injury under OPA.

The Federal Power Act (FPA 1920) regulates the licensing and relicensing of power projects including the construction, operation and maintenance of dams, water conduits, reservoirs, and transmission lines. The issuance of permits requires equal consideration of power and development: energy conservation, protection, mitigation of damage to, and enhancement of fish and wildlife, including spawning grounds and habitat. Relicensing requires consultation with Fish and Wildlife Service, National Marine Fisheries Service and state fish and wildlife agencies and conducting appropriate studies with such agencies. Because of the site-specific nature of such projects, specific species will be impacted and the resulting studies will require consideration of life-history characteristics and habitat needs of the species likely to be impacted. For many species, behavioural testing may be required to generate the necessary life-history and habitat needs information.

The International Environmental Protection Act (IEPA) of 1983 mandates the Agency for International Development (AID) to ensure that proposed and ongoing actions do not inadvertently endanger wildlife species or their habitats or adversely impact biological diversity. In meeting the mandates of the IEPA, behavioural data may be applicable to determine the habitat needs and information about the life-cycle characteristics of species that are potentially affected by AID projects.

The Endangered Species Act (ESA) of 1973 authorises the determination and listing of threatened and endangered species and requires the preparation of recovery plans for listed species. Section 7 of the ESA requires Federal agencies to ensure that actions funded, authorised or carried out by them is not likely to jeopardise the continued existence of listed species or modify their critical habitat and requires the preparation of biological assessment. Section 7 also stipulates a consultation process to consider the impact of pesticides and environmental contaminants in the critical habitat of endangered species, and to limit such uses if necessary. Under the ESA consultation is required when any aspect of the habitat may be altered. Thus, for example, in the case of the Cherokee darter, a threatened species, and the Etowah darter, an endangered species, ''Federal involvement is expected to include the Environmental Protection Agency through the Clean Water Act's provisions for pesticide registration and waste management actions. The Corps of Engineers will consider these species in project planning and operation and during the permit review process (for dredging, wetland management, and dredge disposal), the federal Highway Administration will consider impacts of federally funded bridge and road construction projects when known habitat may be impacted. Continuing urban development within the Etowah river system may involve the Farmers Home administration and the loan programmes. The Soil Conservation Service will consider the species during project planning and under their farmer's assistance programmes. The Forest Service will consider downstream impacts to habitat of the Etowah darter'' (Fish and Wildlife Service 1994).

The ESA places a significant emphasis on behavioural data. Compliance with the ESA requires significant information about the habits and habitat uses of threatened and endangered species. Such information would include species preferences or aversions to environmental variables, response to competitors, predators, and forage as well as life history information related to rearing habitats and reproduction. Alteration in habitat such as changes in flow, water temperature, water quality, turbidity or clarity, substrate, and possibly even, electromagnetic fields and noise levels might have a significant impact on the species that would be evident in the behavioural response of the organism.

In summary, there are numerous potential applications of behavioural data in the environmental regulatory process. The NRDA regulations recognise certain behavioural impairments that lead to reduced survival and diminished long-term viability as *de facto* environmental injury. In addition, through weight of evidence, behavioural data also demonstrate injury through adverse effects to reproduction, development and viability by altering essential life functions which include habitat selection, competition, feeding, predator avoidance and reproduction. Disruption of these essential functions can become apparent ecologically through loss of populations or reductions in reproductive success and viability of the progeny year-class.

In consideration of the site-specific assessments required by regulations such as NRDA, OPA, CWA, and ESA, it is especially important to consider those species that have unique behavioural characteristics that may cause them to be particularly vulnerable to injury by hazardous substances and habitat alteration.

This would include (1) species that form mass assemblages which place them at greater risk than when they are dispersed; (2) species that change habitats periodically through narrow front migrations that concentrate migrants along particular routes; (3) species that exhibit inflexible behaviours that must be performed in correct order at a correct time, under specific environmental conditions; (4) species that are particularly dependent on specific resources such as critical prey; and (5) species that perform 'keystone functions' (e.g. grazing, predation, commensal, host) within their community (Ellis 1984).

11.3 POTENTIAL BEHAVIOURAL MEASURES OF ENVIRONMENTAL INJURY

Behavioural responses directly linked to the maintenance of a local population are likely to be the most valid responses to consider in support of environmental regulations. Populations can disappear because of mortality or reproductive failure. They can also disappear because of emigration (or the failure to return), competition or predation. The following section describes behavioural functions that are applicable to the regulatory process.

11.4 HABITAT SELECTION

The selection of appropriate habitat is critical to the survival of individuals and their offspring. Many aquatic species have specific habitat needs for survival and reproductive success. Habitat selection can be inhibited through physiological impairments in the ability of the organism to perceive and respond to environmental stimuli as well as from altered environmental cues which result when a hazardous substance masks, mimics, or inhibits responses to the habitat (Suttlerlin 1974).

11.4.1 MIGRATION BEHAVIOUR

Migration is a critical aspect of habitat selection in some species. Habitat alteration can create barriers to the migratory path that block the organism's movement to preferred habitat. For example, the seaward migration was halted when the fish encountered oil-contaminated water and gas-supersaturated water (Gray 1990). Even though avoidance of this harmful condition prevented immediate injury to the fish, the inhibition of the seaward migration prevented the fish from reaching foraging areas and probably prevented them from developing in a manner sufficient to sustain reproduction. Similarly, the spawning migration of Atlantic salmon was disrupted by mining effluents contaminated with copper and zinc (Saunders and Sprague 1967). Such migration barrier issues can arise from point source pollution related to OPA, NRDA, CWA and other regulations, as well as changes in other habitat variables such as temperature, water quality, or substrate. Migrations can also be inhibited by chronic exposures that interfere with the physiological mechanisms underlying migratory behaviour. Parr smolt transformation of anadromous salmonids involves a series of hormonal, physiological and behavioural changes that stimulate out-migration and prepare the fish for the transition from fresh to sea water. Environmental perturbations can disrupt smoltification and prevent out-migration of juvenile salmon. Reduced or delayed smolt migration may result in reduced growth, poor recruitment, lowered survival and ultimately fewer returning adult salmon. Studies with coho (*Oncorhynchus kisutch*), chinook (*O. tshawytscha*) and sockeye (*O. nerka*) salmon have shown that metals and other aquatic contaminants inhibit smoltification and impair the ability of fish to survive in seawater (Lorz and McPherson 1976, Nichols *et al.* 1984, Hamilton *et al.* 1986). Coho salmon exposed to sublethal concentrations of copper (5–30 ug/l) exhibited reduced downstream migration, lowered gill ATPase activity (an indication of incomplete smoltification), and poor survival in sea water (Lorz and McPherson 1976).

The consequences of impaired migrations include insufficient growth, insufficient return, and reduced reproduction. An assessment of migration is indicated when a migratory species is absent from areas adjacent to sites of impact where the species was formerly found or where habitat conditions are similar to those of reference sites inhabited by the species. Although the investigation of migration is not applicable to all fish species, such responses should be considered when salmonid habitats are impacted by the release of hazardous substances. Techniques for examining smoltification and associated seaward and homing migrations have been developed to a sufficient degree accurately to quantify the response in a cost-effective manner. The response has been demonstrated to occur in the field as well as the laboratory. Smoltification and associated migratory behaviour are recommended as measures of NRDA injuries.

11.4.2 AVOIDANCE BEHAVIOUR

Adaptive behaviour allows fish to avoid areas containing unfavourable conditions and to occupy acceptable areas. Fish may respond to noxious, aversive habitat conditions in a manner that minimises exposure to them. In contrast, a chemical inducing attractance increases exposure and probability of injury or death. Responses of avoidance and attraction can significantly influence the level or duration of exposure experienced by the organism through contact with water, sediment or diet (Beitinger 1990). For such responses to occur it is necessary that: (1) the substance activate a receptor; (2) the fish interprets this stimulation as being sufficiently noxious to induce locomotory responses to escape; and (3) the chemical concentration gradient provide sufficient directional cues to lead the organism away from the contaminant plume (Sutterlin 1974).

Avoidance behaviour has been observed in the field as well as in the laboratory. In tests of metals characteristic of those found in a river downstream from a large mining extraction operation, telemetry studies conducted at the confluence of the river with an uncontaminated tributary revealed a similar avoidance of the contaminated water within the concentration range that induced avoidance responses in laboratory studies (Woodward *et al.* 1997). Laboratory-field studies with a mixture of heavy metals (copper, chromium, arsenic and selenium) representative of effluent from fly ash settling basins of a coal-burning electrical plant showed that fathead minnows (*Pimephales promelas*) avoided a 73.5 g/l mixture of these metals in a natural stream, and similarly avoided a 34.3 g/l mixture in an artificial stream (Hartwell *et al.* 1987). Sprague *et al.* (1965) found similar avoidance of copper zinc mixtures in the laboratory and in the field.

There is extensive literature on contaminant avoidance studies with fish (Beitinger 1990, Little *et al.* 1985). Generalisations regarding the avoidance of aquatic contaminants by fish are difficult to make because of the variety of species and experimental designs used, as well as variations in the chemical modes of action of the substances studied (Giattina and Garton 1983). Generically similar organic compounds may not induce similar reactions and different species may not respond in the same manner to a compound (Little *et al.* 1985). In his review of published avoidance data for over 75 different chemicals, Beitinger (1990) found that about a third of the chemicals were avoided, whereas the others either failed to elicit a response or induced inconsistent responses. Fish were attracted by some contaminants including detergents (Hara and Thompson 1978), certain metals (Timms *et al.* 1972) and petroleum hydrocarbons (Lawrence and Scherer 1974). This may have occurred as a result of novelty, confusion with natural chemical cues (Sutterlin 1974), or the result of damage to chemoreceptors (Hansen *et al.* 1999a,b). The sensitivity of avoidance/attraction responses ranged from less than 3% of the LC_{50} (concentration lethal to 50% of the test population) for the herbicide, 2,4-D, to more than a 1000 times the LC_{50} for the chlorinated biphenol, arochlor 1254 (Little *et al.* 1985). Brown trout (*Salmo trutta*) avoided concentrations as low as 10% of the average concentration of a metals mixture measured in the Clark Fork River in Montana (DeLonay *et al.* 1995).

Avoidance measures are also applicable to other environmental variables that may change as a result of human activity. Avoidance procedures are also applicable to changes in other water quality parameters, such as salinity, hardness, turbidity, as well as habitat variables such as temperature, light, velocity of water flow and deviations in the magnitude of these variables over seasonal or annual cycles. For example, preferred temperatures are often correlated with physiological optima for various metabolic functions (Beitinger *et al.* 1975), and temperature is likely a major variable in habitat selection. Preferred temperatures differ widely among species (Coutant 1977) and conspecifics (Reynolds and Thomson 1974). Contaminants can also influence temperature selection by fish (Peterson 1976, Beitinger 1990).

The avoidance response remains an effective measure of injury under NRDA. Avoidance reactions to contaminants can be detrimental because organisms may be displaced from preferred habitats to suboptimal areas where they may face greater competition and predation pressure or inadequate resource availability (Atchison *et al.* 1987). Thus, avoidance-driven displacements can result in locally reduced biomass, loss of species diversity, diminished year-class strength, and loss of productivity and diversity in that area. Because of acceptance of avoidance behaviour as a measure of injury, data on the avoidance response has been applied in several NRDA investigations (Lipton *et al.* 1997). Field telemetry studies confirm laboratory data in showing significant avoidance, and thus injury to natural resources.

Behavioural avoidance studies should be considered when a suspected hazardous substance is known to cause avoidance responses and when fish are absent from sites adjacent to a suspected or known source of a hazardous substance. Avoidance data should also be collected to monitor mitigation and habitat recovery efforts.

11.5 COMPETITION

Aquatic organisms often share the same habitat, and competition among species or age classes may result when resource needs overlap. One species may gain competitive advantage over another by more efficient exploitation of those resources, or through aggressive interactions (Schoener 1974). Changes in prey selection (Werner and Hall 1974), spatial distribution (Finger 1982), or diurnal patterns of activity (Reynolds and Casterlin 1976) may segregate competitors in time and space. Contaminant exposure may indirectly influence competition between and within species by reducing a resource such as food, or vegetative cover. Loss of food, such as a contaminant intolerant prey, may cause a shift in food preference and intensify competition. Decreases in natural food supply have been shown to cause an increase in intraspecific aggression and the size of territories (Kalleberg 1958). Contaminant-induced changes in behaviour may also influence competition by reducing the efficiency with which one species exploits resources. Also delayed development may result in a life stage emerging at a time when competition with other species is more intense. Either case would result in reduced availability or accessibility of a resource and result in reduced growth, delayed development and impaired reproduction. Contaminants also alter aggressiveness among fish. Aggression insures access to vital resources and distributes individuals over space (Keenleyside 1979). Aggression among bluegill increased during sublethal exposure to a metals mixture (Henry and Atchison 1979), and to organophosphorus (Henry and Atchison 1984, Fairchild and Little 1999) and pyrethroid insecticides (Little *et al.* 1993b). Coho salmon became more aggressive during sublethal exposure to an organophosphorus pesticide (Bull and McInerney 1974). The loss of spawning territories occurred as Atlantic salmon became less aggressive in territorial defence following pesticide exposure (Symons 1973).

The consequences of altered aggression include loss of access to resources, increased stress, injury, increased energy expenditure and displacement to marginal habitats. An assessment of exploitive and interference competition is recommended when abundance or distribution of a native species declines in the presence of a non-indigenous species. Linkages of competition with resource injury is not sufficiently understood relative to direct impacts on growth, viability or population status. Thus, the evaluation of competition is limited to controlled laboratory tests which could provide weight of evidence, particularly when competitive species are present at an impacted site, or when there is an indication of aggressivity such as abandoned nesting areas or physical trauma.

11.6 PREDATOR–PREY INTERACTIONS

Predator-prey interactions measure the ability of fish to feed, as well as their ability to avoid being eaten by other organisms. Predator-prey interactions provide a good example of a behavioural measure of injury because deficits in either response will have immediate implications for the organism's survival (Brown *et al.* 1987). Feeding and prey vulnerability have been used to examine sublethal contamination because predator and prey may be behaviourally affected by toxicants (Sandheinrich and Atchison 1990).

11.6.1 FORAGING BEHAVIOUR

Survival of early life stage fish will decline if the development of foraging behaviour is delayed or inhibited. Impaired feeding behaviour also correlates with reduced growth, which may lengthen a fish's period of vulnerability to predation (Werner and Hall 1974) and impair overwinter survival (Oliver *et al.* 1979). Because changes in the ability of a fish to detect, pursue, capture and consume prey will affect growth and survival, these behavioural variables may provide an ecologically relevant measure of contaminant-induced injury. The experimental environment for such studies can vary in complexity depending on the amount of ecological realism one hopes to achieve, but it can also be readily adapted to standard toxicity testing procedures (Mathers *et al.* 1985).

A diversity of chemical substances have been shown to have an effect on various aspects of the feeding sequence, including detection of prey (Lemly and Smith 1987), prey selection (Bryan *et al.* 1995), latency to respond to prey

(Confer and Blades 1975), prey capture (Little *et al.* 1990), handling time and ingestion of prey (Sandheinrich and Atchison 1990), as well as general motivation to feed (Little *et al.* 1990). Thus, several variables of feeding behaviour can be measured in toxicity studies (Sandheinrich and Atchison 1990). The sensitivity of feeding behaviour as an index of sublethal exposure among rainbow trout (*O. mykiss*) exposed to various agricultural chemicals ranged from less than 0.3% to 50% of the LC50 (Little *et al.* 1990). Inhibited motivation to feed seemed to be the predominant effect at higher concentrations, whereas reduced feeding efficiency and reduced strike frequencies were the predominant impairments at exposure to lower concentrations of the agricultural chemicals. Feeding, measured as frequency of prey capture, was sensitive in defining no-effect concentrations of aluminium in acid-exposed brook trout (*Salvelinus fontinalis*) and was among the most sensitive indices of sublethal pH and aluminium exposure (Cleveland *et al.* 1989). Sublethal exposure to other metals (Atchison *et al.* 1987, Cleveland *et al.* 1989), pesticides (Little *et al.* 1993b, Fairchild and Little 1999), dioxins (Mehrle *et al.* 1988) and petroleum hydrocarbons (Woodward *et al.* 1987) have been shown to inhibit the feeding response in fish.

Feeding behaviour has been recommended to be included as a weight of evidence of resource injury in NRDA investigations and should also be useful in determining water quality criteria as well as the sublethal toxicity of effluents. The procedures for quantifying contaminant impacts on feeding behaviour have been developed for numerous species. The response is clearly affected by a range of contaminants and the impact of inhibited feeding behaviour on growth, development, viability and subsequent reproductive success is clearly understood. Evaluation of the response is effectively assessed in laboratory investigations. Feeding responses in free-ranging fish is largely limited to an assessment of gut content.

11.6.2 PREDATOR AVOIDANCE

Practically all organisms are vulnerable to predation during some portion of their life cycle and predation is a frequent cause of mortality in the field. Predation tests measure the ability of prey to escape predation. Increased vulnerability to predation may occur when a toxicant alters the ability of fish to detect or respond to predators (Brown *et al.* 1985). A common laboratory approach for measuring predation combines predators with contaminant-exposed and unexposed prey. Census of the surviving control and exposed prey are made when approximately 50% of the prey population have been captured (Little *et al.* 1985). Structural complexity of the experimental setting, including the addition of refuges for prey, is used to increase the environmental realism of the test. A range of contaminants have been shown to increase predation-induced mortality. Little *et al.* (1990) reported that different types of toxicants at concentrations as low as 2% of the lethal concentration significantly increased the predation of exposed prey. The response did not consistently follow a dose-response relationship, however, because the behavioural aberrations resulting from the exposure sometimes made the prey less conspicuous to the predators due to inactivity or reduced mobility. In tests with methyl parathion, Little (unpublished data) found that a 50% decrease in brain cholinesterase was associated with a 40 % decrease in swimming activity and heightened rates of predation. Assessment of predator-prey behaviours during toxicity tests provides not only a sensitive measure of toxicant effect, but also a measure of injury, since predation-induced mortality is affected by exposure.

Contaminant exposure may also cause aberrations or inhibitions of other defensive responses that protect the organism from predation. For example, schooling was consistently diminished following exposure of fathead minnows to the range of chemicals having different toxicological modes of action (Drummond *et al.* 1986).

Predator avoidance is most effectively evaluated in controlled investigations. Investigations of free-ranging organisms are largely limited to analysis of predators and the appearance of grossly atypical responses among affected organisms that would preclude an increased vulnerability to predation. Future advances in dietary DNA analysis will greatly increase understanding of prey selection and increase the utility of predation tests as measures of resource injury. It is recommended that laboratory observations of predation avoidance be included as weight of evidence measures of resource injury. Measures of predator avoidance are also applicable to tests of contaminant effluents and other changes in water quality.

In addition, measures of predation should be considered when assessing the impact of nonindigenous species on native organisms.

11.7 REPRODUCTION

Reproductive success is of fundamental importance to the maintenance of natural populations. In addition to selecting an environment suitable for the development of eggs and larvae, many organisms must establish and defend nesting areas, attract mates, and care for young. Because reproduction is a vulnerable stage of the life cycle, spawning and egg viability have become important endpoints in chronic toxicity studies. Relatively little is known of the impact of contaminants on reproductive behaviour. Alteration of any behaviour in the spawning sequence could inhibit reproduction (Schoder and Peters 1988). For example, male fathead minnows must prepare a quality nest site to attract gravid females then, after spawning, assume parental responsibilities for maintaining the nest site and guarding the eggs. Fathead minnows have been shown to reduce or even cease these spawning behaviours after exposure to

1.0 mg/l lead (Weber 1993) and greater than 37 g/l copper (Pickering *et al.* 1977) immediately prior to spawning. Premature spawning occurred among Pacific herring exposed to benzene (800 ug/l) which resulted in decreased survival of eggs and larvae. Male shore crabs exposed to crude oil, naphthalene, or binaphthyl failed to respond to female sex pheromones for up to 11 days (Takahashi and Kittredge 1973).

Laboratory demonstration of disrupted reproductive behaviour resulting from exposure to hazardous substances released in the environment is recommended as evidence of natural resource injury. This would include any response in the spawning sequence including courtship, nest preparation and defence, viability of gametes, behaviours associated with the release of eggs or larvae, and parental care.

11.8 BEHAVIOURAL MAINTENANCE FUNCTIONS

Essential life functions are affected by the impairment of underlying behavioural patterns including neuromotor functions, responsiveness to physical, chemical and social environmental stimuli, development of experientially derived responses and social responses.

Clinical signs of behavioural toxicity are grossly apparent responses that are of a magnitude of disruption that would be readily viewed as abnormal. Such responses would include loss of equilibrium, convulsive or tremorous movements, abnormal postures, or when the appearance of the organism in the habitat is atypical relative to diurnal rhythms or location.

Neurotoxicity is frequently evident as changes in the form, frequency or posture of locomotor behaviour. For example, the first indication of toxicity when bluegill received pulsed doses of the pyrethroid insecticide, es-fenvalerate, were tremorous tail movements as fish initiated movement (Little *et al.* 1993b). Sublethal parathion exposure caused convulsive movements in rainbow trout (Brewer *et al.* 1999). Swimming behaviour is well studied as a measure of toxicant effect and is one of the most common behavioural measures in toxicity investigations. Changes in swimming behaviour resulting from sublethal contaminant exposure may impair the performance of essential life functions. Swimming can be measured in many ways, including the frequency and duration of movements, as well as the form and pattern (Little and Finger 1990). Descriptions of swimming form and pattern include frequency and angle of turns, distance of movement, linear and angular velocity, position in the water column and swimming posture (Kleerekoper 1974). The measurement of these swimming responses are usually limited to the laboratory where swimming responses are effectively measured in automated biomonitoring systems (Baatrup and Bailey 1998, Miller *et al.* 1982, Smith and Bailey 1988). However, anticipated improvements in telemetry technology should enable similar evaluations in the field.

Behavioural changes, such as altered swimming behaviour, usually occur much earlier than mortality when fish are exposed to contaminants. A review by Little and Finger (1990) revealed that the lowest behaviourally effective toxicant concentration that induced changes in swimming behaviour of fish ranged from 0.1% to 5.0% of the LC50. When observations were made over time, behavioural changes commonly occurred 75% earlier than the onset of mortality. Swimming behaviour is affected by many chemical stressors, including metals, organochlorines and industrial contaminants. Concentrations that induced behavioural changes often were considerably lower than the concentrations that caused mortality (Little and Finger 1990). Swimming behaviour is also often affected before reductions in growth are detected. For example, swimming activity was significantly reduced in rainbow trout after a 96-h exposure to an organophosphorus defoliant at concentrations that only affect growth after 30 days (Little *et al.* 1990). Development of locomotory responses, frequency of movements and duration of activity were significantly inhibited in brook trout at lower aluminium concentrations under acidic conditions than those that affected survival or growth (Cleveland *et al.* 1991). Hypoactivity and hyperactivity, as well as deviations in diurnal rhythms, may disrupt feeding and increase vulnerability to predation (Steele 1983). Heightened activity may increase an organism's vulnerability to predation, whereas reductions in swimming activity lessen the chance of encountering prey by reducing effective search areas (Laurence 1972).

The appearance of grossly abnormal locomotory responses in the field (in absent evidence of disease, parasitism or physical trauma) should be examined as an injury deriving from exposure to environmental chemicals. Locomotory responses such as frequency and form are most effectively evaluated in controlled laboratory studies. Locomotory data be useful as weight of injury evidence of natural resource injury during evaluations of toxicity. Measures of locomotory behaviour provide effective weight of evidence documentation in support of injury to natural resources, as well as in defining water quality criteria and in determining sublethal effluent concentrations. Measures of locomotory behaviour could also be very effective in the evaluation of mitigation and habitat recovery activities.

11.9 FUTURE DIRECTIONS

The site-specific nature of human intrusions and disruption of aquatic habitats places a premium on information about life history and habitat use in aquatic organisms. Such information will be necessary to judge the impacts that are presently occurring in natural habitats and will also provide guidance as to the mitigation or rehabilitation of injured habitats. Although much may be known about commercially important organisms, often surprisingly little is known about the habits and needs of aquatic species with which to make informed management decisions. As increasing emphasis is placed on the restoration of habitat it will be especially important to identify biological measures to gauge the success of the restoration. As it is frequently practised, remediation often reflects an engineering effort, the goals of which have little to do with habitability requirements for the organisms that once lived there or might live there again. Consider for example that a clean-up criterion for copper contamination would be unsuccessful if it were based on drinking water criteria, because such concentrations are aversive to fish, and habitats contaminated to that level would be lost to the species. Behavioural criteria should be especially important since these reflect the habits and habitats of species most affected.

Whereas some environmental regulations are primarily focused on the toxicity of pure chemical substances, in the absence of other chemicals or other environmental variables, the site-specific focus of many regulations forces consideration of the environmental hazard in a matrix of multiple stressors. The interactions of several stressors can be additive in their injury to an aquatic organism. The ability of an organism to adapt to a constant challenge of multiple stressors may be overwhelmed if the magnitude or duration of a particular stressor increases. Behavioural data may provide an integrated measure of such impacts. In order to accurately understand the environmental risks and ensure that appropriate restoration goals are met will require a comprehensive assessment of a variety of environmental stresses on various life stages of impacted organisms.

The standardisation of behavioural procedures will significantly facilitate their use in the regulatory process (Little 1990). Behavioural researchers often follow similar procedures such that general protocols can be gained from the literature. However, only recently has the standardisation of behavioural methods been initiated by the American Society for Testing and Materials (ASTM 1998a,b). It is anticipated that such efforts will yield a suite of standardised methodologies.

Behavioural measurements must often be conducted in the laboratory, and this has hampered field validation. Significant improvements in telemetry technology are anticipated to advance the study of behaviour in the field. The behavioural early warning systems developed for laboratory use in the 1970s may well be feasible in free-ranging organisms within the next decade.

In conclusion, the development of environmental regulation in the United States has ranged from numerical criteria to site-specific assessment. Although behavioural measures have not been extensively used in environmental regulation, there is precedence for their application, particularly in NRDA. The potential is great for extensive application of behavioural measures in support of regulations that have a site-specific or species-specific focus since the regulatory issues often hinge upon the behavioural characteristics of the organism and its response to changes in its habitat. The implementation of effective recovery plans for many endangered species is not possible without consideration of behaviour.

REFERENCES

ASTM, Annual Book of ASTM Standards (1998a) ASTM Guide E 1711-95, *Standard Guide for Measurement of Behaviour During Fish Toxicity Test*, American Society for Testing and Materials, West Conshohocken, PA, pp. 1224–1240.

ASTM, Annual Book of ASTM Standards (1998b) ASTM Guide E 1604-94, *Standard Guide for Behavioural Testing in Aquatic Toxicology*, American Society for Testing and Materials, West Conshohocken, PA, pp. 1001–1008.

Atchison GJ, Henry MG and Sandheinrich MB (1987) Effects of metals on fish behaviour: A review. *Environmental Biology of Fishes*, **18**, 11–25.

Baatrup E and Bayley M (1998) Animal locomotor behaviour as a health biomarker of chemical stress. *Archives of Toxicology* (Suppl.) **20**, 163–178.

Beitinger TL (1990) Behavioural reactions for the assessment of stress in fishes. *Great Lakes Research*, **16**, 495–528.

Beitinger TL, Magnuson JJ, Neill WH and Schaffer WR (1975) Behavioural thermoregulation and activity patterns in the green sunfish, *Lepomis cyanellus*. *Animal Behaviour*, **23**, 203–210.

Besch WK, Kemball A, Meyer-Waarden K and Scharf B (1977) A biological monitoring system employing rheotaxis of fish. In *Biological Monitoring of Water and Effluent Quality*, Cairns J Jr, Dickson KL and Westlake GF (eds), ASTM STP 607, American Society for Testing and Materials, Philadelphia, PA, pp. 56–74.

Blaxter J and Hallers-Tjabbes CT (1992) The effect of pollutants on sensory systems and behaviour of aquatic animals, Netherlands. *Journal of Aquatic Ecology*, **26**, 43–52.

Brewer SK, Little EE, DeLonay AJ, Beauvais SL and Jones SB (1999) The use of automated monitoring to assess behavioural toxicology in fish: linking behaviour and physiology. *Environmental Toxicology and Risk Assessment. (1998) Eighth Volume*, 163–178. ASTM STP 1364, Henshel DS, Black MC and Harrass MC (eds), American Society for Testing and Materials, West Conshohocken, PA.

Brown JA, Johansen PH, Colgan PW and Mathers RA (1985) Changes in the predator-avoidance behaviour of juvenile guppies (*Poecilia reticulata*) exposed to pentachlorophenol. *Canadian Journal of Zoology*, **63**, 2001–2005.

Brown JA, Johansen PH, Colgan PW and Mathers RA (1987) Impairment of early feeding behaviour of largemouth bass by pentachlorophenol exposure: a preliminary assessment. *Transactions of the American Fisheries Society*, **116**, 71–78.

Bryan MD, Atchison GJ and Sandheinrisk MB (1995) Effects of cadmium on the foraging behaviour and growth of juvenile bluegill, *Lepomis macrochirus*. *Canadian Journal of Aquatic Science*, **52**, 1630–1638.

Bull CJ and McInerney JE (1974) Behaviour of juvenile coho salmon (*Oncorhynchus kisutch*) exposed to sumithion (fenitrothion), an organophosphate insecticide. *Journal of the Fisheries Research Board of Canada*, **31**, 1867–1872.

Cleveland L, Little EE, Ingersoll CG, Wiedmeyer RH and Hunn JB (1991) Sensitivity of brook trout to low pH, low calcium and elevated aluminum concentrations during laboratory pulse exposures. *Aquatic Toxicology*, **19**, 303–317.

Cleveland L, Little EE, Wiedmeyer RH and Buckler DR (1989) Chronic no-observed-effect concentrations of aluminum for brook trout exposed in low calcium dilute acidic water. In *Environmental Chemistry and Toxicology of Aluminum*, Lewis TE (ed.), Lewis Publishers Inc., Chelsea, MI, pp. 229–246.

Confer JL and Blades PI (1975) Reaction distance to zooplankton by *Lepomis gibbosus*. *Vereinigung für Theoretische und Angewandte Limnologie. Verhandlungen*, **19**, 2493–2497.

Comprehensive Environmental Response Compensation and Liabilities Act (1980). 26 United States Code 4611-4682 Public Law 95-510 94 Statute 2797. 42 United States Code 9601-9626. Public Law 98-802, 97 Statute 485 (1983).

Coutant CC (1977) Compilation of temperature preference data. *Journal of the Fisheries Research Board of Canada*, **34**, 739–745.

DeLonay AJ, Little EE, Lipton J, Woodward DF and Hansen JA (1995) Behavioural avoidance as evidence of injury to fishery resources: applications to natural resource damage assessments. In *Environmental Toxicology and Risk Assessment Fourth Volume* ASTM STP 1262, LaPoint TW, Price FT and Little EE (eds), American Society for Testing and Materials, Philadelphia, PA, pp. 268–280.

Diamond JM, Parson MJ and Gruber D (1990) Rapid detection of sublethal toxicity using fish ventilatory behaviour. *Environmental Toxicology and Chemistry*, **9**, 3–12.

Drummond RA, Russom CL, Geiger DL and DeFoe DL (1986) Behavioural and morphological changes in fathead minnows, *Pimephales promelas*, as diagnostic endpoints for screening chemicals according to modes of action. In *Aquatic Toxicology, 9th Aquatic Toxicity Symposium*, Poston TM and Purdy R (eds), ASTM STP 921, American Society for Testing and Materials, Philadelphia, PA, pp. 415–435.

Drummond RA and Carlson RW (1977) Procedures for measuring cough (gill purge) rates of fish. EPA-600/3-77-133. Diluth, Minn. U.S. EPA Environmental Research Laboratory.

Ellis DV (1984) Animal behaviour and pollution. *Marine Pollution Bulletin*, **15**, 163–164.

Endangered Species Act (1973) 16 United States Code 1531-1544, 87 Statute 884.

Fairchild JF and Little EE (1999) Use of behavioural endpoints to determine protective concentrations of the insecticide, fonofos, for bluegill. *Environmental Toxicology and Risk Assessment, ASTM STP 1364*, Henshel DS, Black MC and Harrass MC (eds), American Society for Testing and Materials, West Conshohochen, PA.

Federal Insecticide, Fungicide, and Rodenticide Act (1947) 7 United States Code 136 Public Laws 92-516 (1972), 94-140 (1975), 95-396 (1978) and 96-539 (1980).

Federal Power Act (1920) 16 United States Code 791–828. Public Law 99-495, 100 Statute 1243 (1986).

Federal Water Pollution Control Act (1948) 33 United States Code 1251-1376, 62 Statute 1155.

Fish and Wildlife Service (1994) Endangered and threatened Wildlife and Plants: Determination of threatened status for the Cherokee darter and endangered status for the Etowah darter. Final Rule. 50 CFR Part 17 RIN 1018-AC01. Federal Register 12/20/1994.

Finger TR (1982) Interactive segregation among three species of sculpins (*Cottus*). *Copeia*, **1982**, 680–694.

Giattina JD and Garton RR (1983) A review of the preference-avoidance responses of fishes to aquatic contaminants. *Residue Reviews*, **87**, 43–90.

Gray RH (1990) Fish behaviour and environmental assessment. *Environmental Toxicology and Chemistry*, **9**, 53–68.

Hamilton SJ, Palmisano A, Wedemeyer GA and Yatsutake WT (1986) Impacts of selenium on early lifestages and smoltification of fall chinook salmon. *Transactions of the 51st Wildlife and Natural Resources Conference*, pp. 343–356.

Hansen JA, Marr JCA, Lipton J, Cacela D and Bergman HL (1999a) Differences in neurobehavioural responses of chinook salmon (*Oncorhynchus tshawytscha*) and rainbow trout (*O. mykiss*) exposed to copper and cobalt: (1) behavioural avoidance. *Environmental Toxicology and Chemistry*, **18**, 1972–1978.

Hansen JA, Rose JD, Jenkins RA, Gerow KG and Bergman HL (1999b) Differences in neurobehavioural responses of chinook salmon (*Oncorhynchus tshawytscha*) and rainbow trout (*O. mykiss*) exposed to copper: (2) neurophysiological and histological effects on the olfactory system. *Environmental Toxicology and Chemistry*, **18**, 1979–1991.

Hara TJ and Thompson BE (1978) The reaction of whitefish, *Coregonus clupeaformis*, to the anionic detergent sodium lauryl sulphate and its effects on their olfactory responses. *Water Research*, **12**, 893–897.

Hartwell SI, Cherry DS and Cairns J Jr. (1987) Field validation of avoidance of elevated metals by fathead minnows (*Pimephales promelas*) following *in situ* acclimation. *Environmental Contamination and Toxicology*, **6**, 189-200.

Henry MG and Atchison GJ (1979) Influence of social rank on the behaviour of bluegill, *Lepomis macrochirus*, exposed to sublethal concentrations of cadmium and zinc. *Journal of Fish Biology*, **15**, 309-315.

Henry MG and Atchison GJ (1984) Behavioural effects of methyl parathion on social groups of bluegill (*Lepomis macrochirus*). *Environmental Toxicology and Chemistry*, **3**, 399-408.

Henry MG and Atchison GJ (1991) Metal effects on fish behaviour- advances in determining the ecological significance of responses. In *Ecotoxicology of Metals: Current Concepts and Applications*, Newman MC and McIntosh A (eds), Lewis Publishers, Boca Raton, FL, pp. 131-144.

International Environmental Protection Act (1983) 22 United States Code 215, 97 Statute 1045.

Kalleberg GH (1958) Observations in a stream tank of territoriality and competition in juvenile salmon and trout. *Institute of Freshwater Research Drottningholm Report*, **39**, 55-98.

Keenleyside MHA (1979) *Diversity and adaptation in fish behaviour. Zoophysiology*, **11**, Springer-Verlag Berlin, 208 pp.

Kleerekoper H (1974) Effects of exposure to a subacute concentration of parathion on the interaction between chemoreception and water flow in fish. In *Pollution and Physiology in Marine Organisms*, Vernburg FJ and Vernberg WB (eds), Academic Press, New York, pp. 237-245.

Laurence GC (1972) Comparative swimming abilities of fed and starved larval largemouth bass (*Micropterus salmoides*). *Journal of Fish Biology*, **4**, 73-78.

Lawrence M and Scherer E (1974) Behavioural responses of whitefish and rainbow trout to drilling fluids. *Fisheries Marine Service Research Developmental Technical Report* 502.

Lemly AD and Smith RJF (1987) Effects of chronic exposure to acidified water on chemoreception of feeding stimuli by fathead minnows, *Pimephales promelas*: mechanisms and ecological implications. *Environmental Toxicology and Chemistry*, **6**, 225-238.

Lipton J, Little EE, Marr JCA and DeLonay AJ (1997) Use of behavioural avoidance testing in natural resource damage assessment: *Environmental Toxicology and Risk Assessment: Biomarkers and Risk Assessment (5th volume) ASTM STP 1306*, Bengtson D and Henshel DS (eds), American Society of Testing and Materials, Philadelphia, PA.

Little EE and Brewer SK (2001) Neurobehavioural toxicity in fish. In *Target Organ Toxicity in Marine and Freshwater Teleosts*, Schlenk D and Benson WH (eds), Taylor and Francis Publisher, London, pp. 139-174.

Little EE, Fairchild JF and DeLonay AJ (1993a). Behavioural methods for assessing the impacts of contaminants on early life stage fishes. *American Fisheries Society Symposium*, **14**, 67-77.

Little EE, Dwyer FJ, Fairchild JF, DeLonay AJ and JL Zajicek (1993b) Survival and behavioural response of bluegill during continuous and pulsed exposures to the pyrethroid insecticide, ES-fenvalerate. *Environmental Toxicology and Chemistry*, **12**, 871-878.

Little EE (1990) Behavioural toxicology: stimulating challenges for a growing discipline. *Environmental Toxicology and Chemistry*, **9**, 1-2.

Little EE and Finger SE (1990) Swimming behaviour as an indicator of sublethal toxicity in fish. *Environmental Toxicology and Chemistry*, **9**, 13-19.

Little EE, Archeski RD, Flerov BA and Kozlovskaya VI (1990) Behavioural indicators of sublethal toxicity in rainbow trout. *Archives of Environmental Contamination and Toxicology*, **19**, 380-385.

Little EE, Flerov BA and Ruzhinskaya NN (1985) Behavioural approaches in aquatic toxicity: a review. In *Toxic Substances in the Aquatic Environment: An International Aspect*, Mehrle PM Jr, Gray RH and Kendall RL (eds), American Fisheries Society, Water Quality Section, Bethesda, MA, pp. 72-98.

Lorz HW and McPherson BP (1976) Effects of copper or zinc in freshwater on the adaptation to sea water and ATPase activity, and the effects of copper on migratory disposition of coho salmon (*Oncorhynchus kisutch*). *Journal of the Fisheries Research Board of Canada*, **33**, 2023-2030.

Marcucella H and Abramson CI (1978) Behavioural toxicology and teleost fish. In *The Behaviour of Fish and other Aquatic Animals*, Mostofsky DI (ed.), Academic Press, New York, pp. 33-77.

Mathers RA, Brown JA and Johansen PH (1985) The growth and feeding behaviour responses of largemouth bass (*Micropterus salmoides*) exposed to PCP. *Aquatic Toxicology*, **6**, 157-164.

Mehrle PM, Buckler DR, Little EE, Smith LM, Petty JD, Peterman PH, Stalling DL, DeGraeve GM, Coyle JJ and Adams WJ (1988) Toxicity and bioconcentration of 2,3,7,8-tetrachlorodibenzodioxin and tetrachlorodibenzofuran in rainbow trout. *Environmental Toxicology and Chemistry*, **7**, 47-62.

Miller DC, Lang WH, Graeves JOB and Wilson RS (1982) Investigations in aquatic behavioural toxicology using a computerized video quantification system. In *Aquatic Toxicology and Hazard Assessment: Fifth Conference*, Pearson JG, Foster RB and Bishop WE (eds), STP 766, American Society for Testing and Materials, Philadelphia, PA, pp. 206-220.

Morgan WSG (1979) Fish locomotor behavior patterns as a monitoring tool. *Journal of the Water Pollution Control Federation*, **51**, 580-589.

Natural Resource Damage Assessments: Final Rule. 1986. Federal Register 51, 27674-27753.

Nichols JW, Wedemeyer GA, Mayer FL, Dickhoff WW, Gregory SV, Yatsutake WT and Smith SD (1984) Effects of freshwater exposure to arsenic trioxide on the parr-smolt transformation of coho salmon. *Water Research*, **1**, 143-149.

Oil Pollution Act (1990) 33 United States Code 2701, 104 Statute 484.

Oliver JD, Holeton GD and Chua KE (1979) Overwintering mortality of fingerling small-mouth bass in relation to size relative energy stores and environmental temperature. *Transactions of the American Fisheries Society*, **108**, 130-136.

Olla BL, Pearson WH and Studholme AL (1980) Applicability of behavioural measures in environmental stress assessment. *Rapports et Procès-Verbaux des Réunions Commission Internationale pour l'Exploration Scientifique de la Mer Mediterranée Monaco*, **179**, 162-173.

Poels CLM (1977) An automatic system for rapidly detection of acute high concentrations of toxic substances in surface water using trout. In *Biological Monitoring of Water and Effluent Quality*, Cairns J, Dickson KL Jr and Westlake GF (eds), ASTM STP 607, American Society for Testing and Materials, Philadelphia, PA, pp. 85-95.

Peterson RH (1976) Temperature selection of juvenile Atlantic salmon, *Salmo salar*, as influenced by various toxic substances. *Journal of the fisheries Research Board of Canada*, **33**, 1722-1730.

Pickering Q Brungs W and Gast M (1977) Effect of exposure time and copper concentration on reproduction of the fathead minnow (*Pimephales promelas*). *Water Research*, **11**, 1079-1083.

Rand GM (1985) Behaviour. In *Fundamentals of Aquatic Toxicology: Methods and Applications*, Rand GM and Petrocelli SR (eds), Hemisphere Publishing, New York, pp. 221-256.

Reynolds WW and Casterlin ME (1976) Activity rhythms and light intensity preferences of *Micropterus salmoides* and *M. dolomieui*. *Transactions of the American Fisheries Society*, **105**, 400-403.

Reynolds WW and Thomson DA (1974) Responses of young gulf grunion, *Leuresthes sardina*, to gradients of temperature, light, turbulence and oxygen. *Copeia*, **1974**, 747-758.

Saunders RL and Sprague JB (1967) Effect of copper-zinc mining pollution on a spawning migration of Atlantic salmon. *Water Research*, **1**, 419-432.

Sandheinrich MB and Atchison GJ (1990) Sublethal toxicant effects on fish foraging behaviour: empirical vs mechanistic approaches. *Environmental Toxicology and Chemistry*, **9**, 107-120.

Schoener TW (1974) Resource partitioning in ecological communities. *Science*, **185**, 27-39.

Schoder JH and K Peters (1988) Differential courtship activity of competing guppy males (*Poecilia reticulata* Peters; Pisces: Poecilidae) as an indicator for low concentrations of aquatic pollutants. *Bulletin Environmental Contamination and Toxicology*, **40**, 396-404.

Smith EH and Bailey HC (1988) Development of a system for continuous biomonitoring of a domestic water source for early warning of contaminants. In *Automated Biomonitoring: Living Sensors as Environmental Monitors*, Horwood E, Gruber DS and Diamond JM (eds), Chichester, England, pp. 182-205.

Sprague JB, Elson PF and Saunders RL (1965) Sublethal copper-zinc pollution in a salmon river A field and laboratory study. *International Journal of Air and Water Pollution*, **9**, 531-543.

Steele CW (1983) Effects of exposure to sublethal copper on the locomotor behaviour of the sea catfish, *Arius felis*. *Aquatic Toxicology*, **4**, 83-93.

Sutterlin AM (1974) Pollutants and the chemical senses of aquatic animals — Perspective and review. *Chemical Senses and Flavor*, **1**, 167-178.

Symons PEK (1973) Behaviour of young Atlantic salmon (*Salmo salar*) exposed to or force-fed fenitrothion, an organophosphorous insecticide. *Journal of the Fisheries Research Board of Canada*, **30**, 651-655.

Takahashi PT and Kittredge JS (1973) Sublethal effects of the water soluble component of oil: chemical communication in the marine environment. In *The Microbial Degradation of Oil Pollutants*, Ahearn DC and Meyers SP (eds), Louisiana State University Press, Publication LSU-SG-73-01, Baton Rouge, LA, pp. 259-264.

Timms AM, Kleerekoper H and Matis J (1972) Locomotor response of goldfish, channel catfish, and largemouth bass to a 'copper-polluted' mass of water in an open field. *Water Resources Research*, **8**, 1574-1580.

Toxic Substances Control Act (1976) 15 United States Code 2910-2671. Public Law 94-469, 90 Statute 2003.

Weber DN (1993) Exposure to sublethal levels of waterborne lead alters reproductive behaviour patterns in fathead minnows (*Pimephales promelas*). *Neurotoxicology*, **14**, 347-358.

Werner EE and Hall DJ (1974) Optimal foraging and the size selection of prey of the bluegill sunfish (*Lepomis macrochirus*). *Ecology*, **55**, 1042-1052.

Woodward DF, Little EE and Smith LM (1997) Toxicity of five shale oils to fish and aquatic invertebrates. *Archives of Environmental Contamination and Toxicology*, **16**, 239-246.

12

Development of Behavioural Ecotoxicology Experiments in Precollege Science Classrooms

CARMEN AGUILAR[1,3], DAVID H. PETERING[1], RANDALL RYDER[2] AND DANIEL N. WEBER[1]

[1]Marine and Freshwater Biomedical Sciences Center, University of Wisconsin-Milwaukee, WI, USA

[2]Department of Curriculum and Instruction School of Education, University of Wisconsin-Milwaukee, WI, USA

[3]Great Lakes Water Institute, University of Wisconsin-Milwaukee, WI, USA

12.1 INTRODUCTION

Nonmammalian organisms have a distinguished history serving as models in biomedical research (Mottet and Laondolt 1987). The understanding of the mechanisms of development at the molecular biological level stem largely from studies with nontraditional organisms such as *Xenopus laevis* (frog), *Drosophila melanogaster* (fruit fly) and *Xenorhabdus elegans* (worm) (Bullock 1984, Cohen 1985, Zettergren *et al.* 1991). The major vertebrate model for development is now the common aquarium zebrafish (Westerfield 1995). Similarly, modern knowledge about the processes of nerve impulse conduction developed out of the study of the squid axon and frog neurons (Bransford and Stein 1984, Paul 1991). This chapter is devoted to describing a novel approach to the middle school Life Science course that centres on the use of aquatic organisms as experimental subjects in the classroom laboratory and on experiments that link toxicology with basic science concepts. Integral to this approach is a network of support for teachers that is necessary if they and their students are successfully to conduct these experiments. Our objective is to describe the features of this model of inquiry-based learning and how it can be propagated among school districts.

Behavioural Ecotoxicology. Edited by Giacomo Dell'Omo. © 2002 John Wiley & Sons Ltd

12.1.1 BACKGROUND

A decade ago studies showed that many American children were not well grounded in the knowledge or the methods of science (American Association for the Advancement of Science [AAAS] 1989, Mullis and Jenkins 1988, IAEEA 1989). At that time the number of college students planning to major in science and mathematics had been dropping for two decades (Tobias 1990). Concerns were voiced that textbook science did not adequately portray the practice of everyday science. Also, close ties between science and society were missed in the precollege classroom, and the focus on student memorisation of information had to be replaced by an emphasis on concepts and processes (AAAS 1989). In a word, students needed to 'do' science, not just learn about it.

There are several reasons to direct one's attention at the middle school educational experience. The middle school period marks a number of crucial transitions for children. They enter puberty and begin the trek to biological and psychological adulthood. They encounter more intense, focused course work that starts to lay the intellectual foundation for their lives as adults. More specifically, middle school offers students their first extended, unified course in Life Science. Popular culture targets this age group for these reasons and in so doing enters into full-scale competition for the minds of young people. The open curiosity of younger children about the natural world can quickly be replaced by disinterest in the face of peer and media pressure. Without overstating the situation, middle school provides a crucial opportunity to engage the minds of young people with the wonders and satisfaction of understanding the living world around them.

Tobacco, alcohol, drugs and pollution are all issues of personal and group hygiene that middle school children are facing in their lives. Children are also becoming aware of the global problems of overpopulation and environmental degradation. An understanding of life science can aid the student in comprehending and evaluating related personal and community health problems.

According to the National Science Standards scientific inquiry is the key to coordinating science content and making it functionally useful for children as they grow into adulthood. In their first course in life science, students should begin to learn how to conduct various types of inquiry. These involve basic skills in observing and describing things or events and understanding the facets of experimentation required for finding new information and making new discoveries. In the process, children should be introduced to methods of data acquisition, treatment and analysis, and, ultimately, they should begin to wrestle with the evaluation and explanation of experimental results that require both a strong knowledge base and high level reasoning.

12.2 THE MODERN LIFE SCIENCE CURRICULUM AND THE NEED FOR A NEW PARADIGM

The AAAS report, *Science for All Americans* (1989) suggests the following areas have to be addressed in precollege education if science literacy is to improve in this country: (1) The gap between current school or textbook science and the science of everyday life must be closed. (2) The connections between science and society must be made more explicit within the classroom setting. Students should have an understanding of the relationship among science, technology, and society. (3) Teaching practice needs to be strengthened by providing teachers with hands-on research experiences. Research experiences will strengthen practice by moving them away from overdependence on textbooks, providing instructional methodologies grounded in real-life research experiences. (4) Students need to be directed away from memorising information as a means of understanding the connections between the concepts and processes of science. Traditional teaching practices impede progress toward scientific literacy because they place limits on the process of creating understandings in the context of 'doing' science. This chapter, using behavioural ecotoxicology as a vehicle for science education, outlines a programme developed at the University of Wisconsin-Milwaukee that addresses these concerns.

During the middle school years students are beginning concentrated inquiry into various subjects. In this environment, few teachers are provided with either adequate preparation time or the substantial science budgets necessary to bring significant new experiments and other activities into their curricula. We propose a model in which the creation of experiments and related technology-based materials is complemented by the development of an effective support structure for the scientific, technical, and pedagogical aspects of implementing science experiments in the classroom.

For more interactive learning to occur, teachers will need to orchestrate a broader range of instructional experiences than they do presently use, providing students with opportunities for scientific investigations within a problem-solving framework (Applebee *et al.* 1989). To assist teachers in this lofty goal and, in so doing, attend to the first gap identified by the AAAS (1989) report, we argue that long-term connections between teachers and students with scientists need to be created. Such connections will not only address the weaknesses of teachers in environmental health but also provide personal contact between students and those who do science. For all these reasons our educational approach centres on teachers as well as students.

12.2.1 A NEW APPROACH

Scientific inquiry, in particular, offers an ideal opportunity for the development of disciplined, reasoned thinking in a setting of exploration. Such inquiry needs to be centred on authentic first-hand experiences. Such projects can engage students in actual science experiences, essentially the work of experts with its characteristic energy and openness. Projects encourage critical thinking, connect classroom to personal experiences, making science relevant to everyday life, and prepare students for lifelong learning (Stokes 1991, National Institute of Environmental Health Sciences [NIEHS] 1994). Project-based instruction provides a realistic presentation of the complexities and uncertainties of science along with real-life problems in general. A project-based form of instruction allows students to learn as scientists learn, within a supportive but rigorous community of mentors, peers and skilled technicians in the form of the teacher, cooperating scientists, fellow students and support personnel (Pollard *et al.* 1992). Thus, our approach address another gap identified by the AAAS (1989) report, learning to see connections in science without simply memorising facts (Aguilar *et al.* 1999).

12.2.2 ENVIRONMENTAL HEALTH SCIENCE AND USE OF AQUATIC ORGANISMS FOR CLASSROOM STUDY

Environmental health science is based on interests at the macroscopic level that have both scientific and social importance, i.e., contaminants in the water we drink, the food we eat, and the air we breathe. As it probes the underlying mechanisms related to environmentally based health and disease, it naturally relates to the environment which children have been intensely exploring on their own. An essential first step in the development of a systematic understanding of biology is to focus on living systems and their characteristics in relationship to their environment.

The attractiveness of nonmammalian, particularly aquatic organisms for the classroom mirror to a significant extent their utility for scientists. Useful attributes in different examples include simplicity of the living system, ease of care and handling, the large number of offspring, rapid development outside a placenta, and genetic and physiological similarity to mammals. In particular, it is possible safely and humanely to bring into the classroom directly or with the aid of visual technologies a variety of nonmammalian organisms that can stimulate, inspire, and deepen the understanding of children in their first encounter with the life sciences. Furthermore, use of aquatic organisms as experimental subjects naturally leads children into discussions of the need for using model organisms in research and for the ethical use and handling of living things.

12.2.3 BIOETHICAL CONSIDERATIONS

Lack of knowledge on how to handle such potentially awkward situations as animal death or perceptions by the students that the animals are suffering may hamper the teacher from introducing live animals into the classroom. It may seem obvious that life sciences are dedicated to understanding living systems.

Yet, paradoxically, controlled investigation of living animals in the precollege classroom is uncommon. Instead, dead animals are routinely examined and issues important to the intellectual development of the children, i.e., the humane treatment of animals while conducting critically needed experiments on the basic mechanisms of life, are avoided. To appreciate the organism and its environment, to get a 'sense for the organism', a classroom should be filled with living creatures. Animal reproduction, development, and individual and group behaviours are best examined with living organisms. It is clear, however, that few schools have either the facilities or skilled personnel to house and utilise mammals for such studies. In fact, classrooms are generally precluded from using mammals due to strict federal guidelines for animal care, as well as the limitations of resources in schools. In contrast, fish and other aquatic organisms are readily kept in static, aerated freshwater aquaria and are ideal for introductory observation and manipulation. Thus, we argue that not only should live animals be used while students confront important bioethical issues but that aquatic organisms are excellent alternatives, especially for studies in behavioural ecotoxicology.

12.3 WHY STUDY BEHAVIOURAL ECOTOXICOLOGY IN A LIFE SCIENCE CLASSROOM?

When the AAAS (1989) listed as a critical gap in modern precollege science curricula the fact that there is insufficient connections between the knowledge gained in class with real-world problems, it touched upon a true concern of precollege students, i.e., the 'What-Does-This-Have-To-Do-With-Me' syndrome. We argue that using behavioural ecotoxicology in these classrooms is an excellent method of attacking this issue.

12.3.1 WHY STUDY BEHAVIOUR?

Because behaviour acts as a link between an organism's internal physiology and external environment, i.e., specific changes in physiological functions responding to specific environmental changes cause definable and predictable behaviours, it is an easy way for young minds to visualise some simple connections between biology, chemistry and earth sciences. Therefore, behavioural observation in the classroom facilitates the understanding of many of the processes that define a living animal such as reproduction, neural and hormonal integration, respiration, digestion, and locomotion (at least at some life history stage).

Observing the whole, intact animal allows the student to analyse how these various biological processes produce an individual that is capable of functioning in and adapting to a given environment, including environments that contain toxic substances. Expanding upon this, these students should also see themselves as a whole, intact entity, as well. Importantly, because behaviour is a readily, easily, observable phenomenon that is a response to both biotic (e.g., social context) and abiotic (e.g., pollution) factors, integrating various scientific topics is enhanced.

Animal health status results from the interaction of individual genetic potential with a multitude of environmental factors. While the former varies from individual to individual, the specific factors, both positive and negative, affect whole populations and are subject to human impact. Using simple behavioural observations of living animals and especially seeing how behaviours can change or mature naturally over time, assists the student into a deeper understanding of their own development. By getting first-hand knowledge of how environmental contaminants alter behaviour of animals and, by implication, the development of their own behaviour, students acquire a context to understand the intricacies of nature and the dangers of pollution.

Although human behaviour clearly differs in specific aspects from fish behaviour, the underlying physiological mechanisms in each organism are fundamentally similar; behaviour merely reflects the coordination of those mechanisms within a social and environmental context to which the individual must adapt and to which the student can relate. Fish, then, can be viewed as an example of the effects due to environmental degradation as well as a model for human environmental health. The use of a *model system* is a key scientific concept for students to learn about critical thinking and models developed to understand several different systems. As humans, we learn about ourselves and the world around us by studying other organisms. Observing altered behaviours of animals exposed to toxic substances, impress upon the students that they, too, may be affected by environmental contamination. Without relying on complex equipment used in modern science, studying animal behaviour opens a window of understanding for the student on how the world around him/her operates.

12.3.2 WHY TOXICOLOGY?

Precollege classes provide introductions to biology, physical and earth sciences, and health precisely because of the developmental and behavioural concerns mentioned above. As they are learning about the underlying basis of healthy living, students are also being admonished about the adverse health effects of smoking, alcohol and drugs. Although they may be aware of 'the environment' and issues of environmental degradation, they will best understand it if issues of health become focused on their personal needs. Thus, a natural organising focus for novel, integrative units in science is environmental health. This subject covers both positive and negative interactions between organisms and their environment. It also integrates a number of disciplines, biology (relating to the organism), chemistry (the compound or pollutant), geological (compound availability), and environmental health science, i.e., environmental toxicology, and current issues are inherently interdisciplinary (how the compound enters the organism and how can this influence health). There are also physical aspects considered, because exposure to ultraviolet or gamma radiation are continuing problems and because of the distribution of pollutants from source to humans through water, the atmosphere, soil and sediment.

The emphasis on interdisciplinary science is important for urban schools which teach large numbers of children who confront daily an inhospitable environment. Some of the problems often start during the child's early development, e.g., poor maternal and infant nutrition and care, exposure to a variety of toxic substances, such as lead, in the immediate environment resulting in aberrant behaviour. It is also an important topic for students living in other environments, in which exposure to toxic substances occurs, for example, through the consumption of fish or in the farm environment. While such environmental factors can be responsible for poor health, they can also provide a stimulus for students and teachers to understand constructive interactions between the environment and its member organisms including humans. Thus, inclusion of human environmental health science and related topical issues in precollege curricula has the potential to break down some of the artificial barriers of the disciplines and meet the student with a realistic mix of the sciences and related real-world problems.

12.4 THE PROGRAMME OUTLINE

12.4.1 OVERVIEW OF THE PROGRAMME

We have built an integrated programme that elevates the quality of precollege life science course work (Figure 12.1) and, in so doing, fills another of the gaps identified by the AAAS (1989) report, i.e., strengthening teacher competence in science. The programme has a number of components that focus on the introduction of inquiry-based experiments into the curriculum. First and foremost, the programme offers a set of experiments to enrich and enliven the life science curriculum. To use them effectively, teachers need support mechanisms such as workshops, scientific interactions, website and other Internet-based opportunities, and audiovisual materials from the NIEHS Center. Secondly, a science support facility designed as part of our programme and operated by the school system is needed to provide teachers with continuing, on-site technical and scientific support. Thirdly, teachers must be provided with the scientific background and skills necessary to bring scientific experimentation and exploration into the classroom.

At the same time, teachers will become familiar with a series of laboratory-based curricular modules that are student-friendly and will be used in their own classrooms to teach students about human environmental health. Additionally, as another form of external support, teachers are familiarised with long-distance learning techniques, which link them to scientists and other personnel throughout the year.

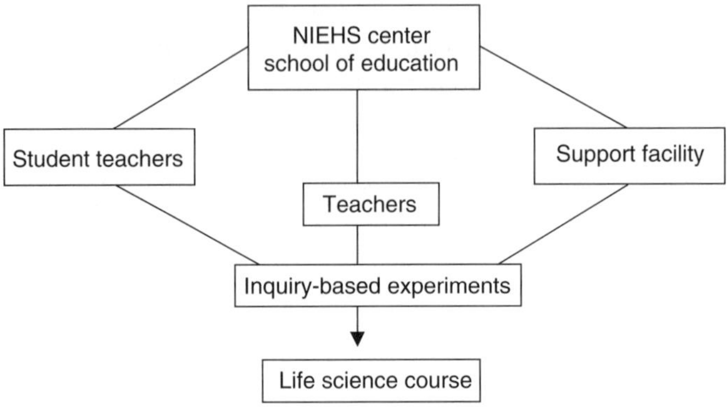

Figure 12.1 Components of model programme for middle school life science.

Information in environmental health is provided by a course entitled 'An Introduction to Environmental Health Science'. The emphasis in this course is on the presentation of overviews of several important topics, beginning with the environmental context (e.g., air, water, food, and inner city), the chemical or physical nature of the environmental factor(s) under consideration, and the mechanisms by which such factors impinge on human health. Much of the information given to teachers is in a format digestible by students.

Representative topics include:

1. Chemical context for living in a first world nation: urban and rural issues.
2. Air pollution: sources and health effects.
3. Water pollution: PCBs and dioxins, food chain concentration, and toxins in edible fish.
4. Reproductive toxicology of environmental oestrogenic substances: biology and introduction to epidemiology with a focus on PCBs.
5. Fate of chemicals in the environment: metabolism and remediation.
6. Pesticides and herbicides: neurotoxicology of farm, garden and home chemicals.
7. Toxicology of heavy metals: focus on lead and mercury.
8. Modifiers of toxic responses: individual genetic variation and nutrition.

For teachers with relatively little background, the course follows the logical path: what is the environment? How does our technological society with its intense reliance on chemicals interact with the environment? How does the quality of the environment relate to personal health? What is meant by the term *chemical hygiene*? Can we apply these concepts and ideas to urban air and water pollution, referring both to scientific studies and the key role of regulatory bodies such as the US Environmental Protection Agency (USEPA)? How do the issues in second and third world countries compare with our own? Teachers receive copies of the Scientific American book, '*Managing Planet*

Earth' (1990). They also get a notebook of readings that amplify on course material so that they have a set of high quality, professional resources for reference. Most importantly, they are invited to address questions that arise during the year to the Center staff at the University.

Issues of purchase costs and bioethics are also dealt with for participating teachers. The grant that supports this programme has, built into it, provisions for start-up funds for the teachers, e.g., fish, aquaria and general laboratory supplies. Thus, teachers can begin experiments as soon as they identify how to coordinate it with their established curriculum.

12.4.2 BEHAVIOURAL ECOTOXICOLOGY MODULES: DEVELOPMENT AND IMPLEMENTATION

In the modules described in Table 12.1, a number of basic facets of science naturally come into play. First, there is a simple fascination with living systems. Whatever the particular experiment relates to, reproduction, feeding, etc., questions will naturally arise about what is occurring in these biological processes. To begin to recognise patterns in these processes, the students learn how to observe the organisms and what to observe. By discussing their observations, students find that some people are seeing different activities from others. They, like any real scientist, can now begin to define how they will make reliable, reproducible, complete observations. The use of behavioural observations and analyses will also aid the student in learning additional educational skills, i.e., collection, organisation and interpretation of new information, particularly new information with special relevance to their own lives. All data are collected on specially prepared sheets to facilitate understanding of how the observations could be interpreted. The teacher introduces simple statistical concepts, such as averages, errors (random, systematic), and distributions, to the students. Finally, the idea that the environment affects complex biological processes will be apparent as variables in the experimental system are systematically changed.

Table 12.1 Behavioural ecotoxicology laboratory modules

Module/Biological Concepts	Environmental Health Issue	Experimental Approach
Dietary uptake	Bioaccumulation	*Daphnia* feeding
Reproduction	Lead poisoning	Observe changes in egg development in fish and frogs
Behavioural development	Toxic metal exposure	Observe fish, frog larvae posthatch development
Reflexive reactions	Sensitivity to pollutants	Dose–response, age-related sensitivity to directional stimulus

Teachers are an integral part of module development. Each scientist-designed module is tested, edited and revised through teacher feedback to make sure that it is age-appropriate. The text of each module lists the multiple goals of the experiment, which are typical for any authentic scientific experiment. Besides the detailed description of all of the experimental manipulations, the module contains a glossary of terms, and discussion questions. There is a suggested time-frame for the experiment, an instructional design for the students, an outline showing how the teachers can present the content of the module to students, and miscellaneous resource materials about the experimental animal and environmental health issues of the toxicants to be used in the module.

Conceptually, we have developed a series of behavioural ecotoxicology modules that discuss specific behaviours and underlying behavioural mechanisms and if done sequentially, integrate with each other to provide a holistic understanding of behavioural ecotoxicology. This holistic understanding of nature in general and behavioural ecotoxicology in particular not only involves learning new information but also correcting previous misconceptions. This relates to critical thinking and problem solving as new information reinforces the understanding of the concepts (Mayer 1988). Therefore, these modules are designed to both 'unlearn' old misconceptions while, at the same time, create opportunities to experience new ideas, to learn how one can learn, i.e., the very process of science.

12.5 ENVIRONMENTAL HEALTH MODULES

12.5.1 BIOACCUMULATION AND FEEDING BEHAVIOUR

In a short, highly visual, experiment, the concept of bioconcentration of fat-soluble chemicals is illustrated using the small aquatic organism *Daphnia magna*. A red dye is added to a suspension of *Daphnia magna*; after a period of time, the organism turns red and remains so when the dye is removed from its growth medium. At the conclusion of this module, students are able to visualise the accumulation of toxic substances by the use of a red dye and *Daphnia magna*. In turn, students will become aware of the movement and concentration of substances within the food web solely as a function of feeding behaviour, i.e., food can be a vector of toxic compounds and feeding behaviour can affect the rate and amount of toxicant uptake. Using this as an introductory module, students are introduced to the role of behaviour in studying toxicology and to the concept of bioaccumulation. Often, students will observe the young, newly hatched *Daphnia* still in the brood sac. By noticing that the adult has the red dye in its gut but the young do not, another aspect of bioaccumulation is learned - maternal transfer. This, of course, is critical to humans with such pollutants as lead, PCBs and methylmercury (Gestenberger *et al.* 1993). How such transfers affect the behaviour of children suddenly becomes more real.

12.5.2 BEHAVIOURAL DEVELOPMENT

Frogs are good models for illustrating embryo, larval and juvenile behaviour. Because frog eggs can be fertilised as a group, large numbers of synchronously developing frogs can be made available for observation. Since the embryos grow in culture and not *in utero* as in mammals, the stages of development can be examined in detail. Although humans do not metamorphose as frogs, we still display distinct developmental stages. Thus, the frog can *model* human growth and development, even from a behavioural perspective. After all, the perturbation of development, both physical and behavioural, by hormones and toxic chemicals seen in the frog is also a human phenomenon. In *Xenopus laevis* and *Rana pipiens* exposure to chemicals such as alcohol and lead produces changes in locomotor activity and sensitivity to directional stimuli. For example, students observe that while alcohol induces sluggishness among the larvae, lead induces hyperactivity. In addition, lead causes changes in the ability of the newly hatched tadpoles to respond to a tap on the aquarium; response time is longer and length of response shorter than control tadpoles (Zettergren *et al.* 1991). The observations made by the teacher and students lead to a discussion of how neurotoxins work and that this mechanism may be different dependent upon the particular class of chemicals being introduced. Appreciation of the development of behaviour from fertilisation of the egg by the sperm to the appearance of the adult is central to understanding the range of behavioural responses to environmental contaminants and, perhaps most importantly, the role age plays in determining the intensity of those responses. Observation of this process leads to a compelling interest in the underlying mechanisms of differentiation and development. Most exciting is that the above experiments were student generated, i.e., we had not developed a specific module for this (although we now are preparing such a module). What began as a module examining the *developmental* effects of alcohol and lead (mitotic activity, limb development, etc.) became a springboard for additional observations because the students themselves were already attuned to behavioural alterations as important indicators of environmental contamination. Thus, critical thinking skills were both developed and applied to new situations.

12.5.3 REPRODUCTIVE BEHAVIOUR

This module provides an added layer of complexity to interpreting behavioural observations. Each of the previous behavioural ecotoxicology modules relied upon actions of *single* organisms. Now students are asked to understand how the intricate interplay of *multiple* organisms in a social context can be affected by environmental contaminants. The underlying concept is that they, too, are social animals and how they interact with others can be altered by even short-term exposure to pollutants.

In this module students observe the effects of small concentrations of waterborne lead on the reproductive behaviours of fathead minnows (*Pimephales promelas*) (Weber *et al.* 1991, Weber 1993, Alados and Weber 1999) and by extension consider the behavioural effects of lead poisoning in people. The teacher should help the student understand that although fish behaviour is clearly different from their own, the physiological mechanisms that define a living animal are fundamentally the same. A videotape is given to each teacher along with the module that shows the normal and abnormal behaviour of the fish. This visual review of fish behaviour helps students and teachers identify specific behavioural categories for their observations and help them feel confident in recording their data. Because of the simplicity of the experimental set-up, students easily modify the original module to create self-designed experiments that not only yield new insights for them into the nature of behavioural ecotoxicology but demonstrate to the teachers that these modules are important in the development of higher critical thinking skills. In one classroom, students asked the question as to whether previously lead-exposed fish could be 'cured' if they were put into clean water. By then actually conducting that experiment in behavioural ecotoxicology, they were able to examine how contaminants may have long-term effects even after short-term exposures. The same group of students generated additional potential experiments, e.g., what would happen if a lead-exposed female with eggs mated with a control male?

Middle school students are also passing from one stage to another. Yet, how do animals, including humans, know when to pass from one life history stage to another? Therefore these experiments are directly relevant to them because they observe the maturation process within a few weeks. As the photoperiod is increased and water temperature reaches its optimal value, colour and structural changes of male and female fish are recorded. Comparisons between control and lead-exposed systems can be observed within a maximum of 2–3 weeks and often within one week. Through this short time frame and with daily changes in physical appearance and behaviour, student attention and interest is maintained. When breeding pairs of minnows are exposed to lead, specific changes in reproductive behaviour and physical development occur within a few days. Students may wish to see the difference between exposing fish before they are reproductively ready (i.e., <15 h photoperiod) vs. when they have begun to oviposit (>15 h photoperiod). The timing of exposure to an environmental contaminant has important health implications for children both born and unborn; these groups may be more susceptible to pollution-induced behavioural alterations than adults. The fact that scientists are asking the same questions lends credibility to student-derived questions and hypotheses.

Other stimuli, besides light, e.g., pheromones or visual cues from other fish, may affect breeding status. Again the module can be modified so that students can test for this by using different 'barriers' (opaque and solid vs. with small holes, clear and solid vs. with holes) designed to separate breeding pairs and identify environmental cues used to facilitate breeding within the entire population. A daily record is then maintained to follow the breeding success of each group (e.g., changes in male coloration, size, secondary sex structures, nest preparation/maintenance, and interaction with the female). The student should categorise observations as number of events, intensity and duration, sequence, winner/loser (if appropriate) of specific interactions.

12.6 DEVELOPING CRITICAL THINKING THROUGH SOCRATIC QUESTIONING

The modules we have developed contribute to the development of problem-solving skills through Socratic questioning, a method that promotes critical thinking through classroom discussion (Paul 1991). This form of questioning helps students evaluate their own thinking, also in comparison with that of their peers, and promotes consideration of multiple ideas and the inter-relationships between those ideas. In this form of discussion the teacher promotes critical thinking with four types of questions: origin, support, conflicts and other thoughts, and implications and consequences. Using the module on behavioural ecotoxicology of minnows exposed to lead as the experimental paradigm (see sample module in Box), a variety of questions can be developed that will fulfil the goals of enhancing critical thinking skills.

In the *origin* question, the teacher asks students to focus on the source of information used in the response to a question. Such questions allow the student to learn how to find information on his/her own, an important part of any research project. Examples of such areas of discussion include:

- Where would you go to find information on the history of lead mining in Wisconsin?
- Where would you find information on the health aspects of lead poisoning?
- What sources would you use to understand how minnows reproduce in the wild?

When the teacher uses *support* questions that focus on how information was gained or conclusions made, the student is faced with thinking through the philosophical basis for experimental design. This class of questions, along with those in the *origin* category, encourages students to refer (and therefore, review) their laboratory notebooks and journals for concrete answers. Examples of such areas of discussion include:

- Why were observations recorded on the first day at the beginning of the experiment?
- Why are the fish reproductively active mostly when the lights go on?
- Develop a hypothesis to explain some of your observations on altered egg development due to lead exposure.
- How do your observations on fish egg development compare with the effects that lead has on human development?

New information may *conflict* with established perceptions. The teacher tests the strength of previously held 'truths' by challenging those beliefs or conclusions. Examples of such areas of discussion include:

- What are some arguments some people may use to counter the concept that fish are good models of human behaviour?
- If not all fish respond to lead the same way, how can you use your data to understand the behavioural toxicity of lead on animals?
- Creation of debate: science and policy.

Students need to learn that there are *implications and consequences* to their research. To accomplish this final step, the teacher should design questions that will place student information, ideas and conclusions into a larger context. Clearly, it is in the discussion of the meaning of the results about the effect of environmental contamination on learning and behaviour that the children's view of this problem can move from the classroom laboratory into the real world. For example, full examination of the meaning of low level lead on childhood behaviour can be undertaken. Examples of such areas of discussion include:

- Female minnows are fractional spawners. What does that mean and what ecological benefit does this confer to the minnow population?
- How does lead alter the success of this reproductive strategy in terms of how often (frequency) eggs are released (oviposited) and the number of eggs oviposited?
- Identify how changes in behaviour might adversely affect long-term fish population survival even if the fish do not die from lead poisoning?
- How was parental care among minnows affected by lead exposure?
- Long-term lead exposure has been shown to cause hyperactivity in children. How does this knowledge help to explain some of the observed fish behaviour? (Hint: Look at the 'Patrol' data — why are control fish behaving differently from lead-exposed fish?).

The following example includes relevant sections from one of the modules used in the classroom.

Box Sample Module: Issues in environmental health: module on the effects of lead on reproductive behaviour in fish

In this module students will examine the effects of lead uptake by fathead minnows and will consider the general effects that lead poisoning have on people. This experiment will take place over 5 weeks and will require students to conduct the experiment and engage in learning activities following the experiment. The module is organised in four sections:

1. A brief description of the goals and objectives.
2. A description of the materials and preparation required for the experiment.
3. How to conduct the experiment itself.
4. An instructional design for the students that includes activities.

1. Goals and objective:

The **goals** of this module are to:

1. Experience the scientific process (formulating hypotheses, designing experiments, recording, analysing, and interpreting data) by investigating a "real-life" problem that has practical implications for the student.
2. Maintain an experimental aquarium and care for the animals.
3. Learn and apply the concepts of controlled experiments, blind study, and treatment groups.
4. Learn basic terminology used in environmental health sciences.
5. Practice safe laboratory techniques, especially if potentially hazardous chemicals are present.
6. Apply the term "model" to scientific inquiry.

The **specific objectives** of this module are to:

1. Observe the effects on reproductive behaviour of a specific environmental contaminant, lead, to which some students may have been exposed.
2. Observe the various components of one, specific complex behaviour and to determine how each action is important to the successful accomplishment of the overall activity.
3. Develop skills in preparing stock solutions for chemical exposure.
4. Define potential ecological implications of the results.
5. Assess reproductive success in terms of egg oviposition and development.
6. Examine the relationship of altered behaviour in fish to the potential effects in the lives of the students.

2. Materials and preparation required for the experiment

- The experimental aquarium
- The control aquarium
- Preparing mature, adult minnows for reproductive behavioural observations
- Maintenance of the aquariums and monitoring of the experiment

3. Conducting the experiment (required time: 2–3 weeks)

- Recording student observations
- Collecting eggs

4. What needs to be done at the end of the experiment

Appendix, glossary, discussion questions, suggested planning calendar, data sheet, additional worksheets on humans and lead.

12.7 INSTRUCTIONAL STRATEGIES FOR CURRENT LABORATORY MODULES

The problem-solving strategy contains the following elements: (1) identifying the problem, (2) defining the problem, (3) exploring strategies to solve the problem, (4) acting on ideas to solve the problem, and (5) looking for the effects of one's thought processes. This problem-solving strategy presents steps to aid students in their understanding of the process of problem-solving. It has two notable strengths. First, it is a practical approach for increasing students' metacognitive awareness of the problem-solving process. Secondly, it offers a practical means to integrate critical thinking within the subject matter. Curricula for this project will incorporate this strategy in the following manner.

1. Identifying the Problem. The first step will assist students in identifying that a problem exists. For the student lacking background information or problem solving strategies, recognising the existence of a problem may be a difficult task. For example, at the outset of the experiment on learning behaviour in fish, the teacher may ask the students to define learning and ask how one could test whether fish or any other organisms are able to learn. The teacher can also inquire whether environmental conditions can affect learning. Speculations can be made that begin the thought and experimental processes involved in scientific inquiry.

2. Defining Problems. The problem will be defined by having the teacher (1) activate students' background knowledge, (2) relate that information to the perceived problem, then (3) isolate those factors that seemingly identify the problem. In relationship to the general question and speculations about learning, the teacher will pose for the first time the experimental situation of fish consuming live prey of several sizes. How will they confront a new part of their environment, namely, this variable, live prey? Can this experiment demonstrate learning? Hypotheses can be articulated about these questions. Suppose the fish had been exposed to lead (or some other agent) that might cause a learning deficit, how would the experiment be extended to determine whether lead-treated fish had difficulty in learning?

3. Exploring Alternative Approaches. In their effort to define approaches to solve problems, Bransford and Stein (1984) note that experienced problem solvers adhere to a conscious and systematic approach that includes the following steps:

1. Breaking down a complex problem into its various parts.
2. Working a problem backward.
3. Reworking the problem to solve first a simpler, specific situation.

To promote students' ability to draw upon or locate concepts relevant to the problem at hand, the curriculum will include the following steps:

1. Ask students to identify procedures for locating concepts that may assist in solving the problem. Emphasise to students that identifying the concepts that apply to a problem is an important step in solving a problem.
2. Discuss with students concepts they believe will assist in solving the problem. Have students explain to the class why they selected these concepts.
3. Model to students the steps followed in trying to identify concepts that would help solve a problem. To aid the students and teachers in the methods used to detect learning, special reading materials will be provided.

4–5. Acting on the Plan and Looking at the Effects. The final two steps consist of acting on identified problem-solving strategies and carefully examining to see the outcomes of these strategies. So, the students will move into the laboratory and do experimental observations, examine their data for reproducible information, and then evaluate their hypotheses in light of the actual results.

More general topics for schoolwide discussion will be based on areas of current public interest and discussion as exemplified by articles in newspapers and magazines. For example, a current topic of intense interest focuses on proposed effects of various federal budget scenarios on the capacity of the USEPA to carry out its regulatory functions. Such articles present the opportunity to formulate study materials, which integrate environmental health issues with basic inquiry into the role of government in our industrialised society.

12.8 EVALUATION OF THE PROGRAMME

The evaluation procedures for this project is composed of both formative (evaluation during the project) and summative (evaluation at the end of the project) phases. The formative phase, which includes an assessment of the teacher and student population and the planning and development of the project, determines whether or not each project component is operational and achieving its desired objectives. The methodology used in this phase depends heavily on the evaluator participating in all activities and conducting periodic interviews of the relevant parties during project events. Anecdotal materials such as curricula plans and itineraries are also examined. The summative phase takes place at or near the end of each event in the project and supplies data on whether or not the project goals were met and the degree to which the project activities should be altered to improve the project outcomes.

The project evaluator uses the Stake Model to provide an overall framework for the evaluation (Stake 1975). The project evaluator is an integral part of the development and planning activities from the beginning. The intent is not only to utilise the evaluator's expertise in terms of evaluation, but to tap the evaluator for timely advice throughout the project duration. Action Planning is be utilised to ensure the implementation of the project activities into the classroom (Shroyer and Enochs 1987).

ACKNOWLEDGEMENTS

This work was supported by NIEHS grants #ES08271 and #RR14267 to DH Petering and RJ Ryder. The authors appreciate the contributions of LD Zettergren in the development of the frog behavioural model.

REFERENCES

Aguilar C, Petering D, Ryder R, Weber D and Zettergren L (1999) *Water Quality, Toxicology, and Environmental Education*, Society of Environmental Toxicology and Chemistry, Midwest Chapter, La Crosse, WI, 25-26 March.

Alados CL and Weber DN (1999) Lead effects on the predictability of reproduction behavior in fathead minnows (*Pimephales promelas*): a mathematical model. *Environmental Toxicology and Chemistry*, **18**, 2392-2399.

American Association for the Advancement of Science (1989) *Science for All Americans*, American Association for the Advancement of Science, Washington, DC.

Applebee A, Langer J and Mullis IVA (1989) *Crossroads in American Education*. Educational Testing Service, Princeton, NJ.

Bransford JD and Stein BS (1984) *The Ideal Problem Solver: A Guide for Improving Thinking, Learning and Creativity*, WH Freeman and Co., New York.

Bullock TH (1984) Comparative neuroscience holds promise for quiet revolutions, *Science*, **225**, 473-478.

Cohen G (1985) Oxidative stress in the nervous system. In *Oxidative Stress*, Sies H (ed.), Academic Press, New York, pp. 396-399.

Gerstenberger SJ, Pratt-Shelley J, Beattie MS and Dellinger JA (1993) Mercury concentrations of walleye (*Stizostedion vitreum*) in 34 Northern Wisconsin lakes. *Bulletin of Environmental Contamination and Toxicology*, **50**, 612-617.

International Association for the Evaluation of Educational Achievement (1989) *Science Achievement in Seventeen Countries*, Pergamon Press, New York.

Managing Planet Earth: Readings from Scientific American Magazine (1990), W.H. Freeman and Company, New York.

Mayer R (1998) *Teaching & Learning Computer Programming*. Erlbaum, Hillsdale, NJ.

Mottet KN and Laondolt ML (1987) Advantages of using aquatic animals for biomedical research on reproductive toxicology. *Environmental Health Perspectives*, **71**, 69-75.

Mullis IVS and Jenkins LB (1988) *The Science Report Card: Elements of Risk and Recovery. Trends and Achievement Based on the 1986 National Assessment of Education Progress (NAEP)*. Educational Testing Service, Princeton, NJ.

NIEHS (1994) *Estrogens in the Environment III: Global Health Implications*, Symposium sponsored by the National Institute of Environmental Health Sciences, 9-11 January.

Paul RW (1991) Dialogical and dialectical thinking. In *Developing Minds: A Resource Book for Teaching Thinking*, Costa AL (ed.), Association for Supervision and Curriculum Development, Alexandria, VA.

Pollard HB, Dhariwal K, Adeyemo OM, Markey CJ, Caohuy H, Levine M, Markey S and Youdim MBH (1992) A Parkinsonian syndrome induced in the goldfish by the neurotoxin MPTP. *FASEB Journal*, **6**, 3108-3116.

Shroyer G and Enochs L (1987) Strategies for assessing the unique strengths, needs and visions of rural science teachers. *Journal of Research in Rural Education*, **4**, 39-43.

Stake RE (1975) *Program Evaluation, Particularly Responsive Evaluation (Occasional Paper No. 5)*, The Evaluation Center, Western University.

Stokes W (1991) Standard guide for conducting the frog embryo teratogenesis assay-*Xenopus* (Fetox) In *Annual Book of ASTM Standards*, American Society for Testing and Materials, Philadelphia, pp. 1-11.

Tobias S (1990) They're not dumb. They're different. *Change*, **July/August**, 11-30.

US Department of Education and National Science Foundation (1992) *Statement of Principles*. Washington, DC.

Weber DN, Russo AS, Seale DB and Spieler RE (1991) Waterborne lead affects feeding abilities and neurotransmitter levels in juvenile fathead minnows (*Pimephales promelas*). *Aquatic Toxicology*, **21**, 71-80.

Weber DN (1993) Exposure to sublethal levels of waterborne lead alters reproductive behavior patterns in fathead minnows (*Pimephales promelas*), *Neurotoxicology* **14**(2-3), 347-358.

Westerfield M (1995) The Zebrafish book: A guide for the laboratory using Zebrafish (*Danio revio*). Institute of Neurosciences, University of Oregon Press, Eugene, Oregon.

Whitehead AN (1929) *The aims of education*, Macmillan, New York.

Zettergren LD, Conrad CA, Petering DH, Weber DN and Goodrich MS 1991) Immunochemical and immunohistochemical studies of cadmium associated proteins in *Rana* tadpoles, *Toxicology Letters*, **59**, 221-228.

Index

Note: Page references in *italics* refer to Figures; those in **bold** refer to Tables